Light Scattering in Semiconductor Structures and Superlattices

NATO ASI Series

Advanced Science Institutes Series

A series presenting the results of activities sponsored by the NATO Science Committee, which aims at the dissemination of advanced scientific and technological knowledge, with a view to strengthening links between scientific communities.

The series is published by an international board of publishers in conjunction with the NATO Scientific Affairs Division

A	**Life Sciences**	Plenum Publishing Corporation
B	**Physics**	New York and London
C	**Mathematical and Physical Sciences**	Kluwer Academic Publishers
D	**Behavioral and Social Sciences**	Dordrecht, Boston, and London
E	**Applied Sciences**	
F	**Computer and Systems Sciences**	Springer-Verlag
G	**Ecological Sciences**	Berlin, Heidelberg, New York, London,
H	**Cell Biology**	Paris, Tokyo, Hong Kong, and Barcelona
I	**Global Environmental Change**	

Recent Volumes in this Series

Volume 269—Methods and Mechanisms for Producing Ions from Large Molecules
edited by K. G. Standing and Werner Ens

Volume 270—Complexity, Chaos, and Biological Evolution
edited by Erik Mosekilde and Lis Mosekilde

Volume 271—Interaction of Charged Particles with Solids and Surfaces
edited by Alberto Gras-Martí, Herbert M. Urbassek,
Néstor R. Arista, and Fernando Flores

Volume 272—Predictability, Stability, and Chaos in N-Body Dynamical Systems
edited by Archie E. Roy

Volume 273—Light Scattering in Semiconductor Structures and Superlattices
edited by David J. Lockwood and Jeff F. Young

Volume 274—Direct Methods of Solving Crystal Structures
edited by Henk Schenk

Volume 275—Techniques and Concepts of High-Energy Physics VI
edited by Thomas Ferbel

Series B: Physics

Light Scattering in Semiconductor Structures and Superlattices

Edited by

David J. Lockwood and Jeff F. Young

National Research Council
Ottawa, Ontario, Canada

Plenum Press
New York and London
Published in cooperation with NATO Scientific Affairs Division

Proceedings of the NATO Advanced Research Workshop on
Light Scattering in Semiconductor Structures and Superlattices,
held March 5-9, 1990,
in Mont-Tremblant, Quebec, Canada

Library of Congress Cataloging-in-Publication Data

NATO Advanced Research Workshop on Light Scattering in Semiconductor
  Structures and Superlattices (1990 : Mont-Tremblant, Québec)
    Light scattering in semiconductor structures and superlattices /
  edited by David J. Lockwood and Jeff F. Young.
       p.   cm. -- (NATO ASI series. Series B, Physics ; v. 273)
    "Proceedings of the NATO Advanced Research Workshop on Light
  Scattering in Semiconductor Structures and Superlattices, held March
  5-9, 1990, in Mont-Tremblant, Québec, Canada"--T.p. verso.
    Includes bibliographical references and indexes.
    ISBN 0-306-44036-9
    1. Semiconductors--Optical properties--Congresses.
  2. Superlattices as materials--Optical properties--Congresses.
  3. Light--Scattering--Congresses. 4. Phonons--Congresses.
  I. Lockwood, David J. II. Young, Jeff F. III. North Atlantic
  Treaty Organization. Scientific Affairs Division. IV. Title.
  V. Series.
  QC611.6.O6N36   1990
  537.6'226--dc20   91-27996
                                                         CIP

ISBN 0-306-44036-9

© 1991 Plenum Press, New York
A Division of Plenum Publishing Corporation
233 Spring Street, New York, N.Y. 10013

Printed in the United States of America

SPECIAL PROGRAM ON CONDENSED SYSTEMS OF LOW DIMENSIONALITY

This book contains the proceedings of a NATO Advanced Research Workshop held within the program of activities of the NATO Special Program on Condensed Systems of Low Dimensionality, running from 1985 to 1990 as part of the activities of the NATO Science Committee.

Other books previously published as a result of the activities of the Special Program are:

Volume 148   INTERCALATION IN LAYERED MATERIALS
              edited by M. S. Dresselhaus

Volume 152   OPTICAL PROPERTIES OF NARROW-GAP LOW-
              DIMENSIONAL STRUCTURES
              edited by C. M. Sotomayor Torres, J. C. Portal,
              J. C. Maan, and R. A. Stradling

Volume 163   THIN FILM GROWTH TECHNIQUES FOR LOW-
              DIMENSIONAL STRUCTURES
              edited by R. F. C. Farrow, S. S. P. Parkin,
              P. J. Dobson, J. H. Neave, and A. S. Arrott

Volume 168   ORGANIC AND INORGANIC LOW-DIMENSIONAL
              CRYSTALLINE MATERIALS
              edited by Pierre Delhaes and Marc Drillon

Volume 172   CHEMICAL PHYSICS OF INTERCALATION
              edited by A. P. Legrand and S. Flandrois

Volume 182   PHYSICS, FABRICATION, AND APPLICATIONS OF
              MULTILAYERED STRUCTURES
              edited by P. Dhez and C. Weisbuch

Volume 183   PROPERTIES OF IMPURITY STATES IN SUPERLATTICE
              SEMICONDUCTORS
              edited by C. Y. Fong, Inder P. Batra, and S. Ciraci

Volume 188   REFLECTION HIGH-ENERGY ELECTRON
              DIFFRACTION AND REFLECTION ELECTRON
              IMAGING OF SURFACES
              edited by P. K. Larsen and P. J. Dobson

Volume 189   BAND STRUCTURE ENGINEERING IN
              SEMICONDUCTOR MICROSTRUCTURES
              edited by R. A. Abram and M. Jaros

Volume 194   OPTICAL SWITCHING IN LOW-DIMENSIONAL SYSTEMS
              edited by H. Haug and L. Banyai

Volume 195   METALLIZATION AND METAL-SEMICONDUCTOR INTERFACES
              edited by Inder P. Batra

Volume 198   MECHANISMS OF REACTIONS OF ORGANOMETALLIC
              COMPOUNDS WITH SURFACES
              edited by D. J. Cole-Hamilton and J. O. Williams

SPECIAL PROGRAM ON CONDENSED SYSTEMS OF LOW DIMENSIONALITY

Volume 199   SCIENCE AND TECHNOLOGY OF FAST ION CONDUCTORS
              edited by Harry L. Tuller and Minko Balkanski

Volume 200   GROWTH AND OPTICAL PROPERTIES OF WIDE-GAP II–VI
              LOW-DIMENSIONAL SEMICONDUCTORS
              edited by T. C. McGill, C. M. Sotomayor Torres,
              and W. Gebhardt

Volume 202   POINT AND EXTENDED DEFECTS IN SEMICONDUCTORS
              edited by G. Benedek, A. Cavallini, and W. Schröter

Volume 203   EVALUATION OF ADVANCED SEMICONDUCTOR MATERIALS
              BY ELECTRON MICROSCOPY
              edited by David Cherns

Volume 206   SPECTROSCOPY OF SEMICONDUCTOR MICROSTRUCTURES
              edited by Gerhard Fasol, Annalisa Fasolino,
              and Paolo Lugli

Volume 213   INTERACTING ELECTRONS IN REDUCED DIMENSIONS
              edited by Dionys Baeriswyl and David K. Campbell

Volume 214   SCIENCE AND ENGINEERING OF ONE- AND
              ZERO-DIMENSIONAL SEMICONDUCTORS
              edited by Steven P. Beaumont and Clivia M. Sotomayor Torres

Volume 217   SOLID STATE MICROBATTERIES
              edited by James R. Akridge and Minko Balkanski

Volume 221   GUIDELINES FOR MASTERING THE PROPERTIES OF
              MOLECULAR SIEVES: Relationship between the Physicochemical
              Properties of Zeolitic Systems and Their Low Dimensionality
              edited by Denise Barthomeuf, Eric G. Derouane,
              and Wolfgang Hölderich

Volume 239   KINETICS OF ORDERING AND GROWTH AT SURFACES
              edited by Max G. Lagally

Volume 246   DYNAMICS OF MAGNETIC FLUCTUATIONS IN
              HIGH-TEMPERATURE SUPERCONDUCTORS
              edited by George Reiter, Peter Horsch, and Gregory C. Psaltakis

Volume 253   CONDENSED SYSTEMS OF LOW DIMENSIONALITY
              edited by J. L. Beeby

Volume 254   QUANTUM COHERENCE IN MESOSCOPIC SYSTEMS
              edited by B. Kramer

PREFACE

Just over 25 years ago the first laser-excited Raman spectrum of any crystal was obtained. In November 1964, Hobden and Russell reported the Raman spectrum of GaP and later, in June 1965, Russell published the Si spectrum. Then, in July 1965, the forerunner of a series of meetings on light scattering in solids was held in Paris. Laser Raman spectroscopy of semiconductors was at the forefront in new developments at this meeting. Similar meetings were held in 1968 (New York), 1971 (Paris) and 1975 (Campinas). Since then, and apart from the multidisciplinary biennial International Conference on Raman Spectroscopy there has been no special forum for experts in light scattering spectroscopy of semiconductors to meet and discuss latest developments. Meanwhile, technological advances in semiconductor growth have given rise to a veritable renaissance in the field of semiconductor physics. Light scattering spectroscopy has played a crucial role in the advancement of this field, providing valuable information about the electronic, vibrational and structural properties both of the host materials, and of heterogeneous composite structures. On entering a new decade, one in which technological advances in lithography promise to open even broader horizons for semiconductor physics, it seemed to us to be an ideal time to reflect on the achievements of the past decade, to be brought up to date on the current state-of-the-art, and to catch some glimpses of where the field might be headed in the 1990s.

It was in this spirit that we set about organising the Workshop at Mont Tremblant. The enthusiasm of others for holding this meeting greatly supported us in our endeavours and, indeed, demonstrated the need for such a Workshop. In the end, the Workshop proved to be a tremendous success, exceeding all our expectations, thanks largely to the strong contributions from each and every one of the participants at the formal presentations and the informal evening round-table discussions. The participants worked very hard at this meeting, as is simply demonstrated by the following facts: (i) despite the arduous morning, afternoon, and evening schedule already set out, participants organised another impromptu discussion session on the Thursday night just before the banquet; and (ii) the Mont Tremblant Club management and staff noted to their amazement that the participants consumed more coffee than the previous record holders, an Alcoholics Anonymous group, and commented that they had never seen a business group work so hard at their vacation spot!

Following in the tradition of the earlier meetings mentioned above, we hope these Workshop Proceedings will serve not only as a record of the meeting but also as a lasting reference work in the light scattering field. Papers are presented here in the same order they were given in the individual sessions, but we have reordered some of the Workshop sessions to place related material next to each other. We have also introduced the Proceedings with a jointly written review chapter that links together some of the information contained in the opening introductory lecture by A. Pinczuk, the discussion session lead by M. Cardona, and the after dinner speech by E. Burstein on " 'Looking Back': Reminiscences of the 'Good Old Days' ".

We are grateful to the NATO Scientific Affairs Division and the National Research Council, Physics Division, for making this meeting possible through their financial support, and to D.E. Tunney, J.-A. Zahab, B. Legault and M. Cole of the Physics Division, NRC for their assistance in the preparations for the meeting and in producing these proceedings. We are also grateful to the many people who provided valuable ideas for the programme in response to our request in the initial announcement letter, our manuscript referees, session chair persons, and, in particular, to our programme advisory committee members M. Cardona and A. Pinczuk who valiantly responded to our request to play special roles in the meeting. We also were pleased to have as our special guest and after dinner speaker, E. Burstein, the only person at the Workshop who also attended the historic meeting in Paris 25 years ago. Finally, we thank all of the participants for their considerable efforts both at the Workshop and in preparing their manuscripts for these Proceedings.

David J. Lockwood
Jeff F. Young

CONTENTS

Inelastic Light Scattering from Semiconductors...1
  E. Burstein, M. Cardona, D.J. Lockwood, A. Pinczuk,
    and J.F. Young

OPTIC PHONONS

Acoustic, Optic and Interface Phonons: Low Symmetry Superlattices........................19
  M. Cardona

Raman Scattering in α-$Sn_{1-x}Ge_x$ Alloys...33
  J. Menéndez, K. Sinha, H. Höchst, and M.A. Engelhardt

Phonon Spectra of Ultrathin GaAs/AlAs Superlattices..39
  E. Molinari, S. Baroni, P. Giannozzi, and S. de Gironcoli

Resonant Raman Scattering in GaAs-AlAs Multiquantum Wells
    Under Magnetic Fields...53
  J.M. Calleja, F. Meseguer, F. Calle, C. López, L. Viña,
    C. Tejedor, K. Ploog, and F. Briones

Optical Phonons and Raman Spectra in InAs/GaSb Superlattice...............................63
  G. Kanellis and D. Berdekas

Raman Scattering Studies of Optical Phonons in GaAs/AlAs
    and $GaAs/Al_xGa_{1-x}As$ Superlattices (abstract only)....................................79
  Z.P. Wang, H.X. Han, and G.H. Li

Analysis of Raman Spectra of GeSi Ultrathin Superlattices
    and Epilayers..81
  M.W.C. Dharma-wardana, G.C. Aers, D.J. Lockwood,
    and J.-M. Baribeau

Interface Roughness and Confined Vibrations..103
  B. Jusserand

ACOUSTIC PHONONS

Interaction of Light with Acoustic Waves in Superlattices
    and Related Devices..123
  J. Sapriel and J. He

Localised and Extended Acoustic Waves in Superlattices
    Light Scattering by Longitudinal Phonons...139
  B. Djafari Rouhani and E.M. Khourdifi

STRAIN RELATED EFFECTS

Optical Phonon Raman Scattering as a Local Probe of Si-Ge
Strained Layers..159
J.C. Tsang, J.L. Freeouf, and S.S. Iyer

Strain Characterization of Semiconductor Structures and Superlattices.........................173
E. Anastassakis

Characterization of Strain and Epitaxial Quality in Si/Ge Heterostructures....................197
D.J. Lockwood and J.-M. Baribeau

Raman Scattering Characterization of Strain in (001) and (111)
GaSb/AlSb Single Quantum Wells and Superlattices and in
Metastable Ge_xSn_{1-x} Alloys...219
G.P. Schwartz

MICRO RAMAN/SMALL STRUCTURES/IMPURITIES

The Raman Line Shape of Semiconductor Nanocrystals......................................229
P.M. Fauchet

Raman Scattering of III-V and II-VI Semiconductor
Microstructures..247
M. Watt, A.P. Smart, M.A. Foad, C.D.W. Wilkinson,
H.E.G. Arnot, and C.M. Sotomayor Torres

Towards Two Dimensional Micro-Raman Analysis of
Semiconductor Materials and Devices (abstract only)......................................257
B. Wakefield and W.J. Rothwell

Raman Spectroscopy for Characterization of Layered
Semiconductor Materials and Devices..259
H. Brugger

Raman Spectroscopy of Dopant Impurities and Defects
in GaAs Layers..275
J. Wagner

Raman Microprobe Study of Semiconductors...291
S. Nakashima

MAGNETIC SUPERLATTICES AND II-VI MATERIALS

Surface Modes in Magnetic Semiconductor Films and Multilayers...........................311
M.G. Cottam and S. Gopalan

Vibrational, Electronic, and Magnetic Excitations
in II-VI Quantum Well Structures..323
A.K. Ramdas and S. Rodriguez

Zinc Blende MnTe as Efficient Confinement Layers in ZnTe
and CdTe Single Quantum Well Structures..341
A.V. Nurmikko

Raman Scattering Study of CdTe/CdMnTe Superlattices...................................353
L. Viña, F. Calle, J.M. Calleja, F. Meseguer, L.L. Chang,
J. Yoshino, and M. Hong

TIME RESOLVED STUDIES

Nonequilibrium Electrons and Phonons in GaAs
and Related Material...367
J.A. Kash

Subpicosecond Raman Study of Hot Electrons
and Hot Phonons in GaAs...383
D. Kim and P.Y. Yu

Time-Resolved Raman Studies of the Transport Properties
of Excitons in GaAs Quantum Wells.................................393
K.T. Tsen and O.F. Sankey

Non-Equilibrium Phonon Dynamics in Ge and GeSi Alloys..................401
J.F. Young, D.J. Lockwood, J.-M. Baribeau,
P.J. Kelly, A. Othonos, and H.M. van Driel

Time-Resolved Raman Measurements of Electron-Phonon Interactions
in Quantum Wells...421
J.F. Ryan and M.C. Tatham

RELATED PHENOMENA

Resonant Three-Wave Mixing via Subband Levels in Quantum Wells:
Theoretical Considerations...441
E. Burstein and M.Y. Jiang

N-Layer Superlattice Phonons..451
L. Dobrzynski, A. Rodriguez, J. Mendialdua, D.J. Lockwood,
and B. Djafari Rouhani

Far-Infrared and Raman Studies of Semiconductor Superlattices...........461
T. Dumelow, A.A. Hamilton, K.A. Maslin, T.J. Parker,
B. Samson, S.R.P. Smith, D.R. Tilley, R.B. Beall,
C.T.B. Foxon, J.J. Harris, D. Hilton, and K.J. Moore

Optical Properties of Periodically δ-Doped GaAs (abstract only)..........477
J.M. Worlock, A.C. Maciel, M. Tatham, J.F. Ryan,
R.E. Nahory, J.P. Harbison, and L.T. Florez

Nonlinear Response of Virtual Excitations in Semiconductor Superlattices...479
P. Hawrylak

Sequential Resonant Tunneling in Superlattices: Light Scattering
by Intersubband Transitions..491
S.H. Kwok, E. Liarokapis, R. Merlin, and K. Ploog

Elementary Excitations in Low-Dimensional Semiconductor Structures.......499
S. Das Sarma

ELECTRONIC EXCITATIONS

Electronic Properties of Parabolic Quantum Wells........................525
L. Brey, N.F. Johnson, J. Dempsey, and B.I. Halperin

Electronic Raman Scattering from Modulation Doped Quantum Wells..........543
D. Richards, G. Fasol, U. Ekenberg, and K. Ploog

Micro-Raman Spectroscopy for Large In-Plane Wave Vector
 Excitations in Quantum-Well Structures.....................................561
 G. Abstreiter, S. Beeck, T. Egeler, and A. Huber

Inelastic Light Scattering by the High Mobility Two-Dimensional
 Electron Gas...571
 A. Pinczuk, D. Heiman, S. Schmitt-Rink, C. Kallin,
 B.S. Dennis, L.N. Pfeiffer, and K.W. West

Concluding Remarks..587
 M. Cardona

Group Photograph..593

Participants...595

Author Index...599

Subject Index..601

INELASTIC LIGHT SCATTERING FROM SEMICONDUCTORS

E. Burstein,[a] M. Cardona,[b] D.J. Lockwood,[c] A. Pinczuk,[d] and
J.F. Young[c]

[a]Physics Department, University of Pennsylvania, Philadelphia, PA
19104-6396, U.S.A.
[b]Max-Planck-Institut für Festkörperforschung, 7000 Stuttgart 80
Germany
[c]National Research Council, Ottawa, ON K1A 0R6, Canada
[d]AT&T Bell Laboratories, Murray Hill, NJ 07974-2070, U.S.A.

INTRODUCTION

As mentioned in the Preface of these Proceedings, one important aspect of the
Workshop was to review recent advances in the field of inelastic light scattering from
elementary excitations in semiconductors. To that end, the workshop organizers requested
an introductory lecture from A. Pinczuk that would cover developments in what could be
called the artificially structured materials era, which has occupied the last decade or so. But
rather than just review recent work for this introductory chapter, the organizers felt a more
comprehensive history of research developments would be appropriate, particularly as this
had not been attempted before. E. Burstein's lecture on early developments in light
scattering spectroscopy of semiconductors, together with M. Cardona's encyclopaedic
knowledge of the field, naturally lead to their co-option in such a task. What follows is a
general account of key and other important developments in the subject to date, biased to
some extent by the knowledge and preferences of the contributors. We apologize in
advance for any inadvertent omission of other major relevant research work.

EARLY DEVELOPMENTS

The subject of inelastic light scattering was first investigated theoretically by
Brillouin,[1] who in 1922 published a study of light scattering by density waves in liquids,
followed a year later by Smekal's fundamental work on two-level atoms.[2] As a result of
his prediction that elementary excitations would produce shifts in the frequency of the
scattered light, the inelastic scattering observed experimentally by Raman[3] in 1928 was
often called the Smekal-Raman effect, but is now more commonly known as Raman
scattering. In 1925, Kramers and Heisenberg[4] developed the old-style quantum theory of
Smekal to derive a scattering formula from classical wave theory by means of the
correspondence principle. The Kramers-Heisenberg intensity relation was of fundamental
importance in the development of quantum mechanics. Then, in 1927, Dirac[5] rederived
these results using his quantum theory of radiation.

Light Scattering in Semiconductor Structures and Superlattices
Edited by D.J. Lockwood and J.F. Young, Plenum Press, New York, 1991

The announcement of Raman's discovery in 1928 led to an intense flurry of activity in light scattering spectroscopy, as well as the Nobel Prize in Physics for Raman in 1930. Raman had investigated inelastic light scattering from vibrations in liquids and solids, and Landsberg and Mandelstam[6] almost simultaneously announced the same effect in quartz. As a result of this latter work, Raman scattering is referred to as combination scattering in Russian literature. Other notable advances at this time were: an interpretation by Rocard[7] of the Raman effect as a modulation of the electric dipole moment by vibrational modes of the oscillating molecule; Wood[8] re-examined luminescence spectra dating from 1906 and found features due to Raman scattering in quartz, and he also introduced the term anti-Stokes scattering; in 1930 Robertson and Fox, and Ramaswamy independently reported the first-order Raman spectrum of diamond,[9] which could also be considered as the first Raman measurement on a "semiconductor"; the second-order Raman scattering from NaCl was published by Rasetti[10] and interpreted in terms of a second-order polarizability; in 1933 Born and Blackman[11] introduced an anharmonicity mechanism for second-order infrared and Raman processes. The work of this period culminated in the seminal paper by Placzek,[12] published in 1934, wherein he presented a phenomenological theory for vibrational and electronic Raman scattering in terms of the first- and higher-order polarizabilities. The first experimental observation of electronic Raman scattering (in gaseous NO) had been reported earlier in 1930 by Rasetti.[13]

Not long after the experimental discovery of the Raman effect, Gross[14] found a triplet light-scattering spectrum in liquids. The frequency shift of the outer components was much smaller than in previous vibrational scattering, consistent with the expected acoustic wave scattering predicted earlier by Brillouin and independently by Mandelstam.[15] These two lines comprise what is now commonly called the Brillouin (or Mandelstam-Brillouin) spectrum. The central elastic component was explained by Landau and Placzek[16] as scattering from nonpropagating density fluctuations.

Other key stepping stones and milestones along the way in this early pre-laser period include: the discovery by Raman and Negungadi[17] of Raman scattering from a soft mode associated with the α-β phase transformation in quartz; the Born-Bradburn theory[18] of second-order Raman in alkali halides couched in terms of the polarizability of atoms and the dependence on the change in distance between pairs of atoms; Krishnan's detailed study[19] of the first- and second-order spectrum of diamond; the analysis by Couture and Mathieu[20] of the "polarization" character of the scattering for different orientations of crystals and for different directions of the incident and scattered light; the observation by Couture-Mathieu and Mathieu[21] of Raman scattering from longitudinal optical (LO) phonons as well as transverse optical (TO) phonons in ZnS; and Poulet's explanation[22] of the I_{LO}/I_{TO} intensity ratio in ZnS (or other piezoelectric crystals) in terms of atomic displacements and electro-optic contributions to the Raman tensor (later referred to as the Faust-Henry coefficient). There was also much fine experimental work on the second-order Raman spectrum of alkali halides (see, for example, Refs. 19, 23-25). At the end of this pre-laser period, the seminal theoretical work of Loudon[26] on light scattering from solids was published. Loudon developed a microscopic model for Raman scattering in non-polar and polar crystals, summarized the Raman tensors for all crystal classes, discussed the electro-optic tensor in terms of local electric fields, demonstrated the cancelation of electron and hole contributions to Raman scattering in the zero wave vector limit, etc. Like Placzek's earlier polarizability theory, this work had far-reaching consequences.

POST-LASER DEVELOPMENTS

The introduction of the laser light source in the 1960s revolutionized the experimental methods of light scattering spectroscopy. New developments in grating spectrometry and Fabry-Perot interferometry soon followed. These new devices coupled with extremely sensitive photomultiplier pulse-counting and (later) multichannel detector systems allowed many new discoveries to be made, including, for example, the very weak scattering of light by magnons.[27] In this part of the review we consider developments mainly in the semiconductor field, but key milestones in related areas are also noted. The particularly recent subjects of quantum wells, superlattices and time-resolved studies are discussed separately.

1962—1971

The importance of the laser as a light scattering source was vividly demonstrated in the early 1960s. Stimulated Raman,[28] Brillouin,[29] and Rayleigh[30] scattering were soon discovered, and a historical account of the early theoretical and experimental research into these stimulated effects has been given by Bloembergen.[31] In 1964, Hobden and Russell[32] reported on the first application of a laser as a source to excite the Raman spectrum of a semiconductor (GaP), and Jones and Stoicheff[33] observed the inverse Raman effect. Then in 1965, Maker and Terhune[34] pioneered the technique of what is now called coherent anti-Stokes Raman spectroscopy (CARS); Terhune et al.[35] observed the hyper-Raman effect (a three-photon process) in water; Russell[36] observed photographically the He-Ne laser excited first- and second-order spectrum of CaF_2 and GaP and, significantly, the third-order spectrum of GaP; Krishnan and Krishnamurthy[37] analyzed the GaP spectrum; Burstein and Ganesan[38] developed a method for determining the Raman matrix element from the electric-field-induced infrared absorption in diamond; Russell[39] recorded the laser Raman spectrum of an opaque material, Si, for the first time in backscattering; Birman[40] derived selection rules for two-phonon Raman scattering; Kleinman[41] made use of polarization effects to determine which Raman-active irreducible representations are present in two-phonon spectra; and Henry and Hopfield[42] observed laser Raman scattering from lower-branch polaritons in GaP.

In 1966, Damen et al.[43] developed the now standard polarization notation used to describe Raman spectra when investigating the phonons in ZnO, in which they also observed polariton Raman scattering;[44] Anastassakis et al.,[45] from a measurement of electric-field induced infrared absorption in diamond, deduced the magnitude of the atomic-displacement Raman tensor to be $| a | = 4 \times 10^{-16}$ cm^2 (later, Grimsditch and Ramdas[46] found from direct measurement $| a | = 4.4 \pm 0.3 \times 10^{-16}$ cm^2); Wolff[47] studied theoretically Thomson and Raman scattering by mobile electrons in crystals and suggested the possibility of Raman scattering involving Landau levels; Wolff's theory was extended by Yafet[48] to predict a spin-flip Raman transition involving virtual transitions between the conduction and valence bands of InSb; Mooradian and Wright[49] observed phonon Raman scattering in the III-V compounds GaAs, InP, AlSb, and GaP, and also observed LO phonon-plasmon coupled modes in GaAs,[50] the first of many such studies; Leite and Porto[51] measured resonant phonon Raman scattering in CdS; Ganguly and Birman[52] analyzed the exciton enhancement of Raman scattering by optical phonons; and Giordmaine and Kaiser[53] observed Raman scattering from intense coherently-driven lattice vibrations in calcite (a forerunner of the pump-probe technique applied later to time-resolved studies).

Developments in 1967 included measurement of the Raman spectrum of Ge[54] and electronic light scattering from impurity levels in Ge;[55] the spontaneous scattering from magnetic levels in InSb[56] including transitions between Landau levels and the spin-flip transition; phonons in ZnSe, ZnTe and InSb;[57] electric-field induced scattering by odd-parity soft optical phonons in $KTaO_3$ and $SrTiO_3$;[58] and soft phonons in $BaTiO_3$ including the first use of backscattering in a transparent crystal.[59] There followed in 1968 an experimental and theoretical analysis of Raman scattering in SiC polytypes,[60] which involved the first considerations of zone folded modes and can be considered a natural precursor to the artificial superlattice work of the 1980s; a study of light scattering from single-particle electronic excitations;[61] the observation of extremely strong spin-flip Raman scattering by donor and acceptor impurities in CdS;[62] and the observation of space-charge electric-field induced Raman scattering by LO phonons in the surface depletion region of InSb.[63] The first of a new series of international conferences on light scattering in solids was held in New York in September 1968, and many important developments in the light scattering field were reported in the proceedings of that meeting.[64]

In 1969 there were reports on phonons in Se;[65] the polariton formulation of exciton enhanced Raman scattering by LO phonons;[66] multiple LO phonon peaks due to resonant Raman scattering in CdS and the relationship to hot luminescence;[67] followed, in 1970, by reports on multiple LO phonon peaks in various zincblende and wurtzite semiconductors;[68] the Franz-Keldysh mechanism for space-charge electric-field induced scattering by LO phonons;[69] inelastic light scattering from semiconductor plasmas in a magnetic field;[70] strain-induced shifts and splittings of the first-order Raman line in Si;[71] stimulated spin-flip Raman scattering in InSb;[72] and electric-field induced Raman scattering in diamond.[73]

The second international conference on light scattering in solids was held in Paris in July 1971 and again many new advances were reported.[74] Other significant advances reported elsewhere include measurement of the Raman spectrum of graphite;[75] examination of morphic effects due to external forces on the infrared and Raman spectra of optical phonons;[76] a theory of wave-vector dependent exciton resonant-Raman scattering mediated by the Fröhlich interaction;[77] the observation of resonant light scattering by single particle spin-flip and collective charge-density excitations in GaAs;[78] a theory of interband electronic Raman scattering in semimetals and semiconductors;[79] a report on spin-flip Raman scattering from conduction electrons in CdSe and ZnSe in a magnetic field;[80] an investigation of wave-vector dependent (Fröhlich) Raman scattering by odd-parity LO phonons in Mg_2Si, Mg_2Ge, and Mg_2Sn;[81] and measurement of the Raman spectrum of α-Sn.[82]

1972—1990

After ten years of post-laser Raman spectroscopy, applications in solids had become routine and the number of studies in semiconductor systems had mushroomed. Because so much work has been done, only highlights of the later developments can be given here. In-depth reviews of this more recent period of research can be found in the series of Springer-Verlag volumes on "Light Scattering in Solids" edited by M. Cardona and G. Güntherodt.[83]

In 1972, Brillson and Burstein,[84] in a calculation of electric-field induced Raman scattering by LO phonons in an external magnetic field, showed that the intraband-Fröhlich-interaction Franz-Keldysh matrix element is proportional to the average separation of the electron and hole in the intermediate state; Geurts and Richter[85] observed interband electronic Raman scattering in InSb; and Yu and Shen[86] used a tunable dye laser to obtain the excitation profile for electric-field induced Raman scattering by LO phonons in InSb. Also observed were electric-field induced resonant Raman scattering by LO phonons in CdS,[87] resonant Raman scattering in Ge at the E_1 and ΔE_1 gaps,[88] and near-resonance spin-flip electronic Raman scattering in InSb.[89] Brenig et al.[90] investigated spatial dispersion effects in resonant Brillouin scattering from polaritons. In 1973, Compaan and Cummins[91] observed resonant Raman scattering by odd-parity LO phonons at the 1s yellow exciton in CdS; Doehler et al.[92] carried out measurements at 2 μm (i.e., in transmission) of Raman scattering by phonons and electronic impurity levels in Ge; Cardona[93] elaborated on Raman scattering as a form of modulation spectroscopy; and Evans et al.[94] observed Raman scattering from surface polaritons in a thin GaAs film on Al_2O_3, which lead to a subsequent theoretical analysis by Chen et al.[95] in 1975. Scattering by coupled phonon-electron excitations was observed in p- and n-type Si and quantitatively interpreted.[96,97] The line shapes exhibited Fano profiles with asymmetry parameters dependent on laser frequency. This work lead to the determination of the sign of the

Raman tensor of Si.[98] Vogt, in 1974, applied the hyper-Raman technique to solids, investigating first CsI[99] and later, odd-parity and "silent" phonons in $SrTiO_3$.[100]

The third international conference on light scattering in solids was held in Campinas in July 1975 and the maturity of the bulk-semiconductor light scattering field was evident from papers presented at that meeting.[101] In 1975, a definitive study of Raman scattering from phonons in the chalcopyrite archetype $AgGaS_2$ was reported along with an interpretation of chalcopyrite optic modes based on the Bettini lattice dynamical model,[102] and the first observation of Raman scattering from the upper-branch polariton in a semiconductor (ZnO) was also reported.[103] Zone folding in GaSe polytypes was studied in 1976,[104] as was resonant two-photon electronic Raman scattering in CuCl,[105] the use of resonant Raman scattering by LO phonons as a probe of the surface space-charge field at n- and p-type InAs surfaces,[106] and the lattice dynamics of the ordered vacancy compound $HgIn_2\square Te_4$, where the Brewster-angle Raman scattering technique was fully exploited for an opaque semiconductor.[107] Scattering by intervalley density fluctuations was observed in n-type Si in 1977.[108]

Light scattering by guided wave polaritons in GaSe was investigated theoretically[109] and experimentally[110] in 1978, and conditions were established for observing inelastic light scattering by intersubband charge-carrier excitations at inversion and accumulation layers.[111] Then in 1980 there were investigations of the coupling of intersubband charge-density excitations with LO phonons and the microscopic mechanisms for Raman scattering by intersubband single-particle non-spin-flip excitations and by intersubband charge-density excitations,[112] resonant Raman scattering at InAs surfaces in MOS junctions (observation of coupled LO phonon - intersubband charge density mode),[113] and electronic Raman scattering by shallow donors in CdTe and GaAs and its formulation in terms of exciton-mediated light scattering.[114] Very recently, double and triple resonances have been observed in the scattering by phonons in semiconductors[115] and absolute resonant cross sections have been interpreted using excitons as intermediate states.[116] Also, in the past couple of years, considerable experimental information has been obtained by measuring resonant Raman scattering by phonons in a strong magnetic field.[117]

SEMICONDUCTOR SUPERLATTICES

Phonons

The mini-Brillouin-zone (MBZ) of a superlattice can always be approximately considered as a folded version of that of the bulk material. The concept of folding, however, is particularly useful for branches whose bulk counterparts overlap and differ relatively little from each other. This is usually the case for most of the acoustic branches of the bulk materials. We can then describe the corresponding dispersion relations of superlattices as an average of the bulk ones, folded back to the MBZ, and with minigaps opening at the zone center and edges. The concept of "folding", however, is not very useful for the bulk optical modes, which are usually well separated in frequency in the two components (e.g., Ge-Si, GaAs-AlAs). In this case the concept of "confinement" becomes more useful: optical vibrations are localized to either one of the components and are non-dispersive throughout the MBZ (except for the so-called interface modes, see below). The distinction between folded and confined modes is not merely academic. Surely one can "fold" confined modes to the first MBZ but this folding is of no use since the modes are non-dispersive. Note that for superlattices composed of a thickness d_1 (d_2) of material 1 (2) the equivalent q-vector of folded acoustic modes at the center of the MBZ is

$$q = \frac{2\pi}{d_1 + d_2} \cdot m, \quad m = 1,2,3,... \tag{1}$$

while that of confined ones is

$$q = \frac{\pi}{d_1} \cdot m \quad \text{in medium 1}$$

$$q = \frac{\pi}{d_2} \cdot m \quad \text{in medium 2}$$

(2)

These wavevectors are different unless $d_1 = d_2$.

The concepts of folding and confinement were used as being more or less equivalent since the first attempts at observing confined optical modes were made in 1977.[118] We have already mentioned above that the concept of folding has been used earlier[60] to describe light scattering in SiC polytypes, materials for which folding is appropriate to either optical or acoustic phonons.

The first conclusive observation of folded acoustic modes in superlattices was probably made by Colvard et al.[119] Their theory has been worked out at an early stage by Rytov,[120] in connection with the propagation of seismic waves through stratified media. Minigaps had been observed earlier by Narayanamurti et al.[121] by acoustic transmission spectroscopy and, more recently in amorphous superlattices by Santos et al. with the same techniques.[122]

Shifts in optical modes versus layer thickness, resulting from confinement, were reported in Ref. 118 and later by several Japanese groups.[123,124] Strong evidence for the formation of good quality superlattices, however, was first obtained through the observation of several optic modes under resonant conditions.[125,126] Even stronger evidence for these "confined" optic modes was obtained under non-resonant conditions for the GaAs/Ga$_{1-x}$Al$_x$As system by Jusserand et al.[127] In order to avoid the difficulties involved in the theoretical treatment of "phonons" in mixed crystals such as GaAl$_x$As$_{1-x}$ Sood et al.[128] investigated (GaAs)$_l$(AlAs)$_m$ systems. These authors clearly spelled out the difference between "confinement" and "folding" which has been unequivocally used since then. They also investigated the polarization selection rules for these superlattices on and off resonance. In the latter case they found that for parallel polarizations (in backscattering) modes with m even (Eq. (2)) are seen. They identified the relevant electron-phonon coupling as the Fröhlich interaction (we would like to recall with respect and admiration the recent death of H. Fröhlich on January 23, 1991). For crossed polarizations parallel to cubic axes they observed modes with m odd and were able to prove that they were induced by the deformation potential interaction, equivalent to that at the top of the bulk valence bands. These selection rules settled unambiguously the question of boundary conditions for confined phonons in polar materials: like in the covalent counterparts (e.g., Ge-Si) the vibrational amplitude, not the electrostatic potential, must vanish at the interfaces.

Modes which have now become known as electrostatic interface modes were observed by Merlin et al.[126] who assigned them to long wavelength LO and TO modes propagating with in-plane $\bar{q}$ vectors, with frequencies given by the solution of $\langle \varepsilon \rangle = 0$ and $\langle \varepsilon^{-1} \rangle = 0$ ($\langle \ \rangle$ represents the arithmetic average over the two media). More detailed observations were performed by Sood et al.[129] who attributed them to "electrostatic interface modes". In the in-plane propagation limit and for $q \to 0$, these modes, which had been calculated by several authors,[130] coincide with those of Merlin et al.[126]

We note that the interface modes observed in Ref. 129 correspond to large in-plane wavevectors which cannot be produced in a near-backscattering process. Hence the contribution of interface roughness or other defects was invoked. This conjecture received support from the observation that the scattering by interface modes was screened out for strong laser power densities.[131] Observations of similar modes have been also made by electron energy loss spectroscopy.[132] Only recently the dispersion relations of interface modes with a well-defined $\vec{q}$ has been measured by scattering on a bevel cut at the edge of the superlattice[133] and also on superlattices with an in-plane periodicity generated by a second, MBE-grown, vertical superlattice.[134]

The detailed nature of the interface modes has been the source of considerable confusion since their original observation in Refs. 126, 129. Its understanding is important for the treatment of the electron-phonon interaction. It has more recently become clear that interface modes are not "additional" vibrational modes but conventional modes of the superlattice considered as a crystal with a large primitive cell, affected by long-range electrostatic fields. This understanding has been reached through simple lattice dynamical models[135] and also through detailed lattice dynamical calculations based on force constants and effective charges carried over from the bulk.[136] The interface phenomena correspond to long-range electrostatic field effects for propagation with an in-plane component of q. Hence only infrared-active modes (for the space group of the superlattice) exhibit such effects. Such are the confined modes with m odd, in particular that with m = 1 (interface effects for m = 3 and larger are usually negligible). The confined m = 1 LO and TO modes for $\vec{q}$ along the growth axis evolve into interface modes when $\vec{q}$ is tipped towards the layer planes.

Most of the experiments on "confined" and "folded" phonons in superlattices have been performed for systems grown along the [001] axis. In these systems only LO modes are allowed for backscattering on the planes. TO modes can, however, be seen if the laser is incident on the superlattice edge.[137] More recently, experiments on superlattices grown along lower symmetry axes, such as [111], [110], and [012] have been performed.[136,138,139] In many of these cases LO, TO, LA, and TA modes can be observed. In all cases they map well onto bulk dispersion relations using Eqs. (1) and (2) or, in the case of confined phonons, a slight modification obtained by adding to m a number δ ($0 < \delta < 1$) to allow for a slight penetration into the forbidden layer.[140] These low-symmetry growth directions yield a large variety of possible space groups, especially in the case of the Ge/Si system.[141]

Multiphonon peaks are also observed in superlattices.[142] They usually correspond to overtones and combinations of modes near $\vec{q} = 0$ (confined modes) with m even, although in the case of short period superlattices m = 1 (interface) modes are also seen.[143] These peaks are particularly easy to observe near outgoing resonance, i.e., when the scattered frequency equals that of an edge exciton. Such resonances, including those for one-phonon scattering, were first observed and treated theoretically by P. Manuel et al.[144] However, no detailed understanding of the nature of these resonance and the resulting scattering cross sections is available to date. This is also the case for the folded acoustic modes whose Raman polarizability is known to be related to the photoelastic constants of both constituents.[145]

Among recent important developments in the field of Raman scattering by phonons in superlattices we mention the observation of confined modes with a fairly sharp bulk $\vec{q}$ vector in the mixed crystal layers of GaAs/Ga$_x$Al$_{1-x}$ superlattices.[146] These experiments prove that, in spite of the disorder, optical modes with a well defined $\vec{q}$ vector exist in bulk alloys of the Ga$_x$Al$_{1-x}$ type.

Because of the available knowledge on the effect of strain on Raman phonons of bulk materials[147,148] it is possible to obtain the degree of strain in lattice mismatched superlattices using Raman spectroscopy.[149] The technique is particularly appropriate for Ge/Si superlattices because of the large mismatch (~4%) between the lattice constants of both constituents. Nevertheless, not only Ge-Si but also III-V and II-VI systems have been investigated.

Infrared investigations of superlattices have not been as productive as their Raman counterparts. However interesting results are beginning to appear.[150,151] Infrared spectroscopy enables one to obtain information about TO modes in [001]-oriented superlattices,[151] which is hard to obtain by Raman spectroscopy. The quantitative analysis of the data, however, requires cumbersome fits with the optical response functions of multilayer systems.[150]

We conclude by mentioning a number of review articles on phonons in superlattices which the reader may find useful.[152]

Electronic Excitations

Interest in light scattering by free carriers confined at semiconductor surfaces and interfaces was stimulated by a communication of Burstein et al. presented at the 14th International Conference on the Physics of Semiconductors.[111] This work considered the mechanisms of inelastic light scattering in semiconductors and pointed out that with resonant enhancements of the scattering cross sections, as demonstrated in experiments carried out with conventional lasers,[79,153-155] the method has the sensitivity required to observe elementary excitations of two-dimensional electron systems in semiconductors. This proposal led to the first observations of resonant light scattering by free electron systems in modulation doped GaAs-AlGaAs heterostructures.[156,157] During the last ten years light scattering research of free carriers in semiconductor quantum wells and superlattices has been extensive. Much of this work has been reviewed in Refs. 158-160.

Within the framework of the effective-mass approximation the light scattering mechanisms of free carriers at semiconductor interfaces are similar to those in the parent 3D systems. The basic light scattering processes that apply in bulk semiconductors were considered in Refs. 47, 48, and 161-164; and the first experiments, in III-V compounds, were reported in Refs. 50, 56, 61, and 165. This pioneering work stimulated many light scattering studies of free carriers in bulk semiconductors that have been reviewed in Refs. 158 and 166-168. From these studies light scattering emerged as a very flexible experimental method to study the single particle and collective excitations of semiconductors. The collective excitations are plasmons or coupled plasmon-LO phonon modes. In the 3D systems the single particle excitations are spin-density modes and in the case of multivalley semiconductors, like n-type silicon, there are also intervalley density fluctuations.

The large resonant enhancements of the light-scattering cross sections required to observe the excitations of low density (10^{10} - 10^{12} cm^{-2}), 2D electron systems are predicted for photon energies near the optical transitions between the valence and conduction states that contribute to the effective mass of the free carriers.[111,112] In the case of direct gap III-V semiconductors free electrons occupy states of zone-center conduction band minima. For these carriers and for shallow donor levels the relevant optical resonances occur at the fundamental E_0 and spin-orbit split $E_0 + \Delta_0$ gaps. For free holes in states of the valence band maxima and for shallow acceptor levels the most important resonance is at the fundamental gap E_0. In the case of GaAs-AlGaAs heterojunctions and quantum wells the optical gap energies are $E_0 \simeq 1.5$ eV and $E_0 + \Delta_0 \simeq 1.90$ eV, easily accessible with tunable visible and infrared lasers. In the case of free holes in Si the relevant resonance is $E_0 \simeq 3.4$ eV.[111,169] Light scattering by free electrons at Si interfaces should show resonant enhancements near the zone-boundary optical gap at $E_2 = 4$ eV. The intensities of collective modes display resonant enhancements at all the optical gaps.[111-113]

Most of the light scattering studies of free electron systems with reduced dimensionality have been carried out in the semiconductor multilayer structures grown by molecular beam epitaxy. Several successful experiments have also been reported in the case of space-charge layers at metal-insulator-semiconductor interfaces.[113,169,170] The systems investigated most extensively are in the n-type modulation doped GaAs-AlGaAs heterostructures.[158-160] Studies of 2D hole gases have revealed some of the complexities of the energy levels and interactions of these systems.[171,172] Resonant inelastic light scattering has been a very useful spectroscopic tool in studies of periodic doping superlattices (nipi structures),[173-175] of shallow donor and acceptor states in quantum wells,[176,177] and of δ-doping layers.[178,179]

In recent work the light scattering method has also been applied in the investigation of the elementary excitations of one-dimensional electron gases.[180-183] We believe that light scattering will continue to have a broad impact in studies of free carrier systems with reduced dimensionality in artificial semiconductor microstructures and nanostructures. The papers on electronic light scattering presented in these proceedings cover some of the most exciting current research in the field.

TIME-RESOLVED STUDIES

The majority of Raman scattering experiments make use of continuous wave (CW) laser sources and are designed to elucidate the elementary excitation spectrum of a system of interest that is in equilibrium. As most of the contributions to this volume attest, this conventional light scattering technique is an extremely powerful means by which to study the influence of varying material parameters on a system's lattice and electronic excitation spectrum. Experimentally, it is usually the case that measures are taken to ensure the laser light used as a source for the scattering does not disturb the system of interest away from its equilibrium state (i.e., care is taken to ensure that the spectra are independent of the laser intensity).

By either increasing the intensity of the CW laser source, or by using pulsed laser excitation, it is possible to induce changes in the elementary excitation spectra, which then represent the non-equilibrium, laser-excited system.

High-intensity CW laser excitation was used by Pinczuk et al.[184] to study the Raman spectra of electron-hole plasmas in GaAs. By fitting the spectra with a finite-temperature Lindhard-Mermin dielectric response function they were able to identify both optic coupled plasmon-LO phonon modes, and perhaps more significantly, they also observed the acoustic plasma mode characteristic of a multi-component plasma.

Nather and Quagliano[185] performed an extensive Raman study of CW photo-excited plasmas in confined, bulk GaAs layers. They succeeded in obtaining both single particle and collective mode spectra over a large range of plasma densities and temperatures. They also extracted density, temperature and plasma drift velocities from high-quality fits of their spectra using a multi-component dielectric function calculation which (necessarily) included intervalence band transitions.

Q-switched YAG laser pulses were used by Vasconcellos et al.[186] to both excite and probe the electron-hole plasma resulting from two-photon absorption in the bulk of GaAs crystals as early as 1977. By studying the shape of the single particle spectra they were able to deduce the intensity dependent carrier temperature, averaged over the laser pulse width (a few nanoseconds). In 1979, Kardontchik and Cohen[187] employed nanosecond dye laser pulses to both excite and probe electron-hole plasmas in GaP at 2 K. They measured the excitation frequency dependence of both branches of the coupled electron-hole-plasmon-LO phonon modes, but were unable to quantitatively explain the spectra. They also commented on the narrow linewidth of these modes as compared to those observed in comparable density n-type doped GaP samples. Interestingly, in 1986, Yugami et al.[188] reported similar nanosecond studies of coupled mode spectra in photo-excited GaP at 80 K, and concluded that the photo-excited plasma was heavily overdamped.

The examples above demonstrate the utility of pulsed lasers as a means of producing and probing non-equilibrium semiconductor plasmas. However, the full potential of pulsed lasers for Raman scattering is realized only when a relatively intense "pump" pulse is used to induce the non-equilibrium state, and a second "probe" pulse, synchronized and delayed with respect to the pump, is used to scatter from the non-equilibrium system as it returns towards equilibrium. Owing to the coherent nature of the Raman process, the temporal resolution of such experiments is limited (except under resonant conditions) only by the laser pulse duration.

Depending upon the laser pulse duration and intensity, this time-resolved Raman technique can and has been used to study a number of interesting non-equilibrium processes in semiconductors and semiconductor structures. There is an extensive literature[189,190] on the use of nanosecond probe pulses to time-resolve the lattice temperature rise of Si irradiated by high intensity nanosecond pulses which ultimately cause melting of the Si surface. This was done by measuring the temporal evolution of the optical phonon anti-Stokes/Stokes scattering ratio, and making a number of complicated corrections for the temperature dependent parameters which influenced it.

Von de Linde et al.[191] were the first to use picosecond probe pulses to study the dynamics of individual LO phonon mode populations following the injection of an electron-hole plasma in bulk GaAs by means of a picosecond pump pulse. The basic technique introduced in their 1980 paper has been subsequently used by many others in different material systems and on different timescales. In their experiment, the pump pulse injects electrons into the central valley of the GaAs with an excess energy above the conduction band edge of ~ 500 meV. These electrons relax towards the band edge primarily by the emission of small wavevector LO phonons via the Fröhlich interaction. By monitoring the strength of the anti-Stokes component of the LO phonon Raman signal produced by a delayed probe beam (proportional to the mode occupation number), von der Linde et al.[191] were able to observe the buildup of LO phonons produced via this electronic relaxation process, as well as the subsequent decay of the non-equilibrium LO phonon population. The decay time represented a direct measure of the population lifetime of the LO phonons as determined by anharmonic decay processes into lower energy phonons. It was noted that this lifetime determined directly in the time domain, ~ 7 ps at 77 K, was consistent with the lifetime deduced previously from CW Raman linewidth measurements of the LO phonon mode in GaAs.

Kash and coworkers[192] subsequently used subpicosecond laser pulses to extend von der Linde et al.'s measurements to room temperature, at which the LO phonon lifetime was found to be only 3.5 ps. The improved temporal resolution also allowed them to estimate the electron-LO phonon emission time, ~ 165 fs, from the risetime of the non-equilibrium LO phonon signal. More recently Kash et al.[193] used picosecond Raman scattering to study the nature of the Fröhlich interaction in group III-V semiconductor alloys. By monitoring the peak non-equilibrium populations of both GaAs-like and AlAs-like LO phonon modes in AlGaAs, they were able to deduce the relative Fröhlich coupling strengths of the electrons with the two types of LO phonons. They concluded that the overall coupling of the electrons to the LO phonons was unaffected by the alloying and that the relative coupling strength could be accounted for using spectroscopically determined parameters in a calculation of the Fröhlich interaction strengths.

In a separate experiment, Kash et al.[194] compared the wavevector dependence of non-equilibrium LO phonon generation in bulk GaAs, $Al_{0.11}Ga_{0.89}As$, and a 50 nm thick slab of GaAs by time-resolving the anti-Stokes LO phonon signal observed both in forward and back scattering geometries. The absence of a non-equilibrium signal in the forward scattering geometry for bulk GaAs and $Al_{0.11}Ga_{0.89}As$ samples, together with a significant forward scattered signal in the thin GaAs slab, led Kash and coworkers to conclude that the wave vectors of Raman active LO phonons in AlGaAs are well-defined, and the corresponding modes are not spatially localized.

Tsen and Morkoç[195,196] extended the same technique to study the GaAs-like mode in GaAs/AlGaAs multiple quantum well structures. They deduced that, within experimental error, the LO phonon lifetime was the same in 20-nm thick GaAs layers as it was in the bulk. They also concluded that the electron-LO phonon emission time was also very similar to that in the bulk.

Collins and Yu[197] devised an elaborate experiment in which the LO phonon generation rate, as deduced using von der Linde et al.'s technique, was studied as a function of the pump beam frequency. Discontinuities in the generation rate occurred at incident photon energies corresponding to the excitation of electrons into the conduction band with sufficient energy to scatter into the L and X (separate discontinuities) satellite valleys. Theoretical fits to their data were used to extract intervalley deformation potentials.

Young et al.[198] and Genack et al.[199] independently reported the results of optical phonon dynamic studies in the group IV material, Ge. Genack et al.[199] pointed out the fact that the lifetimes obtained in the time-domain were consistently longer than the corresponding inverse CW linewidths, over a range of temperatures up to 300 K, and postulated that this might be due to isotopic disorder in Ge. However, very recent results on an isotopically-pure Ge crystal[200] indicate that this disorder mechanism does not contribute to the optical phonon linewidth in naturally occurring Ge. By modeling the microscopic plasma and phonon dynamics in picosecond-laser-excited Ge, Othonos et al.[201] showed that the temporal evolution of the optical phonon population deduced from the Raman experiments[198] could be quantitatively understood using deformation potentials obtained from the literature.

All of the picosecond experiments described above basically involved monitoring the changing population of a well-defined optical phonon mode through its anti-Stokes Raman efficiency. There are also a number of examples in which the evolution of an entire portion of the spectrum (anti-Stokes and/or Stokes) is time-resolved. In the case of the single-particle spectrum from photo-excited plasmas in GaAs[202] and Si,[203] this information has been used to infer the temporal evolution of the plasma temperature,[202] and relative changes in its density.[203]

Another example of solid state plasma diagnosis using temporally and spectrally resolved Raman signals makes use of the coupling of photo-excited plasmons and LO phonons in zincblende materials. By using a 3 ps probe pulse to monitor the Stokes spectrum near the bare LO phonon line in InP and GaAs as an 80 ps pump pulse injected an electron-hole plasma, Young et al.[204,205] showed that the LO phonon spectrum broadened and reduced in intensity by a factor of ~ 5 times, and shifted up in energy by ~ 20 cm^{-1} for injected plasma densities of $< 5 \times 10^{17}$ cm^{-3}. By fitting these coupled plasmon-LO phonon spectra using a multi-component dielectric function calculation, the temporal evolution of the surface plasma density was deduced and described by a simple diffusion model. These results also explicitly point out a limitation on phonon dynamic studies in GaAs and InP; namely that the photo-excited plasma density must be kept well below 5×10^{16} cm^{-3} in order that the anti-Stokes signal intensity reflects only the occupation number of the mode, and not its renormalization via interaction with plasmons.

Time-resolved Raman scattering from intersubband electronic transitions in GaAs/AlGaAs multiple quantum wells was reported by Oberli et al.[206] They used pump and probe pulses of different frequencies to selectively inject electrons into different conduction subbands and to resonantly enhance the signal from different intersubband transitions. By monitoring the decay of the Stokes signal from the second to third subband transition when injecting carriers into both the first and second subbands, they deduced a second-to-first intersubband scattering time of ~ 325 ps when the subbands were separated by less than an LO phonon energy. They could not resolve the intersubband scattering rate with their apparatus (resolution ~ 8 ps) in thinner GaAs quantum wells in which the subbands were separated by more than an LO phonon energy.

More recently, Tatham et al.[207] used sub-picosecond pulses and different resonance conditions to time-resolve the anti-Stokes, second-to-first intersubband signal in a pump probe experiment similar to that of Oberli et al.[206] Using this technique they successfully resolved the intersubband relaxation rate in samples where the subband spacing exceeded the LO phonon energy. It was found it to be ~ 1 ps, in good agreement with model calculations of LO-phonon assisted intersubband scattering rates, including multiple subband occupancies.

A different type of quantum well experiment was recently reported by Tsen et al.[208] They studied the lateral diffusion dynamics of electron-hole plasmas in GaAs quantum wells by spatially and temporally monitoring the intersubband Raman signal following local excitation. They found that the lateral ambipolar diffusion coefficient was ~ 120 cm^2/s at an ambient temperature of ~ 105 K, and that this diffusion coefficient was not sensitive to the GaAs quantum well width from 10 to 30 nm.

This brief review of time-resolved Raman studies of semiconductors and semiconductor structures is meant only to highlight some of the main techniques and the types of information that can be deduced using them. Much more detail of these techniques and new methods of extracting important dynamical information using time-resolved Raman scattering can be found in the five articles on this topic in these proceedings.

FUTURE PROSPECTS

The above portions of this chapter are intended to serve as background material, laying the foundation for the remaining chapters of this volume that summarize the current state-of-the-art in the field of light scattering in semiconductor structures. In comparing the articles that follow with examples from the historical review in this chapter, it should be clear that this field is now being driven to a large extent by the rapidly advancing semiconductor growth and fabrication technologies. As these technologies continue to expand at an increasing rate, their role in determining the directions of the light scattering field will continue to grow in the coming decade. To keep pace with these technological developments, advances in our fundamental understanding of the light scattering process are essential in order to take full advantage of the information provided by this powerful technique.

From the materials point of view, there will definitely be developments in smaller bandgap compound semiconductor materials compatible with opto-electronic/fibre optic applications. Silicon will surely remain the dominant electronic material, and there will be increasing demands to better understand the interfaces of Si with insulators, metals, and other semiconducting materials. For any of these materials, nanofabrication techniques will be applied to produce reduced-dimensional structures from the basic 2-D quantum well structures studied so extensively during the 1980s. In order to use light scattering to effectively study the nature of structures with multiple dimensions less than 100 nm in a variety of material hosts, new experimental methods and a better theoretical framework will have to be developed in parallel. Resonance enhancement will become increasingly important, requiring the development of more tunable light sources (at shorter and longer wavelengths) and associated detectors, as well as a quantitative understanding of resonant mechanisms in these structures. Also of practical importance is a theoretical understanding of selection rules (and the breakdown thereof), the fundamental overlap of Raman scattering with the general field of non-linear optics, and the development of electronic Raman theory to the same level as our current knowledge of phonon Raman scattering.

All of the above relate to the use of Raman scattering to elucidate the equilibrium properties of exotic new structures and materials. As the ultimate application of these structures is in novel electronic and opto-electronic devices, there will also be increasing applications of Raman scattering to help further our understanding of non-equilibrium transport processes in these devices.

Of course, these are all predictions based on extrapolation of past experience, and the most exciting developments to be reported in the year 2000 at a Workshop on Light Scattering in Semiconductors will undoubtedly have to do with developments that cannot be foreseen. What can be foreseen is that the future will be more exciting than we can predict!

REFERENCES

1.   L. Brillouin, Ann. Phys. (Paris) 17, 88 (1922).
2.   A. Smekal, Naturwiss. 11, 873 (1923).
3.   C.V. Raman, Ind. J. Phys. 2, 387 (1928); C.V. Raman, Nature 121, 619 (1928); C.V. Raman and K.S. Krishnan, Nature 121, 501 (1928).
4.   H.A. Kramers and W. Heisenberg, Z. Phys. 31, 681 (1925).
5.   P.A.M. Dirac, Proc. Roy. Soc. London 114, 710 (1927).
6.   G. Landsberg and L. Mandelstam, Naturwiss. 16, 557 and 772 (1928).
7.   Y. Rocard, C.R. Acad. Sci. (Paris) 186, 1107 (1928).
8.   R.W. Wood, Nature 112, 349 (1928); R.W. Wood, Phil. Mag. 6, 729 (1928).
9.   R. Robertson and J.J. Fox, Nature 125, 704 (1930); C. Ramaswamy, Ind. J. Phys. 5, 97 (1930).
10. F. Rasetti, Nature 127, 626 (1931); E. Fermi and Rasetti, Z. Phys. 71, 689 (1931).
11. M. Born and M. Blackman, Z. Phys. 82, 55 (1933).
12. G. Placzek, Handbuch der Radiologie 6 (part 2), 209 (1934).
13. F. Rasetti, Z. Physik 66, 646 (1930).
14. E.F. Gross, Nature 126, 201 (1930).
15. L.I. Mandelstam, Zh. Radio-Fiziko-Khimicheskogo Obshchestva 58, 381 (1926).
16. L. Landau and G. Placzek, Phys. Z. Sowjetunion 5, 172 (1934).
17. C.V. Raman and T.M.K. Negungadi, Nature 145, 147 (1940).
18. M. Born and M. Bradburn, Proc. Roy. Soc. A 188, 161 (1947).
19. R.S. Krishnan, Proc. Ind. Acad. Sci. A 19, 216 (1944); R.S. Krishnan, Proc. Ind. Acad. Sci. A 24, 25 and 45 (1946).
20. L. Couture and J.P. Mathieu, Ann. Phys. (Paris) 3, 521 (1947).
21. L. Couture-Mathieu and J.P. Mathieu, C.R. Acad. Sci. (Paris) 236, 25 (1953).
22. H. Poulet, C.R. Acad. Sci. (Paris) 238, 70 (1954); H. Poulet, Ann. de Phys. 10, 908 (1955).
23. A.I. Stekhanov and M.L. Petrova, Zh. Eksp. Teor. Fiz. 9, 1108 (1944); A.I. Stekhanov and M.B. Eliashberg, Opt. Spectrosc. 10, 174 (1961).
24. A.C. Menzies and J. Skinner, J. Phys. Radium 9, 93 (1948); A.C. Menzies and J. Skinner, J. Chem. Phys. 46, 60 (1949).
25. H.L. Welsh, M.F. Crawford, and W.J. Staple, Nature 164, 737 (1949).
26. R. Loudon, Proc. Roy. Soc. A 275, 218 (1964); R. Loudon, Advan. Phys. 13, 423 (1964).
27. See, for example, M.G. Cottam and D.J. Lockwood, "Light Scattering in Magnetic Solids", Wiley, New York (1986).
28. E.J. Woodbury and W.K. Ng, Proc. IRE 50, 2637 (1962).
29. R.Y. Chiao, C.H. Townes, and B.P. Stoicheff, Phys. Rev. Lett. 12, 592 (1964).
30. D.I. Mash, V.V. Mororov, V.S. Starunov, and I.L. Fabelinskii, JETP Lett. 2, 25 (1965).
31. N. Bloembergen, Am. J. Phys. 35, 989 (1967).
32. M.V. Hobden and J.P. Russell, Phys. Lett. 13, 39 (1964).
33. W.J. Jones and B.P. Stoicheff, Phys. Rev. Lett. 13, 657 (1964).
34. P.D. Maker and R.W. Terhune, Phys. Rev. 137, A801 (1965).
35. R. Terhune, P.D. Maker, and C.M. Savage, Phys. Rev. Lett. 14, 681 (1965).
36. J.P. Russell, J. de Phys. 26, 620 (1965).
37. R.S. Krishnan and N. Krishnamurthy, J. de Phys. 26, 630 (1965).
38. E. Burstein and S. Ganesan, J. de Phys. 26, 637 (1965).
39. J.P. Russell, Appl. Phys. Lett. 6, 223 (1965).
40. J.L. Birman, Phys. Rev. 127, 1093 (1965).
41. L. Kleinman, Solid State Commun. 3, 47 (1965).
42. C.H. Henry and J.J. Hopfield, Phys. Rev. Lett. 15, 964 (1965).
43. T.C. Damen, S.P.S. Porto, and B. Tell, Phys. Rev. 142, 570 (1966).
44. S.P.S. Porto, B. Tell, and T.C. Damen, Phys. Rev. Lett. 16, 450 (1966).
45. E. Anastassakis, S. Iwasa, and E. Burstein, Phys. Rev. Lett. 17, 1051 (1966).
46. M.H. Grimsditch and A.K. Ramdas, Phys. Rev. B11, 3139 (1975).
47. P.A. Wolff, Phys. Rev. Lett. 16, 225 (1966).
48. Y. Yafet, Phys. Rev. 152, 858 (1966).
49. A. Mooradian and G.B. Wright, Solid State Commun. 4, 431 (1966).
50. A. Mooradian and G.B. Wright, Phys. Rev. Lett. 16, 999 (1966).

51. R.C.C. Leite and S.P.S. Porto, Phys. Rev. Lett. $\underline{17}$, 10 (1966).
52. A.K. Ganguly and J.L. Birman, Phys. Rev. Lett. $\underline{17}$, 647 (1966).
53. J.A. Giordmaine and W. Kaiser, Phys. Rev. $\underline{144}$, 676 (1966).
54. J.H. Parker, Jr., D.W. Feldman, and M. Ashkin, Phys. Rev. $\underline{155}$, 712 (1967).
55. G.B. Wright and A. Mooradian, Phys. Rev. Lett. $\underline{18}$, 608 (1967).
56. R.E. Slusher, C.K.N. Patel, and P.A. Fleury, Phys. Rev. Lett. $\underline{18}$, 77 (1967).
57. M. Krauzmann, C.R. Acad. Sci. (Paris) $\underline{264}$ B, 1117 (1967); W. Taylor, Phys. Lett. $\underline{24}$ A, 556 (1967).
58. P.A. Fleury and J.M. Worlock, Phys. Rev. Lett. $\underline{18}$, 665 (1967); P.A. Fleury and J.M. Worlock, Phys. Rev. Lett. $\underline{19}$, 1176 (1967).
59. A. Pinczuk, E. Burstein, W. Taylor, and I. Lefkowitz, Solid State Commun. $\underline{5}$, 429 (1967).
60. D.W. Feldman, J.H. Parker, Jr., W.J. Choyke, and L. Patrick, Phys. Rev. $\underline{173}$, 698 and 787 (1968).
61. A. Mooradian, Phys. Rev. Lett. $\underline{20}$, 1102 (1968).
62. D.G. Thomas and J.J. Hopfield, Phys. Rev. $\underline{175}$, 1021 (1968).
63. A. Pinczuk and E. Burstein, Phys. Rev. Lett. $\underline{21}$, 1073 (1968).
64. G.B. Wright, ed., "Light Scattering Spectra of Solids", Springer-Verlag, New York (1969).
65. A. Mooradian and G.B. Wright, in: "The Physics of Se and Te", W.C. Cooper, ed., Pergamon, New York (1969), p. 266.
66. E. Burstein, S. Ushioda, A. Pinczuk, and D.L. Mills, Phys. Rev. Lett. $\underline{22}$, 348 (1969); D.L. Mills and E. Burstein, Phys. Rev. $\underline{188}$, 1465 (1969).
67. R.C.C. Leite, J.F. Scott, and T.C. Damen, Phys. Rev. Lett. $\underline{22}$, 780 (1969); M. Klein and S.P.S. Porto, Phys. Rev. Lett. $\underline{22}$, 782 (1969).
68. J.F. Scott, T.C. Damen, W.T. Silfvast, R.C.C. Leite, and L.E. Cheesman, Optics Commun. $\underline{1}$, 397 (1970).
69. A. Pinczuk and E. Burstein, in: "Proc. Int. Conf. on Physics of Semiconductors, Boston, 1970", S.P. Keller, J.C. Hensel, and F. Stern, eds., USAFC Tech. Inform. Div., Oak Ridge, TN (1970), p. 266.
70. F.A. Blum, Phys. Rev. $\underline{B1}$, 1125 (1970).
71. E. Anastassakis, A. Pinczuk, E. Burstein, M. Cardona, and F. Pollak, Solid State Commun. $\underline{8}$, 133 (1970).
72. C.K.N. Patel and E.D. Shaw, Phys. Rev. Lett. $\underline{24}$, 451 (1970).
73. E. Anastassakis and E. Burstein, Phys. Rev. $B\underline{2}$, 1952 (1970).
74. M. Balkanski, ed., "Light Scattering in Solids", Flammarion, Paris (1971).
75. L.J. Brillson, E. Burstein, A.A. Maradudin, and T. Stark, in: "Proc. Conf. on Physics of Semimetals and Narrow Gap Semiconductors", D. Carter and R. Bate, eds., Pergamon Press. Oxford (1971), p. 187.
76. E. Anastassakis and E. Burstein, J. Phys. Chem. Solids $\underline{32}$, 313 (1971).
77. R.M. Martin, Phys. Rev. B $\underline{4}$, 3676 (1971).
78. A. Pinczuk, L.J. Brillson, E. Burstein, and E. Anastassakis, Phys. Rev. Lett. $\underline{27}$, 317 (1971).
79. E. Burstein, D.L. Mills, and R.F. Wallis, Phys. Rev.B $\underline{4}$, 2429 (1971).
80. P.A. Fleury and J.F. Scott, Phys. Rev. B$\underline{3}$, 1979 (1971).
81. E. Anastassakis and E. Burstein, Solid State Commun. $\underline{9}$, 525 (1971).
82. C.J. Buchenauer, M. Cardona, and F. Pollak, Phys. Rev. B$\underline{3}$, 1243 (1971).
83. M. Cardona and G. Güntherodt, eds., "Light Scattering in Solids", Springer-Verlag, Heidelberg (1975-1989), Vols. I-V.
84. L.J. Brillson and E. Burstein, Phys. Rev. B$\underline{5}$, 2973 (1972).
85. J. Geurts and W. Richter, in: "Proc. Int. Conf. on Physics of Semiconductors, Edinburgh, 1978", B.L.H. Wilson, ed., Inst. of Physics, London (1979), p. 513.
86. P. Yu and R. Shen, Phys. Rev. Lett. $\underline{29}$, 468 (1972).
87. M.L. Shand, W. Richter, and E. Burstein, J. Nonmetals $\underline{1}$, 53 (1972).
88. F. Cerdeira, W. Dreybrodt, and M. Cardona, Solid State Commun. $\underline{10}$, 591 (1972).
89. S.R.J. Brueck and A. Mooradian, Phys. Rev. Lett. $\underline{28}$, 161 (1972).
90. W. Brenig, R. Zeyher, and J.L. Birman, Phys. Rev. B$\underline{6}$, 4613 and 4616 (1972).
91. A. Compaan and H.Z. Cummins, Phys. Rev. Lett. $\underline{31}$, 41 (1973).
92. J. Doehler, P.J. Colwell, and S.A. Solin, Phys. Rev. B $\underline{9}$, 636 (1974).
93. M. Cardona, Surf. Sci. $\underline{37}$, 100 (1973).

94. D.J. Evans, S. Ushioda, and J.D. McMullen, Phys. Rev. Lett. 31, 369 (1973).
95. Y.J. Chen, E. Burstein, and D.L. Mills, Phys. Rev. Lett. 34, 1516 (1975).
96. R. Beserman, M. Jouanne, and M. Balkanski, in: "Proc. 11th Int. Conf. on Physics of Semiconductors", P.W.N.-Polish Scientific, Warsaw (1972), p. 1181.
97. F. Cerdeira, T.A. Fjeldly, and M. Cardona, Phys. Rev. B 8, 4734 (1973); M. Chandrasekhar, J.B. Renucci, and M. Cardona, Phys. Rev. B 17, 1623 (1978).
98. M. Cardona, F. Cerdeira, and T.A. Fjeldly, Phys. Rev. B 10, 3433 (1974).
99. H. Vogt, Appl. Phys. 5, 85 (1974).
100. H. Vogt and G. Neuman, Phys. Stat. Sol. (b) 92, 57 (1979).
101. M. Balkanski, R.C.C. Leite, and S.P.S. Porto, eds., "Light Scattering in Solids", Flammarion, Paris (1975).
102. D.J. Lockwood and H. Montgomery, J. Phys. C 8, 3241 (1975); D.J. Lockwood, in: "Ternary Compounds 1977", G.D. Holah, ed., Inst. Physics, Bristol (1977), p. 97.
103. J.H. Nicola, J.A. Freitas, Jr., and R.C.C. Leite, Solid State Commun. 17, 1379 (1975).
104. A. Polian, K. Kunc, and A. Kuhn, Solid State Commun. 19, 1079 (1976).
105. N. Nagasawa, T. Mita, and M. Ueta, J. Phys. Soc. Japan 41, 929 (1976).
106. S. Buchner, L.Y. Ching, and E. Burstein, Phys. Rev. B 14, 4459 (1976).
107. A. Miller, D.J. Lockwood, A. MacKinnon, and D. Weaire, J. Phys. C 9, 2997 (1976).
108. M. Chandrasekhar, M. Cardona, and E.O. Kane, Phys. Rev. B 16, 3579 (1977).
109. K.R. Subbaswamy and D.L. Mills, Solid State Commun. 27, 1085 (1978).
110. J.B. Valdez, G. Malter, and S. Ushioda, Solid State Commun. 27, 1089 (1978).
111. E. Burstein, A. Pinczuk, and S. Buchner, in: " Physics of Semiconductors 1978", B.L.H. Wilson, ed., Inst. of Physics, London (1979), p. 1231.
112. E. Burstein, A. Pinczuk, and D.L. Mills, Surf. Sci. 98, 451 (1980).
113. L.Y. Ching, E. Burstein, S. Buchner, and H.H. Wieder, J. Phys. Soc. Japan 49, Suppl. A, 951 (1980).
114. R.G. Ulbrich, N.V. Hieu, and C. Weisbuch, Phys. Rev. Lett. 46, 53 (1980).
115. A. Alexandrou, in: "Proc. 20th Int. Conf. on Physics of Semiconductors", E.M. Anastassakis and J.D. Joannopoulos, eds., World Scientific, Singapore (1990), p. 1835.
116. C. Trallero-Giner, in: "Proc. 20th Int. Conf. on Physics of Semiconductors", E.M. Anastassakis and J.D. Joannopoulos, eds., World Scientific, Singapore (1990), p. 1899.
117. T. Ruf and M. Cardona, Phys. Rev. Lett. 63, 2288 (1989); T. Ruf and M. Cardona, Phys. Rev. 41, 10747 (1990).
118. J.L. Merz, A.S. Barker, Jr., and A.C. Gossard, Appl. Phys. Lett. 31, 117 (1977); A.S. Barker, A.C. Gossard, and J.L. Merz, Phys. Rev. B 17, 3181 (1978).
119. C. Colvard, R. Merlin, M.V. Klein, and A. Gossard, Phys. Rev. Lett. 43, 298 (1980).
120. S.M. Rytov, Sov. Phys. Acoust. 2, 68 (1956).
121. V. Narayanamirti, H.L. Stoermer, M.A. Chin, A.C. Gossard, and W. Wiegemann, Phys. Rev. Lett. 43, 2012 (1979).
122. P.V. Santos, J. Mebert, O. Koblinger, and L. Ley, Phys. Rev. B 36, 1306 (1987).
123. K. Kubota, M. Nakayama, H. Katoh, and N. Sano, Solid State Commun. 49, 157 (1984).
124. A. Ishibashi, Y. Mori, M. Itabashi, and N. Watanabe, J. Appl. Phys. 58, 2691 (1985).
125. G.A. Sai-Halasz, A. Pinczuk, P.Y. Yu, and L. Esaki, Solid State Commun. 25, 381 (1978).
126. R. Merlin, C. Colvard, M.V. Klein, H. Morkoç, A.Y. Cho, and A.C. Gossard, Appl. Phys. Lett. 36, 43 (1980).
127. B. Jusserand, D. Paquet, and A. Regreny, Phys. Rev. B 30, 6245 (1984).
128. A.K. Sood, J. Menéndez, M. Cardona, and K. Ploog, Phys. Rev. Lett. 54, 2111 (1985).
129. A.K. Sood, J. Menéndez, M. Cardona, and K. Ploog, Phys. Rev. Lett. 54, 2115 (1985).

130. R.E. Camley and D.L. Mills, Phys. Rev. B 29, 1965 (1984); R.P. Pokatilov and S.I. Beril, Phys. Stat. Sol. (b) 118, 567 (1983).
131. G. Ambrazevičius, M. Cardona, R. Merlin, and K. Ploog, Solid State Commun. 65, 1035 (1988).
132. P. Lambin, J.P. Vigneron, A.A. Lucas, P.A. Thiry, M. Liehr, J.J. Pireaux, R. Caudano, and T.J. Kuech, Phys. Rev. Lett. 56, 1227 (1984).
133. A. Huber, T. Egeler, W. Ettmüller, H. Rothfritz, G. Tränkle, and G. Abstreiter, Proceedings of the Int. Conf. on Superlattices and Microstructures, Berlin 1990, to be published.
134. H.D. Fuchs, to be published.
135. K. Huang and B. Zhu, Phys. Rev. B 38, 13377 (1988).
136. Z. Popović, M. Cardona, E. Richter, D. Strauch, L. Tapfer, and K. Ploog, Phys. Rev. B 41, 5904 (1990).
137. J.E. Zucker, A. Pinczuk, D.S. Chemla, A. Gossard, and W. Wiegmann, Phys. Rev. Lett. 53, 1280 (1984).
138. E. Friess, H. Brugger, K. Eberl, G. Krötz, and G. Abstreiter, Solid State Commun. 69, 899 (1989).
139. Z. Popović, M. Cardona, E. Richter, D. Strauch, L. Tapfer, and K. Ploog, Phys. Rev. B 43, 4925 (1991).
140. A. Fasolino, E. Molinari, and K. Kunc, Phys. Rev. Lett. 56, 1751 (1986); B. Jusserand and D. Paquet, Phys. Rev. Lett. 56, 1752 (1986).
141. P. Molinas i Mata, I. Alonso, and M. Cardona, Solid State Commun. 74, 347 (1990).
142. A.K. Sood, J. Menéndez, M. Cardona, and K. Ploog, Phys. Rev. B 32, 1412 (1985).
143. D.J. Mowbray, M. Cardona, and K. Ploog, Phys. Rev. B 43, 1598 (1991).
144. P. Manuel, G.A. Sai-Halasz, L.L. Chang, Chin-an Chang, and L. Esaki, Phys. Rev. Lett. 37, 1701 (1976).
145. C. Colvard, T.A. Gant, M.V. Klein, R. Merlin, R. Fischer, H. Morkoç, and A.C. Gossard, Phys. Rev. B 31, 2080 (1985).
146. B. Jusserand, D. Paquet, and F. Mollot, Phys. Rev. Lett. 63, 2397 (1989).
147. E. Anastassakis, A. Pinczuk, E. Burstein, M. Cardona, and F. Pollak, Solid State Commun. 8, 133 (1970).
148. P. Wickboldt, E. Anastassakis, R. Sauer, and M. Cardona, Phys. Rev. B 35, 1362 (1987).
149. F. Cerdeira, A. Pinczuk, J.C. Bean, B. Battlog, and B.A. Wilson, Appl. Phys. Lett. 45, 1138 (1984); G H. Brugger, T. Wolf, H. Jorke, and H.J. Herzog, Phys. Rev. Lett. 54, 2441 (1985).
150. T. Dumelow, A.A. Hamilton, K.A. Maslin, T.J. Parker, B. Samson, S.R.P. Smith, D.R. Tilley, R.B. Beall, C.T.B. Foxon, J.J. Harris, D. Hilton, and K.J. Moore, in these proceedings.
151. Yu.A. Pusep, A.F. Milekhin, M.P. Sinyukov, K. Ploog, and A.I. Toporov, to be published.
152. B. Jusserand and M. Cardona in: "Light Scattering in Solids V", M. Cardona and G. Güntherodt, eds., (Springer, Heidelberg, 1989), p. 49; M.V. Klein, IEEE J. QE-22, 1760 (1986); J. Menéndez, J. Luminesc. 44, 285 (1989); M. Cardona, in: "Lectures on Surface Sciences", G.R. Castro and M. Cardona, eds., (Springer, Heidelberg, 1987), p. 2; M. Cardona, Superlattices and Microstructures 5, 27 (1989); M. Cardona, Superlattices and Microstructures 7, 183 (1990); M. Cardona, in: "Spectroscopy of Semiconductor Microstructures", G. Fasol, A. Fasolino, and P. Lugi, eds., (Plenum, New York, 1990), p. 143.
153. A. Pinczuk, L. Brillson, E. Anastassakis, and E. Burstein, in: "Light Scattering in Solids", M. Balkanski, ed., Flammarion, Paris (1971), p. 115.
154. G. Abstreiter, R. Trommer, M. Cardona, and A. Pinczuk, Solid State Commun. 30, 703 (1979).
155. A. Pinczuk, G. Abstreiter, R. Trommer, and M. Cardona, Solid State Commun. 30, 429 (1979).
156. G. Abstreiter and K. Ploog, Phys. Rev. Lett. 42, 1308 (1979).
157. A. Pinczuk, H.L. Störmer, R. Dingle, J.M. Worlock, W. Wiegmann, and A.C. Gossard, Solid State Commun. 32, 1001 (1979).

158. G. Abstreiter, M. Cardona, and A. Pinczuk, in: "Light Scattering in Solids IV", M. Cardona and G. Güntherodt, eds., Springer-Verlag, Berlin-Heidelberg (1984), p. 5.
159. G. Abstreiter, R. Merlin, and A. Pinczuk, IEEE J. Quantum Electron. 22, 1609 (1986).
160. A. Pinczuk and G. Abstreiter, in: "Light Scattering in Solids V", M. Cardona and G. Güntherodt, eds., Springer-Verlag, Berlin-Heidelberg (1989), p. 153.
161. P.M. Platzman and N. Tzoar, Phys. Rev. 136, A11 (1964).
162. P.M. Platzman, Phys. Rev. 139, A379 (1965).
163. P.M. Platzman and P.A. Wolff, "Waves and Interactions in Solid State Plasmas", Academic Press, New York (1973).
164. D.C. Hamilton and A.L. McWhorter, in: "Light Scattering Spectra of Solids", G.B. Wright, ed., Springer, Berlin-Heidelberg-New York (1969), p. 309.
165. C.K.N. Patel and R.E. Slusher, Phys. Rev. Lett. 21, 1563 (1968).
166. A. Mooradian, in: "Festkörperprobleme IX", Pergamon-Vieweg, Braunschweig (1969), p. 74.
167. M.V. Klein, in: "Light Scattering in Solids", M. Cardona, ed., Springer-Verlag, Berlin-Heidelberg (1973), p. 147.
168. W. Hayes and R. Loudon, "Scattering of Light by Crystals", Wiley, New York (1978).
169. G. Abstreiter, U. Claessen, and G. Tränkle, Solid State Commun. 44, 673 (1982).
170. M. Baumgartner, G. Abstreiter, and E. Bourgert, J. Phys. C17, 1617 (1984).
171. A. Pinczuk, D. Heiman, R. Sooryakumar, A.C. Gossard, and W. Wiegmann, Surface Sci. 170, 573 (1986).
172. D. Heiman, A. Pinczuk, A.C. Gossard, J.H. English, A. Fasolino, and M. Altarelli, in: "Proc. 18th Int. Conf. on the Physics of Semiconductors", O. Engström, ed., World Scientific, Singapore (1987), p. 617.
173. G.H. Döbler, H. Kuengel, D. Olego, K. Ploog, P. Ruden, H.J. Stolz, and G. Abstreiter, Phys. Rev. Lett. 47, 864 (1981).
174. C. Zeller, B. Vinter, G. Abstreiter, and K. Ploog, Phys. Rev. B26, 2124 (1982).
175. G. Fasol, P. Ruden, and K. Ploog, J. Phys. C17, 1395 (1984).
176. T.A. Perry, R. Merlin, B.V. Shanabrook, and J. Comas, Phys. Rev. Lett. 54, 2623 (1985).
177. D. Gammon, R. Merlin, W.T. Masselink, and H. Morkoc, Phys. Rev. B33, 2919 (1986).
178. N. Krischka, A. Zrenner, G. Abstreiter, and K. Ploog, cited in Refs. 159 and 160.
179. J.M. Worlock, A.C. Maciel, M. Tatham, J.F. Ryan, R.E. Nahory, J.P. Harbison, and L.T. Florez, in these proceedings.
180. J.S. Weiner, G. Danan, A. Pinczuk, J.P. Valladares, L.N. Pfeiffer, and K.W. West, Phys. Rev. Lett. 63, 1641 (1989).
181. G. Abstreiter, S. Beeck, T. Egeler, and A. Huber, in these proceedings.
182. T. Egeler, G. Abstreiter, G. Weiman, T. Demel, D. Heitmann, P. Grambow, and W. Schlapp, Phys. Rev. Lett. 65, 1804 (1990).
183. A. Goni, A. Pinczuk, J.S. Weiner, B.S. Dennis, J.M. Calleja, L.N. Pfeiffer, and K.W. West, Int. Symp. on Nanostructures and Mesoscopic Systems, submitted.
184. A. Pinczuk, J. Shah, and P.A. Wolff, Phys. Rev. Lett. 47, 1487 (1981).
185. H. Nather and L.G. Quagliano, J. Luminescence 30, 50 (1985).
186. A.R. Vasconcellos, R.S. Turtelli, and A.R.B. de Castro, Solid State Commun. 22, 97 (1977).
187. J.E. Kardontchik and E. Cohen, Phys. Rev. Lett. 42, 669 (1979).
188. H. Yugami, S. Nakashima, Y. Oka, M. Hangyo, and A. Mitsuishi, J. Appl. Phys. 60, 3303 (1986).
189. A. Compaan, in: Mat. Res. Soc. Symp. Proc. Vol. 35, D.K. Biegelsen, G.A. Rozgonyi, and C.V. Shank, eds., Materials Research Society, Pittsburgh, (1985), p. 65.
190. D. von der Linde, G. Wartmann, M. Kemmler, and Z. Zhu, in: Mat. Res. Soc. Symp. Proc. Vol. 23, J.C.C. Fan and N.M. Johnson, eds., Elsevier, New York, (1984), p. 123.
191. D. von der Linde, J. Kuhl, and H. Klingenberg, Phys. Rev. Lett. 44, 1505 (1980).
192. J.A. Kash, J.C. Tsang, and J.M. Hvam, Phys. Rev. Lett. 54, 2151 (1985).
193. J.A. Kash, S.S. Jha, and J.C. Tsang, Phys. Rev. Lett. 58, 1869 (1987).

194. J.A. Kash, J.M. Hvam, J.C. Tsang, and T.F. Kuech, Phys. Rev. B $\underline{38}$, 5776 (1988).
195. K.T. Tsen and H. Morkoç, Phys. Rev. B $\underline{34}$, 4412 (1986).
196. K.T. Tsen and H. Morkoç, Phys. Rev. B $\underline{38}$, 5615 (1988).
197. C.L. Collins and P.Y. Yu, Phys. Rev. B $\underline{30}$, 4501 (1984).
198. J.F. Young, K. Wan, and H.M. van Driel, Solid-State Electronics $\underline{31}$, 455 (1988).
199. A.Z. Genack, L. Ye, and C.B. Roxlo, SPIE Proc. $\underline{942}$, 140 (1988).
200. H.D. Fuchs, C.H. Grein, C. Thomsen, M. Cardona, W.L. Hansen, and E.E. Haller, Phys. Rev. B $\underline{43}$, 4835 (1991).
201. A. Othonos, H.M. van Driel, J.F. Young, and P.J. Kelly, Phys. Rev. B $\underline{43}$, 6682 (1991).
202. Y. Huang and P.Y. Yu, Solid State Commun. $\underline{63}$, 109 (1987).
203. K.T. Tsen and O.F. Sankey, Phys. Rev. B $\underline{37}$, 4321 (1988).
204. J.F. Young and K. Wan, Phys. Rev. B $\underline{35}$, 2544 (1987).
205. J.F. Young, K. Wan, A.J. SpringThorpe, and P. Mandeville, Phys. Rev. B $\underline{36}$, 1316 (1987).
206. D.Y. Oberli, D.R. Wake, M.V. Klein, J. Klem, T. Henderson, and H. Morkoç, Phys. Rev. Lett. $\underline{59}$, 696 (1987).
207. M.C. Tatham, J.F. Ryan, and C.T. Foxon, Solid-State Electronics $\underline{32}$, 1497 (1989).
208. K.T. Tsen, O.F. Sankey, G. Halama, S.Y. Tsen, and H. Morkoç, Phys. Rev. B $\underline{39}$, 6276 (1989).

ACOUSTIC, OPTIC AND INTERFACE PHONONS:

LOW SYMMETRY SUPERLATTICES

Manuel Cardona

Max-Planck-Institut für Festkörperforschung
Stuttgart, Federal Republic of Germany

INTRODUCTION

Optical spectroscopies (Raman, ir) allow only the observation of excitations very close to the center of the Brillouin zone (BZ). The formation of a superlattice divides, through loss of translational symmetry, the volume of the BZ by an integer (mini-Brillouin zone, MBZ), and the number of $k \simeq 0$ modes is enhanced by the same factor. Thus the information that can be gained through optical spectroscopies may be greatly increased with respect to the bulk constituents. Raman scattering is more powerful in this regard than infrared spectroscopy. The spectroscopic coupling constant to a given mode of the MBZ is related to the fluctuation of differential polarizabilities in the Raman case and to that of dynamical charges e_T^* for infrared spectroscopy. The former is usually much larger than the latter, since the polarizabilities undergo resonances at different (laser) frequencies ω_L in the various constituent materials. The e_T^* fluctuations are very small for superlattices made by replacing isovalent atoms (Ge/Si, GaAs/AlAs, ZnTe/CdTe). They are, however, larger if the substitution is not isovalent (e.g., Ge/GaAs, GaAs/ZnTe). Nevertheless, this case has not yet received attention because of difficulties involved in preparing such superlattices.

The lattice dynamics of semiconductor superlattices and the corresponding results from optical and electron energy-loss spectroscopies have been the object of a number of recent reviews.[1-8] The reader should consult them for background and references. Here we concentrate on recent developments, especially on results for low-symmetry growth directions.

SPACE GROUPS OF DIAMOND AND ZINCBLENDE-TYPE SUPERLATTICES

Semiconductor superlattices of these systems offer a rich variety of symmetry properties depending on composition (here we discuss Ge/Si, A_1C_1/A_1C_2, and A_1C_1/A_2C_2, where A and C are cations and anions of a zinc blende compound), number of layers of each constituent, and direction of growth. The best studied system, and one of the simplest from the point of view of symmetry, is $(GaAs)_m/(AlAs)_n$ grown along [001]. For this case the point group is always tetragonal D_{2d} ($\bar{4}2m$), regardless of m and n. Two space groups are possible, depending on whether $m+n$ is even [D_{2d}^5 ($P\bar{4}m2$), primitive tetragonal], or odd [D_{2d}^9 ($I\bar{4}m2$), body-centered tetragonal]. The super-primitive cell (SPC) of the former is a tetragonal prism, and so is the MBZ. For the latter the SPC and MBZ are more complicated. The MBZ's can be seen in Fig. 3.18 of Ref. 1. The number of atoms per SPC is, in any case, $2(m+n)$. Hence the number of phonons of $k \simeq 0$ (including the bulk-like acoustic ones) is $6(m+n)$.

The symmetry becomes lower if the two constituents have neither a common anion nor a common cation [e.g., $(InAs)_m(GaSb)_n$]. In this case the symmetry operations that transform

Light Scattering in Semiconductor Structures and Superlattices
Edited by D.J. Lockwood and J.F. Young, Plenum Press, New York, 1991

19

z into $-z$ are lost (S_4, $2C_2'$), and the point group becomes C_{2v}. The space groups are C_{2v}^1 ($Pmm2$) for $m + n$ even and C_{2v}^{20} ($Imm2$) for $m + n$ odd. The loss of the $z \to -z$ operation implies the loss of parity for confined wave functions (see next section).

The difference in the lattice parameters of the constituents Δa_0 plays an important role in the growth, structure, and spectroscopic behavior of superlattices. If Δa_0 is small ($\lesssim 1\%$) the transverse lattice constant of the superlattice is well defined and equal to an average of those of the constituents given by elasticity theory (see p. 145, Ref. 1). With increasing Δa_0 or increasing layer thicknesses, transverse lattice constant matching (pseudomorphic growth) breaks down because of the generation of dislocations that make up for the misfit. Raman spectroscopy has been a fruitful technique for the investigation of the state of strain in superlattices. For background and recent results see Refs. 9 and 10. Although the macroscopic lattice parameters of pseudomorphic growing superlattices can be obtained from the elastic compliance constants of the constituents, the position of some of the atoms within the unit-cell is not determined by symmetry and requires independent unit-cell parameters (internal strains). These parameters can, in principle, be determined by x-ray scattering. Such work, however, has not yet been performed. In order to unravel the number of internal parameters needed to specify a given superlattice structure it is helpful to the phonon spectroscopist to consider the number of fully symmetric optic phonons at Γ using factor-group techniques.[11] Each one of these modes leads to a free parameter of the SPC, since the corresponding atom can be displaced in the corresponding directions without reducing the symmetry of the crystal.[12]

Of particular interest from the point of view of symmetry are the $Ge_m Si_n$ superlattices.[13-15] The [001] case, for instance, leads to six different space groups with four different point groups. The latter are for $m = n = 1$ T_d ($\bar{4}3m$, zinc blende); for m and n even D_{2h} (mmm, orthorhombic holohedral); for m and n odd D_{2d} ($\bar{4}2m$, tetragonal); and for $m + n$ odd D_{4h} ($4/mmm$ tetragonal holohedral). The corresponding space groups are obtained by considering that for $m + n = 4\kappa$ (κ integer) the Bravais lattice is primitive; for $m + n = 4\kappa + 2$ it is body centered. In these cases the SPC contains a formula unit ($Ge_m Si_n$). For $m + n =$ odd, however, one needs two formula units per SPC. The space group is D_{4h}^{19} ($I4_1/a\ md$). The D_{2d} point group appears in the D_{2d}^9 ($I\bar{4}m2$) and D_{2d}^5 ($P\bar{4}m2$) space groups, while D_{2h} is found in the D_{2h}^5 ($Pmma$) and D_{2h}^{28} ($Imma$) space groups. The spectroscopist should note that the holohedral groups D_{2h} and D_{4h} have a center of inversion; thus Γ phonons will be either Raman or infrared allowed (or silent) but not both. The infrared activity, however, is expected to be very small.[16] In the other cases, modes can be both infrared and Raman active. The orthorhombic group D_{2h} implies optically biaxial behavior with anisotropy (slow and fast axis) in the superlattice planes. Observation of this anisotropy would require growth on atomically flat substrates, which has not yet been accomplished. However, the anisotropy has been recently calculated and found to be significant for small $m + n$.[17]

For growth along [111] the $Ge_m Si_n$ system leads to five different space groups, including the trivial zinc blende case and trigonal primitive and rhombohedral groups both with point groups D_{3d} (m or n even, holohedral, inversion symmetry) and C_{3v} (m and n odd). As usual, the case of [111] superlattices with zincblende-type constituents is simpler. One obtains only one point group, namely C_{3v} ($3m$) and either a primitive or a centered space group, depending on whether or not $m + n$ is a multiple of three.[18]

Another interesting case is $Ge_m Si_n$ [110]. These orthorhombic superlattices always have inversion symmetry (point group D_{2h}; the zinc blende structure does not belong to them). Three space groups are possible: D_{2h}^5, D_{2h}^7, and D_{2h}^{28}. Parity with respect to $\bar{z} \to -\bar{z}$ ($\bar{z}$ direction of growth) is always a good quantum number. The corresponding [110] $(C_1A_1)_m/(C_2A_1)_n$ superlattices belong to four different space groups: D_{2d}^5 ($m = n = 1$), C_{2v}^7 (m, n even), C_{2v}^1 (m, n odd), and C_{2v}^{20} ($m + n$ odd). Only the first one has $z \to -z$ parity.[19]

The $Ge_m Si_n$ superlattices grown along [112], [120], and [114] are discussed in Ref. 15. The latter two generate a large number of space groups spreading over most crystal systems.

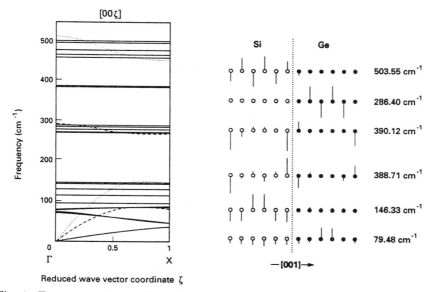

Fig. 1. Transverse-phonon-dispersion curves for a Si_6Ge_6 superlattice with Si bulk transverse optical (TO) and transverse acoustic (TA) branches overlaid as dotted lines and with Ge bulk TO and TA modes as dashed lines. Atomic displacements for selected modes are shown at the right. From Ref. 20.

CONFINED MODES

Confined modes result when the dispersion relations of both constituents do not overlap. This is the case for optical modes when the reduced masses of the constituents are sufficiently different (e.g., Ge–Si, GaAs–AlAs). For long wavelengths acoustic modes overlap (typically up to $k \geq \pi/a_0$), and propagating modes result (acoustic folded modes).

In zinc blende crystals (but not in Ge or Si) a gap usually exists between the top of the acoustic branches and the optical ones. The acoustic branches of one of the materials reach higher frequencies than those of the other. Thus confined acoustic modes may exist in the materials with the higher acoustic frequencies. TA confined modes also exist, corresponding to the top TA modes in cases when the TA and longitudinal acoustic (LA) bands do not mix. (This happens in [001] superlattices for propagation along [001], in particular for the Ge/Si systems.) These nondispersive modes can be observed near the top of the TA bands (see 146 cm^{-1} mode in Fig. 1 [20]).

Quasiconfined (resonant) modes may result even at frequencies at which the dispersion relations of both constituents overlap (the case in which the overlap is between bands of different symmetry is trivial). Such situation is important in the Ge/Si [001] case in which there is actually no gap for $\vec{k} \parallel$ [001] from $\omega = 0$ to the $k = 0$ optical frequencies. Despite the overlap, the boundary conditions are not favorable to a coupling of longitudinal optical (LO) modes of Ge with the LA of Si and nearly confined Ge optical modes obtain. The Raman spectra of such confined modes are displayed in Fig. 2a for a Si_4Ge_{18} superlattice grown on a Ge substrate. Note that the Ge_1 mode is close to the frequency expected for bulk Ge at 4 K (305 cm^{-1}), since the Ge is not strained. The superlattice of Fig. 2b, however, was grown on a Si-Ge mixed crystal of a similar composition. Hence the Si layers are under tension, and the Si_1 mode is at lower frequency than that of the Raman phonon of the Si substrate (also shown in the figure). The modes Ge_ℓ and Si_ℓ can be mapped as usual on the dispersion relations of the (strained) bulk with

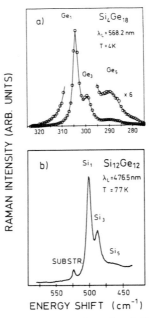

Fig. 2. Raman spectra of confined Ge phonons in a Si_4Ge_{18} and confined Si phonons in a $Si_{12}Ge_{12}$ superlattice. The Ge modes are fitted with a set of Lorentzians (solid lines); the experimental spectrum is given by the open circles. In the spectrum of the Si modes the optical-phonon peak of the Si substrate is observed at about 524 cm^{-1}. From Ref. 21.

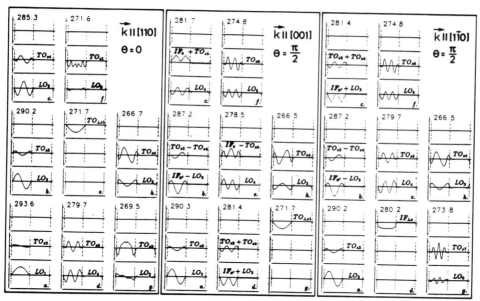

Fig. 3. Displacement vectors of a number of GaAs-like zone-center optical modes for a [110] $(GaAs)_{13}/(AlAs)_{14}$ superlattice. Plotted are the optical relative-displacement vectors of anion-cation pairs along the layer axes; the anion and cation of one pair belong to the same plane perpendicular to the layer axis. Left panel, **k** approaches zero along the [110] direction ($\theta = 0$); middle and right panels, **k** approaches zero along the [001] and [1$\bar{1}$0] directions ($\theta = \pi/2$).

$$k = \frac{\pi}{(n_j + 1)a_0}\ell \tag{1}$$

where n_j is the number of monolayers of material j (Ge or Si) and the summand 1 takes care of the minor spillover of the confined mode into the "forbidden" material.

Another interesting case of confinement despite overlap applies to $(C_1 A_1)/(C_2 A_2)$ zinc blende type superlattices such as InAs/GaAs.[22] In this case the dispersion relations of the bulk nearly overlap for both optic and acoustic modes (see Fig. 4 in Ref. 22): the average atomic masses of both constituents are nearly the same. Despite this, surprisingly well confined optical modes obtain (Fig. 5 of Ref. 22). Two types of interfaces are possible: heavy (In–Sb) and light (Ga–As). For the former the In–Sb reduced mass is considerably larger than that of the bulk constituents, while for Ga–As it is smaller. This fact dampens the vibration of the bulk at the interface, forcing the latter to have nearly zero vibrational amplitude: thus confinement in one of the two constituents results (because of some leakage the modes are referred to as quasi-confined or resonant). In this case, microscopic interface modes, in which mainly In–Sb or Ga–As vibrate, also appear (see INTERFACE MODES below).

Note that for the conventional [001]-grown superlattices in the backscattering configuration one can only observe longitudinal modes. Other growth directions allow the observation of transverse modes. For [110] growth, TO and also LA and TA modes are Raman active in backscattering.[19,23] In Ref. 23 the Ge-like TO_ℓ modes of Si_3Ge_9 and Si_3Ge_{15} are reported. An interesting, unexpected feature is that confined TO modes for $\ell = 1, 2, 3$ are observed: the $\ell = $ even modes should be forbidden by parity, as in the case of [001] superlattices. The reason why the $\ell = 2$ phonon is observed in $(y'x')$ configuration $(x'\|[110], y'\|[1\bar{1}0], z\|[001])$ is not known. A possibility is that it corresponds to the $\ell = 1$ LO phonon activated through inaccuracies in the internal polarizations (the LO phonon is allowed for the $y'z$ polarization).

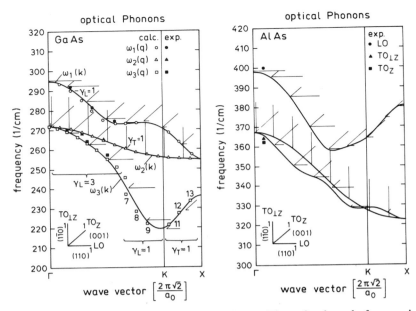

Fig. 4. Experimental (•, ▲, ■) and calculated (○, △, □) confined-mode frequencies as a function of confinement wave vector $k_\ell = [\ell/(m + \gamma_{L,T})](2\pi\sqrt{2}/a_0)$, together with theoretical optical-phonon-dispersion curves of (a) bulk GaAs and (b) AlAs in the Γ-K-X direction and the projections of the corresponding phonon displacements of the cation relative to the nearest anion along the [110], [1$\bar{1}$0], and [001] directions. From Ref. 19.

The [110] GaAs/AlAs system has been investigated in detail in Ref. 19, both experimentally and theoretically. In this system the LO ℓ modes for ℓ odd (even) mix with the TO ℓ' even (odd) modes polarized along [001]. Figure 3 ($\theta = 0$) indicates that $\ell' \simeq \ell + 1$. Figure 4 shows the optical bulk-dispersion relations of GaAs and AlAs and the mapping of the observed and calculated modes for a (GaAs)$_{13}$/(AlAs)$_{14}$ superlattice obtained with the equation

$$k_\ell = \frac{2\pi\sqrt{2}}{a_0(m + \gamma_{L,T})}\ell \qquad (2)$$

The mixture of polarization directions is indicated by the vertical, horizontal and diagonal bars attached to the dispersion curves. The TO$_{y'}$ modes map perfectly with $\gamma_T \approx 1$ and are unmixed. The mapping of the upper mixed LO-TO$_z$ hybridized modes is also straightforward for $\gamma_L \simeq 1$ but the modes for $\ell = 9, 10$ are missing. These modes are nearly degenerate with the TO$_z$ modes at $\vec{k} \simeq 0$ with which they hybridize as shown in the left panel of Fig. 3. We have thus plotted one of them for $k = 0$.

Despite the LO-TO$_z$ mixing which in the superlattice implies a change in ℓ from odd to even, but not in the bulk, the mapping of superlattice modes on bulk modes is rather good. The only "fudging" needed is to use $\gamma_L \simeq 3$ for the first eight modes of the lower TO branch, a fact that enables us to recover one of the last modes of the upper branch. The modes labeled $7 - 9$ are predominantly LO polarized (mode switching as a result of the LO-TO$_z$ anticrossing

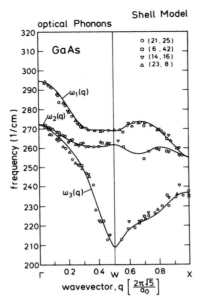

Fig. 5. Optical-phonon-dispersion curves of bulk GaAs in the [012] direction of $\vec{k}$ space, together with calculated frequencies of superlattice confined modes vs. $k_\ell = [\ell/(m+1)](2\pi\sqrt{5}/a_0)$ for all samples studied here.

which takes place in the middle of the BZ). Between K and X mode switching with the LA modes takes place as the frequency increases with increasing k.

For $(\mathrm{GaAs})_m/(\mathrm{AlAs})_n$ [012] only C_2 point-group symmetry is found. This case has also been thoroughly studied, both theoretically and experimentally.[24] Heavy mixing now takes place among the three polarization directions. A mapping of frequencies calculated for four superlattices is shown in Fig. 5. For this mapping the wave vector

$$k = \frac{2\pi\sqrt{5}}{a_0(m+\gamma)}\ell \tag{3}$$

was taken with $\gamma = 1$. Despite the simplicity of the mapping and the heavy polarization mixing, excellent agreement between superlattice and bulk modes is found. Available experimental data[24] confirm these results.

As already mentioned, another interesting effect appears in superlattices that do not have $z \rightarrow -z$ symmetry. This happens in [111] zinc blende superlattices[18] and also in [111] $\mathrm{Ge}_m/\mathrm{Si}_n$ for m and n odd.[15] In this case the envelope function of confined modes is no longer either a sine or a cosine (odd or even with respect to the center of the layer). It has been shown in Ref. 18, however, that in the system C_1A_1/C_2A_1 if one takes the same force constants for both constituents one finds symmetric and antisymmetric envelope functions for the displacements of C_1 and C_2 but not for those of the common anion. Deviations from the above assumptions are seldom going to be large.

25

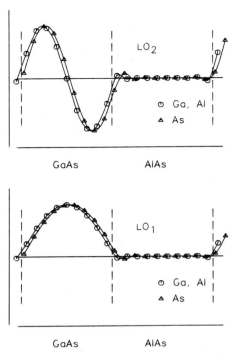

Fig. 6. Displacement pattern of LO_ℓ confined GaAs-like zone-center modes of a [111] $(GaAs)_9/(AlAs)_{10}$ superlattice. The frequencies of these modes are 393.93 cm^{-1} (LO_1) and 391.76 cm^{-1} (LO_2). The layer boundaries (dashed lines) are taken to be the bisector planes between Ga and As layers. From Ref. 18.

As an illustration we show in Fig. 6 the envelope functions calculated for the first two LO modes of a $(GaAs)_9/(AlAs)_{10}$ superlattice under the assumption of equal force constants. Note the symmetry of the cation displacement as opposed to the asymmetry of those of the anion (which is nevertheless small).

Even in the case of the [001] growth direction, relatively low orthorhombic symmetry results for Ge_m/Si_n if m and n are even. Consequently the confined TO modes split into [110]- and [1$\bar{1}$0]-polarized components.[13,14] This splitting is due to the Ge-Si interface and thus decreases rapidly with increasing m and n; it amounts to 60 cm^{-1} for $m = n = 2$ and only 10 cm^{-1} for $m = n = 4$. Unfortunately, one cannot couple to these modes in backscattering. The orthorhombic anisotropy also results in a difference in the strengths of LO Raman spectra for $z(x', x')\bar{z}$ and $z(y', y')\bar{z}$ polarizations ($x'\|[110]$, $y'\|[1\bar{1}0]$). This is shown in Figs. 7 and 8 for Si_2/Ge_2 and Si_2/Ge_4 superlattices as calculated with the bond-polarizability model (polarizability $\alpha_{Ge} = 2\alpha_{Si}$). The $(x', x') - (y', y')$ asymmetry is larger for $m = n = 2$ and somewhat smaller for $m = n = 4$. Note that there is also a (x', x') spectrum that is forbidden in the bulk materials and is related to the $(x', x') - (y', y')$ asymmetry (the Raman tensor components $R_{xx} = R_{x'y'} + R_{x'y'}$; $R_{x'x'} \simeq -R_{y'y'}$). For $m = n = 2$ no Raman-active Ge-Si LO modes are present (only infrared-active). They appear, however, for $m, n \geq 4$.

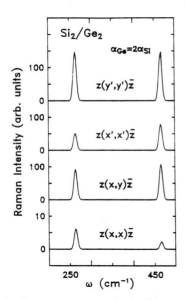

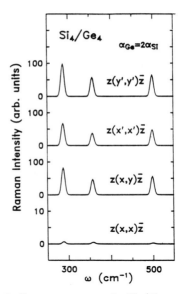

Fig. 7. Raman spectra of a Si_2/Ge_2 super-lattice calculated with the bond-polarizability model for several backscattering polarization configurations. From Ref. 25.

Fig. 8. Raman spectra of a Si_4/Ge_4 super-lattice calculated with the bond-polarizability model for several backscattering polarization configurations. From Ref. 25.

The calculated $(x', x') - (y', y')$ asymmetry should, however, be difficult to observe, since the substrates usually have monolayer steps, which effect a random switching of x' and y'. Observations thus require superlattices grown on carefully prepared substrates, if not step-free at least with a preferential x' orientation. Another difficulty encountered in Ge/Si superlattices concerns the observation of the Ge-Si modes ($350 \, \text{cm}^{-1}$ in Fig. 8). These modes are localized at the interface (microscopic interface modes, see next section) and thus are very sensitive to interface quality. Data obtained so far[26,27] (Fig. 9) suggest heavy alloying between the two components at the interface.

INTERFACE MODES

We have mentioned the existence of interface modes confined to two monolayers around the interfaces (Fig. 10 for [111] GaAs AlAs; also GaAs and InSb modes of GaSb/InAs [22]). These are the microscopic interface modes.[18,28] We show in Fig. 10 microscopic interface modes calculated for a $(GaAs)_9/(AlAs)_{10}$ [111]-grown superlattice.[18] They appear near the top of the LA bands of both constituents.[18] The plotted atomic displacements correspond to the W point of the MBZ, although the modes are localized in the W-X region and beyond. Some experimental indication of these modes has been obtained in the Raman spectra (Fig. 12 of Ref. 18).

Besides these modes one finds in zincblende-type superconductors macroscopic interface modes of electrostatic nature, extending to a depth $\approx \pm 1/k_x$ from the interface (k_x is the in-plane wave vector). They differ from the Kliewer-Fuchs modes of a single slab[29] in that they also propagate along the growth direction z so as to satisfy Bloch's theorem for superlattice periodicity. Their electrostatic potential is usually obtained by solving Laplace's equation

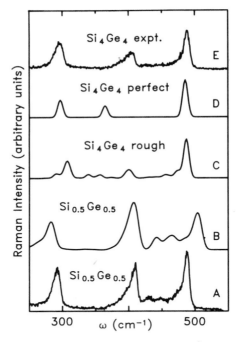

Fig. 9. Measured (at λ_L = 4579 Å) and calculated room-temperature Raman spectra of Si_4Ge_4 (C, D, E) and $Si_{0.5}Ge_{0.5}$ (A, B). The calculations for Si_4Ge_4 were performed for a perfect (D) and a rough (C) interface.[25-27]

$\nabla^2 \phi = 0$ in both media and applying the ordinary electrostatic boundary conditions for E and D.

This treatment is, however, rather crude since it neglects boundary conditions for the mechanical displacement $\vec{u}$. Actually, the macroscopic interface modes are nothing more than regular phonons of the superlattice with an in-plane $\vec{k}$ component k_x and an infrared activity parallel to it. It is easy to see that the electrostatic "interface" modes are nonanalytic for $k \to 0$; i.e., they depend on the angle $\theta = \tan^{-1}(k_x/k_z)$ (see Fig. 1 of Ref. 5). For $\theta = 0$ ($k_x \equiv 0$) they coincide with the $\ell = 1$ LO and TO confined modes. For $\theta = \pi/2$ (in-plane propagation) they have a smaller L–T splitting obtained by solving (1 and 2 represent the two bulk constituents)

$$d_1 \epsilon_1(\omega_{LI}) + d_2 \epsilon_2(\omega_{LI}) = 0; \quad \frac{d_1}{\epsilon_1(\omega_{TI})} + \frac{d_2}{\epsilon_2(\omega_{TI})} = 0 \tag{4}$$

i.e., the LI and TI frequencies for in-plane propagation are obtained as the longitudinal and transverse modes of different effective media with average dielectric functions represented by Eq. (4). For $d_1 = d_2$ both effective media coincide, and $\omega_{TI} = \omega_{LI}$. The $\ell = 1$ confined mode for $\theta = 0$ evolves continuously with increasing θ into the ω_{LI} and ω_{TI} in-plane modes with smaller (zero if $d_1 = d_2$) L–T splitting.

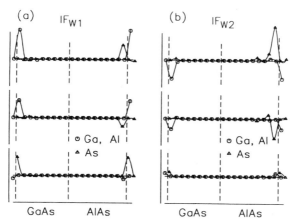

Fig. 10. Displacement pattern of the interface modes IF_{W1} and IF_{W2} with wave vector at the W point on the BZ boundary with frequencies of 227.45 cm^{-1} (a) and 196.62 cm^{-1} (b). Shown are the components of the displacement vectors along [1$\bar{1}$0], [11$\bar{2}$], and [111] (from the bottom to the top). From Ref. 18.

Because of their angular dispersion, interface modes usually cross confined modes with $\ell > 1$. Anticrossings result if the confined modes of higher ℓ are infrared active along $\vec{k}$. This is usually the case only for ℓ odd. Examples of long-wavelength interface modes for $\theta = \pi/2$ are given in the center and right frames of Fig. 3 for a [110] superlattice. In this case of orthorhombic symmetry θ is not sufficient to specify the problem: $k\|[001]$ and $k\|[1\bar{1}0]$ lead to different results. In fact, in the latter case the ℓ even LO modes (ℓ odd TO modes) are not infrared active and thus do not participate in the interface modes. The corresponding crossings can be seen in the angular dispersion (vs. θ) diagrams of Fig. 11 (right-hand side). For $k\|[001]$ LO modes with ℓ even are infrared active because of admixture with TO$_z$ (ℓ odd): all bands anticross as shown in the left-hand side of Fig. 11.

Interface modes have been observed by Raman[30] and electron energy-loss spectroscopy.[31] The scattering wave vector had no m-plane component, so it had to be assumed that the observed interface modes are activated by surface roughness or defects. This corresponds to the observation that the ratio of strengths of these modes to those of confined modes is strongly sample dependent. The ratio also decreases with increasing laser intensity (screening) and magnetic field.[32,33]

We conclude by mentioning an interesting observation concerning interface modes in large-period (500–1000 Å) GaAs/AlAs [001] superlattices with the GaAs layers heavily doped with Si.[34] The doping leads to electron concentrations around 10^{17}/cm^3 in the GaAs layers. In backscattering one observes instead of the LO phonons plasmon-phonon coupled modes ω^+ and ω^-.

With the laser frequency near that of the lowest valence-conduction transitions one observes the ω^+ and ω^- bands plus sharp peaks at the ω_{LO} frequencies of GaAs (substrate) and ω_{TO} of AlAs (see Fig. 12, sample C). Peaks at frequencies that seem to correspond to electrostatic interface modes are also observed. These peaks, forbidden in backscattering, are probably induced by impurity scattering. While they are all positive peaks for sample C, the TO and IF peaks of the AlAs region are observed as dips in sample A which has an ω^+ peaking near the TO frequency of AlAs. This result suggests strong coupling between the TO and IF modes and the broad ω^+ plasmon-phonon peak. Phenomenological fits on the basis of three coupled frequencies (ω_{TO}, ω_{IF} and ω^+) are shown by dashed lines in Fig. 12.[34] The microscopic origin of the coupling has not yet been elucidated.

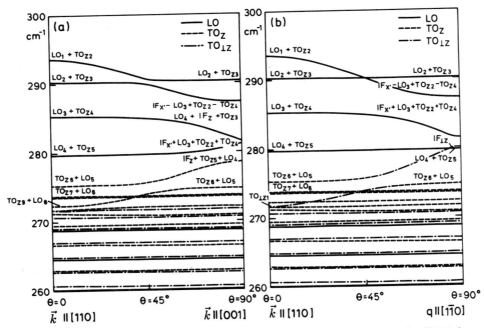

Fig. 11. Calculated angular dispersion of the optical modes of a $(GaAs)_{13}/(AlAs)_{14}$ superlattice for infinitesimal $\vec{k}$: (a) $\vec{k}$ in the $(x'-z)$ plane; (b) $\vec{k}$ in the $(x'-y')$ plane. The solid lines represent modes predominantly polarized along x' (LO), while the dashed and the dashed-dotted lines correspond to predominant TO_z and $TO_{\perp z}$ polarizations. From Ref. 19.

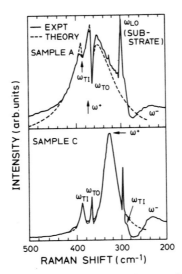

Fig. 12. Raman spectra of two large period GaAs/AlAs superlattices doped with Si (n-type) in the GaAs layers, showing destructive (A) and constructive (C) interference between ω^+ phonon plasmon modes and ω_{TO}, ω_{IF} of AlAs.

REFERENCES

1. B. Jusserand and M. Cardona, in: "Light Scattering in Solids V", Springer, Heidelberg (1989).

2. M. V. Klein, in: "Raman Spectroscopy: Sixty Years on Vibrational Spectra and Structure," Elsevier, Amsterdam (1989).

3. J. Menéndez, J. Lumin. $\underline{44}$, 285 (1989).

4. J. Sapriel and B. Djafari Rouhani, Surface Sci. Rep. $\underline{10}$, 189 (1989).

5. M. Cardona, Superlatt. Microstr. $\underline{5}$, 27 (1989).

6. M. Cardona, "Spectroscopy of Semiconductor Microstructures," Plenum Press, New York (1990).

7. M. V. Klein, IEEE J. Quantum Electron. $\underline{QE-22}$, 1780 (1986).

8. M. Cardona, Superlatt. Microstr., issue dedicated to Frits de Wette (in press).

9. E. Anastassakis, Proceedings of the ECS-Symposium on "Analytic Techniques for Semiconductor Materials and Process Characterization," Berlin, Sept. 1989 (in press).

10. R. M. Abdelouhab et al., Phys. Rev. B $\underline{39}$, 5857 (1989).

11. D. L. Rousseau, R. P. Bauman, and S. P. S. Porto, J. Raman Spectrosc. $\underline{10}$, 253 (1981).

12. E. Anastassakis and M. Cardona, phys. status solidi (b) $\underline{104}$, 589 (1981).

13. M. I. Alonso, M. Cardona, and G. Kanellis, Solid State Commun. $\underline{69}$, 479 (1989).

14. M. I. Alonso, M. Cardona, and G. Kanellis, Solid State Commun. $\underline{70}$, i (1989).

15. P. Molinàs i Mata, M. I. Alonso, and M. Cardona, Solid State Commun. $\underline{74}$, 347 (1990).

16. U. Schmid, N.E. Christensen, and M. Cardona, Phys. Rev. B $\underline{41}$, 5919 (1990).

17. U. Schmid, N.E. Christensen, and M. Cardona, to be published.

18. Z. V. Popović, M. Cardona, E. Richter, D. Strauch, L. Tapfer, and K. Ploog, Phys. Rev. B $\underline{41}$, 5904 (1990).

19. Z. V. Popović, M. Cardona, E. Richter, D. Strauch, L. Tapfer, and K. Ploog, Phys. Rev. B $\underline{40}$, 3040 (1989).

20. R. A. Ghambarri and G. Fasol, Solid State Commun. $\underline{70}$, 1025 (1989).

21. E. Friess, K. Eberl, U. Menczingar, and G. Abstreiter, Solid State Commun. $\underline{73}$, 203 (1990).

22. A. Fasolino, E. Molinari, and J. C. Maan, Phys. Rev. B $\underline{39}$, 3929 (1989).

23. E. Friess et al., Solid State Commun. (in press).

24. Z. V. Popović, M. Cardona, E. Richter, D. Strauch, L. Tapfer, and K. Ploog, Phys. Rev. B $\underline{40}$, 1207 (1989).

25. M. I. Alonso, Ph.D. thesis, University of Stuttgart (1989).

26. M. I. Alonso, F. Cerdeira, D. Niles, M. Cardona, E. Kasper, and H. Kibbel, Appl. Phys. Lett. $\underline{66}$, 5645 (1989).

27. S. Wilke, Solid State Commun. $\underline{73}$, 399 (1990).

28. S-F. Ren, H. Chu, and Y-C. Chang, Phys. Rev. B $\underline{40}$, 3060 (1989).

29. K. L. Kliewer and R. Fuchs, Phys. Rev. <u>144</u>, 495 (1966).

30. A. J. Sood, J. Menéndez, and M. Cardona, Phys. Rev. Lett. <u>54</u>, 5115 (1985).

31. P. Lambin, et al., Phys. Rev. Lett. <u>56</u>, 1227 (1984).

32. G. Ambrazevičius, M. Cardona, R. Merlin, and K. Ploog, Solid State Commun. <u>65</u>, 1035 (1988).

33. D. Gammon, R. Merlin, and H. Morkoç, Phys. Rev. B <u>35</u>, 2552 (1987).

34. A. K. Sood, M. Cardona, A. Fischer, and K. Ploog, in: "Recent Trends in Raman Spectroscopy" 2–6 Nov. 1988, Calcutta, S. B. Banerjee and S. S. Jha, eds., World Scientific Publishers, Singapore (1989).

RAMAN SCATTERING IN α-Sn$_{1-x}$Ge$_x$ ALLOYS

J. Menéndez* and K. Sinha

Department of Physics, Arizona State University, Tempe
Arizona, USA

H. Höchst and M.A. Engelhardt

Synchrotron Radiation Center, University of Wisconsin-Madison
Stoughton, Wisconsin, USA

INTRODUCTION

The Si-Ge system provides a prototypical example of a substitutional semiconductor alloy. Because of its potential technological applications, it has been intensively studied during the past thirty years. In particular, light scattering techniques have been used to investigate the compositional dependence of phonon modes. Most of this work has concentrated on the three main Raman active optical vibrations, assigned to "Si-Si", "Si-Ge", and "Ge-Ge" modes.[1],[2],[3],[4],[5],[6],[7],[8] In spite of this effort, however, the compositional dependence of Raman modes is not well understood. A powerful experimental approach to this problem would be the comparison of Raman spectra for different isovalent alloys such as Si$_{1-x}$C$_x$, Ge$_{1-x}$Si, and α-Sn$_{1-x}$Ge$_x$. Unfortunately, only Ge$_{1-x}$Si$_x$ shows solid solubility for the entire compositional range $0 \leq x \leq 1$. Non-equilibrium growth techniques such as Molecular Beam Epitaxy, however, open up a new alternative for the fabrication of these alloys. Recently, the successful MBE growth of α-Sn$_{1-x}$Ge$_x$ has been reported.[9] In this paper, we summarize recent Raman results on this system and discuss the implications of the results for the understanding of phonons in IV-IV alloys.

EXPERIMENT

The growth of our samples has been explained in detail in Ref. 9. Raman experiments were performed at room temperature with the 7525 Å and 6764 Å lines of

Light Scattering in Semiconductor Structures and Superlattices
Edited by D.J. Lockwood and J.F. Young, Plenum Press, New York, 1991

33

a Kr$^+$ laser. These lines were selected in view of their proximity to the E_1 gap of α-Sn. The scattered light was analyzed with a SPEX 1404 double monochromator and detected with standard photon-counting techniques. A constant flow of helium onto the sample eliminated spurious lines associated with air. The possibility of a laser-induced $\alpha \rightarrow \beta$ transformation required special attention. In bulk α-Sn, the α-β transition occurs near 13 °C.[10] In thin MBE films, the α-phase has been shown to be stable up to 115 °C,[11] and even higher in more recent work.[12] The incorporation of Ge is expected to further raise this transition temperature. Therefore, the problem was not so critical in our samples. Nevertheless, we used low-power laser beams ($\sim$ 100 mW) focused onto the sample by means of a cylindrical lens.

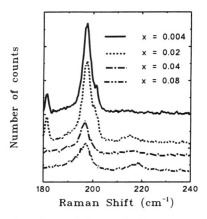

Fig. 1. Raman spectra for four different α-Sn$_{1-x}$Ge$_x$ samples obtained at room temperature with the 7525 Å line of a Kr$^+$ laser.

RESULTS AND DISCUSSION

Figure 1 summarizes the Raman results obtained. Two peaks were observed, the strongest in the 196-198 cm^{-1} range, and a second, weaker peak at frequencies between 215-220 cm^{-1}. These peaks show small linewidths and selection rules consistent with a diamond structured material, confirming the high quality and single-crystal nature of our films. We assign the low-energy line to Sn-Sn vibrations and the high-energy line to Sn-Ge vibrations in tetrahedral alloys. These assignments are based on their common features with the Raman peaks of the intensively studied Ge$_{1-x}$Si$_x$ alloys.[1-8] Firstly, in analogy with results from Ge$_{1-x}$Si$_x$, the peak attributed to Sn-Sn vibrations has a frequency close to the Raman peak of pure α-Sn, as shown in Fig. 1. The peak assigned to Sn-Ge vibration has a frequency - also as expected - between the Raman peaks of α-Sn and Ge (301 cm^{-1}).[13] Its weaker intensity can be understood in terms of the smaller number of Sn-Ge bonds as opposed to Sn-Sn bonds in these Sn-rich alloys. The same argument explains why we are not able to detect the Ge-Ge peak, since the number of Ge-Ge bonds is even smaller. Together with the above analogies with Ge$_{1-x}$Si$_x$ alloys, there are some important differences worth mentioning here. From Fig. 1, we find that the separation between the Sn-Sn and Ge-Sn peaks for the 8%-Ge sample is of the order of 20 cm^{-1}. On the other hand, experimental results[1] for Ge$_{0.92}$Si$_{0.08}$ show a separation between the Ge-Ge and Si-Ge lines of the order of 100 cm^{-1}. It is clear that this separation does not scale with the ionic plasma frequency $\omega_{pl} \propto (Ma^3)^{-1/2}$, where M is the ion mass and a the lattice constant, as has been found for all phonon features along the C - Si - Ge - Sn series.[14]

For a more detailed comparison between Ge$_{1-x}$Si$_x$ and α-Sn$_{1-x}$Ge$_x$ it is necessary to take into account the fact that the peaks in Fig. 1 correspond to films that according to X-ray data[15] are lattice-matched to the CdTe substrate, i.e., expanded laterally with respect to the expected lattice constant of bulk α-Sn$_{1-x}$Ge$_x$ alloys. This effect shifts the Raman frequency by an amount given by[16]

$$\Delta\omega_0 = -2 \ \varepsilon\omega_0 \ [\gamma(1 - C_{12}/C_{11}) + a_s(1 + 2C_{12}/C_{11})/3], \qquad (1)$$

where ω_0 is the bulk Raman frequency, ε the in-plane component of the strain tensor, C_{11} and C_{12} elastic constants, γ the Grüneisen parameter, and a_s the shear phonon deformation parameter given by $a_s = \frac{1}{2}(p-q)/(\omega_0)^2$, where p and q are the anharmonic parameters as used in Ref. 16. Because γ and a_s are not known experimentally, not even for pure α-Sn, we can only obtain semi-quantitative information by making reasonable assumptions as to the value of these parameters.

To proceed with our analysis, we will use $\gamma = 1$ and $a_s = 0.25$ in Eq. 1. The value for γ is reasonable in view of the fact that it is very close to 1 for the Raman phonons of all diamond and zincblende semiconductors.[17] The uniaxial deformation potential a_s has the same value $a_s = 0.25$ for Si[18] and Ge.[19] Since the phonon dispersion curves of α-Sn, Ge, and Si scale extremely well with the ionic plasma frequency, it is reasonable to use for α-Sn the same value of a_s found for Si and Ge. Alternatively, one might use the value $a_s = 0.42$ reported for InSb.[20] This larger value of a_s would only reinforce our conclusions below.

The composition dependence of the bulk Raman modes (for the above choice of parameters) is shown in Fig. 2. Clearly, the Sn-Sn Raman frequency increases as a function of Ge-concentration. This is in sharp contrast with findings in Ge$_{1-x}$Si$_x$ alloys, where the Ge-Ge frequency decreases as the Si-concentration becomes higher.

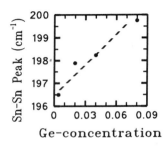
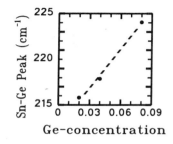

Fig. 2. Composition dependence of Raman frequencies for bulk α-Sn$_{1-x}$Ge$_x$ alloys. The points represent the peak energies from Figure 1 after the strain correction given by Equation 1.

The observed compositional dependence of the Raman modes in our α-Sn$_{1-x}$Ge$_x$ films can be understood in terms of a model that was recently proposed to explain this dependence in Ge$_{1-x}$Si$_x$ alloys.[21] In fact, our α-Sn$_{1-x}$Ge$_x$ results lend further support to the explanation of Ref. 21. According to this model, the frequency of the Raman lines changes as a function of composition due to confinement and strain effects. Confinement effects, well documented from superlattice work, arise because of the small size of the cluster-like regions that contribute to the different Raman peaks. As a result of this, the Raman peaks observed in the alloys correspond to optical phonons with large wavevector in the bulk. These phonons have usually lower frequencies than the k = 0 optical modes, so that the Raman frequencies of the alloy modes tend to be lower than their crystalline counterparts. On the other hand, alloys have a different lattice constant than their parent compounds. This strain effect (not to be confused with the additional strain imposed on MBE films lattice-matched to substrates) gives the second important contribution to the x-dependence of Raman modes. Because the optical branches of these materials have positive Grüneisen parameters, compressive (tensile) strains lead to an increase (decrease) of the Raman frequencies. The combination of confinement and strain has been shown to be sufficient to explain all qualitative findings in the Ge$_{1-x}$Si$_x$ system.[21] In particular, it explains the much stronger composition dependence of the Si-Si peak as opposed to the Ge-Ge peak. For the Si-Si peak, confinement and strain contributions have the same negative sign as x decreases from x = 1, since the Si-rich regions become smaller and the Si-Si bond length is forced to be larger than in bulk Si. Conversely, the two contributions tend to cancel out for the Ge-Ge modes when x increases from x = 0, since confinement will cause a decrease and the compressive strain (lattice constant of the alloy smaller than that of Ge) an increase in the Raman frequency. The Si-Ge peak in Ge$_{1-x}$Si$_x$ can be seen as derived from the optical phonons of a zincblende SiGe compound. The probability of finding large regions with this composition is largest for $x = 0.5$, so that from the point of view of confinement the maximum Si-Ge frequency should occur for $x = 0.5$. Strain effects are responsible for the notorious asymmetry in the composition dependence of this peak. For $x > 0.5$, the alloy lattice constant is smaller than that of zincblende SiGe, so that the Raman frequency is shifted upwards. Conversely, the shift is downwards for $x < 0.5$, because the lattice constant of the alloy is larger than that expected for zincblende SiGe.

The above model provides a natural explanation for our findings in the α-$Sn_{1-x}Ge$ system. The key to this explanation is the much larger strain present in this system as a result of the 14% lattice mismatch between the parent compounds. This is to be compared with 4% lattice mismatch between Si and Ge. As a result of the larger mismatch, the strain effect (which for the Ge-Ge peak in the Si-Ge system is not strong enough to change the sign of its composition dependence) is likely to become the dominant contribution to the composition dependence of the Sn-Sn mode in α-$Sn_{1-x}Ge_x$ alloys. Consequently, it is not surprising that the Sn-Sn frequency be found to <u>increase</u> as the Ge-concentration increases. Similarly, the very large downshift found for the Sn-Ge peak of our samples with respect to the expected value for an α-$Sn_{0.5}Ge_{0.5}$ alloy is a result of the dramatic 7% expansion of the lattice, as opposed to 2% expansion in the $Ge_{1-x}Si_x$ system.

In conclusion, we have investigated the Raman spectrum of α-$Sn_{1-x}Ge$ alloys. Our samples show remarkable similarities with high quality $Ge_{1-x}Si$ alloys, indicating that we have been able to grow these alloys successfully by MBE. The most important qualitative differences between the two systems have been explained in terms of a simple model. A more detailed quantitative analysis of these differences might offer a deeper insight into clustering and relaxation effects in these materials. Experimental as well as theoretical work along these lines is in progress.

ACKNOWLEDGEMENTS
* Supported by NSF under Grant No. DMR-8814918

REFERENCES
1. D.W. Feldman, M. Ashkin, and J.H. Parker, Phys. Rev. Lett. **17**, 1209 (1966).

2. M.A. Renucci, J.B. Renucci, and M. Cardona in <u>Proceedings of the Conference on Light Scattering in Solids</u>, ed. by M. Balkanski (Flammarion, Paris, 1971), p. 326.

3. J.B. Renucci, M.A. Renucci, and M. Cardona, Solid State Commun. **9**, 1651 (1971).

4. W.J. Brya, Solid State Commun. **12**, 253 (1973).

5. J.S. Lannin, Phys. Rev. B **16**, 1510 (1977).

6. D.J. Lockwood, K. Rajan, E.W. Fenton, J.-M. Baribeau, and M.W. Denhoff, Solid State Commun. **61**, 465 (1987).

7. M.I. Alonso and K. Winer, Phys. Rev. B **39**, 10056 (1989).

8. F. Cerdeira, M.I. Alonso, D. Niles, M. Garriga, and M. Cardona, Phys. Rev. B **40**, 1361 (1989).

9. H. Höchst, M.A. Engelhardt, and D.W. Niles, SPIE Proc. Vol. **1106**, 16 (1989).

10. R.W. Smith, J. Less-Common Metals **114**, 69 (1986), and references therein.

11. J. Menéndez and H. Höchst, Thin Solid Films **111**, 375 (1984).

12. M.T. Asom., A.R. Kortan, L.C. Kimerling, and R.C. Farrow, Appl. Phys. Lett. **55**, 1439 (1989).

13. J. Menéndez and M. Cardona, Phys. Rev. B **29**, 2051 (1984).

14. W. Weber, Phys. Rev. B **15**, 4789 (1977).

15. R.C. Bowman, P.M. Adams, M.A. Engelhardt, and H. Höchst, J. Vac. Sci. Technol., in press.

16. F. Cerdeira, A. Pinczuk, J.C. Bean, B. Batlogg, and B.A. Wilson, Appl. Phys. Lett. **45**, 1138 (1984).

17. B.A. Weinstein and R. Zallen, in Light Scattering in Solids IV, ed. by M. Cardona and G. Güntherodt, Topics Appl. Phys. **54** (Springer, Berlin, 1984).

18. E. Anastassakis, A. Cantarero, and M. Cardona, unpublished.

19. F. Cerdeira, C.J. Buchenauer, F.H. Pollak, and M. Cardona, Phys. Rev. B **5**, 580 (1972).

20. M.I. Bell, Phys. Stat. Sol. (b) **53**, 675 (1972).

21. J. Menéndez, A. Pinczuk, J. Bevk, and J.P. Mannaerts, J. Vac. Sci. Technol. B **6**, 1306 (1988).

PHONON SPECTRA OF ULTRATHIN GaAs/AlAs SUPERLATTICES

E. Molinari[a], S. Baroni[b], P. Giannozzi[c], and S. de Gironcoli[c]

[a] CNR, Istituto "O.M. Corbino", Via Cassia 1216, I-00189 Roma, Italy

[b] SISSA-ISAS, Strada Costiera 11, I-34014 Trieste, Italy

[c] IRRMA, PHB-Ecublens, CH-1015 Lausanne, Switzerland

INTRODUCTION

This paper presents some recent results of *ab initio* calculations of the dynamical properties of $GaAs/AlAs$ superlattices (SL's) grown along the (001) direction. In particular we will focus on open problems in the interpretation of the experimental thickness-dependence of the zone center longitudinal optical (LO) SL frequencies in the ultrathin (UT) regime, and clarify the origin of the major discrepancy between the measured Raman spectra and the previous theoretical predictions. This will allow us to draw conclusions on structural properties of such systems, and obtain indications on how LO modes can give useful information for characterization also in UT structures.

The following sections are organized as follows. In the first one, we will quickly review the available experimental and theoretical data on UT SL's, pointing out the main unexplained points. In the second one, we will sketch our theoretical scheme. In the third one, we will then present the predictions of such scheme for the (001) dispersions of bulk $GaAs$ and $AlAs$, which are an essential ingredient for the study of SL's. Our dispersions, in excellent agreement with the available experimental data, show some significant differences from the dispersions used previously, particularly concerning the quasi-flat LO branch of $AlAs$. It will appear that this difference in describing $AlAs$ vibrations is relevant for the interpretation of SL spectra. Finally, we will show results for ideal (perfectly ordered) SL's and for SL's with intermixed configurations (substitutional disorder in the cationic sublattice); the results indicate that a substantial degree of cationic intermixing in the samples, extending beyond the interface planes, is necessary to account for the Raman spectra measured by several groups. Some of the results presented here will appear in Ref. 1.

OPEN PROBLEMS IN THE INTERPRETATION OF SUPERLATTICE RAMAN SPECTRA

In spite of the large theoretical and experimental effort that has been recently devoted to the study of the prototype UT SL's, *i.e.* (001)-grown $(GaAs)_m(AlAs)_n$ with $m, n = 1, 2, 3$, their phonon spectra are much less understood than those of thicker SL's. Indeed, for $m, n \geq 4$ the main mechanism of confinement for optical modes with wavevector $\mathbf{k} = (k_\parallel, k_z)$ parallel to the (001) growth direction ($k_\parallel = 0$), as well as the anisotropic behaviour arising when $\hat{k}$ deviates from (001)—which gives rise to the so called "macroscopic interface modes"— are by now clarified.[2]

For $m, n \leq 3$, instead, the difficulty in interpreting the measured shift of the SL LO modes from the LO frequencies of bulk $GaAs$ and $AlAs$ was apparent already a decade ago. In the pioneering investigation of UT SL's by Barker *et al.*[3], a clear point was that in the optical range simple models (in that case a nearest-neighbor linear chain model) were not capable of explaining Raman data. Fig. 1 and Table 1, reprinted from Ref. 3, are the best illustration of this fact. At that time, the origin of such discrepancy could be attributed to the quality of the

Light Scattering in Semiconductor Structures and Superlattices
Edited by D.J. Lockwood and J.F. Young, Plenum Press, New York, 1991

39

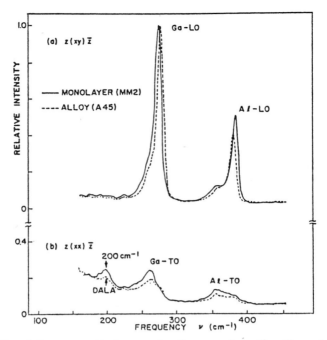

Fig. 1 Raman scattering spectra for a random $Al_{0.45}Ga_{0.55}As$ alloy sample (broken lines) and for a nominal $(GaAs)_1(AlAs)_1$ superlattice (solid lines), taken *(a)* in the crossed and *(b)* in the parallel backscattering configurations. From Barker *et al.*[3].

Table 1 Comparison of LO mode frequencies predicted by the nearest neighbors linear chain model and observed by Raman scattering for the $(GaAs)_1(AlAs)_1$ (monolayer, MM1) superlattice and for the $(GaAs)_2(AlAs)_2$ (bilayer, BB1) superlattice. From Barker *et al.*[3].

I. Monolayer ($MM2$) Mode type	Symmetry	Predicted (cm^{-1})	Observed (cm^{-1})
AlAs LO	Γ_4	372.1	385
[AlAs TO]	$[\Gamma_5]$	[357.8]	[359]
GaAs LO	Γ_4	241.3	275
[GaAs TO]	$[\Gamma_5]$	[262.8]	[261]
LA	Γ_1	201.4	198
II. Bilayer ($BB1$)			
AlAs LO	Γ_4	381.8	391
[AlAs TO]	$[\Gamma_5]$	[360.5]	[355]
AlAs LO	Γ_1	356.1	
GaAs LO	Γ_4	277.3	280
[GaAs TO]	$[\Gamma_5]$	[266.2]	[259]
GaAs LO	Γ_1	236.4	236(?)
LA	Γ_4	201.4	198
LA	Γ_4	120.8	· · ·
LA	Γ_1	118.4	· · ·

samples—which were just starting to be grown and were not yet fully characterized by other independent techniques—or to the severe approximations contained in the model, or to both. A particularly intriguing aspect for the monolayer SL ($m = n = 1$) was the close similarity of its Raman spectrum with that of the corresponding random alloy (Fig. 1).

The experimental trend of LO modes in $(GaAs)_n(AlAs)_n$ as a function of n was first systematically plotted[4] in 1985, and compared with a linear chain model containing the same strong assumptions for the range and type of interactions as Ref. 3. The results, reprinted from Ref. 4, are reported in Fig. 2. From their model, the authors concluded that pure confinement effects were responsible for the observed trend for $n \geq 2$, and that the general picture formulated by Barker $et\ al.$[3], Colvard $et\ al.$[5], and Jusserand $et\ al.$[6] (folding of acoustical and confinement of optical modes) was still valid for their samples down to $n = 2$. The remaining strong discrepancy for $n = 1$ was attributed to a "failure of the elastic model" in that specific case, because the slight difference between the LO peaks of $(GaAs)_1(AlAs)_1$ and the random alloy $Al_{0.5}Ga_{0.5}As$ was taken as a clear indication of the existence of some SL ordering at $n = 1$.

More recently, further experimental work has been done[7-9], but—in spite of the very different sets of samples—the overall picture that can be extracted from Raman measurements has not changed much. A collection of these experimental data for the highest LO $GaAs$-like

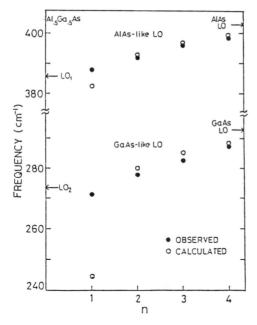

Fig. 2 $GaAs$-like and $AlAs$-like LO phonon frequencies of $(GaAs)_n(AlAs)_n$ superlattices as a function of $n = 1, 2, 3, 4$. Observed frequencies are marked by full circles, calculated frequencies (1st neighbor linear chain model) by empty circles. The arrows on the left-hand scale mark the measured optical frequencies of the alloy $Al_{0.45}Ga_{0.55}As$; the arrows on the right-hand scale mark the experimental zone-center LO frequencies of the two bulk constituents. From Nakayama $et\ al.$[4].

and *AlAs*-like peaks at zone center is presented in Fig. 3. The picture can be summarized as follows: both in the *GaAs*- and in the *AlAs*-like energy ranges, the frequencies of the highest LO modes, ω_{LO1}, almost coincide with the LO modes in the corresponding alloy for the monolayer SL, whereas they smoothly tend to the bulk limit $\omega_{LO}(\Gamma)$ for increasing thickness of the slab.

From the theoretical point of view, a sensible improvement of the models applied to SL's was obtained in the following years: On one side, the linear chain approach—which contains, in principle, no approximation in the mapping of the SL from three dimensions to one dimension, if only propagation with $k = (0, k_z)$ in an ideally ordered crystal is considered—was improved[10] by including longer-ranged interactions, extracted from the *ab initio* bulk force constants, which for *GaAs* were available from the first calculations of Ref. 11. As the observed Raman spectra are usually taken in off-resonance backscattering configurations, the allowed peaks correspond to propagation along k_z, and therefore the one-dimensional model is adequate for their description once the interplanar force constants are chosen correctly. On the other side, various three dimensional models with different accuracy were applied to superlattices[12-16]; some of them have brought a clearer insight in the SL dispersion for propagation with $k_\parallel \neq 0$, as well as in the comparison between "microscopic" and "macroscopic" approaches (see Ref. 2b).

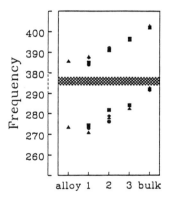

Fig. 3 Collection of *GaAs*- and *AlAs*-like LO Raman frequencies measured by several groups for $(GaAs)_n(AlAs)_n$ superlattices, with $n = 1, 2, 3$. Measured peaks for a random alloy of equivalent composition and for the bulk crystals are also shown for comparison. Triangles are from Ref. 4, diamonds from Ref. 7, squares from Ref. 8, and circles from Ref. 9. Note the break in the frequency scale, separating the *GaAs*-like and *AlAs*-like ranges.

To our knowledge, none of the above-mentioned theoretical models in both categories were able to give conclusive explanations of the experimental trend of Fig. 3, except for a case[9] where the interface parameters were directly fitted to reproduce it with no other physical constraint. The only common feature of all the other models was the fact that no charge rearrangement at the interface was allowed to take place, because the interactions inside the layers were generally assumed to be those of bulk *GaAs* ans *AlAs* and a simple average was chosen at the interface.

The present work started therefore from our opinion, that it was still unclear whether the main discrepancy had to be attributed to theoretical inadequacies or to experimental problems (meaning specifically that the actual samples were substantially different from the nominal samples). In turn, theoretical inadequacies could be either due to insufficient accuracy in the description of $GaAs$ and $AlAs$ already in the bulk phase, or to the unacceptable assumption of negligible charge rearrangement in the SL phase. As we will show below, this last assumption can be proven to be very well supported by the present calculations, where charge transfers in the SL are fully allowed but turn out to be of minor importance for the dynamics. The other possible theoretical inadequacy, instead, is found to be important. Its correction will not be enough to eliminate the discrepancy, which will then have to be searched in the actual structure of the samples. We postpone the discussion of this point to the final part of the present paper.

THEORETICAL SCHEME

We have performed a series of *ab initio* calculations of the phonon spectra of $(GaAs)_n(AlAs)_n$ (001) SL's with $n = 1, 2, 3$. Preliminary results of this research have been presented[17]. The dynamical matrix is obtained within density-functional linear-response theory (for a description of the present linear response approach and its use in phonon calculations see Ref. 18). The macroscopic electric fields arising from long-wavelength LO phonons are dealt with *ab initio*, with no empirical adjustment of the ionic effective charges. The dynamical matrices are calculated at the theoretical equilibrium configurations, obtained allowing full relaxation of the atomic positions in SL's.

The ground state electronic problem is treated within the local density approximation, using norm-conserving pseudopotentials and a large plane wave basis set (12 Ryd cutoff in the kinetic energy). From previous experience[17] we know that pseudopotentials from published tables[19] are appropriate for reproducing phonon dispersions of bulk $GaAs$ and $AlAs$ (which are, of course, calculated at the theoretical equilibrium lattice parameter of each material), but are not as accurate for SL's because they lead to overestimating the lattice mismatch between the two materials, thus introducing an artificial unrealistic strain in the SL. For this reason, we use new norm-conserving pseudopotentials for Ga, Al and As, which are able to give the observed lattice mismatch between $GaAs$ and $AlAs$.[1] Further technical details of our calculation are reported in Ref. 1.

BULK GaAs AND AlAs

Results and Comparison with Previous Theoretical and Experimental Data

Fig. 4 presents our calculated (001) dispersions for $GaAs$ and $AlAs$. For $GaAs$, where neutron-diffraction measurements exist[20a,b], the agreement with experiments is excellent (particularly with very recent data[20b], believed to be very accurate). Both the LO-TO splitting at Γ—hence the effective charges—and the flatness of the TA branch near the zone boundary are accurately predicted by the present calculation.

For $AlAs$, this is the first *ab initio* calculation of the full (001) dispersion. Our results agree well with the experimental data[20c,d], which are, however, very scarce for this material. It is important to notice that we predict that the LO branch of $AlAs$ in the Δ direction is very flat (the width of the dispersion being 5 times smaller than the LO-TO splitting), while in $GaAs$ the width of the dispersion is twice as large as the LO-TO splitting. The reliability of our predictions for $AlAs$ is not only based on the accuracy of our approach and the agreement with the few existing experimental points, but it is also supported by the excellent agreement with experiments found for two very similar compounds, $GaAs$ and $AlSb$, where recent accurate neutron experiments exist.[20-22]

Our results can be compared with the typical dispersions obtained with other approaches. In general, the 1st neighbor force constants are fitted to reproduce the Γ and X experimental frequencies, but the dispersions are not expected to be reliable. This is particularly true for the flattening of the TA branch in the second half of the Brillouin Zone (BZ), which is never reproduced by these models, and for the overestimate of the LO dispersions (see *e.g.* Refs. 3 and 4). This last point is relevant for SL optical modes. The dispersions calculated in Ref. 10 for $GaAs$, using the long-ranged *ab initio* interplanar forces from Ref. 11, substantially improve on both these points. However, when corrected for a remaining underestimate of the

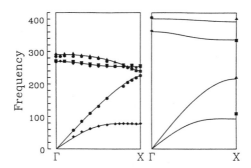

Fig. 4 Calculated phonon dispersions in *GaAs* (left) and *AlAs* (right) along the (001) direction [cm^{-1}]. Experimental points are marked by triangles and circles for L modes, and by squares and diamonds for T modes. Data for *GaAs* are from Ref. 20(a) (circles and diamonds) and 20(b) (triangles and squares); for *AlAs*, they are from Ref. 20(c) (circles and diamonds) and 20(d) (triangles and squares).

Table 2 Interplanar force constants for L- and T- polarized displacements in (a) bulk GaAs, and (b) bulk AlAs. k_m^{ij} is the force constant connecting the plane at $z = 0$ with the plane at $z = m\,a/4$, a being the bulk lattice parameter and $a/4$ the interplanar distance along (001). i,j denote the type of atoms in the two planes respectively ($c \equiv$ cation, $a \equiv$ anion). The force between two cationic planes is different from the force between two anionic planes; hence, for even values of m, i and j have to be specified. Note that for L polarization it is always $k_m = k_{-m}$, whereas for T polarization and m odd it is $k_m \neq k_{-m}$.

(a)	bulk GaAs			(b)	bulk AlAs		
longitudinal		transverse		longitudinal		transverse	
$k_{\pm 1}$ =	-0.90860	k_{+1} =	-1.34470	$k_{\pm 1}$ =	-0.94723	k_{+1} =	-1.34176
		k_{-1} =	-0.17096			k_{-1} =	-0.15590
$k_{\pm 2}^{cc}$ =	-0.14316	$k_{\pm 2}^{cc}$ =	0.05923	$k_{\pm 2}^{cc}$ =	-0.14375	$k_{\pm 2}^{cc}$ =	0.04508
$k_{\pm 2}^{aa}$ =	-0.11387	$k_{\pm 2}^{aa}$ =	-0.04442	$k_{\pm 2}^{aa}$ =	-0.04896	$k_{\pm 2}^{aa}$ =	-0.03191
$k_{\pm 3}$ =	0.00716	k_{+3} =	0.02223	$k_{\pm 3}$ =	0.00474	k_{+3} =	0.01245
		k_{-3} =	-0.05236			k_{-3} =	-0.04288
$k_{\pm 4}^{cc}$ =	-0.00233	$k_{\pm 4}^{cc}$ =	0.01379	$k_{\pm 4}^{cc}$ =	0.00158	$k_{\pm 4}^{cc}$ =	0.00988
$k_{\pm 4}^{aa}$ =	-0.00747	$k_{\pm 4}^{aa}$ =	-0.00942	$k_{\pm 4}^{aa}$ =	-0.00777	$k_{\pm 4}^{aa}$ =	-0.00726
$k_{\pm 5}$ =	0.00595	k_{+5} =	0.00096	$k_{\pm 5}$ =	0.00459	k_{+5} =	0.00032
	0.00595	k_{-5} =	-0.01013		0.00459	k_{-5} =	-0.00756

LO-TO splitting and transferred to $AlAs$ via the "mass and charge approximation"[10], these force constants also yield a LO dispersion larger than the one we find here.[23]   To conclude this part, we must also add that the other (three-dimensional) models yield a large $AlAs$ LO dispersion.[12-16]   This is, however, not surprising because such models are based on fitting parameters, and are therefore not predictive and very hard to use in the absence of clear experimental information.

Calculated Interplanar Force Constants and Validity of the Mass Approximation

Although our calculations are fully three-dimensional, here we also formulate the bulk results in terms of interplanar (rather than interatomic) force constants, in order to provide a practical way of calculating SL phonons for (001) propagation (in the spirit of Ref. 10). The interplanar form will also suitable for comparing the values of our $GaAs$ forces with those of Ref. 11, and for discussing the validity of the "mass approximation".

Our calculated interplanar force constants are given in Table 2a and 2b for $GaAs$ and $AlAs$, respectively. The comparison of Table 2a with Ref. 11b shows that the structure is very similar for both L and T polarizations, although of course quantitative differences exist (e.g. in the range of L forces).[23]

We find, as expected, a strong similarity between our first-principles interplanar forces for $GaAs$ and $AlAs$. This is already an indication that the "mass approximation"—i.e. the assumption that the interactions in the two materials are the same and only the cationic masses differ—should work. The validity of this approximation for $AlAs$ is demonstrated in Fig. 5. The dashed lines ("mass approximation" with the $GaAs$ force constants of Table 2a and $AlAs$ masses) differ from the full calculation ($AlAs$ force constants of Table 2b with $AlAs$ masses) only by a small quasi-rigid shift of the LO branch and an over-estimate of LA. The T dispersion is practically identical. The comparison of these L dashed lines with the $GaAs$ dispersion, reported for convenience in the top panels of the same figure, shows that the same set of interplanar forces (Table 2a) yields a very different dispersion of the LO branch in the two materials by simply changing the masses, and gives good agreement with experiments in both cases. Next we will present an application of this fact to SL's.

ULTRATHIN SUPERLATTICES

Ab initio Results for Ideally Ordered Superlattices

Let us first consider the results for SL's with perfect compositional order in the cationic sublattice. We focus on the highest-lying longitudinal frequencies, ω_{LO1}, which are the most intense Raman modes observable in the usual backscattering configuration, and correspond to the experimental data discussed above.

Fig. 6 shows our results for $\omega_{LO1}(n)$ of ordered SL's as a function of the slab thickness, n. For $n \geq 1$, ω_{LO1} roughly follows the trend predicted for perfect confinement and holding to a good approximation for thicker SL's, i.e. $\omega_{LO1}(n) \sim \omega_{LO}^{bulk}[q_z = \pi/(n+1)]$ (see Refs. 24, 25, 2). For $n = 1$ this relation no longer holds, but ω_{LO1} still fall within the $GaAs$ or $AlAs$ LO bulk bandwidths, showing that some confinement in either material is present even in the thinnest possible SL. An inspection of the displacement patterns associated to these modes[17a] confirms this interpretation. The very small sensitivity of $AlAs$-like LO1 modes to slab thickness is therefore easily understood as being due to the flatness of the bulk LO dispersion.

The main differences between our results and previous calculations are indeed found to be related to differences in the description of the bulk dispersions: (i) the smaller dependence of our highest $AlAs$-like LO modes on n, for $n = 1, 2, 3$ is related to our flatter $AlAs$ bulk LO dispersion all over the Γ-X direction; (ii) the smaller dependence of our highest $GaAs$-like LO modes on n for $n = 2, 3$ is related to our flatter $GaAs$ bulk LO dispersion in the first part of the BZ. In the $GaAs$-like range, the "jump" between $n = 1$ and $n = 2$ is not reduced much with respect to previous calculations.

Let us now compare with Fig. 3, where the available experimental data for $\omega_{LO1}(n)$ are collected. Their smooth thickness-dependence clearly appears to be at variance with theoretical results for ideal SL's both in the $GaAs$- and in the $AlAs$-like frequency ranges. In the $GaAs$-like region, experimental data lie below our predictions for $n = 2, 3$, and slightly above for $n = 1$. In the $AlAs$-like region, they fall consistently below the theoretical results for the SL and also out of the bulk bandwidth.

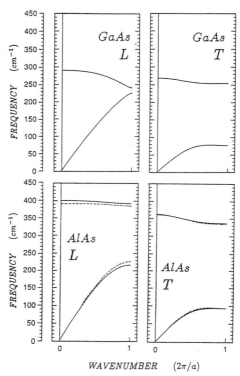

Fig. 5 Validity of the "mass approximation" for the bulk dispersion of *AlAs* when the interplanar force constants of the present work are used. Solid lines: forces from Table 2 for *GaAs* and from Table 3 for *AlAs*, with the appropriate masses. Dashed lines: results of the "mass approximation" for *AlAs* (*i.e.*, force constants of *GaAs* from Table 2, with *AlAs* masses).

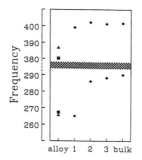

Fig. 6 Calculated frequencies of LO1 modes in perfectly ordered $(GaAs)_n(AlAs)_n$ superlattices, with $n = 1, 2, 3$, and in bulk $GaAs$ and $AlAs$ (diamonds). Calculated frequencies for supercells with cationic distributions simulating $Al_{0.5}Ga_{0.5}As$ alloys are marked by triangles and squares (for further details, see text and Ref. 1). Note the break in the frequency scale, separating the $GaAs$ and $AlAs$ ranges. Compare with the available experimental data in Fig. 3.

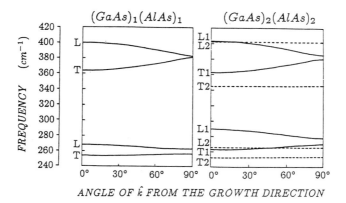

ANGLE OF $\hat{k}$ FROM THE GROWTH DIRECTION

Fig. 7 Anisotropy of optical modes of the $n = 1$ and $n = 2$ superlattices. Zone center frequencies are shown as a function of the angle between the (001) growth direction and the direction $\hat{k}$ of the vanishing wavevector. $\hat{k}$ rotates in the (010) plane from the (001) direction ($\theta = 0°$) to the (110) direction ($\theta = 90°$). L and T labels are meaningful only at $\theta = 0°$. Solid lines are for L1, T1 modes, broken lines for L2, T2 modes.

47

Before discussing the origin of the remaining large discrepancy between theory for ideal SL's and experiments, we comment on the calculated angular dispersions of zone center optical modes for $(GaAs)_1(AlAs)_1$ and $(GaAs)_2(AlAs)_2$. In Fig. 7 the Γ frequencies are shown as a function of the angle θ formed by the direction of the vanishing wavevector with the growth direction. When $\vec{k}$ departs from the (001) growth direction towards the (110) direction, rotating in the (010) plane, θ varies between 0° and 90°. Only for $\theta = 0°$ the L and T polarizations are decoupled, and the usual distinction between L and T modes holds, as well as the labelling with the confinement quantum number: the labels are reported on the left hand side of each panel. In the panel on the right, different lines are used to further distinguish between L1, T1 (solid lines) and L2, T2 modes (dashed lines). Modes above 340 cm^{-1} are $AlAs$-like. Note that the structure of the pictures is similar to the typical results obtained by different models.[12,14] The modes drawn by solid lines (L1, T1) are dispersive and are associated with the "macroscopic interface modes" found by continuum models.[2] We only remark that a non-negligible difference is obtained with respect to Ref. 14 in the $n = 2$ case, particularly in the high frequency range. This is again related to the small confinement energy of $AlAs$-like modes due to the flat bulk $AlAs$ dispersion.

Finally we comment on the application of the "mass approximation" to superlattices. Fig. 8 shows the results of calculations for the (001) L and T dispersion in the $m = n$ UT SL's. We stress that they are obtained by simply diagonalizing linear chain dynamical matrices, the size of which is at most 12×12 in the present cases. Here the interplanar force constants used in the whole SL are the ones calculated for $GaAs$, yielding a good description of the $GaAs$-like region of the SL spectrum, while a downward shift of $AlAs$-like L frequencies with respect to the full calculation is found, consistent with the bulk behaviour in Fig. 5. Note that the flat LO $AlAs$-like confined modes are very close to each other and sometimes practically overlap, as e.g. for the 1st and 2nd mode of the $n = 3$ SL.

It is interesting to note that the Γ-point LO frequencies of Fig. 8, when replotted on top of the results of the full calculations in Fig. 6, show no significant difference, except for the above mentioned quasi rigid shift of all the $AlAs$-like LO frequencies. This result is important, because it definitely rules out any significant role of interface charge rearrangement in the phonon spectra of even ultrathin $GaAs/AlAs$ SL's. If charge transfers would be relevant, the mass approximation would not work at all, especially in these SL's where the relative weight of interface atoms is very high.

The difference between our results for ideal SL's and previous theoretical results, therefore, is *not* due to important chemical effects occurring at the interface, but rather to an improved quality of the present bulk phonon dispersions. This fact gives confidence that interatomic force constant models based on first-principle calculations may prove to be an accurate tool for studying large unit cell systems aimed at simulating alloys of different compositions and thick disordered SL's. Encouraging preliminary results have been obtained along these lines.

Superlattices with Cationic Intermixing

In order to understand whether the remaining discrepancies between theory and experiments are determined by disorder effects, we have also calculated[1] phonon frequencies of selected supercell systems with the purpose of simulating cationic intermixing. We perform a full self-consistent calculation of the electronic ground state and of the dynamical matrix for every configuration, and therefore the computational effort limits us to considering a few supercells of at most 16 atoms. Going beyond this limit requires developing other approximate approaches, such as the one mentioned at the end of the preceding section.

In spite of the limited number of configurations that we can study for the moment (which may affect an accurate determination of the frequencies), we will show that it is possible to get some physical insight on the possible effects of disorder just by looking at a few significant configurations.

Let us first consider the $(GaAs)_2(AlAs)_2$ SL. One can distinguish the situation where disorder leaves some of the planes unaffected (a pure Al and a pure Ga plane exist), and the one where it extends to all the cationic planes. We simulate the first situation by studying a SL with the following sequence of (001) cationic planes: Ga-X-Al-X, where X is a mixed $Ga_{0.5}Al_{0.5}$ plane which we assume to be ordered (2 cations per two-dimensional unit cell). Our results indicate that the $AlAs$-like LO1 frequency practically coincides with the corresponding value for the ordered SL (403 and 402 cm^{-1} for the ordered and disordered cases respectively), whereas the $GaAs$-like mode is lowered by 8 cm^{-1}. The agreement with experimental Raman

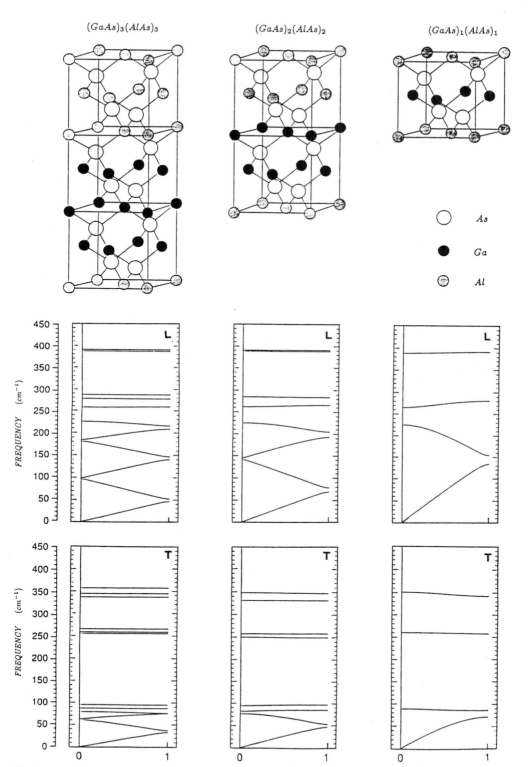

Fig. 8 SL dispersions versus (001) wavevector, in units of π/d (d is the SL period). Here the "mass approximation" is used: $GaAs$ force constants from Table 2 and appropriate Ga, Al, and As masses; see text.

data is therefore slightly better in the *GaAs*-range, but has not improved in the *AlAs*-range. Inspection of phonon eigenvectors shows that ion displacements of *AlAs*-like and (to a minor extent) *GaAs*-like LO1 modes are well confined within the corresponding pure-cation plane. In other words, this kind of *interfacial* disorder affects the LO1 frequencies only through a modification of the relevant confinement length, shifting them closer to the LO1 modes of the $n = 1$ SL. As the *AlAs*-like LO1 frequencies are very close for $n = 2$ and $n = 1$, such a shift is small in that case. While the *AlAs*-like vibrations at lower frequencies are related to the mixed interface planes[17], the *AlAs*-like LO1 mode is practically insensitive to their composition, and it will approximately be at the same frequency as long as a pure *Al* cationic plane exists in the SL. In particular, it will never be shifted out of the bulk *AlAs* bandwidth by this kind of interfacial disorder. A similar analysis also holds for *GaAs*-like modes. In this case, however, the large dispersion of the bulk LO band makes that a rather large shift of the LO1 mode is possible also in the presence of a pure *Ga* plane. A more extended presentation of these results can be found in Ref. 17.

AlAs-like LO1 modes therefore appear to be marginally sensitive to confinement effects and thus are strictly related to the amount of cationic intermixing in the purest planes. Based on their behaviour, we have proposed that the existing experimental data for $n = 2$ not only cannot be assigned to a perfectly ordered superlattice, but are also inconsistent with a model of disorder that leaves some cationic planes completely unaffected.[1,17]

When no pure cationic plane exists, the *AlAs*-like mode is strongly shifted to lower frequencies due to the strong reduction in the associated macroscopic polarization. This trend appears from our results for the $Ga_{0.5}Al_{0.5}As$ alloy (Fig. 6), which we have simulated with a random distribution of cations in a 16-atom supercell or with a chalchopyrite cell. In both cases, the *AlAs*-like frequency moves out of the corresponding bulk bandwidth and turns out to be in agreement with the experimental results for the alloy. Now the shift of the *AlAs*-like frequencies with respect to the ordered cases is very large: a simple understanding of this fact is obtained by recognizing that here the shift mostly reflects the decrease in the macroscopic polarization, a measure of which is given by the LO-TO splitting. A similar interpretation of our data can also be given for the other UT SL's.[1]

Although the present calculations are, of course, limited by the small size of the cells used in simulating disordered configurations, and by the small number of configurations, preliminary results performed with giant supercells and *ab initio* interatomic force constants with the mass approximation essentially confirm the above general picture, and allow to conclude that the large discrepancy between the experimental data of Fig. 3 and the first principles predictions of Fig. 6 are removed when intermixing at interfaces is taken into account in the calculations.[26]

CONCLUSIONS

We have performed a consistent *ab initio* study of the (001) dispersions and interplanar force constants of bulk *GaAs* and *AlAs*, and of ultrathin *GaAs/AlAs* superlattices. We have shown that the bulk dispersions, particularly of *AlAs*, differ from previous expectations, and that such difference is also relevant to the description of SL phonons. Moreover, we have demonstrated that the "mass approximation" for *GaAs* and *AlAs* holds with good accuracy both in the bulk and in the SL, thus ruling out an important role of interface charge rearrangements in determining the SL dynamics. We have provided the ingredients to perform simple and accurate linear chain calculations for L and T propagation in (001) SL's.

Our study of ordered ultrathin SL's, and of some model disordered configurations, indicates that a substantial degree of intermixing extending to all cationic planes is necessary to account for the measured Raman spectra on the samples presently available. *AlAs*-like modes appear to be the most sensitive to such intermixing, and are therefore indicated as important features to be monitored for characterization.

ACKNOWLEDGEMENTS

We are grateful to A. Fasolino and J. Menéndez for many useful discussions. This work was supported in part by the Italian Ministry of University and Scientific Research through the SISSA-CINECA supercomputing project, by the Italian CNR under grants 89.00006.69 and

89.00011.69 (PF Sistemi Informatici e Calcolo Parallelo), and by the Swiss National Science Foundation under grant 20-5446.87. One of us (SB) also acknowledges financial support by the European Research Office of the US Army under grant DAJA 45-89-C-0025.

REFERENCES

1. S. Baroni, P. Giannozzi, and E. Molinari, Phys. Rev. B41, 3870 (1990).

2. For very recent reviews see: (a) B. Jusserand and M. Cardona, in "Light Scattering in Solids V", edited by M. Cardona and G. Güntherodt, Springer, Berlin (1989), p. 49; (b) J. Menéndez, J. of Luminescence 44, 285 (1989).

3. A.S. Barker, Jr., L.J. Merz, and A.C. Gossard, Phys. Rev. B17, 3181 (1978).

4. M. Nakayama, K. Kubota, and N. Sano, Solid State Commun. 53, 493 (1985).

5. C. Colvard, T. A. Gant, M. V. Klein, R. Merlin, R. Fischer, H. Morkoç, and A. C. Gossard, Phys. Rev. B31, 2080 (1985).

6. B. Jusserand, D. Paquet, and A. Regreny, Phys. Rev. B30, 6245 (1984).

7. M. Cardona, T. Suemoto, N.E. Christensen, T. Isu, and K. Ploog, Phys. Rev. B36, 5906 (1987).

8. A. Ishibashi, M. Itabashi, Y. Mori, K. Kawado, and N. Watanabe, Phys. Rev. B33, 2887 (1986).

9. T. Toriyama, N. Kobayashi, and Y. Horikoshi, Jpn. J. Appl. Phys. 25, 1895 (1986).

10. (a) E. Molinari, A. Fasolino, and K. Kunc, in "Proc. 18th Internat. Conf. on the Physics of Semiconductors", edited by O. Engström, World Scientific, Singapore (1987), p. 663; (b) E. Molinari, A. Fasolino, and K. Kunc, Superlatt. Microstruct. 2, 397 (1986).

11. (a) K. Kunc and R.M. Martin, Phys. Rev. Lett. 48, 406 (1982); (b) K. Kunc, in "Electronic Structure, Dynamics and Quantum Structural Properties of Condensed Matter", edited by J. T. Devreese and P.E. Van Camp, Plenum, New York (1985), p. 221.

12. E. Richter and D. Strauch, Solid State Commun. 64, 867 (1987).

13. G. Kanellis, Phys. Rev. B35, 746 (1987).

14. S.F. Ren, H. Chu, and Y.C. Chang, Phys. Rev. B37, 8899 (1988).

15. T. Tsuchiya, H. Akera, and T. Ando, Phys. Rev. B39, 6025 (1989).

16. K. Huang and B.F. Zhu, Phys. Rev. B38, 2183 (1988); F. Bechstedt and H. Gerecke, Phys. Stat. Sol. (b) 154, 565 (1989), and references therein.

17. (a) S. Baroni, P. Giannozzi, and E. Molinari, in "Proc. 19th Int. Conf. on the Physics of Semiconductors", edited by W. Zawadzki, Inst. of Physics of the Polish Academy of Sciences, Warsaw (1988), p. 795; (b) S. Baroni, P. Giannozzi, and E. Molinari, in "Proc. III Int. Conf. on Phonon Physics and VI Int. Conf. on Phonon Scattering in Condensed Matter", edited by S. Hunklinger, W. Ludwig, and G. Weiss, World Scientific, Singapore (1990), p. 722.

18. S. Baroni, P. Giannozzi, and A. Testa, Phys. Rev. Lett. 58, 1861 (1987).

19. G.B. Bachelet, D.R. Hamann, and M. Schlüter, Phys. Rev. B26, 4199 (1982).

20. (a) G. Dolling and J.L.T. Waugh, in "Lattice Dynamics", Pergamon, London (1965), p. 19; (b) D. Strauch and B. Dorner, J. of Phys.: Cond. Matt. 2, 1457 (1990); (c) A. Onton, in "Proc. 10th Int. Conf. Physics of Semiconductors", USAEC, New York (1970), p. 107; (d) B. Monemar, Phys. Rev. B8, 5711 (1973).

21. B. Dorner and D. Strauch, in "Proc. III Int. Conf. on Phonon Physics and VI Int. Conf. on Phonon Scattering in Condensed Matter", edited by S. Hunklinger, W. Ludwig, and G. Weiss, World Scientific, Singapore (1990), p. 82.

22. P. Pavone, Master Thesis, SISSA, Trieste (1989), unpublished.

23. The origin of the slight discrepancies between the present results and Ref. 11 in the description of $GaAs$ is probably related to their use of local pseudopotentials and supercells of limited size. The description of $AlAs$ in Ref. 10—which differs from the present one mainly in its larger LO bandwidth—was obtained using the $GaAs$ force constants of Ref. 11 and adding a Coulomb term to correct the LO-TO splitting; we point out that the discrepancy of the resulting $AlAs$ LO bandwidth[10] with respect to the present one arises not only from the imperfections in the starting $GaAs$ forces[11], but also from the corrective long range term.

24. A. K. Sood, J. Menéndez, M. Cardona, and K. Ploog, Phys. Rev. Lett. 54, 2111 (1985); Phys. Rev. Lett. 54, 2115 (1985).

25. B. Jusserand and D. Paquet, Phys. Rev. Lett. 56, 1751 (1986).

26. E. Molinari, S. Baroni, P. Giannozzi, and S. de Gironcoli, in "Proc. 20th Int. Conf. on the Physics of Semiconductors", edited by E. Anastassakis and J.D. Joannopulos, World Scientific, Singapore (1990), in press; and to be published.

RESONANT RAMAN SCATTERING IN GaAs-AlAs MULTIQUANTUM WELLS

UNDER MAGNETIC FIELDS

J.M. Calleja, F. Meseguer, F. Calle, C. López, L. Viña, and
C. Tejedor

Instituto de Ciencia de Materiales UAM-CSIC
Cantoblanco 28049, Madrid, Spain

K. Ploog

Max Planck Institut für Festkörperforschung
Stuttgart, FRG

F. Briones

Instituto Nacional de Microelectrónica CSIC
Madrid, Spain

INTRODUCTION

Resonant Raman Scattering (RRS) in semiconductor superlattices (SL) and multiquantum wells (MQW) has been extensively studied in recent years in order to obtain information on different aspects of their electronic structure, as well as on their phonon properties. Thus, RRS has been used, among other methods, to investigate the two-dimensional character of excitons in SL's and MQW's,[1] the degree of localization of electron and hole wave functions,[2] the effect of the lattice mismatch-induced strain on electronic states of higher gaps[3] and the electric field effect on the resonance energies and probabilities, including the appearance of initially forbidden transitions.[4] As far phonons are concerned, extensive work has been done on confined longitudinal optical (LO) modes in GaAs-Al$_x$Ga$_{1-x}$As and GaAs-AlAs systems,[5] as well as folded acoustic modes[6] and interface modes,[7-10] including sometimes their resonant behavior. A recent and comprehensive review on light scattering in MQW and SL's can be found in Ref. 11.

As common denominator for most of the previous work, the main points of interest offered by SL's and MQW's, in comparison with bulk materials, are their specific electronic structures (step-like density of states, (DOS), two-dimensional excitons, hole-subbands mixing; etc.) and the appearance of confined, folded and interface vibration modes. On the other hand, conventional optical spectroscopy in external magnetic fields has proven to be a very powerful method to study some aspects of the electronic structure of semiconductor MQW and SL, such as the excitonic properties[12,13] the mixing of the valence subbands[14] and the SL miniband structure.[15,16]

Light Scattering in Semiconductor Structures and Superlattices
Edited by D.J. Lockwood and J.F. Young, Plenum Press, New York, 1991

53

RRS in the presence of a magnetic field has recently been used to enhance sensitivity for the observation of interband transitions.[17] There, the reduction in the dimensionality of the DOS by the external magnetic field leads to the sharpening of the resonant peaks. This also allows for the observation of multiphonon RRS[18] (up to 7LO in bulk GaAs) when the resonance condition for a particular Landau level (LL) is reached. On the other hand, the magnetic field shifts the electron and hole states, so that it can help in the study of incoming and outgoing resonances, as well as the conditions for double resonances. Double resonances have very recently been reported in GaAs[19] when the energy difference between heavy and light hole states of the same LL is tuned by the magnetic field to equal the LO phonon energy. Also a detailed theory of the RRS in the presence of magnetic fields in bulk semiconductors is now available.[20] This theory gives explicit expressions for both the deformation potential and Fröhlich interaction scattering mechanisms, taking into account the hole subbands mixing produced by the magnetic field, and the non-parabolicity of the energy bands.

Moving now to SL's and MQW's, new features are expected to appear in the field dependence of the Raman intensity, due to their specific electronic structures mentioned above. In fact, it has been observed that the magnetic field increases the strength of the Raman resonance of the LO confined modes and especially of the interface modes (IF) in GaAs-$Ga_{1-x}Al_xAs$ SL's.[21]

In this work a study of the magnetic-field dependence of the RRS of GaAs-AlAs MQW is presented. We report the ocurrence of magneto-Raman oscillations similar to those found in bulk GaAs,[19,20] and their usefulness in properly describing RRS in 2D systems is discussed. Double Raman resonances induced by the magnetic field are reported and some of the underlying mechanisms are discussed.

EXPERIMENT

The samples studied are MQW of GaAs-AlAs grown by MBE with well and barrier widths d_1=90 Å, d_2=50 Å (sample A) and d_1=100 Å, d_2=100 Å (sample B), respectively. The sample periodicity and homogeneity have been tested by X-ray diffraction. The experiment was performed in a superconducting magnet (0-13 T) with optical access from the bottom of the cryostat. A Faraday configuration was used with B directed along the SL-axis. A LD-700 dye laser was used as excitation source with a typical power of 1 mW for luminescence and 150 mW for Raman measurements. The laser beam was focused on the sample with a spherical lens, and no heating problems were present since the sample was immersed in liquid He at 2 K. The emitted light was analysed by a double monochromator and a standard photon-counting system. Four different polarization configurations were used ($\sigma\pm$, $\sigma\pm$) by means of achromatic $\lambda/4$ plates.

RESULTS AND DISCUSSION

1. Optical Characterization of the Samples

Photoluminescence and PLE spectra were first recorded in the four polarization configurations for several values of the magnetic field. The luminescent emission was located at 1.587 eV (5.2 meV FWHM) for sample A and at 1.564 eV (1.4 meV FWHM) for sample B. No significant changes in the PL spectra were found for different polarization configurations.

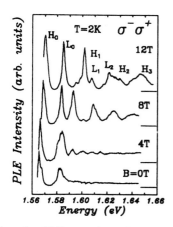

Fig. 1. PLE spectra for sample B at several fields.

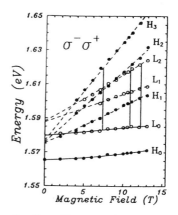

Fig. 2. Fan diagram for sample B.

The PLE spectra of the two samples are very similar, as expected from the nearly identical values of the well widths. The barrier are large enough in all the samples to keep the wells uncoupled. Fig. 1 shows the PLE spectra of sample B at different fields in the $\sigma^-\sigma^+$ configuration. The H_n (L_n) labels refer to the ground- (n=0) and excited states (n≥1) exciton transitions. (They are often considered as transitions between the succesive Landau levels, but due to the non-zero binding energy, the exciton picture seems to be more appropiate, at least at low fields). The heavy or light character of the holes involved in the transitions can be easily determined using the selection rules for circularly polarized light. The fan diagram corresponding to the PLE data is shown in Fig. 2. One can clearly distinguish the small diamagnetic shift of the ground-state transitions (H_0, L_0) from the Landau-like behavior of the excited state transitions. The L_0-H_0 energy difference (about 15 meV) is much smaller than the LO-phonon frequency of GaAs (36 meV, as indicated by the vertical bars in Fig. 2), so that the double resonance condition described in Ref. 19 is not fulfilled by the fundamental hole states. Double resonances can nevertheless occur, as will be shown later, involving higher excited states.

2. Raman Resonances

One of the ways to look for double resonances is to change the magnetic field over a wide range, while the monochromator is set at a frequency fixed 1LO below that of the incident light.[17,19] The result of such a procedure is shown in Fig. 3 for sample A for two values of the incident light frequency. The observed oscillations correspond, in principle, to the succesive incoming (outgoing) resonances occuring every time that the chosen laser frequency (monochromator setting) equals a magneto-optical transition, as the field increases. Whereas this method appears to give good results for bulk GaAs,[17,19] it has a decisive disadvantage in the present case, due to the selective excitation of hot luminescence.[22] This can be observed in Fig. 4, where the Raman lines (the

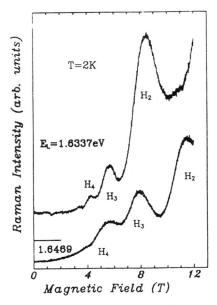

Fig. 3. Magneto-Raman oscillations of the GaAs confined LO mode in sample A for two excitation energies.

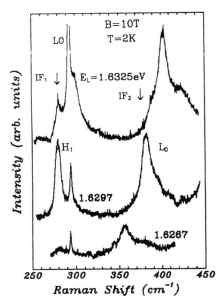

Fig. 4. Raman spectra of sample B at three laser energies, showing the LO and IF phonons together with the hot luminescence bands.

LO-GaAs phonon and the GaAs-like IF_1 and AlAs-like IF_2 interface modes) and the L_0 and H_1 luminescent emissions show a similar resonant enhancement as the incident frequency is changed. This means that the magneto-Raman oscillations shown in Fig. 3 will contain a significant contribution of luminescence, especially under doubly resonant conditions. To avoid this inconvenient, the RRS profiles have been measured as a function of the laser frequency for fixed values of the magnetic field.

The result can be observed in Fig. 5, again for sample A, at 0 and 10T. The peak at the lowest energy corresponds to the incoming resonance at the ground heavy-hole transition (H_0). The remaining peaks are the outgoing resonances at both the ground-state heavy- and light-hole excitons (H_0, L_0) and excited states of the heavy-hole exciton (H_1, H_2). The peak positions are the same than the corresponding ones in the PLE spectra, once the phonon energy is substracted for the outgoing resonances. It is important to notice the decrease of about one order of magnitude of the resonance intensity when going to successively higher-order excited states at constant field. This is the same trend found in Refs. 17 and 19 for bulk GaAs. When comparing the RRS curves for OT and 10T it is not easy to decide about their absolute differences in intensity, due to the strong luminescence that prevents the measurement of the H_0 outgoing resonance at 0 T.

A comparison of the PL, PLE and RRS spectra is shown in Fig. 6 for sample B, in the $\sigma^-\sigma^+$ configuration, at 10T. The curves are presented in a logarithmic scale with rigid vertical shifts for a clear display. The use of circularly polarized light allows for transitions between states with different spin orientations. This is clearly observed in Fig. 6, where a

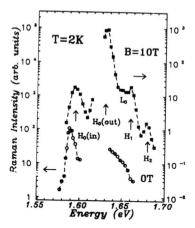

Fig. 5. RRS spectra at 0 T and 10 T for
sample A. Note the difference
in the logarithmic scale for
the two curves.

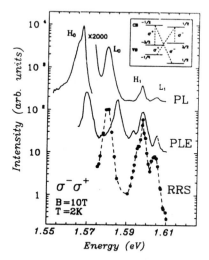

Fig. 6. Comparison between the PL, PLE
and RRS curves at 10T in the
$\sigma^{-}\sigma^{+}$ configuration.

Zeeman splitting appears between the PL peaks (determined by the outgoing polarization, σ^+) and the PLE ones (determined by the incoming polarization, σ^-) of the ground state exciton transitions.[22] The inset of Fig. 6 shows a simplified diagram (in the sense that hole mixing and excitonic effects are ignored) that indicates the participation of different spin states in the σ^+ (PL) and σ^- (PLE) transitions. One observes also in Fig. 6 a clear coincidence of the RRS spectrum (which is plotted as a function of the scattered photon energy) with the PL curve. This again indicates the dominance of the outgoing resonance over the incoming one.

3. Double Raman Resonances (DRRS)

The intensity of the L_0 outgoing resonance of sample B (the peak at 1.58 eV in the RRS curve of Fig. 6) has been studied as a function of the magnetic field. Typical results are shown in Fig. 7 for four different values of the field. One observes a small shift (3 meV) of the resonance maximum to higher energies as the field is increased, which reflects the weak diamagnetic shift of the L_0 transition shown in Fig. 2. Besides, the resonance intensity changes dramatically with the field. This is more clearly observed in Fig. 8, where the intensity of the L_0 outgoing resonance is displayed as a function of the field. The arrows in Fig. 8 are at the same position than the bars in Fig. 2; they indicate the fields at which the difference in energy between the L_2, H_2, L_3 (not shown in Fig. 2) and H_3 transitions, and the L_0 ground state exciton equal the LO-phonon energy. Consequently, the strong peak at 11.4T and the shoulder at 7.5T are interpreted as double resonances[23] where the outgoing channel is L_0 and the incoming one is L_2 (or H_2) and L_3 (or H_3), respectively. In fact, the difference between H_2 and L_2 (or H_3 and L_3) is not very meaningful, since their small difference in energies and the complexity of the valence-band structure[14,19,20] in these systems lead to a strong mixing of the hole states. We will nevertheless use this nomenclature to keep the following discussion in simple terms.

It has been shown[20] that the non-parabolicity of the bands, as described in the Luttinger-Kohn model, leads to the coupling between levels of the heavy-hole (or light-hole) ladder differing 2 units in the LL index, n. According to this result, the peak at 11.4T of Fig. 8 should be assigned to the L_2 in $-$ L_0 out double resonance. The nature of the coupling between levels with $\Delta n = 3$, which would be at the origin of the L_3-L_0 out double resonance, is not explained.

The result shown in Fig. 8 has an essential difference with respect to the 3D double resonance.[19,20] Contrary to the bulk case, our 2D results can be hardly understood in terms of the free particle-phonon interaction as the scattering mechanism, because this would imply the simultaneous jump of the electron and the hole by $\Delta n = 2$ or 3, sharing the phonon energy. In fact, our results strongly suggest the exciton-phonon coupling as the scattering mechanism. This means that the L_2 (or L_3) excitons created by the incoming light would be scattered as a whole by the phonon to the L_0 state. One possible alternative to this mechanism would be a four-band process in the free particle picture, in which the electron and the hole are scattered separately by the phonon. This would be described by terms of order higher than 3 in perturbation theory, which are unlikely to explain the strong observed changes in the resonance intensity.

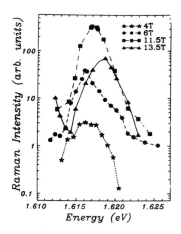

Fig. 7. RRS profiles in the L_0 outgoing region, for different values of the magnetic field.

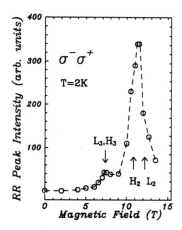

Fig. 8. RRS peak intensity as a function of the magnetic field.

SUMMARY AND CONCLUSIONS

The influence of the magnetic field on the RRS of GaAs-AlAs MQW has been studied. There have been found magneto-Raman oscillations for the GaAs LO mode similar to those reported for bulk GaAs,[19,20] but containing a significant amount of luminescence coming from excited states (Landau levels) transitions. The RRS profile reproduces the PL spectra, thus indicating the predominance of outgoing transitions. Finally, a DRRS has been observed at values of the magnetic field for which the difference between several excited states exciton transitions and the ground state light-hole exciton equals the LO-GaAs phonon frequency. These results suggest the exciton-phonon coupling as the scattering mechanism.

ACKNOWLEDGEMENTS

This work was sponsored in part by CICYT Grant MAT-88-0116-C02.

REFERENCES

1. J.E. Zucker, A. Pinczuk, D.S. Chemla, A. Gossard, and W. Wiegmann, Phys. Rev. Lett. 51, 1293 (1983).
2. J.E. Zucker, A. Pinczuk, D.S. Chemla, A. Gossard, and W. Wiegmann, Surf. Science 174, 175 (1986).
3. C. Tejedor, J.M. Calleja, F. Meseguer, E.E. Mendez, C.A. Chang, and L. Esaki, Phys. Rev. B 32, 5303 (1985).
4. C. Tejedor, J.M. Calleja, L. Brey, L. Viña, E.E. Mendez, W.I. Wang, M. Staines, and M. Cardona, Phys. Rev. B 86, 6054 (1987).
5. M.V. Klein, IEEE J. Quantum Electronics QE-22, 1760 (1986).
6. C. Colvard, T.A. Gant, M.V. Klein, R. Merlin, R. Fischer, H. Morkoc, and A.C. Gossard. Phys. Rev. B 31, 2080 (1985).
7. A.K. Sood, J. Menéndez, M. Cardona, and K. Ploog, Phys. Rev. Lett. 54, 2111 (1985).
8. A.K. Sood, J. Menéndez, M. Cardona, and K. Ploog, Phys. Rev. Lett. 54, 2115 (1985).
9. R. Merlin, C. Colvard, M.V. Klein, H. Morkoc, A.Y. Cho, and A.C. Gossard, Appl. Phys. Lett. 36, 43 (1980).
10. F. Calle, J.M. Calleja, C. Tejedor, F. Briones, J.L. de Miguel, and K. Ploog, Surf. Science 228, 176 (1990).
11. "Light Scattering in Solids V (Topics in Applied Physics, vol. 66)", M. Cardona and G. Güntherodt, eds., Springer, Heidelberg, 1989.
12. J.C. Maan, G. Belle, A. Fasolino, M. Altarelli, and K. Ploog, Phys. Rev. B 30, 2253 (1984).
13. D.C. Rogers, J. Singleton, R.J. Nicholas, C.T. Foxon, and K. Woodbridge, Phys. Rev. B 34, 4002 (1986).
14. F. Ancilotto, A. Fasolino, and J.C. Maan, Phys. Rev. B 38, 1788 (1988).
15. G. Belle, J.C. Maan, and G. Weimann, Solid State Commun. 56, 65 (1985).
16. B. Deveaud, A. Chomettte, F. Clerot, A. Regreny, J.C. Maan, R. Romestain, G. Bastard, H. Chu, and Y.-C. Chang, Phys. Rev. B 40, 5802 (1989).
17. G. Ambrazevicius, M. Cardona, and R. Merlin, Phys. Rev. Lett. 59, 700 (1987).
18. T. Ruf and M. Cardona, Phys. Rev. Lett. 63, 2288 (1989).
19. T. Ruf, R.T. Phillips, C. Trallero-Giner, and M. Cardona, Phys. Rev. B 41, 3039 (1990).
20. C. Trallero-Giner, T. Ruf, and M. Cardona, Phys. Rev. B 41, 3028 (1990).
21. D. Gammon, R. Merlin, and H. Morkoc, Phys. Rev. B 35, 2552 (1987).

22. F. Meseguer, F. Calle, C. López, J.M. Calleja, L. Viña, C. Tejedor, and K. Ploog, presented at the 20th. Int. Conf. on the Physics of Semiconductors (Thessaloniki, 1990).

23. F. Meseguer, F. Calle, C. López, J.M. Calleja, C. Tejedor, K. Ploog and F. Briones, presented at the 5th. Int. Conf. on Superlattices and Microstructures (Berlin, 1990).

OPTICAL PHONONS AND RAMAN SPECTRA IN InAs/GaSb SUPERLATTICES

G. Kanellis and D. Berdekas

Physics Department
University of Thessaloniki
540 06 Thessaloniki, Greece

INTRODUCTION

InAs/GaSb superlattices (SL) of high crystalline quality have been grown[1-2] over the past ten years, and most of the reported work has been focused on their electronic properties [3-4]. To the best of our knowledge, their interesting dynamical properties are still under investigation. A linear-chain model[5] has been applied to long period SL of this kind, giving the basic features of the normal modes of vibration for wavevectors along the SL-axis. Furthermore, 3-dimensional calculations of the phonon dispersion in short-period SL have been reported using either the Bond-Charge Model (BCM)[6] or a Valence Overlap Shell Model (VOSM)[7]. To the best of our knowledge, no experimental data on light scattering from phonons have been reported.

The unit cell masses of InAs and GaSb are almost equal, therefore the optical phonon frequency ranges of the two compounds overlap, in contrast to the well-separated optical phonon branches of GaAs and AlAs, the constituent compounds of the extensively studied GaAs/AlAs SL. On the other hand, the dispersion of the LO branch in InAs is larger than that in GaSb, while for the TO branch the situation is reversed. In the overlapping region, extended modes are expected to appear, instead of the confined ones.

InAs and GaSb do not have any ions in common. As a result, in these systems two different interfaces appear, consisting of Ga-As and In-Sb bonds respectively. This feature, again in contrast to the GaAs/AlAs SL, is of particular interest. Longitudinal optical phonons in GaAs have higher frequencies than those in InAs and GaSb, while in InSb they have lower frequencies. Therefore, the interface modes in the SL under question are expected to have well-separated frequencies above and below the frequency range of the extended modes. In particular, the 1x1 InAs/GaSb SL could equaly well be considered as an GaAs/InSb SL, or as a quaternary compound.

Furthermore, the bulk phonon dispersion is experimentally known for all of the four mentioned binary compounds[8-11]. Hence, detailed investigation of the optical modes of vibration in these systems would be very helpful in studying the atomic interactions of all pairs of ions in the SL and would provide a means for comparison with the interactions in the bulk compounds.

In the next section we give a very brief description of the calculation procedure, and the method and model used, while in the third section we give the symmetry characteristics of these systems and the phonon dispersion for a 1x1 SL. In the fourth section we present the results for the zero wavevector optical modes of vibration in a

Light Scattering in Semiconductor Structures and Superlattices
Edited by D.J. Lockwood and J.F. Young, Plenum Press, New York, 1991

63

16x16 SL and we discuss their confinement characteristics. Finally, in the last section we give the Raman spectrum for a 1x1 SL calculated with the Bond Polarizability Model (BPM).

MODEL AND METHOD OF CALCULATION

The calculation of the phonon dispersion curves is based on a ten-parameter VOSM[12-13], with a modified Valence Force Field. The short range interactions are described by a bond-stretching force constant λ, two bond-bending force constants $kr_1\theta$ and $k'r_1\theta$, in relation to changes in the bond length and in the angles of the bonds, and two more bond-bending force constants $kr_2\theta$ and $k'r_2\theta$, in relation to changes in the angles of the bonds and in the 2nd nearest neighbour distance[14]. The remaining five parameters of the model are: the effective ionic charge Z, the shell charges Y_1 and Y_2 and the core-shell coupling constants k_1 and k_2, for cations and anions, respectively.

The values of the above parameters are obtained by least-square fitting to the experimentally known phonon dispersion of the bulk compounds InAs[8] and GaSb[9] and their values for InAs (GaSb) are found to be: Z=2.00 (kept constant for both

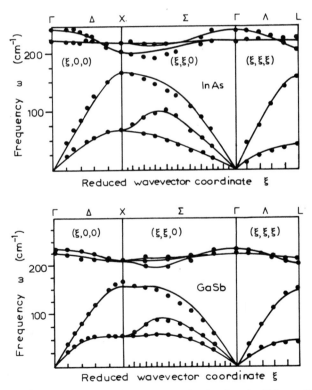

FIG. 1. Phonon dispersion for InAs and GaSb as calculated using the Valence Overlap Shell Model described in the text (solid lines). Points represent measured phonon frequencies.

compounds), Y_1 = 4.930 (4.277) and Y_2 =-2.513 (-3.502) in proton charges, k_1 = 13.790 (15.111) and k_2 = 4.829 (4.448), λ = 1.718 (1.717), $kr_1\theta$ = 0.116 (0.171) and $k'r_1\theta$ = -0.179 (-0.195), $kr_2\theta$ = -0.078 (-0.052) and $k'r_2\theta$ = -0.008 (-0.003) in 10^5 dyn/cm. The calculated dispersion curves for both compounds, using the above values of the parameters, are shown in Fig. 1. It should be noted that in the case of GaSb, the TA (ξ,0,0) branch bends to lower frequencies near the X point, with the highest frequency occuring at about k_x =0.7.

The method used for the calculation of the SL dynamical matrix from the cor-responding dynamical matrices for the constituent compounds, has been described in detail and applied in different cases. Either using a rigid-ion model[15] with the same values of the model parameters for both bulk compounds, or a Keating type model with different values of the force constants for the two compounds[16]. In the case of the shell model, one has to account also for the shell or "electronic" degrees of freedom, includ-ing matrices which describe short and long range, core-core, core-shell and shell-shell interactions. Usually it is assumed that all short range interactions act only through the shells and, therefore, only one short range interaction matrix has to be evaluated for a given SL. On the other hand, the structure dependent Coulomb coefficients, which enter the various pairs of long range interactions, are the same for all these pairs and they constitute the second matrix, which together with the various core and shell charges, will provide the necessary long range interaction matrices for the SL dynami-cal matrix. The above mentioned method is applied for the calculation of these two matrices and the correction in the diagonal blocks for translation invariance is per-formed before the "electronic" degrees of freedom are eliminated.

As it will be discussed next, InAs/GaSb SL may have lattices of P or I type. The transformations from the zinc blende primitive translations to the primitive translations of either kind of SL, are the same with those for Si/Ge SL's along the [0 0 1] direction. These transformations have been presented elsewhere[16].

In the approximation introduced by the above method of calculation, the short range interactions between the pairs In-Ga and As-Sb are taken equal to the average in-teractions of the corresponding pairs In-In and Ga-Ga, and As-As and Sb-Sb, respec-tively. Interactions between the pairs In-Sb and Ga-As are taken equal to the average of In-As and Ga-Sb interactions and the rest of the short range interactions are taken equal to those in the two bulk binary compounds. The lattice constant of the zinc blende underlying lattice is assumed to be equal to 6.07 A.

SYMMETRIES AND PHONON DISPERSION

(InAs)$_n$/(GaSb)$_m$ SL appear in two different orthorombic space groups depend-ing on the number of layers each compound contributes to the SL. If n+m is even, the space group is C^1_{2v} (Pmm2), while if n+m is odd, the space group is C^{20}_{2v} (Imm2).

The crystallographic unit cells for the shortest period InAs/GaSb SL, are given in Fig. 2 together with the phonon dispersion for the 1x1 case. The 1x1 (a) and 2x2 (b) SL have P-type lattice, while the 1x2 (c) SL has I-type lattice. The Ox' and Oy' axes referenced below are in the $x_o + y_o$ and $x_o - y_o$ directions of the underlying zinc blende structure, respectively.

Within the unit cell in P-type lattices, cations occupy a and d Wyckoff sites while anions occupy b and c sites. In the I-type lattices, cations occupy a Wyckoff sites, while anions occupy b sites. All of these sites have mm2 symmetry and each one con-tributes one of A_1, B_1 and B_2 irreducible representations. Hence, for an (InAs)$_n$/(GaSb)$_m$ SL, the long wavelength optic modes of vibration are Raman active and have the following symmetries:

$$[2(n+m)-1](A_1+B_1+B_2).$$

The phonon dispersion for a 1x1 SL is also shown in Fig. 2, for wavevectors parallel (right part) and perpendicular to the SL axis (left part). In the central part of the figure we give the frequency of each mode as the direction of the wavevector changes from one to the other direction.

For this type of SL the point group of the wavevector along the SL axis is the same as for the center of the 1st Brillouin zone (BZ) and the entire phonon branches for $k = (0,0,k_z)$ have the same symmetry properties as their end points at the 1st BZ center. The point group of the wavevector along the Ox'-axis is C_s. The same point group also holds valid for wavevectors in the zOx' plane. The symmetry of the A_1 (longitudinal) and B_1 (transverse) modes becomes Δ_1 and the B_2 (transverse) modes turn into Δ_2 symmetry as the wavevector shows a nonzero k_x component.

The highest frequency A_1 mode lies above the LO branch of InAs and falls into the LO band of GaAs. This mode mainly involves vibrations of the Ga and As atoms and, in SL with thicker layers, becomes an interface mode. In the 1x1 SL, In and Sb participate with four times smaller amplitudes. As the wavevector k changes direction from parallel to perpendicular to the SL axis (remaining very small), the mode remains longitudinal with decreasing frequency, and when k becomes parallel to the Ox'-axis, the displacement pattern becomes (almost) identical to that of the higher frequency B_1 mode for k parallel to the SL axis. B_1 modes involve displacements of the atoms on the Ox'-axis. The remaining A_1 and B_1 modes show little dispersion, while B_2 modes show no dispersion when k changes direction. B_2 modes involve displacements of the atoms on the Oy'-axis.

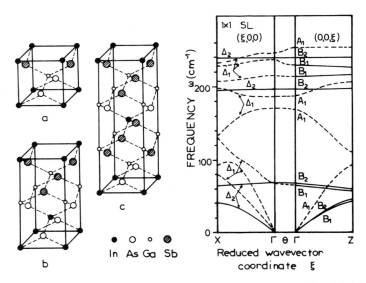

FIG. 2. In the left part of the figure are shown the unit cells for the 1x1 (a), 2x2 (b) and 1x2 (c) InAs/GaSb SL. In the right part of the figure the phonon dispersion for the 1x1 SL is shown. In the right panel, dashed lines represent longitudinal modes while in the central and left panels they represent modes with mixed polarization. Solid lines represent transverse modes polarized along Oy' throughout. For the symmetries of the branches, see text.

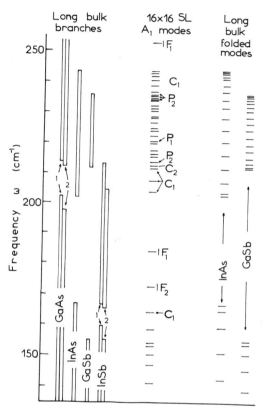

FIG. 3. In the left part of the figure are shown the frequency ranges of the longitudinal bulk branches in the four binary compounds. In the cases of GaAs and InSb, both sets of force constants are used; (1) using those of InAs, and (2) using those of GaSb. In the middle part, the frequencies of SL modes of A_1 symmetry, are shown. In the right part are shown the frequencies of the folded LO and LA modes of bulk InAs and GaSb. The labeling of the modes is explained in the text.

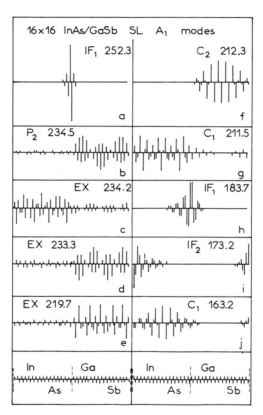

FIG. 4. Displacement patterns for some SL A_1 modes, (a to j). In the lower part of the figure are shown the positions of the ions. For A_1 modes the displacements are parallel to the SL-axis.

INTERFACE, CONFINED AND EXTENDED MODES

As it has been already mentioned above, InAs/GaSb SL show two types of interface bilayers: one formed by the lighter pair of atoms (Ga and As) and a second formed by the heavier pair (In and Sb atoms). Since in both cases the interatomic forces are comparable to the interactions in bulk compounds, the frequencies of the interface modes are expected to appear well above and below the frequency range of the SL modes.

Therefore, for folded optic modes propagating along the SL-axis, the two interfaces act as narrow-band filters for the bulk SL modes. They are situated between the layers of the two different compounds and have frequency bands outside the range of the bulk SL modes. This is indeed the case for A_1 and B_2 folded optic modes with $\mathbf{k} \parallel z$ involving stretching of the interface bonds, since in both cases they lay on the y'Oz plane (Fig. 1a).

For modes of B_1 symmetry, the atomic displacements occur on the Ox'-axis, and only bending of the interface bonds is involved. For this type of vibration the characteristic frequencies of the interface layers fall much closer to or inside the frequency range of the bulk SL modes. As a consequence, folded TO modes of B_1 symmetry "see" a "soft" interface, while the corresponding modes of B_2 symmetry "see" a "hard" interface.

The above remarks are best illustrated by examining the confinement of modes of different symmetries in a SL with thick layers. In Fig. 3 we present the results for the A_1 modes in a 16x16 SL, together with the frequency ranges of LO and LA branches in bulk InAs, GaSb, GaAs and InSb. The above frequency ranges for the latter two compounds have been calculated using both sets of force constants. The differences are not too large and the approximations involved in the calculation of the interactions at the interfaces should correspond to some intermediate results. For comparison, on the same figure we also give the frequencies of the longitudinal bulk modes of InAs and GaSb which are folded to the center of the SL 1st BZ, for comparison. Their relation to the confinement characteristics of the SL modes will be discussed in the following.

As can be seen in Fig. 3, in the higher frequency range the GaSb LO branch overlaps with the corresponding branches of GaAs and InAs. Hence, in this frequency range, only extended modes are expected. It turns out that some modes have very small or even zero amplitudes inside the InAs layer. Displacement patterns for various A_1 modes are shown in Fig. 4. In order to characterize these modes, we have calculated the sum of the squares of the displacements in each layer and divided it by the same sum in both layers. The resulting percentage is taken as a measure of the energy concentrated in each layer. The four modes labeled P_2 on Fig. 3 have over 98% of their "energy" in the GaSb layer. The rest of their "energy" is distributed into the InAs layer. One of these modes is shown in Fig. 4(b). Such modes may be called quasi-confined or pseudo-confined modes. The mode labeled C_2 on Fig. 3, also falls within the overlapping frequency range but only 0.3% of its "energy" is found in the InAs layer. This mode is considered to be confined in the GaSb layer and its displacement pattern is shown in Fig. 4(f). The rest of the modes in this range of frequencies are considered to be extended, although they are mainly localized in one layer, as is shown in Fig. 4(c), (d) and (e). Outside the frequency range of the GaSb LO branch, but still inside the frequency range of the InAs LO branch, the modes are labeled C_1 and are expected to be confined modes[5]. The "energy" leak in the GaSb layer reaches 4% [Fig. 4(g)]. A mode confined in the InAs layer is also found in the higher frequency range of folded LA modes in the SL, as can be seen in Fig. 3, [Fig. 4(j)]. Modes with frequency less than 150cm^{-1}, are well extended modes.

Of particular interest are the three modes localized at the interfaces. Two of them are localized at the Ga-As interface and are labeled IF_1 in the figures. The one with the higher frequency is a sharp symmetric interface mode [Fig. 4(a)] and the other is an asymmetric one [Fig. 4(h)], with frequencies well inside the LO band and near the top of the LA band of bulk GaAs. The third mode is an In-Sb symmetric interface mode with

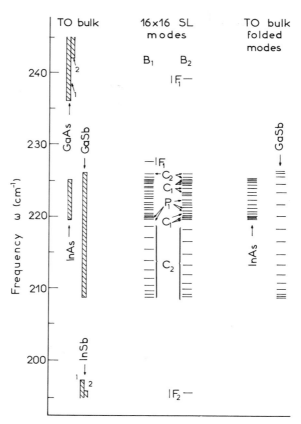

FIG. 5. In the left part of the figure are shown the frequency ranges of the TO bulk branches in the four binary compounds. In the cases of GaAs and InSb, both sets of force constants are used; (1) using those of InAs, and (2) using those of GaSb. In the middle part, the frequencies of SL modes of B_1 and B_2 symmetry, are shown. In the right part are shown the frequencies of the folded TO modes of bulk InAs and GaSb. The labeling of the modes is explained in the text.

frequency at about the middle of the LO band of bulk InSb [Fig. 4(i)]. It should be noted that the frequencies of the interface modes are independent of the layer thicknesses for SL with layers thicker than their length of localization in each layer.

The situation for the transverse modes propagating along the SL-axis is depicted in Figs. 5 to 9. The results for the B_1 and B_2 folded optic modes in the same SL, together with the TO frequency ranges in the four bulk compounds and the frequencies of the TO modes in InAs and GaSb, which are folded to the center of the SL 1st BZ, are shown in Fig. 5. The same information for the folded TA modes is given in Fig. 6, while the displacement patterns for modes of B_1 and B_2 symmetry are given in Figs. 7 to 9. The labeling of the transverse modes follows the same conventions as for the longitudinal ones. The TO branch of InAs has a narrow frequency range which overlaps entirely with that of GaSb, while the corresponding frequency ranges for GaAs and InSb are in higher and lower frequencies, respectively.

The very sharp symmetric Ga-As interface mode of B_2 symmetry [IF_1 in Fig. 5 and Fig. 8(a)] falls near the lower end of the bulk GaAs TO branch and the corresponding, In-Sb interface mode of the same symmetry [IF_2 in Fig. 5 and Fig. 8(j)], which is also very sharp, falls near the upper end of the bulk InSb TO branch. Two Ga-As interface modes of B_1 symmetry are present [IF_1 in Figs. 5 and 6 and Fig. 7(a) and (f)]. The higher frequency one falls in the gap above the TO branches of InAs and GaSb and below the TO branch of GaAs. This is a specific characteristic of that mode, since it involves mainly bond-bending of the Ga-As bond and demonstrates the "soft" behaviour of the interface with respect to B_1 modes. In contrast to this, the "hard" character of the interface with respect to B_2 modes is implied by the high frequency of the corresponding interface mode of B_2 symmetry.

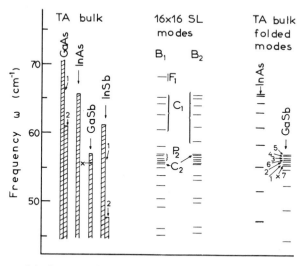

FIG. 6. In the left part of the figure are shown the frequency ranges of the TA bulk branches in the four binary compounds. In the cases of GaAs and InSb, both sets of force constants are used; (1) using those of InAs, and (2) using those of GaSb. In the middle part, the frequencies of SL modes of B_1 and B_2 symmetry, are shown. In the right part are shown the frequencies of the folded TO modes of bulk InAs and GaSb. The labeling of the modes is explained in the text. The numbers on the rightmost column indicate the sequence of folded modes from the X point towards the zone center.

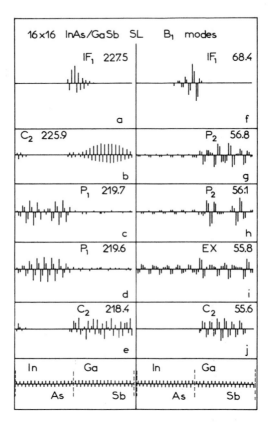

FIG. 7. Displacement patterns for some SL B$_1$ modes, (a to j). In the lower part of the figure are shown the positions of the ions.

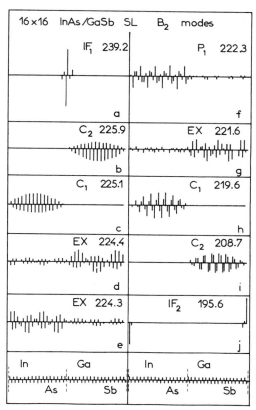

FIG. 8. Displacement patterns for some SL folded transverse optic modes of B_2 symmetry, (a to j). In the lower part of the figure are shown the positions of the ions.

The consequences of this anisotropy, which is reflected on the orthorombic symmetry of the lattice, are in turn reflected on the confinement characteristics of the rest of the modes, as can be seen in Figs. 5 to 9. In the overlaping frequency range of the TO branches, all of the B_1 modes, except two, are extended. The two modes of lower frequencies within this range, labeled P_1 in Fig. 5 and Fig. 7(c) and (d), have over 99% of their "energy" in the InAs layer. Outside this range, apart from the interface mode, the next one of lower frequency, labeled C_2 in Fig. 5 and Fig. 7(b), is confined within the GaSb layer. The next lower one, still above the InAs TO branch, is extended. For the modes of B_2 symmetry, apart from the high frequency interface mode, the two modes above the InAs TO branch, labeled C_2 in Fig. 5, are confined in the GaSb layer. The one of higher frequency is shown in Fig. 8(b). In the overlaping range of frequencies we find eight modes, labeled C_1 in Fig. 5, with over 99.5% of their "energy" in the InAs layer. In Fig. 8(c) and (h) are given the displacement patterns of the higher and lower frequency of these modes, respectively. Another three modes, labeled P_1 in Fig. 5, leak only 0.5-1.5% of their "energy" in the GaSb layer. One of them is depicted in Fig. 8(f). Over the same frequency range, two modes, with a frequency difference of only 0.1 cm^{-1}, have over 90% of their "energy" in different layers, [Fig. 8(d) and (e)].

In the lower, non overlaping range of frequencies of the GaSb TO branch, all of the modes of both symmetries are confined. These are labeled C_2 in Fig. 5. The higher frequency mode of B_1 symmetry, and the lower frequency one of B_2 symmetry, are shown in Fig. 7(e) and Fig. 8(i), respectively.

The results for the folded TA modes of the same symmetries are given in Fig. 6. For modes of this type, involving (almost) in-phase displacements of the interface

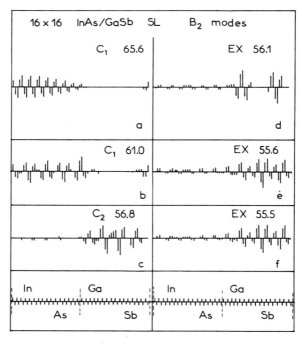

FIG. 9. Displacement patterns for some SL folded transverse acoustic modes of B_2 symmetry, (a to f). In the lower part of the figure are shown the positions of the ions.

bilayer, B_1 interface modes are expected to have higher frequencies than the corresponding modes of B_2 symmetry. This is due to the fact that the former involve bond-stretching while the latter involve bond-bending of the neighbouring bonds. In the higher frequency range, we find a Ga-As interface mode of B_1 symmetry [Fig. 7(f)] and confined modes in the InAs layer, of both symmetries, with frequencies above the GaSb TA branch. Two such B_2 modes are shown in Fig. 9(a) and (b). The latter one involves a comparatively large displacement of the interface layer, with over 18% of its "energy" in that layer. This is an indication that the GaAs interface mode of this symmetry falls in the freqency range of the InAs TA branch and propagates in that layer. In-Sb interface modes of that type are expected to have much lower frequencies. For TA folded modes, the situation is reversed with respect to folded TO modes. In this case, B_2 modes see a "soft" interface, while B_1 modes see a "hard" one. In the lower overlaping frequency range, we find only one confined mode of each symmetry, labeled C_2 in Fig. 6, and shown in Fig. 7(j) for the B_1, and in Fig. 9(c) for the B_2, symmetry. Four more modes of B_1 symmetry have over 98% of their "energy" in the GaSb layer and are labeled P_2 in Fig. 6. Two of them are shown in Fig. 7(g) and (h). B_2 modes of comparable frequencies penetrate more into the InAs layer, [Fig. 9(d),(e) and (f)].

All of the above features could be explained in the following way: The two layers of the SL may be considered as independent oscillators with frequencies ω_1 and ω_2 which are weakly coupled through the interface layers by a force constant g. These frequencies may be taken to be the frequencies of the coresponding bulk branches folded to the center of the SL 1st BZ, as they are given in Figs. 3, 5 and 6. This coupling raises slightly the eigenfrequencies of the oscillators to ω_1' and ω_2'. Assuming a SL mode to have frequency ω_A coinciding with one of the latter two frequencies, the mode appears with maximum amplitude A into that layer (1 or 2) and a much smaller amplitude B into the other layer. It can be shown that, to second order in the coupling g, the ratio of the vibrational amplitudes is,

$$B/A = 2g/[m_A(\omega_A^2 - \omega_B^2)].$$

As can be seen in Fig. 5, the folding of the InAs TO branch in a narrow frequency range results in a dense spectrum for the oscillator "InAs layer", as compared to the spectrum of the oscillator "GaSb layer". Hense, SL modes in the overlaping frequency range have, in general, frequencies closer to InAs TO folded ones. As it has already been discussed, B_1 modes see "softer" interfaces or, for B_1 modes the coupling of the two layers is stronger than for B_2 modes. This explains the stronger localization of the latter modes in the GaSb layer. For folded TA modes, the situation is reversed and B_2 modes show stronger coupling. The oscillator spectrum is more dense for the GaSb layer and B_1 modes localize stronger into that layer. For A_1 modes, the oscillator spectrum is more dense for the GaSb layer [Fig. 3], and modes localize preferably into that layer. In this case the coupling is comparable to that for B_2 modes and, apart from the elastic properties of the interfaces, it must also be influenced by the electric macroscopic field in the SL.

Therefore, it is evident from the above discussion that the interface layers act as polarizers for the transverse phonons propagating along the SL axis, in certain narrow bands of frequencies.

RAMAN SPECTRA

Inelastic light scattering by phonons occurs when the atomic vibrations cause changes of the electronic polarizability of the medium. In order to describe these changes in materials with dominant covalent bonding, such as the SL in question, we adopt the BPM[17]. According to this model, four parameters are assigned to each bond. They are the bond polarizability components α_p and α_v, parallel and perpendicular to the bond axis, respectively, and their derivatives with respect to the bond length (α_p' and α_v', respectively). The values of these parameters can be determined by fitting to experimentally-known Raman spectra or, in the absence of such data as in our case, they can be estimated from other related information. It has been found[18] that the ratio a_p/a_v for many III-V compounds is very close to 1.1 and, moreover, the two components are related by the relation,

$$a_p + 2a_v = 3v_c(\varepsilon_{oo}-1)/16\pi,$$

where v_c is the primitive cell volume and ε_{oo} is the high frequency dielectric constant.

From the above considerations and the values of ε_{oo}, we deduce the values of a_p, a_v (in units of A^3) for the bonds of InAs (GaSb) : $\varepsilon_{oo}=12.2$ (14.4), $a_p=13.3$ (15.9), $a_v=12.1$ (14.4).

The absolute values of the derivatives a_p', a_v' are taken to be equal to the corresponding values of a_p and a_v divided by the bond length. The sign of the above derivatives has been decided after we have calculated the Raman spectra for the 1x1 and 3x3 SL, using all possible combinations of positive and negative signs for each bond according to the following considerations: (1) the experimentally established three-mode behavior for the $In_{1-x}As_{1-y}Ga_xSb_y$ alloy[19] should be clearly present for the higher frequency modes; (2) the Raman activity of the lower frequency modes should be weak, since they "belong" to the acoustical branches of the bulk compounds and are activated in the SL by the zone-folding effect. Moreover, (3) the intensities of the stronger modes of different symmetries should be of the same order of magnitude for the corresponding polarizations, and (4) the intensities of the interface modes should decrease with respect to those of the extended modes, as we go from the 1x1 to larger period SL.

Given the values of the above bond polarizability components and their derivatives, the polarizability derivatives with respect to each normal coordinate for $k=0$, $a'_{ab}(j)$, can be evaluated[17,20] and finally the Raman scattering cross section. For the Stokes part of the spectrum it is given by,

$$I_{\alpha\beta\gamma\delta} = \sum_j a'_{\alpha\beta}(j) \, a'_{\gamma\delta}(j) \, \delta(\omega-\omega_j) \, [n(\omega_j)+1]$$

where j runs over all long wavelength modes of frequency ω_j, $n(\omega_j)$ is the Bose factor which is taken at room temperature ($T=300°K$) and $\delta(\omega-\omega_j)$ is the function,

$$\delta(x) = \varepsilon/[\pi(x^2+\varepsilon^2)]$$

used to broaden the Raman lines. We use the value $\varepsilon=1$ cm^{-1}.

The calculated Raman spectrum for the 1x1 SL is presented in Fig. 10. The intensities are in the same arbitrary units. In the spectrum of A_1 modes (Fig.10, top), the stronger peak at 252.5 cm^{-1} (labeled a), belongs to the mode involving mainly vibrations of the Ga-As atoms. This mode becomes the highest frequency interface mode in thicker SL. The peak labeled (b) in the same figure at 187.0 cm^{-1}, again involves vibrations of the same atoms and becomes the other Ga-As interface mode in thicker SL. Finally the peak labeled (c) at 169.1 cm^{-1}, becomes the In-Sb interface mode in thicker SL.

In the B_1 spectrum (Fig.10, middle), the peak (a) at 228.8 cm^{-1} again becomes a Ga-As interface mode, while the weak peak (b) at 64.5 cm^{-1} might be related to the other interface mode of this symmetry in thicker SL.

In the B_2 spectrum (Fig.10, bottom), the higher frequency peak labeled (a) at 239.4 cm^{-1} and the peak (b) at 195.7 cm^{-1} become, in thicker SL, the Ga-As interface mode and the In-Sb interface modes of that symmetry, respectively.

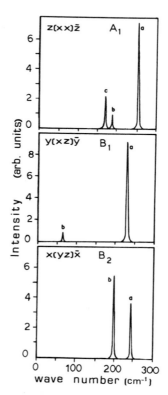

FIG. 10. Raman spectrum for the 1x1 InAs/GaSb SL. The intensity scale is the same for the three symmetries.

CONCLUSION

$(InAs)_n/(GaSb)_m$ SL appear in two different orthorombic space groups depending on the total number of layers in the SL. A Valence Overlap Shell Model has been used to calculate the phonon dispersion in the $n=m=1$ case and to study the optical phonons in the $n=m=16$ case. In the $n=m=1$ case an anticrossing of the phonon branches with Δ_1 symmetry is found, for small wavevectors in the zOx' plane, due to the anisotropy of the optical phonons at the zone center. In the $(\xi,0,0)$ direction another anticrossing between phonon branches of the above symmetry is also present. In the same case, the Raman spectra have been calculated using the Bond Polarizability Model. The four higher frequency modes appear with strong Raman intensities, while another three modes show weak peaks.

In the case of $n=m=16$ SL, interface, confined and pseudo-confined, longitudinal and transverse modes of zero wavevector propagating along the SL axis are found over a wide range of frequencies. Confined modes are found mainly in frequency ranges where the bulk phonon dispersion branches of the constituent binary compounds do not overlap. In overlapping frequency ranges, the phonon symmetry, the orientation of interfaces bonds, and the relative extend in frequency of the corresponding branches of the bulk phonon dispersion, determine an effective coupling between eigenmodes of the two layers. The strength of this coupling determines, in turn, the confinement characteristics of the corresponding SL modes in these frequency ranges.

ACKNOWLEDGEMENTS

We thank Prof. G. Theodorou and Dr. S. Ves for many helpful discussions. This work has been partially financially supported by the Greek Ministry of Research and Technology.

REFERENCES

1.   L.L. Chang and L. Esaki, Surf. Sci., 98, 70 (1980).
2.   L.L. Chang, R. Ludeke, C.A. Chang and L. Esaki, Appl. Phys. Lett., 31, 759 (1977).
3.   E.E. Mendez, L.L. Chang, C.A. Chang, F. Alexander and L. Esaki, Surf. Sci., 142, 215 (1984).
4.   G. Bastard, E.E. Mendez, L.L. Chang and L. Esaki, J. Vac. Sci. and Technol., 21, 531 (1982).
5.   A. Fasolino, E. Molinari and J.C. Maan, Phys. Rev., B33, 8889 (1986).
6.   L. Colombo and L. Miglio, in "Proc. of the Int. Conf. Phonons 89", Heidelberg, 21-25 Aug. 1989, World Scientific, Singapore (1989).
7.   E. Richter and D. Strauch, Solid State Commun., 64, 867 (1987).
8.   N.S. Orlova, Phys. Stat. Sol., b119, 541 (1983).
9.   M.K. Farr, J.G. Traylor and S.K. Sinha, Phys. Rev., B11, 1587 (1975).
10. D.L. Price, J.M. Rowe and R.M. Nicklow, Phys. Rev. B3, 1268 (1971).
11. J.L.T. Waugh and G. Dolling, Phys. Rev., 132, 2410 (1963).
12. K. Kunc and H. Bilz, Sol. St. Commun., 19, 1027 (1976).
13. K. Kunc and O. Nielsen, Comp. Phys. Commun., 17, 413 (1979).
14. G. Kanellis, W. Kress and H. Bilz, Phys. Rev., B33, 8724 (1986).
15. G. Kanellis, Phys.Rev., B35, 746 (1987).
16. G. Kanellis, in "Proc. of the NATO-Advanced Research Workshop on Spectroscopy of Semiconductor Microstructures", 9-13 May 1989, Venezia (Italy) G. Fasol, A. Fasolino and P. Lugli eds., Plenum Press, London (1989).
17. R. Tubino and L. Piseri, Phys. Rev., B11, 5145 (1975).
18. C. Flytzanis and J. Ducuing, Phys. Rev., 178, 1218 (1969).
19. G. Pickering, J. Electron. Materials, 15, 51 (1986).
20. S. Nakashima, H. Katahama, Y. Nakakura and A. Mitsuishi, Phys. Rev., B33, 5721 (1986).

RAMAN SCATTERING STUDIES OF OPTICAL PHONONS IN GaAs/AlAs AND

GaAs/Al$_x$Ga$_{1-x}$As SUPERLATTICES

Z.P. Wang, H.X. Han and G.H. Li

Institute of Semiconductors
Chinese Academy of Sciences
P.O. Box 912
Beijing 100083, China

ABSTRACT

It is well known from the phonon dispersion curves of bulk GaAs and AlAs that though their acoustic branches basically overlap, the optical branches do not. As a consequence, in GaAs/AlAs superlattices the acoustic phonon modes are the folded modes which can propagate through the whole superlattice, and the optical modes are the confined modes which are confined in GaAs and AlAs layers, respectively. (GaAs)$_n$/(AlAs)$_n$ superlattices on (001)-oriented GaAs substrates have point group symmetry D$_{2d}$. There are altogether n GaAs and n AlAs longitudinal optical (LO) and transverse optical (TO) confined modes; LO$_m$ and TO$_m$ with m = 1,2,...,n. The LO$_m$ modes belong to B$_2$ (m odd) and A$_1$ (m even) representations, respectively. Twofold degenerate TO$_m$ modes belong to the E representation.

Raman scattering is an effective tool for measuring the vibrational spectra of superlattices. Actually, almost all experimental results available have been obtained by means of Raman scattering. The first unambiguous Raman observation of LO modes confined in GaAs layers at 180 K and using near-resonance scattering was reported by Colvard et al.[1] in 1981. A typical result for GaAs LO confined modes in GaAs/AlAs superlattices was given by Sood et al.[2] at 10 K under off-resonance or near-resonance conditions. A few LO modes in AlAs layers were observed at 27 K for the first time by Zhang et al.[3] who employed off-resonance scattering. In back-scattering geometry experiments we have observed all of the LO modes confined in GaAs and AlAs layers at room temperature and using off-resonance scattering.[4] It was shown that the Raman activity of the confined LO modes is dependent on the scattering configuration; in the case of x // (1$\bar{1}$0), y // (110) and z // (001), all of the confined LO modes were observed in z(xx)$\bar{z}$ configuration, and no modes in z(xy)$\bar{z}$ configuration.

In a strict back-scattering geometry, TO confined modes with E symmetry are Raman forbidden. To our knowledge, there has only been the report by Sood et al.[2] on TO modes confined in GaAs layers measured at 10 K and under extreme resonance conditions, but no TO modes confined in AlAs layers have been reported yet. However, the effects due to near Brewster-angle incidence and a large aperture of the scattering light collecting lens can create a small wave vector component along the y orientation, and thus induce a Raman activity of TO confined modes. In the near z(xy)$\bar{z}$ back-scattering configuration, in which the interference from the stronger LO modes was avoided, we have observed the TO modes confined in GaAs and AlAs layers at room temperature employing off-resonance scattering.[5] Our results have

Light Scattering in Semiconductor Structures and Superlattices
Edited by D.J. Lockwood and J.F. Young, Plenum Press, New York, 1991

79

shown that Raman scattering measurements of superlattice structures provide an effective method for determining the phonon dispersion curves of the constituent bulk materials. This is particularly important for some materials which are not stable in air, such as AlAs, or which can be prepared only in the form of thin films because of the limitations of the crystal growth technique, such as CdS. For these materials, neutron scattering data are not available. We have provided the first experimental data of LO and TO phonon dispersion curves for bulk AlAs.[4,5]

In the case of GaAs/Al_xGa_{1-x}As superlattices, the behavior of the optical phonon modes becomes complicated. As is well known, in the long-wavelength optical-phonon spectra of Al_xGa_{1-x}As alloys there is the so called 'two-mode' behavior, i.e., GaAs-like and AlAs-like optical phonon modes. The optical phonon dispersion curves are also divided into GaAs-like and AlAs-like branches. The AlAs-like LO branch is separated in energy from the GaAs-like branch of the alloy and from the LO branch of bulk GaAs. But the GaAs-like LO branch of the alloy and the LO branch of bulk GaAs are partly separated and partly overlapped in energy. Thus, we expect that there are three types of LO modes in GaAs/Al_xGa_{1-x}As superlattices: AlAs-like LO modes confined in Al_xGa_{1-x}As layers, GaAs LO modes confined in GaAs layers, and GaAs-like delocalized LO modes. Jusserand et al.[6] first observed the LO odd modes confined in GaAs layers at 80 K under off-resonance scattering conditions. And the LO even modes confined in GaAs layers were observed by Arora et al.[7] at low temperature in near-resonance scattering. In addition to LO odd modes confined in GaAs layers, we have observed AlAs-like LO modes confined in Al_xGa_{1-x}As layers at room temperature in off-resonance scattering and have obtained experimental results for the AlAs-like LO phonon dispersion curve for the Al_xGa_{1-x}As alloy. As to the GaAs-like delocalized LO modes, they can be seen only in the polarized spectra; this is similar to the acoustic-phonon folded modes.

REFERENCES

1. C. Colvard, T. Gant, M. Klein, and A.C. Gossard, J. Physique $\underline{42}$, C6-631 (1981).
2. A.K. Sood, J. Menéndez, M. Cardona, and K. Ploog, Phys. Rev. Lett. $\underline{54}$, 2111 (1985).
3. S.L. Zhang, D. Levi, T. Gant, M. Klein and H. Morkoç, Proc. 10th Int. Conf. on Raman Spectroscopy, p.9-4, (1986).
4. Z.P. Wang, D.S. Jiang, and K. Ploog, Solid State Commun. $\underline{65}$, 661 (1988).
5. Z.P. Wang, H.X. Han, G.H. Li, D.S. Jiang, and K. Ploog, Phys. Rev. B$\underline{38}$, 8483 (1988).
6. B. Jusserand, D. Paquet, and A. Regreny, Phys. Rev. B$\underline{30}$, 6245 (1984).
7. A.K. Arora, E.K. Suh, A.K. Ramdas, F.A. Chambers, and A.L. Moret, Phys. Rev. B$\underline{36}$, 6142 (1987).

ANALYSIS OF RAMAN SPECTRA OF GeSi ULTRATHIN SUPERLATTICES

AND EPILAYERS

M. W. C. Dharma-wardana, G. C. Aers, D. J. Lockwood and
J.-M. Baribeau

Division of Physics, National Research Council of Canada
Ottawa, Ontario, Canada K1A 0R6

1. INTRODUCTION

Molecular beam epitaxy (MBE) techniques[1,2] are now capable of fabricating artificial layered structures with virtually "one-atomic-layer" accuracy. Thus one may construct a structure by depositing a few atomic layers of Ge on a silicon substrate, or build a superlattice $(Ge_mSi_n)_p$ where m atomic layers of Ge are followed by n atomic layers of Si to form a super-cell, which is repeated p times along the growth direction. A versatile method for investigating the structural properties of these epilayers (ELs) and superlattices (SLs) is Raman spectroscopy. In earlier papers[3,4] we studied the low-lying acoustic phonons ($\Delta\omega <$ 100 cm^{-1}) in such structures and found them to be sensitive to the overall periodicities, boundary conditions, and average properties of the lattices. In a more recent paper[5] we have studied the optical phonons, which yield information regarding local structure, interface smudging, and lattice-strain effects. There we calculated the Raman spectra of ideal unstrained $(Ge_mSi_n)_p$ structures, then successively included strain and interface smudging in the theoretical model. In this paper we will review and extend the results of Ref. 5, covering epilayers as well. Since we are concerned with backscattering-like Raman spectra for longitudinal vibrations along the [001] direction, we use a linear chain model[4] to calculate the phonons, while a bond-polarizability approach[5-8] is used to evaluate the Raman scattering intensity.

In section 2 we discuss the details of the linear chain model and the calculation of the Raman intensities. In sections 3 and 4 we present the results of the calculation for the ideal structures and bring out some interesting systematics. In section 5 we consider the effect of strain on the position of the principle Raman peaks for superlattices grown on Si or Ge substrates. The effect of interface smudging is considered in section 6. The results of sections 3-6 show that confinement, strain, and interface smudging need to be considered *together* in interpreting an experimental spectrum. In section 7 we use the information of the previous sections as well as specific calculations and results of annealing studies to interpret observed spectra of specific MBE-grown samples.

2. CALCULATION OF PHONONS AND RAMAN INTENSITIES

In our previous studies[3-5] we used a linear chain model of the total structure

$$Substrate + (Ge_mSi_n)_p + Cap$$

where the substrate was modeled by 500-1000 atomic layers of Si, while the Si cap of 50 Å was modeled by 37 atomic layers of Si. The first atomic layer of the substrate was assumed to be fully anchored, while the surface layer (last layer of cap) was treated as free or anchored, depending on a parameter σ, which specified the degree of anchoring. The $(Ge_mSi_n)_p$ superlattice itself was modeled in each case as m atomic layers of Ge and n atomic layers of Si, repeated p- times. An epilayer with m Ge atomic layers (say) on a substrate can be considered within the same general scheme by setting n = 0 and p = 1. Although optical phonons in SLs are less sensitive to sample-specific details like the presence or absence of a cap, surface boundary conditions, and the presence of a substrate, they become all important in ELs and hence we have retained the sample-specific details in our modeling. Theoretical force constants k_1 to k_4 for bulk silicon[9] and germanium[10] up to fourth-neighbour interactions were used in the linear chain model. For interactions between Si *and* Ge the arithmetic mean of the bulk force constants was used. Table 1 gives the values of the force constants as well as the multiplicative factor f (applied to each bulk force constant). This factor was necessary to reduce the calculated bulk Raman frequency of Si and Ge from 309.8 and 526.3 cm^{-1} to the values of 300.6 and 520.1 cm^{-1}, respectively, observed in our samples at 298 K. In our calculations we do not assume any translational invariance etc but simply diagonalize a matrix that is of dimensionality equal to N_{sub} + N_{sup} + N_{cap}, where N_{sub}, N_{sup}, and N_{cap} are the number of substrate, SL, and cap atomic layers, respectively. For the $\Delta\omega$ > 90 cm^{-1} regime we used N_{sub} = 500 unless otherwise stated.

Following Ref. 5 we have adopted the bond-polarizability[5-8] model for calculating the Raman intensities. If the growth direction, i.e. the [001] direction, is designated the z axis, the backscattering along the z direction will lead to Raman signals with $z(xx)\bar{z}$ and $z(xy)\bar{z}$ polarizations[11]. The corresponding Raman intensities are proportional to the modulations $|\delta\chi^{xy}/\delta R_z|^2$ and $|\delta\chi^{xx}/\delta R_z|^2$, respectively, where R_z is the z component of the vector defining the bond and χ is the susceptibility tensor. The bond polarizability Π of Weber et al.[7] is proportional to χ and is written as

$$\Pi_{\alpha\beta}(\vec{R}) = \frac{\vec{R}_\alpha \vec{R}_\beta}{R^2} \alpha_\parallel(R) + \left(\delta_{\alpha\beta} - \frac{\vec{R}_\alpha \vec{R}_\beta}{R^2} \right) \alpha_\perp(R) \qquad (2.1)$$

where $\vec{R}$ is a vector along the bond and α,β designate x,y and z components. The bond polarizabilities parallel $(\alpha_\parallel)$ and perpendicular $(\alpha_\perp)$ to the bond depend on the bond length R. When atomic planes parallel to the x-y plane vibrate along the z axis, R_z changes and $\Pi_{\alpha\beta}$ is modulated. Menéndez et. al.[12] have recently used this form of the bond-polarizability model to interpret a Raman study of a (Ge_4Si_4) SL. A strictly equivalent model introduced earlier by Bell[8] uses the mean polarizability P and the anisotropy γ and is written.

$$\Pi_{\alpha\beta}(\vec{R}) = \left[P(\vec{R})\vec{I} + \gamma(\vec{R}) \left(\frac{\vec{R}\vec{R}}{R^2} - \frac{1}{3}\vec{I} \right) \right]_{\alpha\beta} \qquad (2.2)$$

where $\vec{I}$ is the unit tensor. Zhu and Chao[13] used this form of the bond-polarizability model in their studies on folded acoustic modes in SLs. Only one-phonon Raman scattering is considered here. The bond distance $\vec{R}$, or more precisely $\vec{R}_{ij}$, connecting the i-th atom with the j-th nearest bonded atom, can be expressed as $R_{ij} = \left(\vec{r}_i + \vec{u}_i\right) - \left(\vec{r}_j + \vec{u}_j\right)$, where $\vec{r}_i, \vec{r}_j$ denote the positions of the unperturbed atoms and $\vec{u}_i, \vec{u}_j$ are the atomic displacements. Only terms linear in these displacements are retained in calculating the change in the susceptibility on changing R_z. Since we are interested in Raman scattering due to longitudinal vibrations along the z direction, reduction of Eq(2.1) or (2.2) leads us to intensities I_{xx} and I_{xy} in the simple form

$$I_{xx}\left(\omega_j\right) \propto \left| \sum_{n=1} e^{iqz_{2n}} \left(a^{xx}_{2n-1} U^j_{2n-1} - a^{xx}_{2n+1} U^j_{2n+1} \right) \right|^2 \qquad (2.3)$$

$$I_{xy}\left(\omega_j\right) \propto \left| \sum_{n=1} e^{iqz_{2n}} \left(a^{xy}_{2n-1} U^j_{2n-1} + a^{xy}_{2n+1} U^j_{2n+1} - 2a^{xy}_{2n} U^j_{2n} \right) \right|^2$$

where U^j_i is the displacement of the i-th *atomic layer* along the [001] direction for the j-th phonon mode. Note that the U_i occurring here refers to an average over the *atomic* displacements u_i taken over the atoms in the plane perpendicular to the linear chain. The phonon modes and displacements were calculated from the linear chain model. The surface layer can be taken to be the first layer. The effect of the finiteness of the photon wave vector q can be approximately included via the phase factor $\exp(iqz_{2n})$, which contains the position z_{2n} of the 2n-th atomic layer along the linear chain. Since the experimental optical modes of $(Ge_mSi_n)_p$-type structures tend to be broad, the doublet splitting of folded modes introduced by the phase factor is negligible and can in general be ignored. However, we have retained this effect in our calculations and estimated q using the average refractive index of the material. The polarizability constants a^{xx} and a^{xy} of each layer are proportional to α_{xx} and α_{xy} of Table 1. These, in turn, can be related to α_1, α_q, and $\alpha_{25'}$ of Cardona[14] in the case of bulk materials.

Table 1. Parameters used for Ge_mSi_n SL calculations. The bulk force constants[9,10] for Ge and Si, k_1 to k_4 up to fourth-neighbour interactions, are given in units of 10^5 dynes/cm^2. A scale factor f is used to bring the calculated bulk Raman frequencies (309.8 and 526.3 cm^{-1}) for Ge and Si to the experimental values (300.6 and 520.1 cm^{-1}) at 298 K. The Raman tensor components,[14] relevant to back-scattering along the z axis are α_{xx} and α_{xy}.

	k_1	k_2	k_3	k_4	f	α_{xx}	α_{xy}
Ge	1.010	0.059	0.016	0.004	0.9426	9.93	4.07
Si	1.128	0.074	0.018	0.004	0.9762	3.32	1.59

We note also that for Ge_mSi_n-type SLs only the $z(xy)\bar{z}$ polarization produces a significant intensity. Hence the Raman intensities depend on the single parameter a^{xy}, which may be thought of as an adjustable constant, deviating where necessary to some extent from the bulk values. For comparison with experimental spectra the calculated spectra were broadened (parameter Γ) to give a Lorentzian form such that

$$I(\omega) \propto \sum_j \frac{\Gamma/\pi}{\left(\omega - \omega_j\right)^2 + \Gamma^2}\left[I_{xx}(\omega_j) + I_{xy}(\omega_j)\right]\left[\frac{n(\omega_j)+1}{\omega_j}\right] \quad (2.4)$$

Here ω_j is the frequency of the j-th mode, and $n(\omega_j)$ is the usual Bose factor.

3. PHONONS AND RAMAN SPECTRA OF IDEAL Ge_m EPILAYERS

In this section we present results of calculations of Raman spectra for *ideal* Ge_m ELs. The calculation assumes a silicon substrate modeled by 500 Si layers, m atomic layers of Ge, viz. Ge_m, and 37 atomic layers of Si to form a 50Å cap. We study the prominent spectral features seen for frequency shifts $\Delta\omega$ greater than, say, 100 cm^{-1}.

Figure 1(a) shows the calculated spectra for m = 3, 4 and 6 atomic layers of Ge obtained with a broadening Γ of 3 cm^{-1} at each peak. The broadening in an actual experimental spectrum would be perhaps 6 cm^{-1}, but in Fig. 1 we have retained a smaller value to expose the "structure" of the spectral features. Thus the Ge_3 EL spectrum shown as

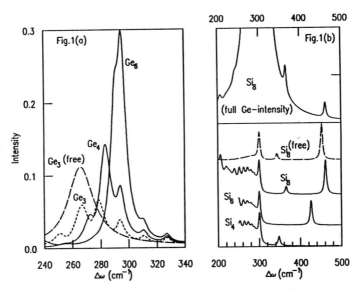

Fig. 1. (a) Calculated spectra for Ge_m epilayers on Si substrates.
(b) Calculated spectra for Si_m epilayers on Ge substrates.

a short-dashed line in Fig.1(a) is strongly oscillatory with six distinct peaks in the 240-340 cm^{-1} region. The long-dashed line labelled Ge$_3$ (free) shows the calculation for a Ge$_3$ layer on a Si substrate but *without* a capping layer. In this case the oscillatory structure has disappeared, and instead we have a smooth spectrum with the main intensity shifted to lower frequencies. The oscillatory structure shown for capped Ge$_3$ and Ge$_4$ are resonant optical phonon modes whose origin is similar to the mechanism given in Ref. 3 for acoustic phonons. The positions of the two main peaks in the Ge$_4$ capped structure is given in Table 2. In the present calculations *the bulk bond polarizabilities* have been used. There is no reason to assume that the bulk bond polarizabilities are applicable to epilayers, and in fact the work of Ref. 5 (and also Ref. 3 and 4) strongly suggests that this may not be the case. Nevertheless, the calculations of this section have been carried out using the values given in Table 1 unless stated otherwise. The frequency of the most intense peaks are given in Table 2 together with a full width at half maximum (FWHM) which refers to the full width of the envelope of the total structure. Thus we see that the very broad, weak spectrum of a Ge$_2$ layer slowly becomes narrow and moves upwards to the bulk value as we go to Ge$_8$. The effect of strain represented by an increase of all force constants of 10%, shown in Table 2, will be taken up in the context of Raman spectra of (Ge$_m$Si$_n$)$_p$ structures.

In Fig. 1(b), we show the calculated spectra of ideal Si$_m$ epilayers on Ge substrates, capped with 20 atomic layers of Ge. The upper panel of Fig. 1(b) shows that the Ge optical phonon at 300 cm^{-1} dominates the intensity of the Si layers. Hence in the lower panel we have calculated the spectra assuming (for the sake of display) a very low Ge-bond polarizability. No peaks are obtained for $\omega > 300$ cm^{-1} for Si$_m$, $m < 4$. For Si$_8$ two peaks are obtained. Also, as shown for the case of Si$_8$, the oscillatory structures due to resonant phonon effects are suppressed when the capping is removed and a simple Si$_m$ epilayer (denoted as Si$_8$-(free)) on a Ge substrate is considered. Unlike the case of Ge$_m$ ELs, the ideal spectra of Si$_m$ ELs remain narrow. This is due to the more complete confinement of the Si$_m$ vibrations compared with Ge$_m$ where the optical branch overlaps with the acoustic branch of the Si substrate and cap. Details of the principal peak positions are given in Table 2. In the case of Si growth on Ge substrates the Si layers are expanded and, in effect, the force constants are reduced in magnitude. The effects of strain (e.g., 10% reduction of force constants) and interlayer mixing will be discussed after a study of the spectra of SLs.

4. PHONONS AND RAMAN SPECTRA OF IDEAL Ge$_m$Si$_n$ STRUCTURES

In this section we present results of calculations of Raman spectra for ideal Ge$_m$Si$_n$ structures, i.e., structures with no strain or interface smudging. The calculation assumes a silicon substrate modeled by an adequate number of atomic layers (500 unless stated

Table 2. Calculated peak positions for Ge$_m$ ELs on Si substrates and capped with 50 Å of Si and for Si$_m$ ELs on Ge substrates capped with Ge. The effect of 10% strain is also shown.

	m	2	3	4	5	6	8
Ge$_m$	$\Delta\omega$	257	267,278	284	288	295	298
	FWHM	81	32	19	15	11	9
	10%	260	283,296	297	303	307	311
Si$_m$	$\Delta\omega$	-	-	347	396	426	365,462
	-10%	-	-	336	378	407	350,440

Table 3. Calculated Raman peaks in ideal Ge_mSi_n SLs. Only four peaks (cm^{-1}) are given.

Ge_m \\ Si_n	1	2	3	4	5	6	7	8
1	417	468,192	489,294	500,359 139	506,402 229	509,431 292,109	512,450 338,189	514,464 372,247
2	409,210	463,253	487,295 266	499,352 230	505,397 260	509,427 294,201	512,448 338,238	513,462 368,265
3	402,260	462,275 111	486,297 225	498,349 264	505,395 277	509,426 296,241	512,447 332,267	513,462 366,278
4	401,278 125,108	462,285 176	486,298 251,108	498,348 279,155	505,395 286,91	509,426 297,260	512,447 331,278	513,462 366,286
6	401,291 220,88	462,293 237,137	486,299 273,202	498,348 291,227	505,395 293,243	509,426 298,277	512,447 330,291	513,462 366,293

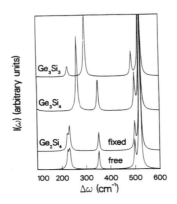

Fig. 2.   Three typical Ge_mSi_n spectra, calculated with a broadening Γ of 3 cm^{-1} at each peak. The substrate peak is at 520 cm^{-1}.

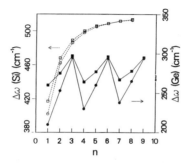

Fig. 3. Confined modes of Si and Ge. The open squares and circles show the highest Si-Si optical phonon frequency in Ge_1Si_n and Ge_3Si_n, respectively, as a function of Si slab thickness n. The filled squares and circles show the highest Ge-Ge optical phonon peak frequency in Ge_3Si_n and Ge_2Si_n SLs, respectively, as a function of n.

otherwise), p repetitions of Ge_mSi_n with p taken to be 24, followed by typically 37 atomic layers of Si to form a 50 Å cap. We study the prominent spectral features seen for frequency shifts $\Delta\omega$ greater than, say, 90 cm^{-1} (e.g., see Fig. 2). Table 3 gives the four main calculated high-frequency features of the Raman spectra of $(Ge_mSi_n)_{24}$ structures.

The Si-Si modes (which have $\Delta\omega > 300$ cm^{-1}) tend to be essentially independent of the Ge environment, as soon as m > 1 (see Fig. 3). This reflects the fact that the bulk Si optical vibrations form a band that does not overlap the Ge vibration bands and are hence essentially *confined*[15,16] to the Si slabs, with no penetration into the Ge layers. In Ge_mSi_n, we find that at n = 4 there are two Si modes, viz. $\Delta\omega_1 \approx 498$ cm^{-1} and $\Delta\omega_2 \approx 349$ cm^{-1}. These correspond to the fully antisymmetric $\overset{\leftarrow}{Si} - \overset{\rightarrow}{Si} - \overset{\leftarrow}{Si} - \overset{\rightarrow}{Si}$ mode and the partially antisymmetric $\overset{\rightarrow}{Si} - \overset{\rightarrow}{Si} - \overset{\leftarrow}{Si} - \overset{\leftarrow}{Si}$ mode (see Fig. 4).

In going from, for example, $(Ge_2Si_6)_{24}$ to $(Ge_2Si_7)_{24}$ the Si-Si-like peaks at 509 and 427 cm^{-1} move upwards to 512 and 448 cm^{-1} due to the decrease in confinement. In addition, a new Si-Si-type mode appears at 338 cm^{-1}, while the Ge-Ge-like peak at 294 cm^{-1} drops down to 238 cm^{-1} (see Fig. 3). The next lowest significant peak in the calculated Raman spectrum of $(Ge_2Si_7)_{24}$ is at 104 cm^{-1} (in Table 3 we have only retained the four main high-energy peaks). Thus an additional Si vibrational mode appears near 335 cm^{-1} when n in Ge_mSi_n reaches 7. Once a new Si mode appears, at an appropriate value of n, its frequency is seen to be more or less independent of m in Ge_mSi_n.

Let us now consider the behaviour of the Ge-Ge optical vibrations. These vibrations are buried in the acoustic continuum of the Si vibrations, and hence their behaviour depends on the number n of the Si layers. If we follow the series Ge_2Si_n we see that the Ge-Ge vibration starts at 210 cm^{-1} for n = 1 and reaches 295 cm^{-1} for n = 3. This behaviour recommences once again at 230 cm^{-1} for n = 4 and reaches 294 cm^{-1} for n = 6. The variation of the partially confined Ge-Ge optic mode frequency as a function of the number of silicon layers shows a three-Si-atom cycle as shown in Fig. 3. This reflects the fact that three silicon atoms are approximately of the same mass as one Ge atom. Thus, for example, the Ge$_2$Si$_9$ system can approximate to a Ge$_2$ ("Ge")$_3$ structure, and hence the Ge-Ge optic mode approaches closer to the bulk value than, say, for a structure like Ge$_2$Si$_{10}$.

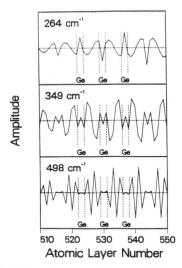

Fig. 4. Vibration amplitudes of the three main peaks of (Substrate) + (Ge$_3$Si$_4$)$_{24}$ + (Cap). The partially antisymmetric Si-Si vibration at 349 cm^{-1} involves displacement of the Ge interface layers. Hence they are called Ge-Si modes in the text. Some Ge interface layers are shown as vertical dashed lines.

Figure 2 shows several Raman spectra calculated using the standard values of the bond polarizabilities (Table 1). The two calculated spectra for Ge$_2$Si$_4$ correspond to free and fixed surface boundary conditions. Of the two peaks at $\Delta\omega$ = 225 and 230 cm^{-1}, the lower one is somewhat sensitive to the surface boundary conditions used. This peak arises from a mode that develops a high amplitude near the surface region, while the peak at 230 cm^{-1} is the expected Ge-Ge vibration for the structure. Thus it is clear that even in studying optical phonons, sample specific effects can sometimes be important.

The results for ideal ELs, viz. Ge$_m$ on Si substrates or Si$_m$ on Ge substrates (Table 2), should be compared with the results for SLs (Table 3). Si$_6$ in an SL shows four distinct peaks, while Si$_6$ as an EL shows only the peak corresponding to the peak at 431 cm^{-1} of the

SLs. The pure optical phonon at 509 cm^{-1} of the SL is not supported by the confined structure. Similarly, all the peaks at lower frequencies, associated with SL periodicity, are lost. The highest Si peak is not regained even with a Si$_8$ slab.

5. THE EFFECT OF STRAIN

Up to this point we have discussed ideal structures of the form (substrate) + (Ge$_m$Si$_n$)$_p$ + (cap) using the force constants applicable to unstrained bulk silicon and germanium. However, if the substrate is silicon, pseudomorphic epitaxial growth requires that the in-plane lattice constant of the Ge layers match the smaller lattice constant of Si. Hence the Ge layers are strained, while the Si layers are generally believed to be free of strain. In fact annealing studies, which will be discussed later, show that even the Si layers may have some strain in them. A complementary situation exists for growth on Ge substrates. By a comparison of the optical phonon peaks observed from incommensurate Ge$_x$Si$_{1-x}$ layers with those of commensurate alloy layers grown on silicon substrates, Cerdeira et al.[17] showed that strain can shift the so-called Ge-Ge, Ge-Si, and Si-Si optical vibration peaks to higher frequencies. Their studies suggest that the Ge-Ge vibrations, as well as the Ge-Si vibrations could be shifted up by as much as 15 cm^{-1} for a fully strained system. The Si-Si peaks are even more sensitive to strain.

In a real SL of the form (substrate) + (Ge$_m$Si$_n$)$_p$ + (cap) we have very little information regarding the microscopic strain profile (interface layer strain may be different from that inside the slabs). Hence we use a simple didactic model. For (Ge$_m$Si$_n$)$_p$ structures grown on Si substrates we compare the spectra calculated using the four bulk force constants for Ge, and then with the bulk values increased by 10%, while keeping the Si force constants unchanged. Such a shift, $k_{Ge} \rightarrow 1.1\ k_{Ge}$, has the effect of moving the main Ge-Ge Raman peak by about 12 cm^{-1} and is of the desired magnitude as seen in experimental samples. We also do a series of calculations where the Si force constants are scaled *downwards* by 10%, i.e., $k_{Si} \rightarrow 0.9\ k_{Si}$, while the Ge force constants are kept unchanged, as this simulates the behaviour of (Ge substrate) + (Si$_n$Ge$_m$)$_p$ + (cap) structures where the Si slabs are strained.

In Table 4 we show in column A the calculations for (Ge$_m$Si$_n$)$_p$ with strained Ge slabs, while column B refers to strained Si slabs in (Si$_n$Ge$_m$)$_p$. Since all structures of the form Ge$_m$Si$_n$ with m and n > 6 have two Si-Si-like peaks ($\Delta\omega$ > 300 cm^{-1}), except in the case of Ge$_m$Si$_2$, we have chosen to call the highest peak ($\Delta\omega$ > 400 cm^{-1}) the Si-Si peak, while the lower peak has been called the Ge-Si peak. (For example, in Fig. 4 the vibration amplitudes for the Si-Si mode at 498 cm^{-1} and the "Ge-Si" mode at 349 cm^{-1} for Ge$_3$Si$_4$ are shown.) The reason for this nomenclature will become clearer by the time we discuss interface smudging effects. In the case of Ge$_m$Si$_2$ only one Raman peak occurs for $\Delta\omega$ > 300 cm^{-1}. The vibrations associated with this peak at about 462 cm^{-1} (see Table 3) are not confined entirely to the Si slabs but also involve the Ge interface layers. Hence this mode cannot be classified as either a pure Si-Si mode or a Ge-Si mode, and entries in Table 4 for Ge-Si and Si-Si peaks in Ge$_2$Si$_2$, Ge$_4$Si$_2$, and Ge$_6$Si$_2$ are thus put in parentheses and entered twice.

In Table 4, column A for $\delta(\Delta\omega)$ (Ge-Ge) shows that the strain shift of the Ge-Ge peak depends both on m and n. Larger Ge slabs (larger m) show a bigger strain shift. The Ge-Si peak is less affected by an increase in k_{Ge} than the Ge-Ge peak. But significantly, the value of the strain shift depends on the silicon slab size n, going from 4 cm^{-1} to a maximum of 6 cm^{-1} and then down to 3 cm^{-1} for n = 6. Further, even the Si-Si vibration is affected, to a lesser extent. Columns B in Table 4 show the effect of strained silicon slabs (i.e., $k_{Si} \rightarrow 0.9\ k_{Si}$) on the three principle peaks. This simulates the case of an epitaxial

Table 4. Effect of strain on the main Raman peaks of $(Ge_mSi_n)24$ SLs. Structures on Si substrates are modeled by a 10% increase in the Ge force constants ($k_{Ge} \rightarrow 1.1\ k_{Ge}$), and the resulting upward (positive) shifts (cm^{-1}) are given in columns labeled A. Structures on Ge substrates are modeled by a decrease of Si force constants ($k_{Si} \rightarrow 0.9\ k_{Si}$), and the resulting downward (negative) shifts are given in B columns. The absolute positions can be obtained by adding the values to the unstrained peak positions given in Table 3. The lower frequency Si-Si peak is called the Ge-Si peak. (In the case of Ge_mSi_2, since there is only one Si-Si peak the same values are entered twice.) The effect of interface smudging and strain is also given for the $(Ge_6Si_6)24$ SL.

m , n	$\delta(\Delta\omega)$ Ge-Ge		$\delta(\Delta\omega)$ Ge-Si		$\delta(\Delta\omega)$ Si-Si	
	A	B	A	B	A	B
2 , 2	8.7	-3.0	(4.0)	-(19)	(4.0)	-(19)
2 , 2	8.2	-4.0	8.0	-12	1.4	-25
2 , 6	6.4	-8.0	3.0	-19	0.5	-26
4 , 2	13	-0.5	(4.4)	-(19)	(4.4)	-(19)
4 , 4	11	-2.4	6.0	-12	1.5	-24
4 , 6	10	-4.4	3.0	-19	0.5	-26
6 , 2	14	0.0	(4.4)	-(19)	(4.4)	-(19)
6 , 4	14	-0.5	6.0	-12	1.6	-24
6 , 6	13	-2.0	3.0	-19	0.5	-26
smudged 6 , 6						
x=0.15	12	0.0	2.6	-40	0.3	-30
x=0.30	11	4.0	4.6	-53	0.9	-32

Si_nGe_m structure grown on a Ge substrate and hence the Ge layers are assumed to be unstrained. The shifts (decrease in $\Delta\omega$) are larger than those obtained in strained Ge grown on Si substrates (columns A). Also, the Ge-Ge peaks are somewhat influenced by the strain in the Si slabs.

These results show that if one set of layers is strained while the other is not (e.g., column A: Ge layers are strained while Si layers are not) it does not necessarily follow that the shift in, say, the Ge-Ge peak unambiguously leads to an estimate of the strain in the Ge slabs. The frequency shifts due to strain depend on the local environment (i.e., values of m, n) in which the vibrating unit is embedded. Our studies on the effect of interface smudging, to be discussed in the next section, will further emphasize this, and hence it seems that extreme caution is necessary in interpreting Raman spectra of commensurate or incommensurate alloy structures, random structures, or structures grown on alloy buffers, or in comparing them with the spectra of SLs. The numbers of Ge-Ge, Si-Ge, or Si-Si bonds and the local strain environment found in a local cluster in a random alloy are unknown, and hence conclusions about strain, confinement etc. are open to a high degree of error.

The effect of strain on a Ge_6Si_6 system where the interface layers are smudged due to admixture of the other component (15 and 30%) is also shown in Table 4. The effects are particularly drastic in the B columns since in this case the smudging and the strain work in the same direction. We discuss interface smudging in the following section.

6. THE EFFECT OF INTERFACE SMUDGING

In the previous sections we studied the behaviour of optical phonon spectra for ideal Ge_mSi_n SLs as a function of m, n, and strain. We assumed that the interfaces are ideal in that we pass from a pure Ge atomic layer to a pure Si atomic layer or vice versa at an interface. Under practical growth conditions the two interface layers are likely to become modified because of kinetic and other processes associated with epitaxial growth. Hence, to simulate experimental conditions it becomes necessary to incorporate interface "smudging" effects. The linear chain model really treats vibrations of whole planes of atoms. Hence, instead of the ideal atomic masses, force constants, and polarizabilities, we need the "smudged" values that correspond to average values over whole planes perpendicular to the direction z defined by the linear chain model. In practice the averages extend only over a characteristic length in the x-y plane corresponding to the damping distance of the interactions in the planes. At present no microscopic information of this sort is available, and hence we resort to simple models. To begin with let us assume that smudging affects only one atomic layer on both sides of an interface. That is, if we consider a Ge/Si interface, the Ge atom will be replaced by an atom of mass m_h, while Si will be replaced by an atom of mass m_l, where h and l designate the "heavy" and "light" atomic layers of the interface.
Hence we write

$$(Ge_mSi_n)_p \rightarrow (m_h \, Ge_{m-2} \, m_h \, m_l \, Si_{n-2} \, m_l \,)_p$$

where we now have $m_h \, m_l$ and $m_l \, m_h$ interfaces instead of Ge-Si and Si-Ge interfaces, respectively. The masses m_h and m_l are taken to be

$$m_h = (1 - x)m_{Ge} + xm_{Si}$$

$$m_l = (1 - x)m_{Si} + xm_{Ge} \qquad \text{with } x \leq 0.5 \qquad (6.1)$$

The value of the intermixing or smudging parameter x will depend on the growth conditions and could vary from interface to interface. However, we assume that esentially reproducible atomic layer epitaxy is used for preparing the Ge_mSi_n SLs. Hence, x may take values within a narrow range near some average value x. In fact, we will attempt to show that Raman data allow us to make an estimate of x.

To study the effect of interface smudging on the Raman spectra we examine in detail a specific model system, viz. $(Ge_6Si_6)_{24}$ smudged to become $(m_h \, Ge_4m_h \, m_l \, Si_4m_l \,)_{24}$, as a function of the fraction of intermixing x that determines the masses m_h and m_l. We have also linearly intermixed the force constants and polarizabilities of the smudged layers, just as for the masses. In these calculations we assume as usual 500 atomic layers of silicon substrate (or Ge substrate as the case may be) and 37 atomic layers (50 Å) of Si cap (or Ge cap), although their presence does not appreciably affect the position of the optical phonon peaks studied here.

Table 5. Effect of interface smudging on the main Raman peaks of the system (Si substrate) + $(Ge_6Si_6)_{24}$ + (cap). The Ge_6Si_6 system becomes, on smudging, $(m_h Ge_4m_h m_l Si_4m_l)_p$, where the masses m_h and m_l are controlled by the smudging parameter x, (Eq. (6.1)), which is the fractional intermixing. Peak positions are given in cm^{-1}.

x	$\Delta\omega$(Ge-Ge)	$\Delta\omega$(Ge-Si)	$\Delta\omega$(Si-Si)
0.0	294	427	509
0.10	300	411	506
0.15	301	405	505
0.20	302	399	504
0.30	307	381	501
0.50	317	375	501

In Fig. 5 we show how the spectrum of the ideal $(Ge_6Si_6)_{24}$ structure evolves as x is increased. The calculated spectra look very like the actual spectra of experimental samples, with a three-mode behaviour.[18] The spectrum with 10% intermixing (x = 0.1) is very similar to the typical experimental spectrum with a mode near 417 cm^{-1} loosely described as a Ge-Si mode. Figure 6 shows the vibration amplitudes for the case x = 0.15, with some of the interface layers marked by dashed vertical lines. The 405 cm^{-1} mode clearly involves excitation of the "Ge-Si" interface atoms. Table 5 shows the evolution of the three peaks as a function of x, the fraction of intermixing.

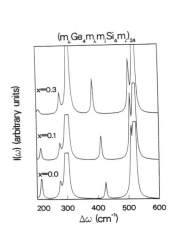

Fig. 5. The spectrum of $(Ge_6Si_6)_{24}$ as a function of x.

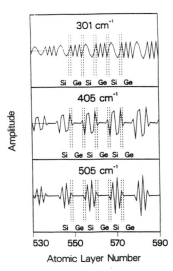

Fig. 6. Vibrational amplitudes for the case x = 0.15.

These calculations do not include strain. Thus, a 15% intermixing would give rise to an *unstrained* Ge-Si-like mode near 405 cm^{-1}. The Ge-Ge-like peak is raised to 301 cm^{-1}, i.e., it is higher than the bulk value. This *increase* is due to the admixture of lighter Si atoms and not due to strain. Similarly, the so-called Ge-Si mode is more properly identified as a silicon slab mode (partially antisymmetric mode) weighed down by the admixture of heavier Ge atoms. This 405 cm^{-1} mode would move upwards, typically to 408 cm^{-1} (see Table 4) if lattice strain is included and could constitute the so called "Ge-Si"-like mode of experimental spectra. It is clear that the Ge-Si-like mode is really a $\overset{\rightarrow}{Si} - \overset{\rightarrow}{Si} - \overset{\leftarrow}{Si} - \overset{\leftarrow}{Si}$ mode bounded by the two heavy silicons, i.e., m_l atoms (or Si layers with Ge admixture), at the two ends forming the interface layers.

The implication of the results of Table 5 are that Ge-Ge-like modes appearing at a higher frequency than that of the bulk Ge-Ge mode (viz. 300 cm^{-1}) do not necessarily imply strain. While partially confined, unstrained, unsmudged Ge-Ge modes fall below 300 cm^{-1} (see Table 3), the corresponding peaks in nonideal superlattices or alloys will tend to lie *above* the bulk value because of the admixture of lighter Si atoms into the smudged interface planes. Similarly, the so-called Ge-Si-like peak could lie above, below, or near the extended Ge-Si mode at 417 cm^{-1}, typical of the m = 1, n = 1 zinc-blende structure (GeSi)$_p$. In fact, it may be misleading to relate the experimental peak near 390-425 cm^{-1} to the Ge-Si zinc-blende peak, as will be seen later, in our discussion of experimental spectra. Similar caution is needed in interpreting the position of the Si-Si peak in terms of confinement and strain alone. Confinement lowers the Si-Si peak from the bulk value of 520 cm^{-1} (see Table 3). But so does admixture of germanium, as seen from Table 5. If the Si layers are lattice matched to a Ge substrate the strain also lowers the Si-Si peaks (see Table 4), and hence the interpretation of alloy spectra as well as SL spectra requires a careful treatment of local m and n in the relevent Ge$_m$Si$_n$ structure, its strain, confinement, and "interface" smudging.

In this discussion of interface smudging we have treated the interface of Si on Ge in essentially the same way ("symmetric model") as the interface of Ge grown on Si. However, since the growth kinetics of Si deposition on Ge is different from Ge deposition on Si, the interfaces on the two sides need not be similar ("asymmetric model"). As examples for this we consider $(m_h Ge_{4m_h} m_l Si_{4m_l})_{24}$ and $(m_h Ge_4 GeSiSi_{4m_l})_{24}$ where the latter case is asymmetric and assumes, for instance, that Si growth on Ge leads to a perfect interface, while Ge growth on Si leads to a smudged interface. This is of course an extreme case and should lead to the largest effects of symmetric/asymmetric modeling. Calculations for the case x = 0.1, for example, show that the "Ge-Si"-like mode shifts from 411 cm^{-1} in the symmetric case to 418 cm^{-1} in the asymmetric case. Also, the intensity of the mode decreases. Both these changes are consistent with simple ideas of decrease of confinement, decrease of mass of the Si-like vibrating units, and decrease in the number of Ge-Si-type bonds. Of the other peaks, viz. Ge-Ge-like and Si-Si-like peaks, the latter is more sensitive and shifts upwards by about 1 cm^{-1}. The shift in the Ge-Si-like peak of about 6 cm^{-1} (or less) should be detectable in careful modeling of good experimental data, but the asymmetric treatment introduces extra parameters into the simulation. In the present calculation we have used the simple symmetric model, noting that the experimental "Ge-Si-like" peaks are fairly broad (we used a broadening parameter $\Gamma = 5$ cm^{-1} in modeling the experimental spectra; see section 7).

The simple two atomic layer interface smudging model

$$Ge_m Si_n \rightarrow m_h Ge_{m-2m_h} m_l Si_{n-2m_l}$$

is not necessarily a universally applicable scheme, since each sample must be studied within a "sample"-specific scheme. For systems with very small values of m and n, e.g., Ge$_3$Si$_3$, a whole slab (i.e., three atomic layers in all) may get smudged, and calculations have to be carried out for each case. In the next section we give examples of the analysis of experimental spectra using the ideas and results presented above.

7. ANALYSIS OF EXPERIMENTAL SPECTRA

Several laboratories including ours have studied Ge_mSi_n ultrathin SLs grown on silicon substrates, germanium substrates, and Ge/Si alloy substrates and using different growth temperatures. X-ray diffraction data and cross-sectional transmission electron microscopy (XTEM) on such Ge_mSi_n SLs show overall good epitaxial growth for suitably low values of m or n, depending on the substrate and the growth conditions. However, Raman spectra of these ultrathin SLs show a characteristic three-mode behaviour. Other weaker spectral features, some of which correspond to folded modes, and other expected features of these SLs are also observed. The almost ubiquitous appearance of the three-mode behaviour, with a Ge-Si peak around 390-425 cm^{-1}, shows that there is considerable intermixing and interface smudging in most experimental samples. Also, a "nominal" Ge_mSi_n structure, where m and n are the target values, may have a distribution of other values, m' and n', which may differ considerably from the target values, due to various kinetic processes and uncertainties associated with MBE growth.

In analysing an experimental spectrum we proceed by noting that, say, if the substrate is silicon, then (a) positions and intensities of the high-energy silicon-like peaks will give an *indication* of the actual values of n in the sample; (b) the position of a given peak is affected by at least three factors, viz. quasi-confinement, strain, and layer smudging; (c) the so-called Ge-Ge peak (near 300 cm^{-1}), or the Ge-Si peak (near 415 cm^{-1}), being dependent on many factors, does not give clear indications of the actual values of m and n in the sample; (d) the observation of SL peaks is a better indicator of the dominant values of m and n in the sample. Our approach is to compare the experimental spectrum with the calculated spectra of the target structure and structures close to it with and without strain and interface smudging. The movement of the peaks under annealing is also a useful guide to the existence of strain in various slabs of the SL. The experimental spectrum is then constructed as a weighted sum of theoretical spectra arising from a few dominant structures. In other words, we assume that the SL is a concatenation of several structures, arranged in series, and in some random order determined by the complexities of the growth kinetics. In addition the spectrum is an average for the structures sampled within the area of the beam. In practice, if the growth conditions are optimal one of the structures will predominate, together with perhaps some other minority structure. Hence fitting to two structures may be adequate. In order to simplify the modeling of such a system we have assumed that the total spectrum can be considered the weighted sum of the spectra of the two structures calculated independently.

In fitting the experimental spectra to the calculations the parameters of the majority structure (defined by the crystal growth parameters) usually needed no adjustment, as in the case of the Ge_2Si_6 structure to be discussed presently. In the case of the minority structure adjustable parameters arise in modeling the smudged layers (e.g., the intermixing fraction x and the degree of strain). The majority structure itself may require smudged interfaces. Hence, assuming symmetric smudging of the interfaces, we have two adjustable parameters per structure to get the peak positions properly aligned. The intensities are fitted by modifying the bulk bond polarizabilities of Si and Ge. The type of agreement between experiment and theory obtained here needed only a modest effort, but any further improvements seem to be very difficult. However, there is no clear guarantee that the chosen parameters and structures are unique, although our experience is that the freedom of choice is limited not only by the experimental spectrum, but also by the need for consistency with other characterization data.

Samples Grown on Si (001) and Ge (001) Substrates

Superlattices of the type Ge_2Si_2, Ge_2Si_4, Ge_2Si_6, Ge_4Si_4, and Ge_4Si_8 were grown on silicon (001) substrates, usually at a growth temperature of $400 \pm 25°C$, and capped with a 50-100 Å Si epitaxial layer. The Ge and Si growth rates were set at 0.4 Å/s for all layers. For epitaxial growth on Ge (001) substrates, a growth temperature of 350°C was used. Structures corresponding to Si_2Ge_6, Si_2G_{12}, Si_2Ge_8, and Si_4Ge_8 were grown (see Ref.

19). The Raman spectra were measured at room temperature in a helium atmosphere, in a pseudo-backscattering configuration.[11] Laser light at 458 or 468 nm was used to excite the spectra, which were analysed with a Spex 14018 double monochromator.

In all the SLs grown on Si (001) substrates, typical Ge-Ge-like peaks were found to lie between 297 and 306 cm^{-1}, while the Ge-Si-like peak varied from 412 to 418 cm^{-1}. The position of the Si-Si-like peak was usually more informative and corresponded to the value predicted from Table 3, but usually additional peaks, which implied the existence of strain or other values of n, were observed. Also, in many samples enhanced intensity features corresponding to the folded acoustic modes could be identified. The SLs grown on Ge (001) substrates showed Ge-Ge-like peaks in the range 296-300 cm^{-1}, while the Ge-Si-like peak position varied from 389.7 to 393.6 cm^{-1}. We confine our detailed discussion to two selected samples, both of which were deemed to be "good" crystals if judged in terms of double crystal x-ray diffraction (DCXD) data, by XTEM, and other characterization criteria.[19] To be more specific, the two samples showed pseudomorphic growth, with continuous layers. The SL grown on (001) Si showed planar growth near the substrate/SL interface, while some waviness was apparent in layers away from the interface, probably indicating that two-dimensional growth is not easily maintained for SL growth on Si at 400°C. In the case of the SL grown on the (001) Ge substrate, planar growth was generally observed despite some bending of the atomic layers near the substrate/SL interface (for XTEM photographs etc. see Ref. 19). It should be borne in mind that if the crystal deviates strongly from planar growth the linear chain model should not provide a good description of the system.

$(Ge_2Si_6)_{48}$ Grown on Silicon (001)

This nominal $(Ge_2Si_6)_{48}$ sample was grown by MBE and capped with a 50 Å silicon layer. The experimental Raman spectrum is shown in the bottom panel of Fig. 7. The high-frequency spectrum ($\Delta\omega \geq 100$ cm^{-1}) shows strong features near 199, 295, 415, and at 512.5 cm^{-1} (see Fig. 8 for the last peak). The strong peak at 520.1 cm^{-1} originates from the substrate. Broad weaker features are seen at 100-140, 230-270, 330-350, and 425-440 cm^{-1}.

The spectrum calculated for the *unstrained* structure (Si substrate) + $(Ge_2Si_6)_{24}$ + (cap), with free surface boundary conditions ($\sigma = 0$) gives Raman peaks at 103, 201, 294, 427, and 509 cm^{-1} (see Table 3, entry under Ge_2Si_6). These peak positions were not affected by changing the surface boundary conditions. Note that the calculations are done for p = 24 SL periods, while the experimental spectrum has p = 48. This has the effect of giving a somewhat lower intensity to the calculated folded mode intensities, while the peak positions remain unchanged. Also, we have used 900 layers to simulate the substrate, which is several micrometres thick and optically penetrated up to several thousand ångströms. The use of a small number of substrate atoms has the effect of giving a broader distribution about the 520 cm^{-1} bulk peak and also giving a poor representation of the intensity relationship between the calculated Si-Si SL and substrate peaks. These limitations should be borne in mind in comparing the relative intensities of experimental and theoretical spectra, although, of course, the peak *positions* are not affected.

In Fig. 7. we give a reconstruction of the measured Raman spectrum (bottom panel) of sample 1, nominally Ge_2Si_6 on Si (001), from theoretical spectra. The dashed line in the top panel is the theoretical spectrum of $(Ge_2Si_6)_{24}$ on 900 atomic layers of Si substrate and capped with 27 atomic layers of Si. The solid line is the same spectrum with the Raman polarizability a^{xy} of Si in the SL enhanced, and the Si-Si force constants increased by 1.4%. The dashed line in the middle panel is the theoretical spectrum of (substrate) +$(Ge_2Si_6)_{24}$ + (cap) with 7.5% intermixing at the interfaces and an increase of k_{Si} by 3%. The solid line is the same, with enhanced a^{xy} of the SL Si layers. The dashed line of the bottom panel is the experimental spectrum, while the solid line is the theoretical model spectrum obtained by adding 75 and 25% of the solid curves of the middle and top panels, respectively.

The experimental spectrum contains perhaps only a shoulder at 509 cm^{-1}, while the well-formed experimental Si-Si peak (see Fig. 8) is at 512.5 cm^{-1}. From Table 3 we might conclude that the structure probably contains Si$_7$ and Si$_8$ slabs as well as the Si$_6$ slabs targeted by the growth conditions. Thus we may expect to see spectral features corresponding to Ge$_{2\pm1}$Si$_{6\pm1}$ or Ge$_{2\pm1}$Si$_{6\pm2}$ in this spectrum. Alternatively, if the nominally strain-free Si slabs had a small amount of strain ($\approx$1% change in k$_{Si}$) the calculated frequency of 509 cm^{-1} for ideal Ge$_2$Si$_6$ would shift to the observed value of 512.5 cm^{-1}. Comparison of the observed features of the experimental spectrum (bottom panel of Fig. 7) with the predictions of Table 3 for unstrained systems, as modified by Tables 4 and 5 for strain and smudging, enables us to explain most of the features in the experimental spectrum. For simplicity we will consider the main spectral features. The strong, well-formed low-frequency peak at 199 cm^{-1} in the experimental spectrum is consistent with the theoretically predicted SL peak (a "folded mode") at 201 cm^{-1} from (Ge$_2$Si$_6$)$_{24}$. It is also found that the 199 cm^{-1} peak is strongly affected by annealing, which would be expected for such a folded mode. A closely related structure, e.g., Ge$_1$Si$_7$, has a peak at 189 cm^{-1}. Hence, admixture of lighter Si atoms due to smudging of Ge$_1$Si$_7$ could also give rise to a frequency near 200 cm^{-1}. On the other hand, the structures (Ge$_2$Si$_8$)$_p$ and (Ge$_2$Si$_7$)$_p$ have peaks at 174 and 238 cm^{-1}, respectively. The absence of such features in the experimental spectrum enables us to conclude that Ge$_2$Si$_7$- and Ge$_2$Si$_8$-type structures are not present in the sample. Similar considerations enable us to eliminate Ge$_3$Si$_7$- and Ge$_3$Si$_5$-type structures. On the whole, we conclude that the experimental sample contains the target structure Ge$_2$Si$_6$, as well as structures derived from Ge$_2$Si$_6$ by interface smudging, together with a possibility of some Ge$_1$Si$_7$ or Si$_8$ regions.

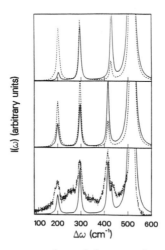

Fig. 7.   Reconstruction of the experimental Raman
spectrum of sample 1, nominally Ge$_2$Si$_6$ on
Si (001), from theoretical spectra (see text).

Now we consider the Ge-Ge-like peak observed at 295 cm^{-1}. The unstrained, unsmudged $(Ge_2Si_6)_p$ and $(Ge_1Si_7)_p$ structures should have peaks at 294 and 189 cm^{-1}, respectively (see Table 3). Thus the 295 cm^{-1} peak is essentially entirely due to the $(Ge_2Si_6)_p$ species, with a small upward shift of ≈ 1 cm^{-1}. This upward shift may be entirely due to strain effects or due to interface smudging effects (admixture of the lighter Si raises the Ge-Ge frequency). The fact that the SL peak, ideally at 201 cm^{-1}, is observed at the *lower* frequency 199 cm^{-1} strongly suggests interface smudging rather than strain. Table 4 shows that the Ge_2Si_6 structure is less sensitive to strain than the other structures given there. If the observed 1 cm^{-1} shift is entirely due to strain, it still implies that the Ge layers have a strain of less than 5% of the expected strain for epitaxial Ge on Si. These considerations suggest that the strain in the Ge layers has become negligible because of interface smudging effects. Such a relaxation would be most effective in Ge_2Si_n systems since both Ge layers would become smudged layers. The presence of a strong Ge-Si-like peak at 415 cm^{-1} also suggests that interface smudging is present. A calculation involving 7.5% intermixing (i.e., $x = 0.075$) of the interface layers gives a spectrum with peaks at 199, 297, 415 and 507 cm^{-1}, in good agreement with the experimental peaks at 199, 295, and 415 cm^{-1}, while the calculated 507 cm^{-1} peak is not in agreement with the 512.5 cm^{-1} peak of the experimental spectrum.

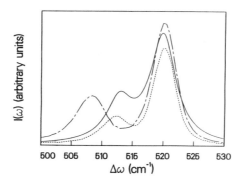

Fig. 8. Part of the Raman spectrum of sample 1. The solid line is the theoretical Raman spectrum of Fig. 7 with some modifications (see text). Dashed line: unannealed experimental spectrum. Chained line: annealed spectrum. After a 15 min anneal at 750°C the superlattice Si peak drops from 512.5 to 508.2 cm^{-1}. The relative intensities for the three curves are arbitrary.

A possible explanation of the origin of the 512.5 cm^{-1} peak is provided by the results of annealing studies. In Fig. 8 we show how the Si-Si peak at 512.5 cm^{-1} shifts when the SL is annealed. The 15 min anneal at 750°C brings the peak down to 508.2 cm^{-1}, i.e., essentially to the calculated value of 507-509 cm^{-1}. We may interpret the annealing process as contributing to the *removal* of some residual strain in the (nominally unstrained) Si layers. The *assumption of the existence of a small amount of strain in the silicon layers*, sufficient to *increase* the Si-Si force constant by 3%, enables us to understand all the features of the Raman spectrum as well as the behaviour of the system under annealing. The theoretical curve shown in Fig. 8 will be described later in this paper, while a more detailed analysis of the annealing studies will be reported elsewhere.

We may now attempt to "synthesize" the observed spectrum as being made up of contributions from unsmudged (Ge_2Si_6) layers and from those with 7.5% intermixed interface layers. Such a theoretical "simulation" spectrum, where the amount of ideal interfaces is taken to be 25%, is shown in Fig. 7 (bottom panel). This simulation spectrum is arrived at in the following way. The calculated spectrum of the ideal unstrained Ge_2Si_6 structure, using the standard values of the Raman polarizabilities (Table 1), is shown as a dashed curve. A broadening parameter $\Gamma = 5$ cm^{-1} is used throughout Fig. 7. The intensity of the Si-Si mode at 427 cm^{-1} in the calculated spectrum is quite weak compared with that of the Ge-Ge mode. Hence we increased the value of a^{xy} for Si from 1.59 to 3.5 and retained the standard value of a^{xy} for Ge at 4.07. The enhanced spectrum is shown as a solid line in the top panel of Fig. 7. The Si-Si force constants in this "enhanced" spectrum have been increased by 1.4% to move the 509 cm^{-1} peak to 512.5 cm^{-1}, as justified by annealing data. This also moves the 427 cm^{-1} peak to 429.5 cm^{-1} but has little effect on the other peak positions. The dashed curve in the middle panel of Fig. 7 is the calculated spectrum of Ge_2Si_6 smudged (x = 0.075) to form the structure $(m_h)_2(m_l \, Si_4 m_l)$ where m_h is a Ge-like layer with 7.5% admixture of Si, and m_l is a Si-like layer with 7.5% admixture of Ge. The solid line in the middle panel shows the same case with a^{xy} for Si enhanced to 3.5. In both these curves the Si-Si force constants k_{Si} have been enhanced by 3%, while the force constants of the smudged layers m_l and m_h have been reduced from the bulk values (Table 1) by 2 and 4%, respectively.

The theoretical simulation spectrum given as the solid line in the bottom panel is obtained from the weighted sum of the enhanced, slightly strained Ge_2Si_6 spectrum (solid line, top panel) using a weight of 0.25, with the spectrum of the enhanced, slightly strained 7.5% smudged system (solid line, middle panel) taken with a weight of 0.75. This reconstruction of the experimental spectrum is in good agreement with the results for the average composition and the total thickness of the structures obtained from XTEM and DCXD.[19]

In this reconstruction we have had to modify the bond-polarizability factors in order to get the correct relative intensities of the major peaks. We did this in a limited manner without attempting a detailed optimization. Further, the 199 cm^{-1} peak, being essentially a folded SL peak, would in any case come out to be weak due to our use of a $(Ge_2Si_6)_{24}$ structure instead of the experimental $(Ge_2Si_6)_{48}$ structure. Similarly, the width of the substrate peak near 520 cm^{-1} is broader than the experimental peak because of the use of a small number of layers for the substrate. Finally, since bond polarizabilities are related microscopically to the band structure of the material in question, it is probably not surprising that the a^{xy} factors needed for a $Ge_m Si_n$ structure are different from those obtained in bulk materials.

The theoretical curve given in Fig. 8 is essentially the same as the theoretical reconstruction shown in the bottom panel of Fig. 7, with some significant differences. Since in Fig. 8 we are comparing the 520 cm^{-1} substrate peak with the Si-Si SL peak, we have attempted to get the relative intensities "visually" correct in the following way. Instead of using 900 layers of substrate with 24 periods of the superlattice and $\Gamma = 5$ cm^{-1}, as was done for Fig. 7, we used 1150 substrate atomic layers with 18 periods of the SL and $\Gamma = 2.5$ cm^{-1} to construct the theoretical spectrum of Fig. 8. These changes reduce the calculated width of the 520 cm^{-1} peak of the substrate and also increase its intensity. In effect these changes are an attempt to get a reasonable representation of the substrate within the limitations of our computer. We stress that the above leads to minimal effects on the peak *positions*. The synthesized spectrum is, as before, composed of 25% of a Ge_2Si_6 structure and 75% of the smudged structure.

The surprising conclusion regarding the nominal $(Ge_2Si_6)_p$ structure studied here is that the Ge layers, having become smudged, carry little strain, while the Si layers have acquired a small amount of strain sufficient to drive the unstrained Si-Si peak near 509 cm^{-1} to the observed 512.5 cm^{-1} peak. Further annealing studies are necessary to determine if the Ge peaks would relax in a manner consistent with Ge layers being relatively free of strain, or whether a different interpretation of the experimental spectrum would become necessary.

It should be noted that XTEM data show the present sample to be a pseudomorphic structure. This is not inconsistent with the present conclusion that the Ge$_2$ layers (which have become m$_h$ layers) carry little strain. The usual estimate of strain in the Ge layers is based on the mismatch between the equilibrium bond lengths of *bulk* Ge and Si. The equilibrium bond lengths, relevent to just two interface atomic layers of Ge need to be determined, possibly by *ab initio* density functional calculations, before realistic estimates of strain in the atomic layer structures can be obtained.

(Si$_4$Ge$_8$)$_{24}$ Grown on a Ge (001) Substrate

The dashed curves in the bottom panels of Figs. 9 and 10 show the experimental spectrum of a nominal $(Si_4Ge_8)_{24}$ structure grown on a Ge (001) substrate and capped with approximately 20 layers of Ge. The spectrum shows well-formed peaks at about 96.4, 298, 390, and 472 cm^{-1} and weaker peaks at ~ 195 , 235, and 418 cm^{-1}.

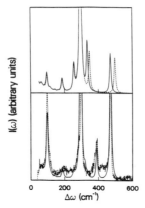

Fig. 9. Reconstruction of the experimental spectrum of sample 2, nominally (Si$_4$Ge$_8$)$_{24}$ grown on Ge (001).

The theoretical spectrum for the unstrained ideal (Ge substrate) + (Si$_4$Ge$_8$)$_{24}$ + (Ge cap) structure is shown as a dashed line in the top panel of Fig. 9. The solid curve is obtained by introducing strain alone (no smudging) into the Si layers so that the main Si-Si peak at 500 cm^{-1} drops to 472 cm^{-1}. The dashed line in the bottom panel is the observed spectrum while the solid line is the theoretical model spectrum (see Fig. 10 for details). A similar lowering of the 500 cm^{-1} peak can be achieved by a mixture of strain and smudging, or smudging alone. However, all these approaches lead to the presence of an extra peak located at 300-350 cm^{-1}, depending on the type of strain + smudging mixture used. Strong smudging of the Ge interface layers also tends to move the Ge-Ge peak into the > 300 cm^{-1}

range because silicon admixtured layers are lighter. However, the experimental spectrum presents no evidence for a feature in the 300-350 cm^{-1} region, and hence structures based on Si$_4$Ge$_8$ cannot be used to explain the presence of the 472 cm^{-1} peak *unless* the intensity of the partially antisymmetric mode ($\approx$300-350 cm^{-1}) could be diminished without at the same time reducing the intensity of the fully antisymmetric mode ($\approx$472-500 cm^{-1}). After a careful study of many possible smudged and strained structures derived from the target structure (Si$_4$Ge$_8$)$_p$ we found that the best agreement with the experimental Raman spectrum could be obtained by considering two dominant structures A and B. The actual physical structure is considered to be made up of 80% of A and 20% of B. The structure A is essentially the target structure (Si$_4$Ge$_8$)$_p$ except that the interface layers are smudged due to 5% intermixing, giving the form $(m_l Si_2 m_l \, m_h Ge_6 m_h)_p$, where m_l contains 95% of Si and 5% of Ge. Similarly m_h contains 95% of Ge and 5% of Si. The m_h and Ge layers are assumed unstrained as the system is epitaxial with Ge (001). The assumption that the Ge layers are unstrained, seems to be in accordance with the results of annealing studies as well. The Si layers as well as the m_l layers contain about 9% strain in the sense that the Si-Si force constants used were 91% of the bulk value.

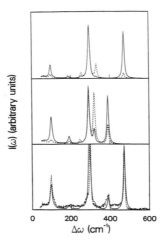

Fig. 10. Reconstruction of the (Si$_4$Ge$_6$)24 SL
spectrum.

The calculated Raman spectrum for this system (structure A), viz. (Si$_4$Ge$_6$)24 on a Ge substrate with 5% smudging of interfaces and strained Si layers, using bulk bond-polarizability factors a^{xy}, is shown as a dashed line in the top panel of Fig.10. The solid line is the Raman spectrum with modified a^{xy}. The middle panel shows the Raman spectrum of structure B, viz. $[(m_l)_4 m_h Ge_4 m_h]24$ on a Ge substrate, where all the Si atomic layers are intermixed (31%) with Ge, while the m_h layers are 50% Ge. Dashed and solid lines are calculated with bulk and modified a^{xy} values respectively. The bottom panel (dashed line) shows the experimental spectrum; the solid line is the theoretical spectrum made up of 80% of (top panel, solid curve) structure A and 20% of (middle panel, solid curve) structure B.

As in the case of the Ge_2Si_6 SL discussed previously, we find that the bulk bond polarizabilities do not give a proper description of the observed spectra. The final form adopted involves an enhancement of a^{xy} for the SL Si layers from the bulk value of 1.59 to 6.0, while the value of a^{xy} for the smudged silicon layers m_l was taken to be 3.5. The polarizability factor for the heavy layers (i.e., m_h layers) had to be lowered to 1.0 (from the usual Ge value of 4.07), and this had the effect of suppressing the intensity of the peak near 320 cm^{-1} to negligible values. The Raman spectrum for the structure A calculated with these modified bond-polarizability factors is shown as a solid line in the top panel of Fig. 10. The second structure, B, contributing 20% to the experimental spectrum, is best modeled as $[(m_l)_4(m_hGe_6m_h)]_p$. That is, all four Si atomic layers have undergone intermixing to give an average m_l containing 69% Si and 31% Ge. The heavy layers m_h contain 50% Ge and 50% Si. The smudged Si layers (i.e., m_l layers) are strained to the extent of about 11% in the sense that the force constants used were 89% of the bulk values. The Raman spectrum of the structure B, calculated using the bulk bond-polarizability factors is shown as a dashed line in the middle panel of Fig. 10. The spectrum calculated with modified values (viz. a^{xy} for $m_l = 3.5$, a^{xy} for $m_h = 1.0$, and a^{xy} for Ge = 4.07, which is the bulk value) is shown as the full line (middle panel). The bottom panel shows the Raman spectrum of the synthesized structure, made up of 80% of structure A and 20% of structure B. This analysis of the structure of the experimental sample is in good agreement with the overall length and Si/Ge composition determined by other characterization methods (XTEM, DCXD).[19]

CONCLUSION

In this paper we have examined the theoretically predicted Raman spectra of Ge epilayers on Si substrates, Si epilayers on Ge substrates, and $(Ge_mSi_n)_p$-type superlattices. We used a linear chain model to calculate the (001) phonons and a bond-polarizability approach to calculate the Raman intensities. The study of the ideal structures enabled us to bring out interesting systematics in the behaviour of the Si-Si and Ge-Ge modes under quasi-confinement. Resonant-phonon peaks asociated with *optical modes* are predicted to be observable for capped epilayers, since the epilayer sandwiched between the substrate and the cap is analogous to a Fabry-Perrot cavity. If no capping is present, and if clean surfaces could be experimentally observed, the resonant features disappear, but effects due to surface modes should appear.

The study of the effect of strain and interface smudging showed that these effects have to be included concurrently with confinement in any discussion of experimental spectra. In particular, a discussion based on confinement and strain alone, ignoring the effect of interface smudging, could be seriously in error. That is, smudging shifts the Si-Si- and Ge-Ge-like peaks in addition to bringing out the so-called Ge-Si-like peak. The same type of considerations become even more imperative in discussing or comparing alloy spectra, since the local environment of the vibrating groups in an alloy is not known. Finally, we have given examples of how an experimental spectrum may be modeled or simulated using spectra calculated for strained and smudged structures derived from the nominal structure $(Ge_mSi_n)_p$. We also showed that annealing studies could be very helpful in resolving some puzzling features and give valuable indications regarding the distribution of strain. The relative magnitudes of the experimental intensities strongly suggest that the bond polarizabilities are modified in these ultrathin Ge_mSi_n superlattices, possibly reflecting changes in band structure. This type of modeling of the experimental Raman spectrum would lead to useful characterization of these $(Ge_mSi_n)_p$ type superlattices, providing some quantitative estimates of interface mixing, strain, and deviation of the actual structure from the target structure.

REFERENCES

1. J. Bevk, J. P. Mannaerts, L. C. Feldman, B. A. Davidson, and A. Ourmazad, Appl. Phys. Lett. 49, 286 (1986).
2. E. Kasper, H.-J. Herzog, H. Dämbkes, and G. Abstreiter, in: "Layered Structures and Epitaxy," Materials Research Society Proceedings, Vol. 56, M. Gibson et al., ed., MRS, Pittsburgh (1986).
3. D. J. Lockwood, M. W. C. Dharma-wardana, G. C. Aers, and J.-M. Baribeau, Appl. Phys.Lett. 52, 2040 (1988).
4. G. C. Aers, M. W. C. Dharma-wardana, G. P. Schwartz, and J. Bevk, Phys. Rev. B 39, 1092 (1989).
5. M. W. C. Dharma-wardana, G. C. Aers, D. J. Lockwood, and J.-M. Baribeau, Phys. Rev. B 41, 5319 (1990).
6. M. A. Maradudin and E. Burstein, Phys. Rev 164, 1081 (1967).
7. W. Weber, S. Go, K. C. Rustagi, and H. Bilz, Proceedings of the 12th International Conference on the Physics of Semiconductors, Stuttgart 1974, Teubner, Stuttgart (1974).
8. R. J. Bell, Methods in Comput. Phys. 15, 216 (1976).
9. A. Fleszar and R. Resta, Phys. Rev. B 34, 7140 (1986).
10. K. Kunc, Table 5.1 in: "Electronic Structure, Dynamics and Quantum Structural Properties of Condensed Matter," J. Devreese and P. Van Camp, eds., Plenum, New York (1984).
11. D. J. Lockwood, M. W. C. Dharma-wardana, J.-M. Baribeau, and D. C. Houghton, Phys. Rev. B 35, 2243 (1987).
12. J. Menéndez,. A. Pinczuk, J. Bevk, and J. P. Mannaerts, J. Vac. Sci. Tech. B6, 1306 (1988).
13. B.Zhu and K. A. Chao, Phys. Rev. B 31, 4906 (1987).
14. M. Cardona, Table 2.7 in: " Light Scattering in Solids II," M. Cardona and G. Güntherodt, eds., Springer-Verlag, Berlin (1982).
15. B. Jusserand, D. Paquet, and A. Regreny, Superlatt. Microstr. 1, 61 (1985).
16. A. Fasolino and E. Molinari, J. Phys. (Paris) Colloq. 48, C5-569 (1987).
17. F. Cerdeira, A. Pinczuk, J. C. Bean, B. Batlogg, and B. A. Wilson, Appl. Phys. Lett. 45, 1138 (1984).
18. E. Kasper, H. Kibbel, H. Jorke, H. Brugger, E. Freiss, and G. Abstreiter, Phys. Rev. B 38, 3599 (1988).
19. J.-M. Baribeau, D. J. Lockwood, M. W. C. Dharma-wardana, N. L. Rowell, and J. P. McCaffrey, Thin Solid Films 183, 17 (1989).

INTERFACE ROUGHNESS AND CONFINED VIBRATIONS

Bernard Jusserand

CNET - Laboratoire de Bagneux
196 Avenue Henri Ravera
92220 Bagneux, France

INTRODUCTION

The lattice dynamics of superlattices based on two different pure constituents, like for instance GaAs/AlAs, InAs/GaSb or Si/Ge, is now very well understood and most of the theoretical predictions have been successfully verified by Raman scattering, at least on GaAs/AlAs structures which are available at a very high quality[1].

The optical phonon branches in GaAs and AlAs are well separated in frequency and the optical vibrations of one material cannot propagate into the other one. As a consequence, the optical vibrations in the GaAs/AlAs superlattices are divided into two families whose eigendisplacements are strongly confined either in the GaAs or in the AlAs layers; the penetration depth into the other compound being less than one monolayer. Their frequencies are thus very sensitive to boundary conditions. Assuming perfect interfaces, one can identify the eigenfrequencies of the modes confined in GaAs layers to those of bulk GaAs at a given set of wavevectors k_p :

$$k_p = \frac{p\pi}{(n_1 + 1)a} \tag{1}$$

Similar conclusions apply to the optical vibrations confined in the AlAs layers. The LO dispersion in GaAs and AlAs being downwards, the confined phonon energies lie below the bulk GaAs (AlAs) one and decrease with decreasing layer thicknesses. This provides a one-to-one correspondence between the confined frequencies and the layer thicknesses. Equation 1 allows one to define an effective thickness $(n_1 + 1)a$ which is common to all the confined vibrations in perfect layers. These predictions have been checked by different groups[2] with reasonable success. However, a general tendency is observed; the dispersion deduced from Raman scattering on superlattices is flatter than the one obtained by neutron scattering on the bulk constituents. In other words, the effective thickness associated to a confined mode increases when the associated local wavevector (see Equation 1) increases.

The main approximation involved in these models is to consider atomically flat interfaces separating the two pure constituents. In real samples, even of the best quality, the interfaces are never flat due to the growth statistics. Such roughness, which extends over a few monolayers around the nominal interface, can be reasonably neglected for superlattices with moderately small individual layer thickness (> 50 Å). In shorter period structures, however, this roughness leads to quantitative departures from the predictions of the lattice dynamics of perfect samples. This paper will be devoted to a description of Raman scattering experiments focussed on this problem and to a quantitative analysis of the dependence of the phonon frequencies on the spatial characteristics of the roughness, both along the growth axis and in the interface plane. This will lead us to reconsider in detail the lattice dynamics of bulk mixed crystals and of perfect superlattices containing layers of these alloys. The effect of interface roughness will be finally treated before we conclude.

Experimental Results

The optical vibrations confined in GaAs and AlAs are particularly sensitive to the interfacial imperfections because their frequencies are mainly determined from boundary conditions at each interface. This was first demonstrated[3] through the variation of the confined frequencies as a function of the growth conditions in nominally identical superlattices. We show in Fig. 1 Raman spectra obtained in the $z(x,y)\bar{z}$ configuration on five nominally identical samples grown by Molecular Beam Epitaxy (MBE) at different temperatures T_s ranging from 510°C to 680°C. We know from X-ray diffraction measurements that the macroscopic parameters of the samples are almost identical : d(GaAs)=28Å and d(AlAs)=12Å and independent of temperature. On the other hand, the intensity of the X-ray satellites decreases significantly with increasing T_s. A similar observation is made on the Raman lines associated to the folded acoustic vibrations, which reflect the long range order along the growth axis in a very similar way to the X-ray satellites : the frequencies of the folded acoustic lines remain unchanged but their intensities significantly decrease. Both results suggest, in agreement with some independent knowledge of the growth process, the existence of an increasing interfacial roughness with increasing T_s.

The effect on the optical vibrations is in a way much stronger : we observe a significant down-shift with increasing T_s of all the lines associated with optical phonons confined either in the GaAs or the AlAs layers. In the AlAs energy range, a single line is mainly observed on all the samples associated to the fundamental confined vibration (p=1 in Eq.1). A small line, tentatively assigned to mode 3, is also observed on the sample grown at 510°C but disappears at higher temperature. In the GaAs frequency range, three lines are observed on all samples, corresponding to mode 1,3,5 according to the now well established selection rules. They all shift towards lower frequency and moreover their relative distance strongly varies : they become more or less equidistant at 680°C while the splitting between line 3 and 5 amounts to almost twice the one between line 1 and line 3 at low T_s. Similar shifts in the confined optical vibrations have also been observed on superlattices during a series of successive thermal annealing stages[4].

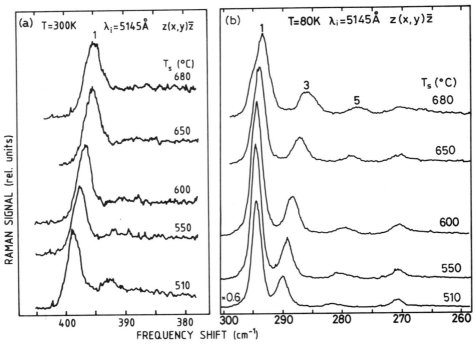

Fig. 1 . Raman spectra in the AlAs-type (a) and GaAs-type (b) frequency range for five GaAs/AlAs superlattices with the same nominal parameters, but grown at different temperature T_s.

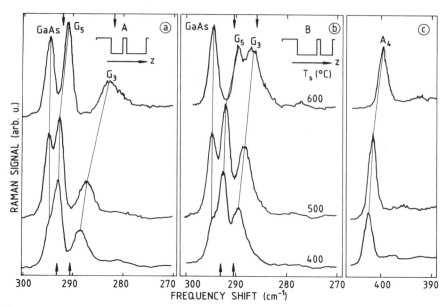

Fig. 2 . a) Raman spectra in the GaAs-type frequency range for three different samples with the same composition sequence A but grown at different temperatures. b) same as a) except that the composition sequence is B. c) same as a) but in the AlAs-type frequency range. In a) and b) the bottom (top) arrows indicate the mode frequencies calculated assuming a perfect (segregation-type) profile.

The dependence of the optical phonon frequency on the growth conditions becomes very strong for very short period superlattices. We show in Fig.2 recent results obtained on samples specially designed to get detailed information on the imperfections of the profile. As shown in the inset, the supercell of each sample contains 4 different layers. Two different sequences have been grown :

$(AlAs)_4(GaAs)_3(AlAs)_1(GaAs)_5$ sequence A
$(AlAs)_4(GaAs)_5(AlAs)_1(GaAs)_3$ sequence B

which are equivalent up to an inversion of the growth direction. The Raman spectra thus must be identical when assuming a perfect realization of the structure and the observed differences are spectroscopic signatures of departures from the nominal composition profile.

Samples are grown by MBE and the superlattice growth temperature ranges from 400 to 600°C. We first characterize the samples using X-ray diffraction and Raman scattering on folded acoustic phonons. We do not observe any systematic change as a function of T_s or of the layer sequence. The measured periods scatter between 3.6 and 3.8 nm, in good agreement with the 3.7 nm nominal value. Moreover, the second X-ray satellite (respectively folded acoustic doublet) is unusually strong with respect to the first one. This demonstrate[5] the presence of an aluminium rich peak inside the GaAs layer which corresponds to the nominal AlAs monolayer. The ratio of satellite intensities only shows small variations from sample to sample.

On the contrary, strong variations are observed in the Raman spectra on confined optical vibrations. We show in Fig.2 the Raman spectra of vibrations confined in GaAs (a and b) or AlAs (c) layers obtained on the samples grown at T_s=400, 500 and 600°C. The spectra in the GaAs energy range contain three main peaks associated respectively with the GaAs buffer layer and the two different GaAs layers in the supercell, labelled G_3 and G_5 according to their nominal thickness. The spectra in the AlAs energy range contain one main peak associated with the thicker AlAs layer labelled A_4. All these layers are very thin, thus preventing the observation of confined modes other than the fundamental one. We are also not able to resolve in these samples the contribution of the AlAs monolayer. Let us first emphasize that the perturbation on the optical vibrations due to a single AlAs monolayer inserted in the GaAs well is much stronger than it is on electronic properties[6]. Each GaAs layer on each side behaves almost independently, which means that the AlAs monolayer acts as an almost infinite barrier.

The Raman spectra qualitatively differ from those which we observed[7] on $Ga_{1-x}Al_xAs/AlAs$ superlattices, i.e. when the aluminum is homogeneously distributed in the well instead of forming a composition spike. This is strong evidence of our ability to grow an aluminum spike in the monolayer range. This allows us to analyze independently the signals coming from each GaAs layer.

In both series of samples the positions of the peaks G_3, G_5 and A_4 strongly varies as a function of T_s. Morover, the GaAs peaks, but not the AlAs one, display a dramatic dependence on the layer sequence. We indicate in Fig.2a, b the frequencies of the GaAs-type confined modes, calculated assuming a perfect profile. The measured frequencies always lie well below the predicted ones and the difference increases with increasing T_s. The same trend is observed on the single AlAs-type mode. Let us now focus on the difference between sequence A and B. The effective thickness of both GaAs layers is clearly very sensitive to the thickness of the AlAs layer which lies immediately undermeath and will be hereafter called the underlying AlAs layer. We plot in Fig.3 the measured line frequencies as a function of these two parameters. One can notice in Fig.3 that this sensitivity is larger in the nominally 3 monolayer than in the nominally 5 monolayer thick GaAs well because the frequency change for a given effective thickness variation is larger. Moreover, a tendency to the saturation in the underlying AlAs thickness dependence is observed at 4 monolayers, the behavior presumably approaching a "bulk-like" limit. On the contrary, the sensitivity to the growth temperature is strongly reduced for small underlying AlAs thicknesses.

On the other hand, the effective thickness of the nominally 4-monolayers-thick AlAs is the same in both sequences, i.e. after either 3 or 5 monolayers of GaAs. We show, however, in Fig.4 some recent Raman spectra in the AlAs energy range obtained on similar samples grown under similar conditions but with thicker GaAs layers and an aluminum spike corresponding to either 1 or 1/2 monolayer of AlAs. In these samples, a clear signal coming from the spike is observed and its frequency depends of the amount of aluminum deposited. This latter result suggests that the AlAs layer properties may also depend on the underlying GaAs thickness.

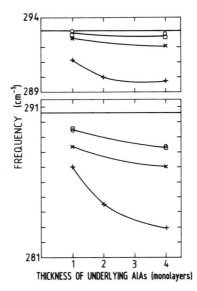

Fig. 3 . Measured frequencies of the GaAs-type fundamental vibrations plotted as a function of the nominal thickness of the underlying AlAs layer. In the upper (lower) part, the results correspond to a nominal GaAs thickness of 5 (3) monolayers. In both parts, the calculated nominal frequency is represented by a horizontal line as it is independent of the underlying AlAs thickness. The experimental results obtained on samples grown at 600, 500, 450 and 400°C are represented by plusses, crosses, open squares and circles respectively. The full lines are a guide to the eye.

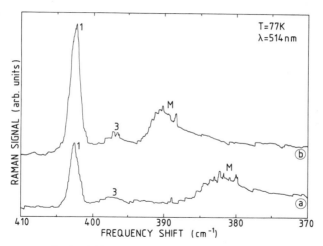

Fig. 4 .   Raman spectra in the AlAs-type frequency range for two different samples with the
composition sequence $(AlAs)_4(GaAs)_{10}(Ga_{1-x}Al_xAs)_1(GaAs)_8$ and x=0.5 (a) or 1. (b)

The general shape of the spectra is not modified in other respects from sample to sample. Even when the phonon frequencies are very different from the ones calculated using the nominal parameters, we do not observe any significant line broadening. Moreover, line splittings (as recently reported in Ref.2) seem to be absent in our samples. This suggests that the imperfections in these samples are caused by "gradual interfaces" ; the in-plane statistics of the roughness should be such that the optical phonons only see the average in-plane concentration and thus experience an ordered, eventually gradual composition profile along the growth axis, while the translational invariance in the layer planes is preserved. Assuming a gradual profile makes one-dimensional models sufficient to analyze the Raman scattering results. From the data corresponding to the AlAs energy range, the information is too scarce to separate the respective effects onto the confinement of an admixture of gallium and of a change in the composition profile. On the contrary, the observation of three lines in the GaAs energy range makes it possible to determine the composition profile in the gallium-rich parts of the structure with good accuracy. We will present in what follows a detailed 1D calculation of the Raman spectra in superlattices with gradual interfaces and we will introduce, on the basis of a 3D lattice dynamics calculation including the interface roughness, a novel microscopic criterion governing the emergence of this behavior.

The main difficulty in performing both tasks arises from the alloy vibrational behavior. Let us recall some experimental results on this point. The Raman spectra on $Ga_{1-x}Al_xAs$ alloys display the well-known two-mode behavior : two optical phonon Raman lines coexist on the whole concentration range which lie respectively in the range of the pure GaAs and AlAs optical phonons and only slightly shift as a function of the alloy composition. Both this two-mode behavior and the strong confinement in GaAs/AlAs superlattices qualitatively originates from the same feature : the large energy splitting between the optical bands in GaAs and AlAs. A virtual crystal approximation to describe these alloy vibrations is therefore meaningless, contrary to what prevails for the electronic properties.

Raman scattering experiments on bulk alloys provide useful but limited information on the alloy lattice dynamics. Inserting these alloys in superlattices allow us to increase this amount of experimental knowledge. Fig.5 displays some Raman spectra in the GaAs energy range obtained on GaAs/GaAlAs superlattices with similar parameters except for the aluminum concentration in the alloy layers which ranges from 100% to 5%. The frequency of the modes confined in the GaAs layers remain almost unchanged as long as the GaAs-type LO mode of the alloy constituents lies at sufficiently higher frequency. When it is no longer the case, the mode slightly shifts towards higher frequency before to transform into a propagative optical vibration with an energy in the range of overlap of the pure GaAs and GaAlAs alloy optical energy band. This whole behavior is typical of a quantum well and the alloy layers appear to act as very effective barriers for the phonons confined in the GaAs layers with energies above the Raman frequency in the bulk alloy.

107

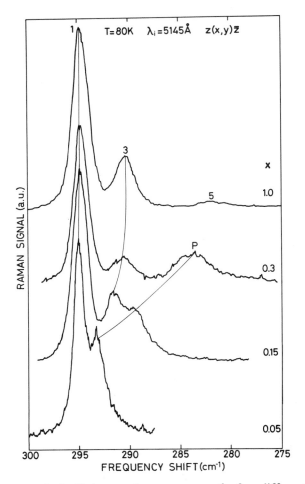

Fig. 5 .   Raman spectra in the GaAs-type frequency range for four different GaAs/Ga$_{1-x}$Al$_x$As samples with the same parameters except for the aluminum concentration x in the alloy layers.

Let us now consider the case of GaAlAs/AlAs superlattices and again the GaAs-type optical vibrations in these structures. Fig.6 displays the associated Raman spectra in three samples with an aluminum concentration in the alloy ranging from 0 to 30%. The spectrum at 15% is rather similar to the one at 0%, while for higher aluminum concentration the different lines become poorly resolved. A general shift of the spectra is observed, which maps onto the one in the corresponding bulk alloy. Moreover, three different peaks remain observable. We attribute these optical phonon lines in the thin alloy layer spectra to confined GaAs-type vibrations, comparable to those observed in pure GaAs thin layers; the AlAs layers acting in both cases as very effective barriers. The distance between the peaks appears to decrease with increasing x, while their width increases. This experiment[7] is the first experimental proof of the dispersive character of the GaAs-type vibrations in the GaAlAs alloys. This observation, and the barrier properties which we described in the previous paragraph, strongly suggest that the GaAs-type optical band in the GaAlAs alloys behaves, at least for small Al concentrations, much like the one in a pure compound with effective parameters. We will present in the next section some methods of approximation to define these effective parameters and compare the related predictions to the experimental results on bulk or thin alloy layers. This will justify the use of these parameters in some of the superlattice dynamics calculations presented here.

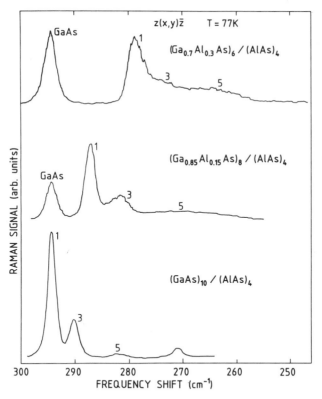

Fig. 6 .   Raman spectra in the GaAs-type frequency range for three different $Ga_{1-x}Al_xAs$/AlAs samples with the same parameters except for the aluminum concentration x in the alloy layers.

Lattice Dynamics of GaAlAs Mixed Crystals

The vibrational properties of the GaAlAs alloy have been analyzed theoretically for a very long time with the aim of qualitatively predict the two-mode behavior and to quantitatively reproduce the Raman scattering results. The favorable feature in this task is the almost identical spring constants of the GaAs and AlAs compounds which essentially differ by the mass of the atom III site. For the remainder of the paper we shall assume that these forces are the same. The first method which was successful in predicting the two-mode behavior is the Random Element Isodisplacement method[8]. It can be reformulated[9] using an effective ordered mass on the disordered site, which is a real quantity but depends on the energy in a rather similar way to the "CPA mass". We will therefore describe directly the Coherent Potential Approximation (CPA), its properties and its predictions.

The CPA description of the alloy[10-13] consists in a fictive ordered crystal which depends only on a frequency dependent isotropic complex mass assigned to the randomly occupied cation sites. This mass is self-consistently determined when one assumes the force constant matrix to be independent of the site occupation and if one demands that a given random occupied site imbedded in the CPA effective medium produces no extra scattering on the average. The CPA crystal being an ordered one, its properties, like phonon densities of states, dispersion curves and Raman activities, can be calculated as for pure compounds. However, some of these notions must be carefully redefined due to the peculiar properties of the CPA mass : it is a complex quantity and varies as a function of the energy, which respectively ensures the CPA crystal to include some disorder effects and to be at each energy the "best" ordered description of the real disordered crystal. Whereas we use a 3D lattice dynamic based on the Overlap Valence Shell Model of GaAs[14] to analyze the experimental results, we prefer to illustrate the CPA predictions on a much simpler system: the GaAlAs linear chain with only nearest neighbor interaction. We thereby retain the main results and in particular the two-mode behavior.

We show in Fig.7 the energy variation of the full density of states, of some partial conditional density of states and of the real and imaginary of the CPA mass at a few different compositions. The two-mode behavior appears in Fig.7 through the existence of two different energy ranges above $200 cm^{-1}$ where the density of states is non vanishing. The position and width of these ranges vary only slightly as a function of x. Their respective GaAs-type and AlAs-type character can be further demonstrated using the conditional partial density of states, calculated on the disordered site. It is only large in the lower energy (higher energy) band when the disordered site is occupied by a gallium (aluminum) atom. The associated vibrations thus display a rather confined behavior.

The imaginary part of the CPA mass is also non-vanishing in the range of allowed energies in the alloy, where the alloy fluctuations can easily scatter elastically any given CPA mode. Thus its amplitude qualitatively maps onto the integrated density of states. The variation of the real part is of greater interest for our purpose. It displays strong variation in each energy gap, so to be successively close to the virtual crystal mass, the gallium and the aluminum one in the three allowed energy bands. This behavior is in qualitative agreement with our experimental observations that the GaAlAs well and barriers act **in the GaAs-type energy range** as if the alloy **coincides in this range** with an effective pure compound with an effective mass close to that of gallium. A more precise comparison can be done on the basis of the phonon dispersion curves. Due to disorder, the translational invariance in the crystal is destroyed and dispersion curves should be expected to disappear. The CPA, being an ordered approximation, allows us to reintroduce this notion[15].However, due to the imaginary part of the CPA mass, the spectral density of states at any value of the wavevector is no longer a set of δ-functions but a continuous function displaying a few peaks which remain rather narrow in GaAlAs. The deduced dispersion curve is thus a thick one, as illustrated in Fig.8 for a few compositions. The zone center properties reproduce well the experimental results on bulk GaAlAs : the frequency decreases with increasing x and an asymmetric line shape develops. Furthermore, using this dispersion curve (in other words using the CPA mass on the disordered sites of the superlattice to perform the superlattices lattice dynamics calculation), we are able[7] to quantitatively explain the Raman scattering results on thin GaAlAs wells presented in Part 1, which provides a new, stringent test of the ability of the CPA to describe the mixed crystal vibrations.

The validity of the CPA for describing the barrier effect of a GaAlAs mixed crystal is not yet well established. The energy range of interest then lies outside the allowed energy bands and, in particular, just above the GaAs-type energy band. In this energy range, the CPA mass

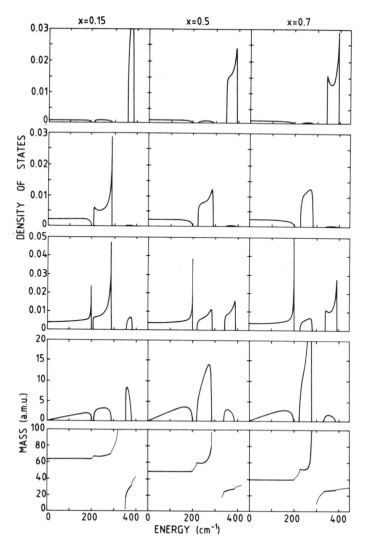

Fig. 7 .   From bottom to top, real and imaginary parts of the CPA mass, full density of states, partial conditional densities of states for gallium occupied and aluminum occupied atom III sites, calculated as a function of energy for three different compositions of a $Ga_{1-x}Al_xAs$ mixed linear chain.

is real and usually display a divergence, which reflects the vanishing of the corresponding element of the site Green function. The physical meaning of this divergence has never been investigated, as far as we know, because it does not play any role in the analysis of the bulk alloy properties. Using this mass in the GaAs/GaAlAs superlattice lattice dynamics, as previously done for GaAlAs/AlAs, seems to give reasonable results. We show in Fig.9 the spectral density of states at the zone center and the corresponding Raman activity for a few superlattices comparable to the structure of Fig.5. The agreement is reasonable due to the following qualitative features : i) the CPA mass is real at these energies, which prevents any disorder-induced broadening of the peaks and ii) its real part is quite different from the gallium one which ensures a strong confinement, as long as the aluminum concentration is sufficiently large. For very small values of x, the CPA mass decreases towards the gallium one and the modes become less confined.

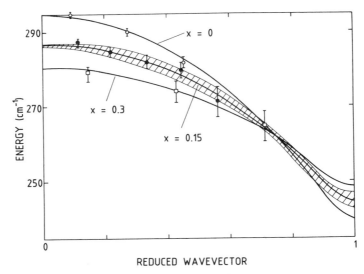

REDUCED WAVEVECTOR

Fig. 8 .   Calculated dispersion curves along the superlattice axis of the GaAs-type LO phonon
in bulk $Ga_{1-x}As_xAs$ mixed crystals, with x=0, x=0.15 and x=0.30. For x=0.15, the
hatched surface reflects the thickness of the dispersion curve, as explained in the
text. Open circles, closed circles and open squares correspond to experimental
frequencies on the samples of Fig.6, plotted according to Eq.1.

In a more quantitative way, the very steep variation of the effective mass predicted by
the CPA at these forbidden energies should be tested by carefully analyzing the confinement
energies in GaAs/GaAlAs superlattices. The few experimental results available up to now can
be nevertheless well reproduced quantitatively on the basis of a more crude model where the
effective mass on the alloy sites is taken constant and real in the small range of interest. In
Ref.3, we fitted its value to the alloy GaAs-type Raman frequency. The previous analysis brings
some theoretical support to this empirical approach, provided one chooses the zone-center
CPA mass as the effective one. We will show in the next section a successful illustration of
this method to calculate the confined optical vibrations in superlattices with gradual interfaces.

We are unfortunately far from treating by the CPA the "modulated alloy", i.e. a
superlattice with an aluminum concentration continuously varying along the growth axis.
Nevertheless, we can extract a lot of new knowlegdge on the alloy lattice dynamics from the
CPA, which was up to now successfully verified by experiments on alloy superlattices. An
alternative approach to the bulk alloy lattice dynamics was not described in this part : the
supercell calculation with a random tossing of the individual site occupations. Some comparison
with the effective mass description was already published[11]: significant differences only take
place close to the band edges due to the effect of large, unfrequent clusters. One should notice
that these energy ranges are of great importance in the superlattice lattice dynamics. Such
supercell methods can be easily extended to modulated alloys, except for computational
limitations. They should thus show a significant development in the near future. We will present
in what follows a first example where mixed interfacial layers are treated.

Confined Vibrations and Interface Roughness

A review of Raman scattering results where the confined optical frequencies depend on
the growth conditions was given at the beginning of this paper. We emphasized that the shape
of the spectra strongly suggest an interpretation in terms of gradual interfaces. This can be
justified if the correlation length of the composition fluctuations in the planes perpendicular
to growth axis is much smaller than the "coherence length" of the confined phonons, which

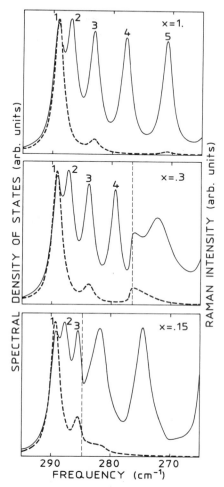

Fig. 9 . Spectral density of states (SDOS) of the GaAs-type LO vibrations (full line) and the associated Raman intensity (dashed line), calculated at a vanishingly small wavevector along the superlattice axis for three different GaAs/Ga$_{1-x}$Al$_x$As structures with the same parameters except for the aluminum concentration x in the barrier. We used the bulk CPA masses in the barrier alloy layer. The vertical dashed line indicates the corresponding LO frequency in the bulk alloy with the same composition.

remains to be defined. Before introducing such a definition and discussing the characteristics of samples where interface terracing should be observed by Raman scattering, we will first quantitatively analyze the presently available experimental results on the basis of lattice dynamics models including a continuously varying mass.

The simplest version of these models was introduced in Ref.3. The parameters of a linear chain with only nearest neighbor interactions is fitted into the dispersion along the (001) direction of the LO phonon in pure GaAs. In a second step, the mass on the atom III site is modified in order the reproduce the GaAs-type LO frequency in the alloy for each value of x. For any given composition profile along the growth axis, we can then diagonalize the dynamical matrix and determine the frequencies of the modes confined in the GaAs-rich parts of the samples (and only these ones). Assuming a gradual profile, for instance an erf-profile, with a varying broadening parameter d_o, we are able to fit this parameter for each spectrum. This fit is only significant when several lines are observed as it is the case of the samples of Fig.1 in the GaAs energy range (but not in the AlAs one). One can then check that the chosen profile is not unreasonable and determine the variation of the parameter d_o as a function of growth conditions. This method is illustrated in Fig.10 where the predicted confined frequencies (modes 1,3 and 5) are shown as a function of d_o for two different choices of the nominal thickness of the well (i.e. when d_o=0). Each frequency decreases with increasing d_o. In order words, the effective thickness seen by each vibration decreases. The comparison with the experimental frequencies is satisfactory because the frequencies of modes 1,3 and 5 can be fitted using the same d_o. Moreover, from the effective thickness scale shown on the right part of the figure for each confined vibration, we get the following result : this thickness is no longer constant from mode to mode as it was in the case of the abrupt profile. It increases with increasing confinement shift. This striking result can be simply understood in a perfect confinement picture, where one assumes that a confined vibration of energy Ω can extend in the part of the sample where the aluminum concentration x is lower than x_o such that $\omega_{LO}(x_o)=\Omega$. Fig.11 shows how reasonable this approximation is.

Though some very useful quantitative data onto an independently chosen profile can be extracted from the Raman spectra, it is, however, impossible to draw the profile from this single information. It is, for instance, very difficult to separate the respective contribution of the two interfaces limiting the GaAs well. In order to circumvent this difficulty, we designed samples with the concentration sequences shown in the insets of Fig.2. In these samples with very thin layers of GaAs and AlAs, we only observe the fundamental confined mode, which prevents testing the composition profile as extensively as before. However, on the basis of the observed dependence of the frequencies on the underlying layer thickness, we get a convincing evidence of the dominant role played by the GaAs on AlAs interface roughness. Instead of an erf-profile, we now use a "segregation profile"[16] to model the vibrations. This model was developed to explain chemical surface analysis results of the deposition of GaAs on AlAs and AlAs on GaAs at 600°C. Using these profiles, shown in Fig.12 for both sequences A and B, we get an excellent description of the Raman spectra on samples grown at 600°C. The predicted frequencies are shown at the top of Fig.2. Our observation of a reduction of the roughness of the GaAs on AlAs interface at lower growth temperature should stimulate the understanding of the MBE growth process.

The previous description of the roughness in terms of gradual interfaces remains valid as long as the spatial fluctuations are smaller than the "coherence length" of confined optical phonons. The rest of the paper will be devoted to a definition of this quantity and to a description of the lattice dynamics of superlattices when the gradual interface approximation fails, i.e. in the presence of terraced interfaces.

For this purpose we need to build a supercell which is larger than the basic one, not only along the nominal superlattice axis, but also in the layer plane. As we are interested in the confined modes, which probe a single layer, we can assume without any loss of generality that equivalent interfaces have the same in-plane statistics. We therefore retain the same periodicity along the growth direction as in the perfect structure. Due to limitations in the available computation time and memory, we must use rather crude approximations to account for the in-plane statistics, which result in the model structure shown in the inset of Fig.13. We consider

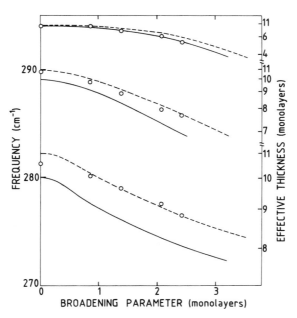

Fig. 10 . Frequency of modes 1,3 and 5 confined in nominally 10 (full lines) or 11 (broken lines) monolayers thick GaAs layers with rough interfaces, calculated as a function of the broadening parameter of the erf-profile. The open circles indicate experimental frequencies deduced from the spectra of Fig.1. The effective thickness corresponding to each confined frequency is shown on the right vertical axis.

one-dimensional fluctuations on a single side of the considered layers while the structure remains perfect along the third direction at this interface (it looks like corrugated cardboard) and along both directions at the other one, which remains atomically flat. Varying the amplitude along z and the distribution along x of the corrugation allows us to handle many different situations.

In Fig.13, we restrict ourselves to a very simple case where the fluctuation is restricted along z to a single interfacial monolayer and looks along x like a lateral superlattice. Two parameters then define the interface : its average aluminum content and its lateral period. This model provides the simplest representation without any empirical parameter of a thin layer with a non-integer thickness. Moreover, it opens the possibility to vary the lateral terraces. When the mass distribution in the supercell is defined, we build up[17] the force constant matrix at a given wavevector in the Brillouin minizone from the bulk matrices at the wavevectors in the full Brillouin zone which are now equivalent due to the modulation.

We show in Fig.13 the predicted frequency of the fundamental vibration confined in a GaAs layer of a thickness $n_1+(1-x)$ monolayers as a function of x for $n_1=2$ and several different in-plane periods a+b between 2 and 40 monolayer thickness, i.e. between 6 and 120Å. a and b are the respective numbers of adjacent Ga and Al atoms in the single interfacial layer (see the inset). The average aluminum content x in the interface layer then takes simple rational values b/(a+b), ranging from 0 to 1. While a single composition x=0.5 can be considered for a+b=2, a series of nine different ones is reproduced for a+b=10. The results display similar trends for other nominal thicknesses, except that the absolute energy variations decrease with increasing n_1.

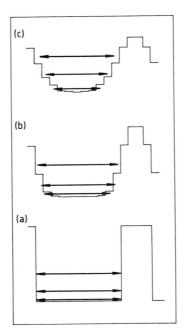

Fig. 11 . Aluminum concentration profiles along the superlattice axis calculated for the structures of Fig. 10 according to an erf-profile with a broadening parameter of 0(a), 1.5(b) and 2.4 (c) monolayers. For each profile, the three arrows correspond to modes 1,3 and 5 confined in the gallium-rich parts of the samples. Their length indicates the effective thickness associated with their frequency. Their vertical position indicates the composition of the bulk alloy whose GaAs-type LO frequency coincides with their frequency.

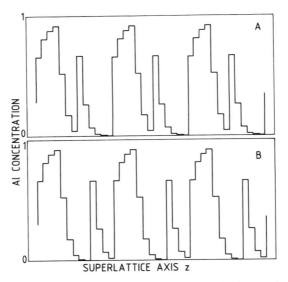

Fig. 12 . Aluminum concentration profiles along the superlattices axis calculated for the sequences A and B of Fig. 2 according to the segregation model of Ref.16.

116

Let us consider the points on the lowest curve of Fig.13. They correspond to Al concentrations 0.1, 0.25, 0.5, 0.75 and 0.9 and the in-plane atomic distribution in the interface layer is such that isolated Ga or Al atoms are separated by a distance comparable to the one in a 1D random alloy of the same composition. These points give us an estimate of the effect of an intermixed layer with short range (alloy-like) disorder. Between x=0.5 and x=1., the frequency of the confined vibrations remains remarkably close to that obtained with exactly two monolayers. Only one Al atom on each fourth site (x=0.25) almost completely pushes away the vibration from the interfacial layer. This results is in good agreement with the predictions of the 1D model with gradual interfaces.

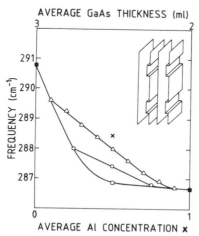

Fig. 13. Frequencies of the fundamental GaAs-type LO vibration calculated for the $(GaAs)_{3-x}(AlAs)_{2+x}$ model superlattice shown in the inset, as a function of the average aluminum concentration x in the in-plane modulated interface atomic layer. Different periods of the in-plane modulation are considered : 2 (open square), 4 (open circles), 10 (open triangles) and 40 (cross) monolayers. The corresponding average thickness (3-x) of the GaAs layers is indicated in the upper scale. The full lines are a guide to the eye.

We can also look in Fig.13 for the variation of the fundamental frequencies as a function of the size of the terraces for a given average concentration. Consider the case where x=0.5 and vary a=b from 1 to 20. The frequency then slightly shifts towards higher values while the eigendisplacement is weakly modified. This result is illustrated in Fig.14 where the amplitude of the eigendisplacement along z on each site in the supercell is shown for a+b=2 and a+b=10. In both cases the eigendisplacement pattern remains very similar to the one in a perfect 5-monolayers-thick GaAs, while the gallium atoms in the interfacial layer are almost at rest. Neither lateral localization nor line splitting is therefore predicted up to this high value of the terrace size. This result is consistent with the absence of any reported observation of such splitting. However it is surprising by comparison with the behavior of electrons confined to a GaAs layer with rough interfaces[18]. Well defined electronic levels, associated to the lowest quantized levels in each thickness, were indeed predicted to appear when the terrace extension exceeds to following dimension :

$$\lambda = \frac{d\sqrt{\frac{m_z}{m_{xy}}}}{p\sqrt{\frac{2\Delta d}{d}}} \qquad (2)$$

where m_z and m_{xy} are the effective mass along z in the bulk constituent (which governs the confinement in quantum wells with infinite barriers) and the one along the layer plane in the superlattice; d and Δd are the nominal thickness and its fluctuation and p is the index of the quantized level. This expression originates in the comparison between the additional confinement energies due to either the reduction of the layer thickness or the lateral localization in the thicker parts. Assuming an isotropic mass, i.e. starting from a cubic crystal and assuming that the confinement does not significantly modify the masses, this length scale does not depend on the actual value of the mass but increases with increasing nominal layer thickness and with decreasing defect amplitude. This model therefore depends on the goemetry of the problem. A naive application of this criterion to the confined vibration which we analyzed previously would let us predict the emergence of lateral localization for very small terraces (respectively 3 and 9 monolayers for $n_1 = 2$ and 5) in complete disagreement with the predictions of our calculation.

We attribute this disagreement to the long range Coulomb forces which strongly affect the dispersive properties of optical phonons around zone center[19]. We show in Fig.15 the dispersion of the GaAs-type optical phonons with displacement mainly oriented along z, calculated at a fixed finite wavevector along z, $k_z = 0.01$ in reduced units, and an in-plane wavevector k_x varying between 0 and 0.3. The dispersion curve of the fundamental mode is rapidly varying close to zone center due to the increasing associated macroscopic polarization. As a result of this large anisotropy, the lateral localization does not appear even in presence of very large terraces. On the contrary, this macroscopic polarization remains very small for the other odd confined modes and is vanishing for the even ones. As a consequence, their in-plane dispersion curves are smooth close to zone center. We thus predict that the corresponding confined modes should be much more sensitive to interface roughness.

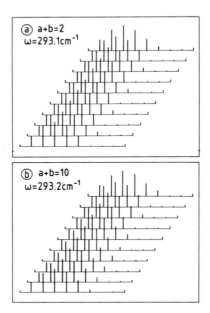

Fig. 14 . Amplitude of the eigendisplacements along z corresponding to the fundamental GaAs-type LO vibrations, shown for the cation sites the supercell of $a(GaAs)_{5.5}/(AlAs)_{2.5}$ superlattice with two different in-plane modulations : a+b=2 (a) and a+b=10 (b).

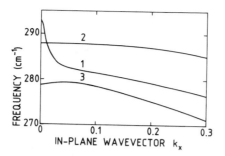

Fig. 15 . Dispersion curves in the layer plane of the higher-frequency GaAs-type optical vibrations calculated for a $(GaAs)_5/(AlAs)_3$ superlattice with perfect interfaces.

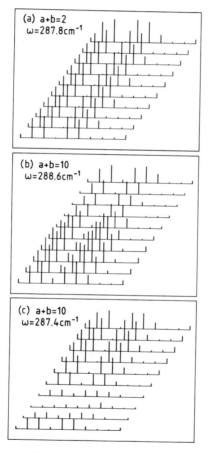

Fig. 16 . Amplitude of the eigendisplacements along z corresponding to the second GaAs-type LO vibrations, shown on the cation sites in the supercell of a $(GaAs)_{5.5}/(AlAs)_{2.5}$ superlattice with two different in plane modulations : a+b=2 (a) and a+b=10 (b and c). In the latter case, two different eigenmodes are displayed, as explained in the text.

We show in Fig.16 some eigendisplacements in the supercell calculated for a corrugated structure with $n_1=5$ and $n_2=3$ and corresponding to the higher even mode (p=2 in Eq.1). We obtain a clear evidence in this case of a mode splitting due to lateral terraces, when they reach a large enough extension. In Fig.16a, we show the eigendisplacement obtained assuming a small in-plane period of 2 monolayers. There is almost no modulation along the x direction, the displacement is hardly distinguishable from the one in the perfect $n_1=5$ GaAs layer. In particular, the displacement of the Ga atom at the center of this perfect well vanishes by symmetry and remains negligible in the locally 6 monolayer thick parts of the rough well. In good agreement with a critical size $\lambda=4$ deduced from Eq. 2, opposite conclusions apply to the sample with larger terraces : two different eigenmodes merge from the p=2 mode of the perfect GaAs layer and exhibit a single node in the vicinity of the center of the layer. The lowest frequency component of the doublet is partially localized in the narrow parts of the GaAs layer and the highest one, more clearly, in the wide parts. A good signature of this difference is again obtained from the displacement of the Ga atom at the center of the narrow parts. Its displacement is vanishingly small for the lowest mode but becomes significant in the wide parts for the highest frequency vibration. The node is now shifted from the gallium side to the neighboring arsenic one.

Similar behavior can be evidenced for the highest index (s>2) confined vibrations, with the qualitative tendency of a decrease in the critical terrace size with an increasing value of s. On the other hand, this critical size for a given value of s increases with increasing nominal thickness n_1. For instance, mode 2 is split in the sample of Fig.16b,c but becomes delocalized when n_1 is increased from 5 ($\lambda=4$) to 10 ($\lambda=12$), the terrace size remaining unchanged. All these variations reflect the change in the perturbative potential and thus in the value of λ.

CONCLUSION

The presence of GaAlAs mixed crystal layers in the GaAs/AlAs superlattices is very difficult and may be impossible to avoid. This makes the lattice dynamics of these structures very difficult to describe. We have attempted to review in this paper the present understanding of this problem. We also emphasized the considerable amount of new information on the lattice dynamics of bulk alloys one can get from their intentional incorporation in superlattices.

ACKNOWLEDGEMENTS

It is a pleasure for us to acknowledge F. Mollot, R. Planel, J.M. Moison, G. Le Roux, J.M. Gérard and J.Y. Marzin for growing or carefully characterizing the samples used in this work and for many helpful discussions.

REFERENCES

1 -   for a review on vibrations in superlattices, see B. Jusserand and M. Cardona, in "Light Scattering in Solids V", ed. by M. Cardona and G. Güntherodt (Springer Heidelberg 1989) p.49.

2 -   G. Fasol, M. Tanaka, H. Sakaki and Y. Horikoshi, Phys.Rev.B38, 605 (1988) and references therein.

3 -   B. Jusserand, F. Alexandre, D. Paquet and G. Le Roux, Appl.Phys.Lett.47, 301 (1986).

4 -   D. Levi, S.L. Zhang, M.V. Klein, J. Kem and H. Morkoç, Phys.Rev.B36, 8032 (1987).

5 -   R. Cingolani, L. Tapfer, Y.H. Zhang, R. Muralidharan, K. Ploog and C. Tejedor, Phys.Rev.B40, 8319 (1989).

6 -   J.Y. Marzin and J.M. Gérard, Phys.Rev.Lett.62, 2172 (1989).

7 -   B. Jusserand, D. Paquet and F. Mollot, Phys.Rev.Lett.63, 2397 (1989).

8 -   I.F. Chang and S.S. Mitra, Adv.Phys.20, 359 (1971).

9 -   D. Paquet, unpublished.

10 -   D.W. Taylor, Phys.Rev.156, 1017 (1967).

11 -   P.N. Sen and W.M. Hartmann, Phys.Rev.B9, 367 (1974).

12 -   R. Bonneville, Phys.Rev.B24, 1987 (1981).

13 - B. Jusserand, D. Paquet and K. Kunc, in "Proceedings of the 17th International Conference on the Physics of Semiconductors", ed. by J.D. Chadi and W.A. Harrison, Springer, New York (1985) p.1165.

14 - K. Kunc and H. Bilz, Solid State Commun.19, 1027 (1976).

15 - P. Soven, Phys.Rev.178, 1136 (1969).

16 - J.M. Moison, C. Guille, F. Houzay, F. Barthe and M. Van Rompay, Phys.Rev.B40, 6149 (1989).

17 - G. Kanellis, Phys.Rev.B35, 746 (1987).

18 - D. Paquet, Superlattices and Microstructures 2, 429 (1986).

19 - R. Merlin, C. Colvard, M.V. Klein, H. Morkoç, A.Y. Cho and A.C. Gossard, Appl.Phys.Lett.36, 43 (1980) ; E. Richter and D. Strauch, Solid State Commun.64, 867 (1987) ; S.F. Ren, H. Chu and Y.C. Chang, Phys.Rev.B37, 8899 (1988).

INTERACTION OF LIGHT WITH ACOUSTIC WAVES

IN SUPERLATTICES AND RELATED DEVICES

J. Sapriel and J. He

Centre national d'études des Télécommunications
196, ave H. Ravera, 92220, Bagneux
France

1. INTRODUCTION

Since the first experimental evidence of folded acoustic modes,[1] many studies have been devoted to light interaction with "acoustic" phonons in superlattices (SL's). Nevertheless there are such a large number of investigated SL samples (corresponding to different layer compositions and thicknesses) and experimental situations (obtained for various laser lines) in the literature[2] that it is worth attempting a unified approach to the problem, which, at least, could minimize the number of parameters used for the description of this interaction phenomenon. Besides, Raman scattering from acoustic phonons is considered as a useful and versatile tool for the investigation of the SL structure (period, composition) and quality (inhomogeneities of period, layer thickness, and composition; sharpness of the interfaces). The results of the study of light interaction with acoustic phonons in SL's can be easily transposed to the problem of light diffraction by ultrasounds in these microstructures . The use of ultrasonic waves (as compared with acoustic phonons of the light-scattering experiments) allows a flexibility by far larger, and the interaction efficiencies are much higher, since one can produce high ultrasonic power densities with a piezoelectric transducer. Acousto-optic Bragg deflectors and modulators can thus be achieved and will be briefly analyzed here.

2 . UNIFIED APPROACH OF LIGHT INTERACTION WITH ACOUSTIC PHONONS

The Propagation of Acoustic Waves in SL

In a superlattice (SL) two models can be used to obtain the phonon dispersion curves along the direction perpendicular to the layers: the elastic model, where each layer is defined by the acoustical properties of the bulk material; and a linear chain model of two alternating diatomic[3-5] sublattices. For a number of monolayers larger than about five in each slab, and as far as the first branches of the dispersion curves are concerned, the two models give the same results.[6] For a given ratio d_{AlAs} / d_{GaAs} of the layer thicknesses the acoustic dispersion curves can be obtained in a standard form, independent of the SL period D, provided one uses the dimensionless parameters $\Omega_p = \omega_p D$ and $K_p = k_p D$ where ω_p is the energy and k_p is the wave vector of the phonon (Fig.1a). Energy gaps at the successive boundaries of the Brillouin zone are related to the acoustic properties of the constitutive layers. So far, neutron scattering has not been used, but it will be shown that, because of the small spreading of the mini-Brillouin zone in the k-vector space, the light-scattering technique can, by itself, allow the determination of the phonon dispersion curves in SL's.

Light Scattering in Semiconductor Structures and Superlattices
Edited by D.J. Lockwood and J.F. Young, Plenum Press, New York, 1991

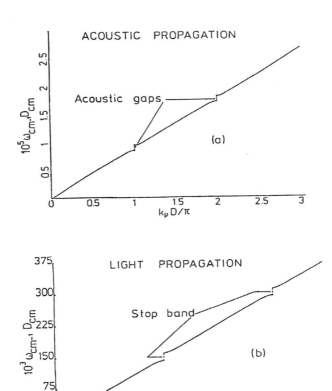

Fig.1. Dispersion curves of (a) acoustic modes in GaAs/AlAs SL's $(d_{AlAs} = 3d_{GaAs})$ and (b) light waves in GaAs/AlAs SL's $(d_{AlAs} / d_{GaAs} = 2/3)$ in the extended Brillouin zone scheme.

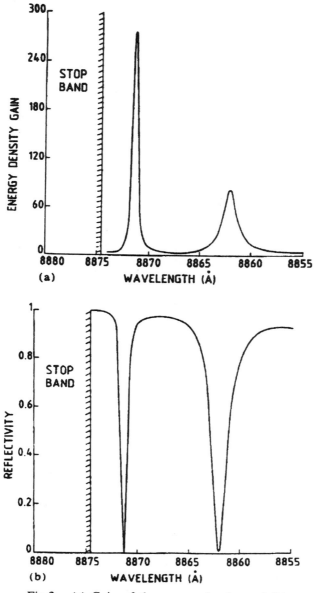

Fig.2. (a) Gain of the energy density and (b)
reflectivity in an SL (d_{GaAs} = 840 Å ,
d_{AlAs} = 560 Å).

The Propagation of Light in SL

The propagation of light of wavenumber $k = 2\pi n/\lambda$ along the SL axis presents strong similarities to that of phonons (Fig.1b). In the extended Brillouin-zone scheme, energy gaps (stop bands) appear whose amplitude depends on the optical properties of the two SL constituents. As will be shown here, these gaps disappear if one considers the damping of the acoustic or optic wave in the different media encountered. The acoustic propagation has been treated in several papers dealing with light scattering in SL's. We'll here emphasize certain new results on the propagation of plane light waves propagating without attenuation along the SL axis.

Let us consider the typical case of an SL bounded by air on one side and by a substrate on the other side, and let us examine the wavelength range near the gap $K = 1$, corresponding to the Bragg reflection on the interfaces. The reflectivity is a function of the light wavelength (Fig.2b). In the stop band the reflectivity is aproximatively 1; out of it, the reflectivity displays a succession of maxima and minima. The minima of reflectivity correspond to maxima of energy density gain[7] in the SL (Fig.2a). Actually, the SL acts as a Fabry-Perot whose resonances correspond precisely to the minima of reflectivity. Such resonances take place at light wavelengths for which the phase change of the Bloch light wave, after the propagation of one cycle in the SL, is equal to a multiple of 2π. One can see that the closer the resonant peak is to the limit of the stop band, the stronger the resonance. This is due to the group velocity value, which decreases and tends to zero for light wavelengths tending to the limit of the stop band. The group velocity being the velocity of the energy propagation, a low group velocity is associated to high-energy densities in the SL.

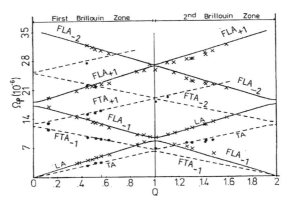

Fig.3. Standard dispersion curves of an SL ($d_{AlAs} / d_{GaAs} = 3$); longitudinal (x) and transverse (•) modes (experimental) . Full and dashed lines are calculated from the elastic model; Q is the reduced scattering wave-vector ($Q = qD/\pi$); $\Omega_p = \omega_{cm^{-1}} D_{cm}$

Acoustic Phonon Dispersion Branches

In an SL, because of the periodicity along the growth axis, the phase matching conditions for a light-acoustic-wave interaction can be assisted by a reciprocal lattice vector, that is

$$\vec{k}_p = \vec{k}_d - \vec{k}_i + (2m\pi/D)\vec{z}_0 \qquad m = 0, \pm 1, \pm 2$$

where $\vec{k}_d$ and $\vec{k}_i$ are the wave vectors of the scattered and incident light, respectively, and $\vec{z}_0$ is the unit vector in the direction of the SL axis. In the backscattering geometry used $\vec{k}_d = -\vec{k}_i$ and $\vec{k}_d - \vec{k}_i = \vec{q} \approx 4\pi n/\lambda$; here n is the refractive index of the light whose wavelength is λ in the SL; $\vec{q}$ is the scattering wave vector.

The integer m is a folding index that labels the different modes; $m = 0$ corresponds to the Brillouin mode (LA or TA). FLA_m and FTA_m correspond to the mth folded longitudinal and the mth folded transverse acoustic modes, respectively. Most of the authors have probed the folded modes corresponding to phonon energies higher than that of the Brillouin mode. Only one paper has clearly reported the existence of a mode at lower energies (more precisely, the one corresponding to $m = -1$) than the Brillouin line.[8]

In Fig.3 we have presented the frequency measurements for GaAs/AlAs SL samples corresponding to different periods but to approximately the same AlAs/GaAs proportion. The synthesis of all the results is given in the form of standard dispersion curves of the lowest LA and TA branches in the acoustic range. One can see the appearance of a folded acoustic branch in the second Brillouin zone, under the Brillouin LA branch.

The lowest frequency modes have been observed with an experimental setup using a Raman spectrometer and a Fabry-Perot interferometer, in tandem. The interaction with longitudinal and transverse acoustic phonons can be observed for phonon energies as low as 1 cm^{-1}. There is an excellent agreement between the experimental results and the calculations obtained with the theory of elasticity (full and dashed lines of Fig.3).

In the (001) oriented SL's, only the LA and the FLA modes are Raman active. The TA and FTA modes are seen only owing to a polarization leakage[9] and are always very weak modes (Fig.4). In Fig.5 is presented a typical Raman spectrum corresponding to a scattering wave vector in the second Brillouin zone.

<u>Intensity Behaviour of the Longitudinal Acoustic Raman Peaks</u>

To obtain a good agreement between the calculated and measured intensities of these modes it is necessary to take into account all the different kinds of modulations in the SL.[10] In

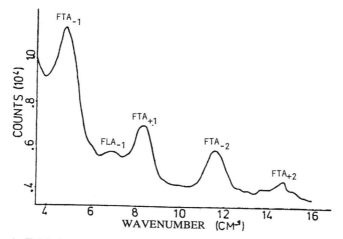

Fig.4. Folded transverse acoustic modes obtained for crossed polarizations of the incident and scattered light.

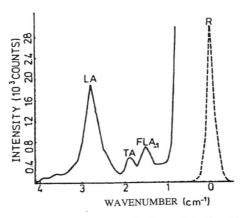

Fig.5. Low-frequency modes in an SL ($Q > 1$).

the intensity calculations we considered an SL made of layers of different acoustic ($\rho \neq \rho', v \neq v'$) optical ($n \neq n'$) and photoelastic ($p \neq p'$) properties. Both light and acoustic waves were treated as Bloch functions. The results of the calculations for GaAs/AlAs SL's are given in Fig.6 on the LA, the $FLA_{\pm1}$, and the $FLA_{\pm2}$. One can see the strong dependence of the mode intensities versus Q. The Brillouin mode is taken as an intensity reference. Though the Brillouin LA is generally the chief mode, other modes become more intense than the Brillouin line near the boundaries $p\pi/D$ (p integer) of the successive Brillouin zone. Besides, the FLA_{+m} intensity is different than that of the FLA_{-m} (see Fig.7). We give in Fig.6 the results of the mode intensities calculated according to Colvard the model[5] ($\rho = \rho', v = v', n = n', p \neq p'$). The intensities according to reference 5 are actually constant with respect to Q, contrary to our theory and to the experimental observations, and no difference is predicted between FLA_{+m} and FLA_{-m} intensities, in disagreement with the experiments (Fig.7). The difference between the intensities of these modes is accounted for[11] by taking $\rho \neq \rho'$ and $v \neq v'$. Another improvement to the theory[10] consists of taking $n \neq n'$ as shown in Fig.8 for $1 < Q < 2$. The discrepancies between the two cases ($n = n'$ and $n \neq n'$) is particularly pronounced for $Q \approx 2$, since for $Q = 2$ the interfaces between GaAs and AlAs play the role of light reflectors at Bragg conditions.

The technique for the measurement of the photoelastic constants that is based on light scattering (Raman and Brillouin) by SL acoustic photons has been refined and applied to $Ga_{1-x}Al_xAs$ and $Si_{1-x}Ge_x$ materials. The photoelastic constants of $Ga_{1-x}Al_xAs$ with respect to GaAs and those of $Si_{0.5}Ge_{0.5}$ with respect to Si have been measured as a function of the wavelength of the laser excitation.[12] One finds that the photoelastic constant of $Ga_{1-x}Al_xAs$ undergoes a nonlinear variation with the aluminium concentration x (Fig.9) and that the ratio of the photoelastic constants of $Si_{0.5}Ge_{0.5}$ and Si varies strongly (Fig.10) as a function of the laser wavelength.

3. EFFECTS OF ACOUSTIC AND OPTICAL DAMPINGS ON THE LIGHT SCATTERING BY PHONONS IN SUPERLATTICES[13]

In this section the effects of the acoustic attenuation and optical absorption are analyzed and experimentally probed in GaAl/AlAs and $Si/Si_{0.5}Ge_{0.5}$ SL's by means of Raman and Brillouin scatterings. Particularly striking are the damping effects for scattering wave vectors

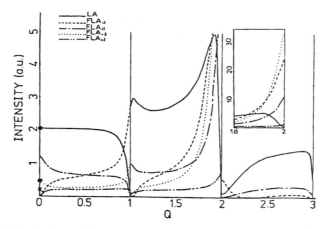

Fig.6. Relative intensities (our model) of the LA and FLA for GaAs/AlAs SL's ($d_{AlAs}/d_{GaAs} = 3$). The values of LA ($\bullet$), $FLA_{\pm 1}$ (*), and $FLA_{\pm 2}$ (x) as calculated in reference 5 are given for comparison.

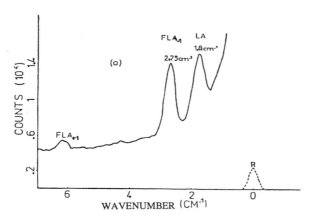

Fig.7. Raman spectrum of a GaAs/AlAs SL's, for $Q = 0.78$. Here, the FLA peak is more intense than the Brillouin LA. Besides, FLA_{+1} is much less intense than FLA_{-1}.

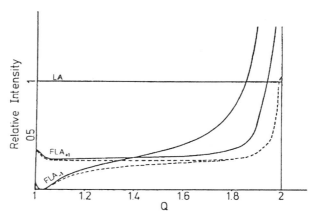

Fig.8. Calculated intensity variations as a function of the reduced scattering wave vector Q in the range $1 < Q < 2$ for
$n \neq n'$ (full lines) and $n = n'$ (dashed lines). In both cases
the acoustic modulation ($\rho \neq \rho'$, $v \neq v'$) and photoelastic
variations are taken into account. For $1.5 < Q < 2$, the
approximation $n = n'$ clearly underestimates the peak
intensity of the $FLA_{\pm 1}$.

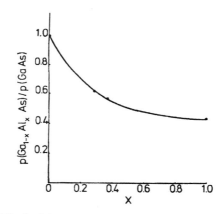

Fig.9. Measured photoelastic constant p_{13}
of $Ga_{1-x}Al_xAs$ as a function of
GaAs p_{13} for $\lambda = 4880$ Å.

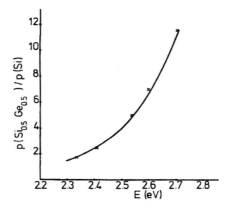

Fig.10. Measured photoelastic constant ratio $p\,(Si_{0.5}Ge_{0.5})/p\,(Si)$ versus the incident photon E energy. For photon energy between 2.3 and 2.7 eV, the variations can be approximated by an exponential (solid line).

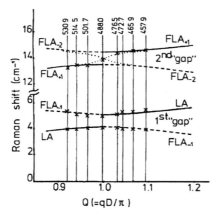

Fig.11. Energies of the acoustic modes and the role of the acoustic attenuation on the acoustic gap in an $Si/Si_{0.5}Ge_{0.5}$ SL.

131

q corresponding to the limits of the successive mini-Brillouin zones. We essentially focus on q values ranging around π/D, which corresponds to Bragg reflection for the acoustic phonons, and near $2\pi/D$, where Bragg reflection occurs for both acoustic and light waves (D is the period of the superlattice along the growth axis z). In these critical regions one can expect an enhancement of the effects of the acoustic attenuation and optical absorption, since both of them significantly change the Bragg reflections. More details on the present study are given elsewhere.[13] We'll successively study the acoustic attenuation effects on the phonon dispersion curves, the acoustic modes broadening as a function of the acoustic and optical damping, and the optical absorption contribution to the relative intensities of the acoustic modes.

The energies of the acoustic modes are measured by light scattering at different laser wavelengths in an $Si/Si_{0.5}Ge_{0.5}$ SL ($d_{(Si)} = d_{(SiGe)} = 140$ Å) and are represented by crosses on Fig.11. The solid lines correspond to the values calculated in the case without acoustic attenuation. At low frequencies ($\omega < 6\,cm^{-1}$) the crosses are in good agreement with the solid line and a gap appears[14] ($\sim 1\,cm^{-1}$). For $\omega > 12\,cm^{-1}$ we found experimental points within "the gap." Actually, at these frequencies the acoustic attenuation can no longer be neglected. The acoustic damping makes the modes at the Brillouin-zone edge shift into the "forbidden gaps." In Fig.12 are represented the calculated dispersion curve (Fig.12a) and attenuation-frequency relation of the acoustic phonons in the considered $Si/Si_{0.5}Ge_{0.5}$ superlattice (Fig.12b). The increase of the attenuation at high frequency changes markedly the feature of the gap at $k_p = 3\pi/D$ (see the inset of Fig.12a). We have used $\alpha_1 = 0$ and $\alpha_2 = A\omega^2$ for the acoustic attenuation in Si and $Si_{0.5}Ge_{0.5}$, respectively. As for the acoustic attenuation α in the SL, it increases with frequency and always displays peaks at $k = n\pi/D$ (Fig.12b), but the peaks are less conspicuous and almost disappear when the attenuation of the constitutive materials is very strong.

The Raman peaks corresponding to the FLA modes undergo a broadening $\Delta\omega$ (FWHM) that is equal to $2\alpha V_g$, $4\alpha_{op}V_g$, and $2\pi V_g/l$, due to the acoustic attenuation α, the optical absorption α_{op}, and the total thickness l of the SL, respectively; $V_g = d\omega/dk$ is the group velocity of the acoustic waves. In Fig.13 are reported the low-frequency spectra of the SL with different excitation laser wavelengths. For low-frequency modes the line width displays a minimum at $Q = 1$ caused by the flatness of the acoustic dispersion curves. This is particularly clear on the upper mode of the low-frequency doublet of Fig.13.

In reference 10 the relative intensities of the different modes have been calculated by taking into account the modulation of the photoelastic, acoustic, and optical properties in the SL, but the optical absorption has been neglected. A good agreement between the theory and the experiments has been obtained for a wide range of scattering wave vectors, not too close to the Brillouin-zone boundaries.[8] However, one can expect a significant effect of the optical absorption on the relative intensities for scattering wave vectors near $Q = 2, 4, 6$. Figure 14 shows the intensity calculated with (solid lines) or without (dashed lines) optical absorption for a GaAs/AlAs SL ($d_1 = 167$ Å, $d_2 = 458$ Å) for a scattering wave vector near the second-Brillouin edge ($Q \approx 2$). When Q is very close to 2, the intensities of the folded modes with respect to the Brillouin line are much reduced by the optical absorption. The Raman spectra corresponding to this Q range are shown in Fig.15. One can see on Fig.14 that when Q tends to 2, FLA_{-2} increases, reaches a maximum and then decreases with respect to the LA (Brillouin line). This is indeed the behaviour predicted when the optical absorption is considered.

4. INTERACTION OF LIGHT WITH ULTRASONIC WAVES; APPLICATION TO ACOUSTO-OPTIC DEFLECTION AND MODULATION OF LASER BEAMS[7]

The preceding study on light interaction with acoustic phonons in SL's along the SL axis can be extended to light diffraction by ultrasonic waves. These waves are generated by piezoelectric transducers and are limited to frequencies lower than 3 GHz ($0.1\,cm^{-1}$). For SL's

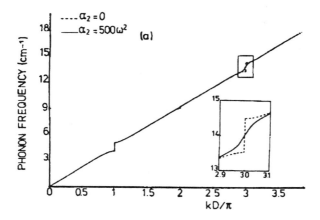

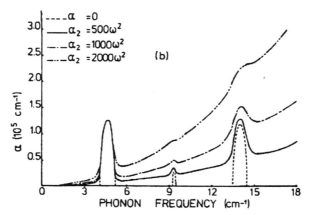

Fig.12. Calculated dispersion curves (a) and
attenuation-frequency relation of the
acoustic phonons (b) in an $Si/Si_{0.5}Ge_{0.5}$
SL (from reference 13). Here the attenuation of Si is neglected and
$\alpha_2(SiGe) = A\omega^2$.

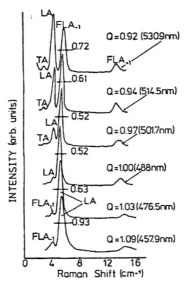

Fig.13. Low-frequency spectra of an
Si/Si$_{0.5}$Ge$_{0.5}$ SL near $Q = 1$
(mini Brillouin-zone edge).

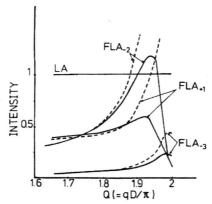

Fig.14. Intensities of the phonon modes for
a GaAs/AlAs SL, calculated with
(solid lines) or without (dashed
lines) optical absorption, near $Q =$
2. The intensity of the Brillouin
(LA) mode is normalized to 1.

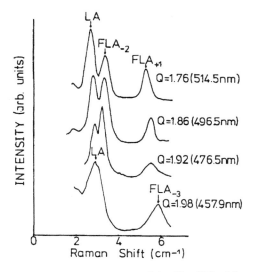

Fig.15. Raman spectra of the SL of Fig.14
for Q values tending to 2.

whose periods are comparable with the light wavelength, the optical properties of the SL's are quite different from those of a bulk medium, since the laser beam propagating along the SL axis will undergo multiple reflections at the different interfaces. From the point of view of propagation of an ultrasonic wave, however, the SL can be considered effectively[15] as a bulk medium.

The coupling of a light beam with a sound wave can take place only when a phase matching condition is satisfied. In a homogeneous bulk medium, the phase-matching conditions is expressed by the momentuum conservation, that is, $k_d = k_i + K$. As the wave vector K of the ultrasonic wave is much smaller than that of light, the direction of propagation of the incident light should be nearly normal to that of the sound (Fig.16a). The frequency shift of the diffracted light with respect to that of the incident light is equal to the frequency of the sound wave.

In an SL, because of the periodicity along the growth axis, the phase-matching condition can be assisted by a reciprocal lattice vector; for example, one can have (see Fig.16b)

$$\vec{k_d} = \vec{k_i} + \vec{K} + (2\pi/D)\vec{z_0}$$

Because of the intervention of the reciprocal lattice vector, the diffraction can take place in a backscattering configuration along the SL axis despite the fact that the wave vector K of the sound wave is much smaller than that of the optical wave. This interaction then occur for light wave vector $k \approx \pi/D$, that is, near the conditions of Bragg reflection of light on the SL interfaces. If one takes advantage of the optical resonance discussed in section 2, the interaction can be strongly increased. This is equivalent to a laser intracavity acousto-optical interaction, which benefits from the high-energy gain localized between the two end mirrors of the laser. As in the case of intracavity devices, the SL must be transparent to the light beam to minimize the optical losses.

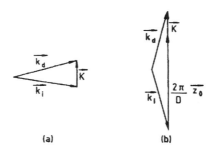

Fig.16. Composition of the wave vectors for acousto-optical interaction in (a) a homogeneous bulk medium and (b) an SL.

As in the case of a homogeneous bulk medium,[16] the diffraction efficiency in the SL (in the limit of small acoustic powers) can be expressed by

$$\eta = \pi^2 P_{ac} M_{SL} L^2 / 2\lambda_0^2 S$$

where P_{ac} is the acoustic power, M_{SL} is the effective figure of merit of the SL, λ_0 is the light wavelength in the air, L is the length of the acousto-optical coupling (total thickness of the SL), and S is the section of the acoustic beam. The increase of the diffraction efficiency due to the optical resonance effect is reflected by the enhancement of the effective figure of merit. Figure 17 shows the effective figure of merit of a GaAs/AlAs SL with respect to that of GaAs, as a function of the light wavelength. One can see that in the resonance condition the effective figure of merit of the SL can be hundreds of times higher than that of GaAs bulk materials (we have considered [001] oriented SL's and longitudinal acoustic waves propagating along the [001] direction).

The effect of optical resonance in the SL permits a strong interaction between the light beam and the ultrasonic wave in a coupling length much smaller than that of bulk media. This allows rapid acousto-optical devices (deflectors and modulators) with broad bandwidth.[17] An acousto-optical modulator built up from an SL structure is schematically drawn in Fig 18. The wavelength and the angle of incidence of the laser beam are adjusted at the resonance condition. The deflection is, therefore, minimum in the abscence of sound waves. When the ultrasound is generated by the piezoelectric transducer, a diffracted beam is produced. The intensity of the diffracted beam depends on the acoustic power. The modulation of the laser beam can then be controlled by the electric signal applied to the transducer.

CONCLUSION

Since the pioneer works on Raman scattering in GaAs/AlAs[1, 18] and Si/SiGe[14, 19] SL's in the acoustic range, the theory of light interaction by acoustic phonons and the interpretation of the light-scattering spectra have undergone several general improvements. Now, one can take into account the acoustic attenuation, the optical absorption, and the total thickness of the SL to account for the intensity and width of the Raman and Brillouin lines. The partial loss of translation symmetry caused by the acoustic attenuation at high acoustic frequency induces the loss of acoustic gaps at the boundaries of the Brillouin zones, and the optical absorption clearly modifies the relative intensities of the modes in these regions. We have also shown that the width of the acoustic modes is sensitive to the group velocity, with striking effects near the Brillouin-zone edges. Besides the interest of light scattering by folded acoustic modes for SL characterization (the structure, quality, optical acoustic, and acousto-optic properties can thus be investigated), the extension of the theory to ultrasonic waves allows the conception of a new generation of fast acousto-optical devices. One can achieve a two-dimensional matrix of acousto-optical microdevices by using such structures.

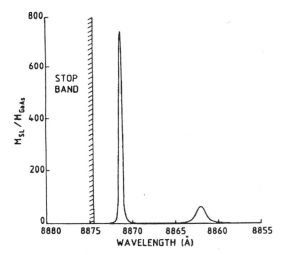

Fig.17. Effective figure of merit M_{SL} of the SL of Fig.2. The maximum corresponds to the optical resonance. The GaAs figure of merit is taken as reference.

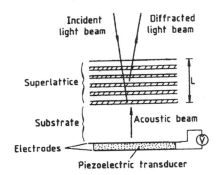

Fig.18. Schematic representation of an acousto-optical modulator based on an SL as the interaction medium. The SL is transparent to the incident photons.

REFERENCES

1. C. Colvard, R. Merlin, M. V. Klein, and A. C. Gossard, Phys. Rev. Lett. 45, 298 (1980).
2. J. Sapriel and B. Djafari-Rouhani, Surf. Sci. Rep. 10, 189 (1989).
3. A. S. Barker, Jr., J. L. Merz, and A. C. Gossard, Phys. Rev. B 17, 3181 (1978).
4. B. Djafari-Rouhani, J. Sapriel, and F. Bonnouvrier, Superlatt. Microstr. 1, 29 (1985).
5. C. Colvard, T. A. Grant, M. V. Klein, R. Merlin, R. Fisher, H. Morkoc, and A. C. Gossard, Phys. Rev. B 31, 2080 (1985).
6. J. Sapriel, B. Djafari-Rouhani, and L. Dobrzynski, Surf. Sci. 126, 197 (1983).
7. J. He and J. Sapriel, Appl. Phys. Lett. 55, 2292 (1989).
8. J. Sapriel, J. He, B. Djafari-Rouhani, R. Azoulay, and F. Mollot, Phys. Rev. B 37, 4099 (1988).
9. J. Sapriel, J. Chavignon, F. Alexandre, and R. Azoulay, Phys. Rev. B 34, 7118 (1986).
10. J. He, B. Djafari-Rouhani, and J. Sapriel, Phys. Rev. B 37, 4086 (1988).
11. B. Jusserand, D. Paquet, F. Mollot, F. Alexandre, and G. Le Roux, Phys. Rev. B 35, 2808 (1987).
12. J. He, J. Sapriel, and H. Brugger, Phys. Rev. B 39, 5919 (1989).
13. J. He, J. Sapriel, and R. Azoulay, Phys. Rev. B 40, 1121 (1989).
14. H. Brugger, H. Reiner, G. Abstreiter, H. Jorke, H. J. Herzog, and E. Kasper, Superlatt. Microstr. 2, 451 (1986).
15. B. Djafari-Rouhani and J. Sapriel, Phys. Rev. B 34, 7114 (1986).
16. J. Sapriel, "Acousto-optics," Wiley, New York, (1979).
17. J. He and J. Sapriel, French Patent 8907458, (1989).
18. J. Sapriel, J. C. Michel, J. C. Toledano, R. Vacher, J. Kervarec, and R. Regreny, Phys. Rev. B 28, 2007 (1983).
19. D. J. Lockwood, M. W. C. Dharma-Wardana, J. M. Baribeau and D. C. Houghton, Phys. Rev. B 35, 2243 (1987).

LOCALISED AND EXTENDED ACOUSTIC WAVES IN SUPERLATTICES LIGHT SCATTERING BY LONGITUDINAL PHONONS

B. Djafari Rouhani* and E.M. Khourdifi

Laboratoire de Physique du Solide
Faculté des Sciences et Techniques, Université de Haute Alsace
68093 Mulhouse Cédex, France

INTRODUCTION

Both the transfer matrix and Green's function methods have been used during the last few years to investigate the acoustic vibrations in superlattices.[1] The extended states which propagate in the whole superlattice form bulk bands which are separated by small gaps (superlattice effect): localised states associated with a perturbation of the perfect superlattice may exist inside these gaps. We have considered the case of defects which conserve the symmetry of translation parallel to the layers, namely a free surface, an interface with a substrate, and a planar defect.

The transfer matrix method which was first applied to study the shear horizontal[2] bulk and surface waves in a superlattice has been extended to sagittal vibrations[3] and more recently to the most general case which involves the three components of the displacement field, as for example in a superlattice composed of cubic materials for a general direction of propagation,[4] or in a piezoelectric superlattice.[5] Due to this coupling between the vibration components, the dispersion curves may show, at each value of $k_{//}$ (wave vector parallel to the layers), not only gaps at the centre and the edge of the reduced Brillouin zone but also inside this zone. The surface modes in these gaps are very dependent upon the composition and thickness of the surface layer which may be different from those in the bulk of the superlattice. In a (001) GaAs-AlAs type superlattice the polarisations of these waves also vary when $k_{//}$ goes from the [100] to [110] direction. Examples of these dispersion curves and the decaying behaviour of the surface modes will be presented.

---

* Present and permanent address : Laboratoire de Dynamique et Structure des Matériaux Molécularies, UFR de Physique, Université de Lille I, 59650 Villeneuve d'Ascq, France.

Light Scattering in Semiconductor Structures and Superlattices
Edited by D.J. Lockwood and J.F. Young, Plenum Press, New York, 1991

139

We also used the transfer matrix method to investigate the existence of localised shear horizontal waves at an interface between a superlattice and a substrate[6] or at a planar defect inside a superlattice.[7] Besides their simplicity, it is known that such waves cannot exist at the planar interface between two homogeneous materials. We find that localised modes can be found in the gaps of the superlattice and even below the bottom of the bulk bands, depending on the relative parameters of the superlattice and the substrate (or the planar defect) and on the nature of the film(s) in contact with the latter. We have also investigated the behaviour of Love waves, as a function of the superlattice thickness and of the wave vector $k_{//}$, when a finite superlattice is deposited on a substrate.

We also derived the dispersion relations of the shear horizontal vibrations in the aforementioned problems by calculating the Green's function of these materials, using the formalism of interface response theory.[8] This method was also applied very recently to the investigation of shear horizontal surface waves on a multi-component superlattice[9]. Besides, the knowledge of these Green's functions makes possible the calculation of local and total densities of states and may be helpful in the derivation of other vibrational properties[8] or in scattering problems (see below).

Most of the Raman experiments in superlattices are performed within a backscattering geometry where the folded longitudinal acoustic phonons are observed. To explain the behaviour of the relative intensities as functions of the scattering wavevector q, it is necessary to take into account the photoelastic[10] and elastic[11] mismatch between the constituents of the superlattice. Using a transfer matrix method we have shown[12] that, when the wave vector q becomes of the order of $\frac{2\pi}{D}$ (where D is the period of the superlattice), it is also necessary to use a Bloch-wave (rather than a plane-wave) description for the propagation of the electromagnetic fields. Besides, when the latter effect becomes negligible our calculation gives rise to a closed form expression for the scattered intensities.

In this paper, we discuss a Green's function approach to this problem using a theory developed for homogeneous materials.[13] This enables us to include another contribution, not considered before, which results from the mismatch between the dielectric constants of the constituents. This effect is associated with the translational displacements of the interfaces during vibration and is the analogue of the surface corrugation contribution in the usual Brillouin scattering.

Both the transfer matrix[1-7] and Green's function[8,9] methods have been extensively described in previous works and therefore we avoid their presentation in this paper. The next section will contain a discussion of the acoustic band structure and localised modes. The last section will be devoted to light scattering by longitudinal phonons.

ACOUSTIC WAVES IN SUPERLATTICES

The superlattice is composed of alternating layers of two (or more) constituents, each of them, α, being characterized by its elastic constants $C_{ijkl}^{(\alpha)}$, its mass density ρ_α and its thickness $d_\alpha = 2h_\alpha$ ($D = \sum_\alpha d_\alpha$ is the period). We call ω the frequency of the wave and $\vec{k}_{//}$ the wave vector parallel to the layers ; then, for given ω and $\vec{k}_{//}$, those wave vectors along the axis x_3 of the superlattice which can be deduced from the bulk dispersion relations are called k_3.

For a general symmetry of the layers in the superlattice and a general orientation of $\vec{k}_{//}$ the three components of the displacement field are coupled together. However, for some particular cases the modes polarised in the sagittal plane (plane containing $\vec{k}_{//}$ and x_3) are decoupled from the shear horizontal waves ; for example, this happens when the films in the superlattice are isotropic[2,3] or [3,4] (0001)-oriented hexagonal slabs, or (001)-oriented cubic slabs with $\vec{k}_{//}$ along [100] or [110] etc... The illustration presented in the following will refer either to a (001) GaAs-AlAs or to a (0001) Y-Dy superlattice.

Bulk Waves

The simplest situation is provided by shear horizontal (or pure longitudinal)waves, which only involve one component of the displacement field. Only one pair of k_3 is associated with given values of $\vec{k}_{//}$ and ω, which can be written[4] as $K + iL$ and $-(K + iL)$. Depending on whether $L = 0$ or $L \neq 0$, the acoustic wave at the frequency ω can propagate in the superlattice, i.e., ω belongs to a bulk band, or is attenuated and ω is inside a gap. These gaps can only appear at the centre and the edges of the Brillouin zone, i.e., for $K = 0$ or $\pm\pi/D$. Indeed, each frequency ω is associated with only one value of $|K|$ and thus each phonon branch necessarily shows a monotonic behaviour with extremes at $|K| = 0$ and π/D.

In the case of sagittal modes there are two pairs of k_3 associated with given $k_{//}$ and ω, which can be written as[4] $\pm(K_1 + iL_1)$ and $\pm(K_2 + iL_2)$. Now, an elastic wave at the frequency ω propagates in the superlattice if $L_1 = 0$ or $L_2 = 0$, while it is attenuated if both L_1 and L_2 are different from zero. Each pair of k_3 (the first for instance) can take four different forms ; it can be

(i)   pure real ($L_1 = 0$),
(ii) pure imaginary ($K_1 = 0$),
(iii) complex but with $K_1 = \pm \pi/D$,   (1a)
(iv) complex with $K_1 \neq \pm \pi/D$.

However in case (iv) the two pairs of k_3 necessarily become

$K+iL$, $-(K+iL)$, $K-iL$, $-(K-iL)$.   (1b)

In the most general case[4] involving the three components of the displacement field, there are three pairs of k_3 associated with given $\vec{k}_{//}$ and ω which take the form $\pm(K_1 + iL_1)$, $\pm(K_2 + iL_2)$, $\pm(K_3 + iL_3)$, and the gaps correspond to frequency domains in which L_1, L_2 and L_3 are all different from zero. Each pair of k_3 can take one of the four forms stated in eq. (1a) ; however, when one pair of k_3 is of the form (iv) and is written as $\pm (K+iL)$, a second pair necessarily becomes $\pm(K-iL)$ while the third pair takes one of the forms (i), (ii), or (iii). Figure 1 gives two examples of the complex band structure in a GaAl-AlAs superlattice showing the combinations of the above-mentioned cases. One can see the presence of direct gaps at the centre and the edge of the reduced Brillouin zone, but also the possibility of indirect gaps inside this zone. This is a consequence of coupling between the components of the displacement field, i.e., the mixing between waves polarised in each constituent as a result of reflection and transmission phenomena at the interfaces. The occurrence of these indirect gaps is independent of the symmetry of the crystals in the superlattice and the orientation of the interfaces and may happen even when the layers are isotropic. Let us also mention that the imaginary parts of the k_3 wave vectors in fig. 1 give the attenuation of the localised waves in the gaps.

In figure 2 we give another representation of the band structure, namely ω versus $k_{//}$, in which small gaps appear in between the bulk bands represented by the shaded areas.

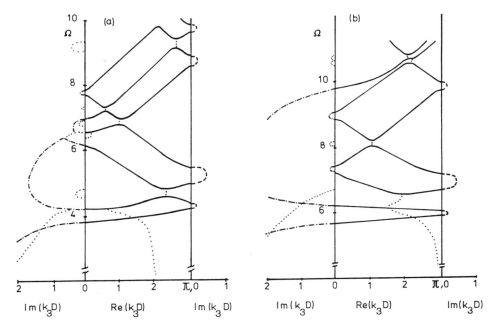

Fig. 1. Complex band structure (ω versus complex k_3) in a GaAs-AlAs (001) superlattice with $d_1 = d_2 = \dfrac{D}{2}$ and (a) $k_{//} D = 4$, (b) $k_{//} D = 6$. $\Omega = \omega D/c_t$ (GaAs) is a dimensionless frequency where c_t (GaAs) is a transverse velocity of sound in GaAs given by $\sqrt{C_{44}/\rho}$. Solid curves : k_3 real (middle panel of the figure). Dash-dotted curves : k_3 imaginary (left panel). Dashed curves : imaginary part of k_3 when its real part is equal to $\dfrac{\pi}{D}$ (right panel). In addition, when k_3 is a complex quantity (eq. (1b)), the dotted curves give both its real and imaginary parts ; in this case the imaginary part is presented in the left panel.

Surface Waves

Localised acoustic waves, associated with the free surface of a semi-infinite superlattice, can be found either below the bottom of the bulk bands or in the gaps of the superlattice.

In the case of pure shear horizontal waves, the surface mode below the bottom of the bulk bands is similar to the well known Love waves which may exist when an adlayer is deposited on a substrate. In the limit $k_{//} D \ll 1$ the existence of this mode requires that the layer at the surface be the "softer" constituent of the superlattice ;[1,2] then this wave has the same slope as the botom of the bulk bands and it can only be distinguished from the latter at the order $k_{//}^2$.

In the case of sagittal waves the mode below the bottom of the bulk bands is the Rayleigh wave, which always exists in the limit $k_{//} D \ll 1$ but may show different behaviours at higher values of $k_{//} D$.[1,3]

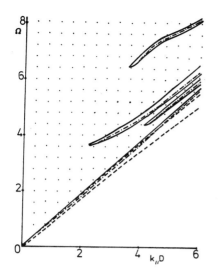

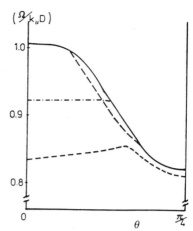

Fig. 2. Bulk bands and surface waves as a function of $k_{//} D$ in a GaAs-AlAs (001) superlattice, when the vector $\vec{k}_{//}$ makes an angle $\theta = 20°$ with the [100] direction. The shaded areas, limited by the solid curves, are the bulk bands. The surface modes are presented either for a GaAs layer (dashed curves) or an AlAs layer (dashed-dotted curves) at the surface. $\Omega = \omega D/c_t$ (GaAs) is a dimensionless frequency.

Fig. 3. Velocities at the bottom of the bulk bands (solid curve) and of the surface waves below them (dashed curves: GaAs at surface; dashed-dotted curve: AlAs at surface), as a function of the angle θ between $\vec{k}_{//}$ and the [100] direction. The magnitude of $k_{//} D$ is taken to be $k_{//} D = 4$. The velocities are normalized to $c_t(\text{GaAs}) = \sqrt{C_{44}/\rho}$.

In the most general case the three components of the vibrations are coupled together and the polarisation of the surface modes is dependent on the orientation of the wave vector $\vec{k}_{//}$. [4] An example of the surface dispersion curves is presented in fig. 2 for a GaAs-AlAs (001) superlattice with $d_1 = d_2 = \dfrac{D}{2}$ and the wave vector $\vec{k}_{//}$ making an angle $\theta = 20°$ with the [100] direction. One can see that the surface waves can be very different depending on the nature of the film which is at the surface (GaAs or AlAs). It is also worth pointing out[1-4] the sensitivity of surface modes in the gaps to the thickness of the surface layer which can be assumed to be different from that of the corresponding films in the bulk.

In a GaAs-AlAs (001) superlattice the polarisation of the bottom of the bulk bands and of the surface waves below it goes from sagittal to shear horizontal when the angle θ between $\vec{k}_{//}$ and the [100] direction varies from 0° to 45°. The velocities of these waves as a function of θ are presented in fig. 3 in the case $d_1 = d_2 = \dfrac{D}{2}$ and for

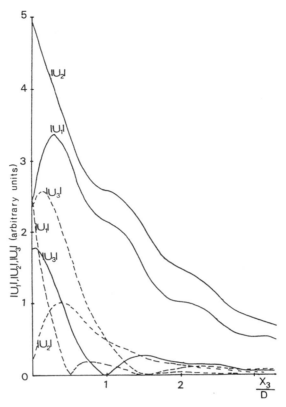

Fig. 4. Modulus of the displacements ($|u_1|$, $|u_2|$, $|u_3|$) versus depth for the two surface modes occurring in fig. 3 for $\theta = 30°$ with a GaAs layer at the surface. The solid (respectively dashed) curves correspond to the mode at higher (resp. lower) frequency. The displacements associated with the two surface modes are normalized so that the $|u_1|$'s take the same value at the surface.

$|k_{//}\,D| = 4$: at intermediary values of θ, there are two surface waves when the superlattice terminates with a GaAs layer ; these modes interact and repel each other when θ is around 30°.

The decay of the surface waves far from the surface is characterized by three attenuation coefficients given by the imaginary parts of the corresponding complex k_3 wave vectors. When the frequency of the surface mode is near a bulk band (particularly inside a narrow gap), at least one of the attenuation coefficients is rather small; conversely, these coefficients become large at a frequency far from the bulk bands. Thus, depending on its frequency, an acoustic surface wave can penetrate over several periods of the superlattice or be very localised in the few layers near the

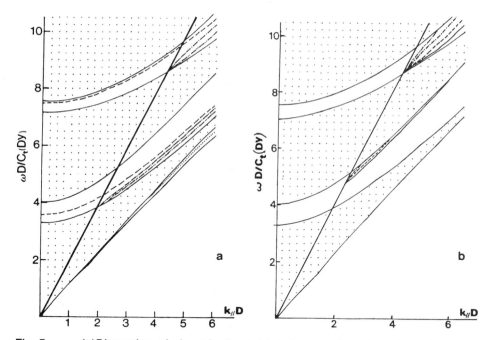

Fig. 5.   (a)Dispersion of shear horizontal interface modes when the substrate is in contact with a Dy film in the superlattice, for different sets of substrate parameters. The velocity of sound in the substrate is taken to be constant($C_t^{(s)}$ = 2C_t(Dy)) and indicated by the heavy line. The dispersion curves are given for the ratio $\gamma = C_{44}^{(s)}/C_{44}$(Dy) equal to ∞ (---), 7(– . – . – .), 3(– .. – .. –), 0.3(.......) and O(———— corresponding to the free surface modes). The branches corresponding to $\gamma = \infty$ are continued inside the bulk bands of the substrate because they are independent of the substrate velocity of sound. The shaded areas are the bulk bands of the superlattice. Dimensionless quantities are reported on both axes.
(b) Same as (a), but the substrate is in contact with a Y film. The dispersion curves are given for γ = 0(————), 2(---), 2.4(– . – . – .) and 5(.....).

surface. On the other hand, the attenuation of the wave may be monotonic or oscillatory depending on whether the real parts of the k_3 are equal to or different from zero.

Figure 4 shows the variations of the displacement field versus the depth x_3 for the two surface modes below the bulk bands of fig. 3, when a GaAs layer is at the surface of the superlattice.

Finally let us mention the investigation of surface acoustic waves on piezoelectric superlattices[5] (in which a dispersive Bleustein-Gulyaev wave can be expected), as well as the study of Rayleigh waves on a superlattice cut perpendicular to the laminations.[14]

Interface between a Superlattice and a Substrate

We have investigated this geometry only for shear horizontal vibrations. Despite their simplicity, the study of these waves leads to the finding of new types of modes; indeed, it is known that the planar interface between two homogeneous elastic media cannot support localised shear horizontal vibrations. We show that such waves exist at the interface between a superlattice and a substrate, with frequencies inside the gaps, or even below the bottom of the bulk bands.

The dispersion relations of acoustic waves when a semi-infinite or a finite superlattice is deposited on a substrate were first derived in ref. 6. An illustration of the shear waves localised at the boundary between a substrate and a semi-infinite Y-Dy (0001) superlattice is given in fig. 5, where the substrate is in contact either with a Y or a Dy film. In fig. 5 the localised modes are represented for several sets of the substrate parameters ($C_{44}^{(s)}$, $\rho^{(s)}$) in such a way that the substrate velocity of sound $c_t^{(s)} = \sqrt{C_{44}^{(s)}/\rho^{(s)}}$ remains constant. Though this representation was chosen for the sake of clarity, it should be pointed out that the dispersion curves are rather sensitive to the value of $C_{44}^{(s)}$ while they are not much affected by the variation of $\rho^{(s)}$ alone. This is due to the dependence of the normal stress boundary condition upon the ratio $\gamma = C_{44}^{(s)}/C_{44}$. In the limit $\gamma \to 0$ one obtains the surface modes of a semi-infinite superlattice in contact with the vacuum. In the other extreme limit, $\gamma \to \infty$, the amplitude of the vibrations goes to zero at the interface and remains vanishingly small in the substrate.

When the substrate is in contact with a Dy film, the frequencies of the interface waves decrease as γ decreases from ∞ (fig. 5a), until the corresponding branches merge into the bulk bands of the superlattice. For lower values of γ, an interface branch is even extracted below the bottom of the bulk bands, merging into these bands at a finite value of $k_{//} D$. When the substrate is in contact with a Y film in the superlattice, only interface waves inside the gaps between the bulk bands may exist: they originate in surface waves for relatively small values of $C_{44}^{(s)}$ and disappear into the bands when $C_{44}^{(s)}$ increases. It is also worth mentioning that the interface modes do not exist if the substrate is made of one of the materials contained in the superlattice.

Besides the transfer matrix method, we obtained the dispersion curves of interface states by calculating the Green's function of the substrate-superlattice system. This also enables us to obtain the local densities of states $n(\omega^2 ; k_{//} ; x_3)$ and the variation $\Delta n(\omega^2 , k_{//})$ of the total density of states with respect to that of the bulk materials (the substrate on one side and the superlattice on the other side). For example, in fig. 5b the localised mode in the first gap becomes an interface resonant state after cutting the sound line of the substrate ; however, this resonance rapidly broadens and becomes spread over the whole gap of the superlattice. Let us also point out that $\Delta n(\omega^2; k_{//})$ shows delta-peak anti-resonances at the band limits of the superlattice ; this means that the phase shift associated with the substrate-superlattice interface perturbation undergoes jumps at these frequencies.

In the case of a finite superlattice deposited on a substrate one can search for Love waves which are localised inside the thin superlattice and decay exponentially into the substrate. When the number N of periods in the superlattice is small, the dispersions and velocities of Love waves are dependent upon N as well as upon the nature of the two layers in the superlattice, which are, respectively, at the interface and at the free

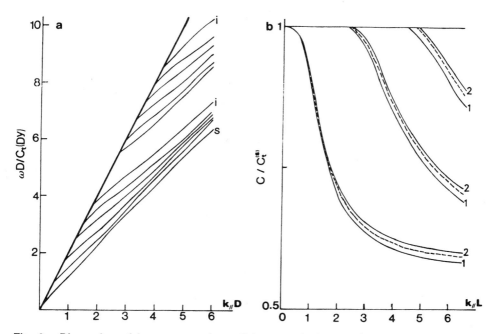

Fig. 6. Dispersion of Love waves for a finite superlattice with Dy layers at both its
ends and N = 5, deposited on a substrate with $C_t^{(s)}$ = 2C_t (Dy) and

$C_{44}^{(s)}$ = 5C_{44} (Dy). The labels i and s refer to the interface and free surface
modes, respectively, discussed in fig. 5.
(b) Velocities of Love waves as a function of $k_{//}$ L, where L is the thickness of
the superlattice. The curves 1 (respectively 2) correspond to a superlattice
with N = 5 and Dy (respectively Y) layers at its ends. The dashed curves are
the result of the effective medium approximation.

surface. In fig. 6a we have represented the dispersion curves assuming that Dy layers
are both at the interface and at the free surface and that the superlattice contains N = 5
layers of Y and N + 1 = 6 layers of Dy. The interface and free surface waves discussed
above can be distinguished in fig. 6a even for such a small number of periods N. The
other branches in this figure correspond to propagating waves in the superlattice;
their number increases with N, leading to the bulk bands of the infinite superlattice in
the limit N → ∞. The case of a superlattice with Y at the free surface and Dy at the
interface provides the possibility of having one surface and one interface branch in the
same gap ; these modes may interact, especially when the number of periods N is
small.

In fig. 6b we have plotted the velocities of the first few Love waves as a function of
$k_{//}$ L (where L is the total thickness of the superlattice) for N = 5 and for two
different choices of the outermost layers. We have also presented the results of the
effective medium approximation (EMA) which are obtained when the superlattice
behaves like an effective homogeneous medium, i.e., when the wavelength is large
compared to the period D. As N increases, the exact results become closer to those of
EMA: therefore, for a given thickness L of the superlattice, one obtains different
Love-wave velocities by varying N and D simultaneously.

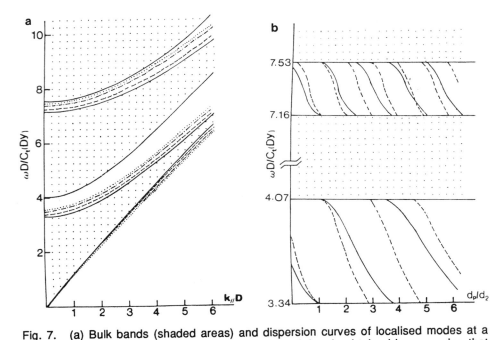

Fig. 7.   (a) Bulk bands (shaded areas) and dispersion curves of localised modes at a planar defect in a Y-Dy superlattice. The defect is obtained by assuming that one layer of Y has a thickness d_p different from those of the normal layers. The localised branches are given for different values of $d_p/d(Y)$: 0.1(.....) ; 0.2($- . - . -$.) ; 0.5(-----).

(b) Full curves : variation of the frequencies of the localised waves presented in (a), at $k_{//} D = 0.5$, as functions of the thickness d_p ; the curves are presented in the first two gaps of the superlattice. Dashed curves : same as for full curves but the planar defect is now obtained by assuming that one layer of Dy has a thickness d_p different from those of the normal layers.

Localised Shear Horizontal Waves Associated with a Planar Defect

In our model the planar defect is obtained by assuming that one layer in the superlattice has different elastic coefficients and/or thickness than the normal layers. From the dispersion relations derived in ref. 7 we have shown that localised waves can exist at a planar defect in a superlattice, even at frequencies below the bulk bands. The occurrence of these modes influences the transmission coefficient, which can be measured, of acoustic waves through the whole structure. Their investigation provides a tool for the determination of acoustic properties of slabs in confined geometries.

Figure 7 (a) gives an illustration of the dispersion of the localised waves associated with a planar defect in a Dy-Y (0001) superlattice with $d_1 = d_2$, by simply assuming that one layer of Y has a thickness d_p different from those (d_2) of the normal Y layers. In this figure the dispersion curves are given for different values of the thickness d_p ranging from 0 to d_2. As a function of d_p, these curves are arranged in increasing order inside the gaps, whereas they are in decreasing order below the bulk bands. At $d_2 = d_p$ one recovers the perfect superlattice and all the branches associated with the planar defect have merged into the bulk bands.

In order to show the variations of the localised modes when $d_p > d_2$, we have plotted, in figure 7 (b), the frequencies of these modes as functions of d_p, at $k_{//} D = 0.5$. One can observe that with increasing d_p these frequencies decrease and the modes merge into the bulk bands, whereas new modes are extracted from higher bands.

Let us also note[7] that at each frequency ω the localised mode is periodically reproduced as a function of d_p. Figure 8 (b) also contains the results corresponding to a Dy-Y superlattice in which one Dy layer has a thickness different from those of the normal layers.

The dispersion curves are also very dependent upon the elastic parameters of the planar defect.[7]

Finally, when a localised mode exists (see fig. 7 (a)) in the long-wavelength limit ($k_{//} D \ll 1$), its velocity is equal to that of the bottom of the bulk bands,[7] the two curves differing only in their curvature. This behaviour is similar to that of the usual first Love mode.

LIGHT SCATTERING BY LONGITUDINAL ACOUSTIC PHONONS

The propagation of an acoustic wave in the superlattice excites periodic variations of strain which in turn induce a modulation of the dielectric tensor ε_{ij} from the photoelastic coupling to elastic fluctuations,

$$\delta\varepsilon_{ij}^{(1)} = -\varepsilon_{ii}\,\varepsilon_{jj} \sum_{k\ell} P_{ijk\ell} \frac{1}{2}\left(\frac{\partial u_k}{\partial x_\ell} + \frac{\partial u_\ell}{\partial x_k}\right) \tag{2}$$

$P_{ijk\ell}$ are the elements of the photoelastic tensor and can be considered as functions of x_3. The coupling of the incident light to the phonons gives rise to a polarisation in the superlattice which creates the scattered field. In this paper we are interested in the propagation of pure longitudinal phonons along the axis x_3 of a superlattice composed of cubic materials with (001) interfaces. In this case we can assume that all the electromagnetic fields (incident, scattered and polarisation waves) are polarised parallel to the x_1 axis and propagate along x_3. Then each medium α ($\alpha = 1,2$) in the superlattice can be characterized by one elastic constant C_α (which means C_{11}), the mass density ρ_α, the dielectric constant $\varepsilon_\alpha = n_\alpha^2$ (where n_α is the index of refraction in medium α) and one photoelastic parameter $p_\alpha = -\varepsilon_\alpha^2\,P_{1133}^{(\alpha)}$.

In each slab $m = (n, \alpha)$, where n is a cell index and α the type of medium, let us define a local coordinate $x_3^{(m)}$ which ranges from $-h_m$ to h_m in the perfect superlattice (the origin of $x_3^{(m)}$ is in the middle of slab m). In the presence of an acoustic wave the boundary between the slabs m and m+1 is subject to a displacement given by $u_3^{(m)}$ ($x_3^{(m)}=+h_m$, t) (or equivalently $u_3^{(m+1)}$ ($x_3^{(m+1)}=-h_{m+1}$, t)). Therefore the dielectric function of the superlattice becomes

$$\varepsilon(x_3, t) = \sum_{(m=n,\alpha)} [\varepsilon_m + \delta\varepsilon_m^{(1)}(x_3^{(m)}, t)]\,[\theta(x_3^{(m)}+h_m) - \theta(x_3^{(m)}-h_m)]$$
$$+\,\delta\varepsilon^{(2)}\,(x_3, t) \tag{3}$$

where $\theta(x_3)$ is the unit step Heaviside function. In eq. (3) the first term represents the dielectric function of the semi-infinite superlattice in contact with a vacuum in the absence of any fluctuations, the second term $\delta\varepsilon^{(1)}$ gives the photoelastic modulation of the dielectric function (see eq. (2)),

$$\delta\varepsilon_m^{(1)} (x_3^{(m)}, t) = p_m \frac{\partial u_3(x_3, t)}{\partial x_3} \tag{4}$$

and the last term $\delta\varepsilon^{(2)}$ is another modulation which arises from the interface (or free surface) translational displacements and is given by

$$\delta\varepsilon^{(2)} (x_3, t) = \sum_m \varepsilon_m (\{\theta[x_3^{(m)} + h_m - u_3^{(m)}(-h_m, t)] - \theta (x_3^{(m)} + h_m)\}$$

$$- \{\theta[x_3^{(m)} - h_m - u_3^{(m)} (h_m, t)] - \theta (x_3^{(m)} - h_m)\}) \tag{5a}$$

or, equivalently,

$$\delta\varepsilon^{(2)} (x_3, t) = \sum_m (\varepsilon_m - \varepsilon_{m+1}) \{\theta(x_3^{(m)} - h_m) - \theta[x_3^{(m)} - h_m - u_3^{(m)}(h_m, t)]\} \tag{5b}$$

The summation in eq. (5), which involves the difference between the dielectric constants of two media in contact at an interface, includes the interface of the superlattice with the vacuum.

The calculation of the emitted electric field $E(x_3, t)$ when the superlattice is submitted to an incident electromagnetic field can be done following the Green's function method presented in ref. 13 for the Brillouin scattering from an homogeneous crystal,

$$E(x_3, t)) = E^{(0)} (x_3, t) + \frac{1}{4\pi c^2} \int dx'_3 \int dt'\ G(x_3, x'_3 ; t\text{-}t')\ X$$

$$\frac{\partial^2}{\partial t'^2} \int dt'' \int \frac{d\Omega}{2\pi} e^{-i\Omega (t'\text{-}t'')} [\delta\varepsilon^{(1)}(x'_3, t') + \delta\varepsilon^{(2)}(x'_3, t')]\ E(x'_3, t'') \tag{6}$$

Here c is the speed of light in the vacuum, $E^{(0)} (x_3, t)$ is the electric field of the wave reflected from the surface in the absence of fluctuations, and G is the Green's function associated with the propagation of an electromagnetic field along x_3 in the vacuum-superlattice system in the absence of acoustic deformation. In the first Born approximation the electric field which appears in the integrand of the right-hand side in eq. (6) can be replaced by the incident field $E^{(i)} (x_3, t)$. After some manipulations, the scattered field $E^{(s)} (x_3, t)$, which is the difference $E\text{-}E^{(0)}$, becomes

$$E^{(s)} (x_3, t) = E^{(1)} (x_3, t) + E^{(2)} (x_3, t) \tag{7}$$

with

$$E^{(1)} (x_3, t) \equiv - \frac{\omega_i^2}{4\pi c^2} \sum_m p_m \int dx_3^{(m)} \int dt'\ G(x_3, x_3^{(m)} ; t\text{-}t') \frac{\partial u_3(x'_3^{(m)}, t')}{\partial x'_3^{(m)}} X$$

$$E^{(i)} (x'_3^{(m)}, t') \tag{8a}$$

150

$$E^{(2)}(x_3, t) \cong - \frac{\omega_i^2}{4\pi c^2} \sum_m (\varepsilon_m - \varepsilon_{m+1}) \int dt' \, [G(x_3, x_3'^{(m)} ; t-t') \, u_3(x_3'^{(m)}, t') \times$$

$$E^{(i)}(x_3'^{(m)}, t')]_{x_3'^{(m)} = h_m} \tag{8b}$$

ω_i is the frequency of the incident light, the fields $E^{(1)}$ and $E^{(2)}$ are, respectively, associated with the modulations $\delta\varepsilon^{(1)}$ and $\delta\varepsilon^{(2)}$, and the summation over m in eq. (8b) includes the vacuum-superlattice interface. The calculation of the integrals in eqs. (8) requires the knowledge of the Green's function $G(x_3, x_3'^{(m)}; t-t')$, of the incident electric field $E^{(i)}(x_3, t)$ and of the displacement field $u_3(x_3, t)$ associated with a phonon (thermal fluctuation) or with the propagation of an acoustic wave.

The latter quantity can be obtained by using, for example, the transfer matrix method presented in refs. 1-4 or 12. In the layer m = (n, α) of the superlattice one can write

$$u_3(x_3^{(m)}, t) = U_+ [a_{\alpha+} e^{ik_{p\alpha}x_3^{(m)}} + b_{\alpha+} e^{-ik_{p\alpha}x_3^{(m)}}] e^{i(k_p nD - \omega_p t)}$$

$$+ U_- [a_{\alpha-} e^{ik_{p\alpha}x_3^{(m)}} + b_{\alpha-} e^{-ik_{p\alpha}x_3^{(m)}}] e^{-i(k_p nD + \omega_p t)} \tag{9}$$

where the subscript p means phonon ; ω_p is the frequency of the acoustic wave ; $k_{p\alpha} = \frac{\omega_p}{v_\alpha}$ where $v_\alpha = \sqrt{C_\alpha/\rho_\alpha}$ is the acoustic velocity of sound in the medium α ; and k_p is the Bloch wave vector in the superlattice associated with the frequency ω_p. In eq. (9) we have included the propagation in both positive and negative x_3 directions in order to take into account the reflection of a wave at the free surface of the superlattice. All the coefficients $a_{\alpha\pm}$ and $b_{\alpha\pm}$ can be determined as functions of ω_p by using the transfer matrix method.[1-4,7] The multiplicative factors U_+ and U_- can then be obtained by writing the two following relations :

(i) the boundary condition at the free surface of the superlattice ; for example when the layer in contact with vacuum is of type $\alpha = 1$ inside the unit cell n = 0, and the superlattice fills the x_3 positive half-space we obtain

$$\sum_{v=\pm} U_v [a_{1v} e^{-ik_{p1}h_1} - b_{1v} e^{ik_{p1}h_1}] = 0 \tag{10}$$

(ii) for a phonon of frequency ω_p, a normalisation condition involving the Bose-Einstein distribution function $n(\omega_p)$,

$$\int dx_3 \, \rho(x_3) \, \omega_p^2 \, |u_3(x_3, t)|^2 \sim n(\omega_p)+1 \quad (\text{or } n(\omega_p)) \tag{11a}$$

for the anti-Stokes (or Stokes) component. In the high-temperature limit this relation becomes

$$\sum_{\nu=\pm 1} \sum_{\alpha=1,2} \rho_\alpha \, |U_\nu|^2 \, \{[\,|a_{\alpha\nu}|^2 + |b_{\alpha\nu}|^2]d_\alpha +$$

$$\frac{1}{k_{p\alpha}} (a_{\alpha\nu} \, b_{\alpha\nu}^* + a_{\alpha\nu}^* \, b_{\alpha\nu}) \, \sin(k_{p\alpha} \, d_\alpha)\} \sim \frac{k_B T}{\omega_p^2} \qquad (11b)$$

Let us point out that in the case of a surface mode of the superlattice, the wave vector k_p is complex ; therefore in eq. (9), and in the following eqs. (10)-(11), we have to keep only one out of the two terms, i.e., the one which corresponds to the decay of the wave function far from the surface.

The expression of the electric field $E^{(i)}$ (x_3, t) in the superlattice can be obtained in a way similar to the above calculation.[12] In the medium α belonging to the unit cell n of the superlattice,

$$E^{(i)} (x_3^{(m)}, t) = \mathscr{E}_1[A_\alpha \, e^{ik_{i\alpha}x_3^{(m)}} + B_\alpha \, e^{-ik_{i\alpha}x^{(m)}}] \, e^{i(k_i n D - \omega_i t)} \qquad (12)$$

where the subscript i means incident ; k_i is the Bloch wave vector associated with the frequency ω_i and can be obtained from the dispersion relation of the electromagnetic waves in the superlattice;[12] $k_{i\alpha} = \dfrac{\omega_i n_\alpha}{c}$; the coefficients A_α, B_α and the multiplicative factor $\mathscr{E}_1$ can be obtained by using the transfer matrix method and then the vacuum-superlattice boundary condition which introduces the transmission coefficient at this boundary and relates $\mathscr{E}_1$ to the incident field $\mathscr{E}_{oi}$ in the vacuum.[12]

Finally, the Green's function $G(x_3, x'_3 ; t\text{-}t')$ can be Fourier-analysed with respect to the time

$$G(x_3, x'_3 ; t\text{-}t') = \int \frac{d\omega}{2\pi} \, G(x_3, x'_3 ; \omega) \, e^{-i\omega (t-t')} \qquad (13)$$

and the expressions of the Green's function elements $G(x_3, x'_3; \omega)$ have been derived in closed form in ref. 15 by using the formalism of interface response theory. Let us point out that in this method the general elements of G are easily obtained once we know its elements in the interface space.

Now the scattered field (eqs. (7)-(8)) can be evaluated by using eqs. (9), (12) and (13) in eqs. (8). The integration over t' in these last equations gives rise to a delta function which determines the frequency of the scattered field as $\omega_s = \omega_i + \omega_p$ (anti-Stokes polarisation). In the following we shall call $k_{s\alpha} = \dfrac{\omega_s n_\alpha}{c}$, and k_s the positive Bloch wave vector in the superlattice associated with ω_s. In practice, ω_p is much smaller than ω_i and ω_s and we have $\omega_s \cong \omega_i$, $k_s \cong k_i$ and $k_{s\alpha} \cong k_{i\alpha}$.

The integration over x'_3 in eq. (8a) can also be easily performed ; then the expression of the component $E^{(1)}$ (x_3, t) of the scattered field in the vacuum side ($x_3 < 0$) becomes :

$$E^{(1)}(x_3, t) = -\frac{\omega_i^2 \, \mathcal{E}_i}{4\pi c^2} e^{-i\omega_s(\frac{x_3}{c}+t)} \left(\left(\sum_{n\geq 0} e^{i(k_i+k_s+k_p)nD} \right) E_+^{(1)} \right.$$

$$\left. + \left(\sum_{n\geq 0} e^{i(k_i+k_s-k_p)nD} \right) E_-^{(1)} \right) \tag{14}$$

where the summations are made over the cells of the superlattice.

In order to express the two factors $E_\nu^{(1)}$ (where $\nu = \pm$) which appear in eq. (14) it is helpful to introduce a few definitions; in particular we need the elements of the Green's function G relating the vacuum-superlattice interface to itself and to the next two interfaces (these expressions can be found in ref. (15)); the three states involved are: $|1->\ \equiv |n=0,\ \alpha=1,\ x_3=-h_1 >$; $|1+>\equiv |n=0,\ \alpha=1,\ x_3=h_1 >$ which is identical to $|2->\ \equiv |n=0,\ \alpha=2,\ x_3=-h_2 >$; and finally $|2+>\equiv |n=0,\ \alpha=2,\ x_3=h_2 >$. We shall call

$$\mathcal{G}_1^{(-)} = <1-|G|1->\ ,\ \mathcal{G}_1^{(+)} \equiv \mathcal{G}_2^{(-)} = <1-|G|1+>,\ \mathcal{G}_2^{(+)} = <1-|G|2+> \tag{15a}$$

Then in each medium of the superlattice ($\alpha = 1,2$) we define the following combinations of these quantities

$$\mathcal{D}_\alpha = \mathcal{G}_\alpha^{(-)} e^{ik_{s\alpha}h_\alpha} - \mathcal{G}_\alpha^{(+)} e^{-ik_{s\alpha}h_\alpha} \tag{15b}$$

$$\mathcal{D'}_\alpha = \mathcal{G}_\alpha^{(+)} e^{ik_{s\alpha}h_\alpha} - \mathcal{G}_\alpha^{(-)} e^{-ik_{s\alpha}h_\alpha} \tag{15c}$$

We also define in each medium $\alpha = 1$ or 2 the following combinations of the wave vectors,

$$K_\alpha^{\lambda\mu} = k_{p\alpha} + \lambda k_{i\alpha} + \mu k_{s\alpha} \qquad (\lambda,\mu = 1\ or\ -1) \tag{16a}$$

and then

$$S_\alpha^{\lambda\mu} = \frac{\sin(K_\alpha^{\lambda\mu} h_\alpha)}{K_\alpha^{\lambda\mu}} \tag{16b}$$

Now the quantities $E_\nu^{(1)}$ in eq. (14) can be written

$$E_\nu^{(1)} = U_\nu \sum_{\alpha=1,2} \frac{p_\alpha k_{p\alpha}}{\sin(k_{s\alpha}h_\alpha)} \{\mathcal{D}_\alpha[a_{\alpha\nu}(A_\alpha S_\alpha^{+-}+ B_\alpha S_\alpha^{--})-b_{\alpha\nu}(A_\alpha S_\alpha^{-+}+ B_\alpha S_\alpha^{++})]$$

$$+ \mathcal{D'}_\alpha [a_{\alpha\nu}(A_\alpha S_\alpha^{++}+ B_\alpha S_\alpha^{-+})-b_{\alpha\nu}(A_\alpha S_\alpha^{--}+ B_\alpha S_\alpha^{+-})]\} \tag{17}$$

where the coefficients U_ν, $a_{\alpha\nu}$, $b_{\alpha\nu}$ (eq. (9)) and A_α, B_α (eq. (12))respectively define the displacement field and the incident electric field in the superlattice.

153

The second contribution $E^{(2)}$ (x_3, t) to the scattered field (eqs. (7)-(8)) takes in the vacuum region ($x_3 < 0$) a form similar to eq. (14) except for a complementary term,

$$E^{(2)} (x_3, t) = - \frac{\omega_i^2 \mathscr{E}_i}{4\pi c^2} e^{-i\omega s(\frac{x_3}{c} + t)} ((\sum_{n \geq 0} e^{i(k_i + k_s + k_p)nD}) E_+^{(2)} +$$

$$+ (\sum_{n \geq 0} e^{i(k_i + k_s - k_p)nD}) E_-^{(2)} + \Delta E^{(2)}) \qquad (18)$$

Here

$$E_v^{(2)} = U_v \sum_{\alpha = 1, 2} \varepsilon_\alpha \{ -\mathscr{G}_\alpha^{(-)} (a_{\alpha v} e^{-ik_p h_\alpha} + b_{\alpha v} e^{ik_p h_\alpha}) (A_\alpha e^{-ik_{i\alpha} h_\alpha} + B_\alpha e^{ik_{i\alpha} h_\alpha})$$

$$+ \mathscr{G}_\alpha^{(+)} (a_{\alpha v} e^{ik_p h_\alpha} + b_{\alpha v} e^{-ik_p h_\alpha}) (A_\alpha e^{ik_{i\alpha} h_\alpha} + B_\alpha e^{-ik_{i\alpha} h_\alpha}) \qquad (19)$$

and

$$\Delta E^{(2)} = \varepsilon_0 \mathscr{G}_1^{(-)} \sum_{v \geq 0} U_v (a_{1v} e^{-ik_p 1 h_1} + b_{1v} e^{ik_p 1 h_1}) (A_1 e^{-ik_{i1} h_1} + B_1 e^{ik_{i1} h_1})$$

$$\qquad (20)$$

where ε_0 is the permittivity of vacuum and the term $\Delta E^{(2)}$ is a contribution coming from the free surface displacement.

Although rather cumbersome, the equations (7), (14) and (18) give closed-form expressions of the scattered field in the vacuum. Depending on the absorption coefficients of light in the constituents of the superlattice, the summations over n in eqs. (14) and (18) may extend far into the superlattice or be limited to the near vicinity of the surface. In the following we present a brief discussion and illustration of the Raman intensities in the backscattering geometry by neglecting the light absorption and by limiting ourselves to the case of bulk acoustic waves or phonons in the superlattice.

Let us recall that in the limit of $N \rightarrow \infty$ the summation $\sum_{n=0}^{N} e^{inx}$ leads to the function $2\pi N \delta(x)$. Therefore, by assuming that in eqs. (14) and (18) the wave vectors k_i, k_s, k_p are chosen to be positive, the major contributions to $E^{(1)}(x_3, t)$ and $E^{(2)}(x_3, t)$ will come from the terms $E_-^{(1)}$ and $E_-^{(2)}$, respectively associated with the wave-vector selection rule

$$k_i + k_s - k_p = \frac{2\pi m}{D} \qquad (21)$$

The intensities of the Raman peaks associated with the Brillouin and folded modes will then be proportional to $|E_-^{(1)} + E_-^{(2)}|^2$, calculated at the frequencies of these modes.

It was established a few years ago[10] that the distribution of scattered intensities in the folded acoustic modes (FLA) is mainly due to the square-wave modulation of the photoelastic constants in the superlattice. However, the acoustic mismatch between the constituents appears[11] to be important in explaining the behaviour of the relative intensities as a function of the diffusion wave vector $q = k_i + k_s \cong 2k_i$. In large-period

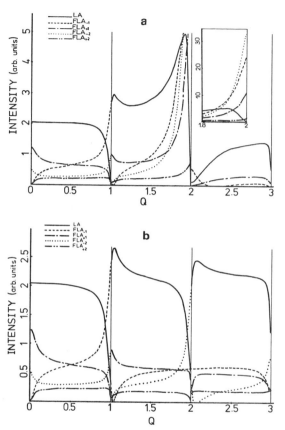

Fig. 8   (a) Intensities of the Brillouin and the first few folded
modes as functions of the diffusion wave vector $(Q=qD/\pi)$ in a
GaAs-AlAs superlattice with $d_{AlAs}/D = 0.74$ and $p_2/p_1 = 0.15$.
The refractive indices of GaAs and AlAs are,respectively,
$n_1 = 4.395$ and $n_2 = 3.37$. For Q close to the boundary of the
second Brillouin zone, several modes become very intense, as
shown in the inset. In this calculation only the photoelastic
contribution $(E^{(1)})$ to the scattered field is taken into account.
(b) Same as (a) but neglecting the dielectric mismatch between
the two layers.

155

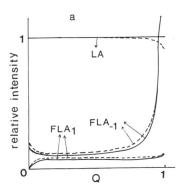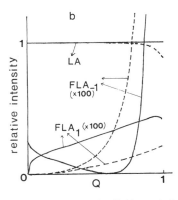

Fig. 9.   Comparison of the relative intensities of the Brillouin (LA) and the first two folded modes (FLA$_{\pm 1}$), taking into account only the photoelastic contribution $E^{(1)}$ (dashed lines) or both $E^{(1)}$ and $E^{(2)}$ contributions (solid lines). All the intensities are normalised with respect to the intensity of LA mode when both contributions are included. Figures (a) and (b) only differ by the magnitude of the photoelastic mismatch in the superlattice, namely $p_2 = p_1/3 = 25$ in (a) and $p_1 = p_2 = 25$ in (b). The other parameters are defined as $\rho_1/\rho_2 = 1.4$, $C_1/C_2 = 1$, $\varepsilon_1 = 19$, $\varepsilon_2 = 16$, $d_1 = d_2$.

superlattices q may even go beyond the first Brillouin zone. In this case we have used the transfer matrix method to calculate the component $E^{(1)}(x_3, t)$ of the scattered field which results from the photoelastic modulation of the dielectric function. We have shown[12] that the scattered intensities (proportional to $|E^{(1)}|^2$ in this case) are significantly modified for $q > \dfrac{\pi}{D}$ if we take into account the Bloch-wave, instead of plane-wave, behaviour of the incident and scattered electromagnetic fields in the superlattice. Let us stress that this behaviour is due to the difference between the dielectric constants of the constituents in the superlattice. An illustration of this effect, taken from ref. 12, is presented in fig. 8.

In this paper, using a Green's function formalism, we have derived the latter field $E^{(1)}(x_3, t)$ in a different way, but also calculated the second contribution $E^{(2)}(x_3, t)$ to the scattered field which results from the translational displacement of interfaces during a longitudinal vibration. The latter field appears because of the dielectric mismatch at the interfaces, or at the free surface in the case of a finite penetration of light. To emphasize its contribution to the total scattered field, we have plotted in fig. 9 the relative intensities of the FLA and Brillouin modes with and without the effect of this field. Two examples are considered which only differ by the magnitude of the photoelastic mismatch between the constituents of the superlattice, namely $p_2 = p_1/3 = 25$ in the first case and $p_1 = p_2 = 25$ in the second; the other parameters are defined as $\rho 1/\rho 2 = 1.4$, $C_1/C_2 = 1$, $\varepsilon_1 = 19$, $\varepsilon_2 = 16$ and may roughly describe the cases of GaAs/AlAs or SiGe/Si superlattices. In the first example (fig. 9a) the results are qualitatively independent of the field $E^{(2)}$, while in the second example (fig. 9b) the relative intensities are significantly affected by this term. (The intensities of FLA$_{\pm 1}$ are very small in the last case due to the absence of photoelastic mismatch in the superlattice.)

The comparison of the experimental Raman intensities with theoretical calculations provides a tool for the determination of the photoelastic parameters of thin-film materials, especially as a function of the wave-length of the light.[16] For this purpose let us now consider a few limiting cases which lead to more simple theoretical expressions. A first simplification is made by assuming that the superlattice behaves like an homogeneous medium as regards the propagation of incident or scattered light. This means that the latter waves are approximated by plane waves (for example, in eq. (12) our assumption implies that $k_{i1} \cong k_{i2} \cong k_i$ and the coefficients B_α are negligible); likewise, the Green's function G becomes that of a semi-infinite homogeneous crystal. Of course the approximation considered here cannot be valid if the scattering wave vector q exceeds the first Brillouin zone. Let us define, as in eqs. (16),

$$K_\alpha^\pm = k_{p\alpha} \pm (k_i + k_s) \qquad \alpha = 1,2 \qquad\qquad (22)$$

Then we obtain

$$I \sim \delta_{k_p - q + 2\pi m/D} \left| \left[a_{1-} (\varepsilon_1 + \frac{p_1 k_{p1}}{K_1^+}) \sin(K_1^+ h_1) - b_{1-} (\varepsilon_1 + \frac{p_1 k_{p1}}{K_1^-}) \sin(K_1^- h_1) \right] + \right.$$
$$\left. e^{\frac{iqD}{2}} \left[a_{2-} (\varepsilon_2 + \frac{p_2 k_{p2}}{K_2^+}) \sin(K_2^+ h_2) - b_{2-}(\varepsilon_2 + \frac{p_2 k_{p2}}{K_2^-}) \sin(K_2^- h_2) \right] \right|^2 \qquad (23a)$$

In this expression the terms containing the photoelastic and dielectric constants are, respectively, associated with the contributions of the $E^{(1)}$ and $E^{(2)}$ fields. However, in principle, it is somewhat contradictory to keep the latter contribution to the scattered field when neglecting the Bloch wave-like behaviour of the light in the superlattice; indeed, both effects result from the difference between the dielectric constants of the media in the superlattice. In a more consistent approximation, we have to attribute the same index of refraction to both constituents ($\varepsilon_1 = \varepsilon_2$); then the terms involving ε_1 and ε_2 in eq. (23a) cancel each other and we obtain

$$I \sim \delta_{k_p - q + 2\pi m/D} \left| p_1 k_{p1} \left[\frac{a_{1-} \sin(K_1^+ h_1)}{K_1^+} - \frac{b_{1-} \sin(K_1^- h_1)}{K_1^-} \right] \right.$$

$$\left. + e^{\frac{iqD}{2}} p_2 k_{p2} \left[\frac{a_{2-} \sin(K_2^+ h_2)}{K_2^+} - \frac{b_{2-} \sin(K_2^- h_2)}{K_2^-} \right] \right|^2 \qquad (23b)$$

This result may be very helpful due to its simplicity, because it only requires the knowledge of the displacement field associated with a bulk phonon in the superlattice (see eq. (9)).

In a further approximation we can assume that the superlattice behaves like an homogeneous medium as regards the propagation of acoustic waves. In this case eq. (23b) leads to the results of ref. 10, namely

$$I_0 \sim \left(\frac{d_1 p_1 + d_2 p_2}{D} \right)^2 \qquad\qquad (24a)$$

for the Brillouin mode and

$$I_m \sim \frac{\sin^2(\pi m d_1/D)}{(\pi m)^2} (p_1 - p_2)^2 \qquad\qquad (24b)$$

for the $FLA_{\pm m}$.

Finally, let us mention that the extension of the above calculation to the case of superlattices composed of more than two constituents can be done without basic difficulties, and even becomes straightforward under the approximations which lead to eqs. (23)-(24). The study of such superlattices may be helpful in the determination of photoelastic constants, in particular if the superlattice is composed of only two materials but with a few combinations of layer thickness.

REFERENCES

1. See for example J. Sapriel and B. Djafari Rouhani, Surface Science Reports 10, 189 (1989).
2. J. Sapriel, B. Djafari Rouhani and L. Dobrzynski, Surface Science, 126, 197 (1983) ; R.E. Camley, B. Djafari Rouhani, L. Dobrzynski and A.A. Maradudin, Phys. Rev. B27, 7318 (1983).
3. B. Djafari Rouhani, L. Dobrzynski, O. Hardouin Duparc, R.E. Camley and A.A. Maradudin, Phys. Rev. B28, 1711 (1983) ; A. Nougaoui and B. Djafari-Rouhani, Surface Science 185, 125 (1987).
4. A. Nougaoui and B. Djafari Rouhani, Surface Science 199, 638 (1988); J. Electron Spectroscopy and Related Phenomena 45, 197 (1987).
5. A. Nougaoui and B. Djafari Rouhani, Surface Science 185, 154 (1987).
6. E.M. Khourdifi and B. Djafari Rouhani, Surface Science 211/212 , 361 (1989).
7. E.M. Khourdifi and B. Djafari Rouhani, Journal of Physics : Condensed Matter, 1, 7543 (1989).
8. L. Dobrzynski, Surface Science, 175, 1 (1986) ; 180, 489 (1987) ; 182, 362 (1987).
9. J. Mendialdua, T. Szwacka, J. Rodriguez and L. Dobrzynski, Phys. Rev. B39, 10674 (1989).
10. C. Colvard, T.A. Gant, M.V. Klein, R. Merlin, R. Fisher, H. Morkoc and A.C. Gossard, Phys. Rev. B31, 2080 (1985).
11. B. Jusserand, D. Paquet, F. Mollot, F. Alexandre and G. Le Roux, Phys. Rev. B35, 2808 (1987).
12. J. He, B. Djafari Rouhani and J. Sapriel, Phys. Rev. B37, 4086 (1988).
13. K.R. Subbaswamy and A.A. Maradudin, Phys. Rev. B18, 4181 (1978).
14. B. Djafari Rouhani, A.A. Maradudin and R.F. Wallis, Phys. Rev. B29, 6454 (1984).
15. L. Dobrzynski, Phys. Rev. B 37, 8027 (1988).
16. J. Sapriel, J. He, B. Djafari Rouhani, R. Azoulay and F. Mollot, Phys. Rev. B37, 4099 (1988) ; J. He, J. Sapriel and H. Brugger, Phys. Rev. B39, 5919 (1989).

OPTICAL PHONON RAMAN SCATTERING AS A

LOCAL PROBE OF Si-Ge STRAINED LAYERS

J. C. Tsang, J. L. Freeouf and S. S. Iyer

IBM T. J. Watson Research Center
IBM Research Division
P. O. Box 218
Yorktown Heights, New York 10598

INTRODUCTION

There has been considerable interest over the last few years in the properties of ultra-thin Si-Ge strained layers and Raman scattering has been widely used to study these materials. The interest in these strained layers has been prompted by proposed technological applications of these materials in high speed electronic devices[1] and theoretical and experimental suggestions that the lowest lying optical transitions in superlattices of these materials can be direct in contrast to the weak, lowest lying, indirect transitions in bulk Si and Ge.[2] For example, it has been argued that direct optical transitions at photon energies near 0.8 eV can be observed in 10 monolayer period, strain symmetrized Ge/Si superlattices grown on Si(100).[2] While questions remain as to whether the fundamental gap in a Si-Ge superlattice can be direct,[3] it is generally agreed that in order to observe this possibility, layer thickness control at the monolayer level, atomically abrupt interfaces between the layers and excellent crystalline quality are all required. These requirements and the general scientific interest in the physics and chemistry of the strained Si-Ge system have placed a significant premium on the characterization of these materials at the atomic level.

Raman studies of Si-Ge multilayers composed of sub-nanometer thickness individual layers of Si and Ge have demonstrated that the Raman spectra show large changes as these layers evolve from thin random alloy layers to pure, single phase layers.[4] These changes have been used to help characterize the Si-Ge layers and to identify the conditions needed for the preparation of homogeneous, single phase layers in these structures.[2,4] In this paper, we consider the electronic origins of the layer thickness and growth temperature dependent changes in the optical phonon Raman spectra in order to understand what Raman spectroscopy measures when it is used to study the local properties of Si-Ge layers at the sub-nanometer level.[5] In particular, we show that E_1 like optical transitions[6] can be observed in ultra-thin Ge layers. The presence or absence of these transitions can produce significant layer thickness and growth temperature dependent changes in the Raman spectra obtained for Ge quantum wells in Si. In addition, there are changes in the Raman spectra of these materials arising from the changes in the number of nearest neighbor Ge and Ge-Si bonds which occur when a Ge_xSi_{1-x} alloy exhibits long range order. These results show why our experimental results are highly sensitive to changes in the local atomic composition of these layers at the sub-nanometer thickness level. This is in spite of the fact that the wavelength of

Light Scattering in Semiconductor Structures and Superlattices
Edited by D.J. Lockwood and J.F. Young, Plenum Press, New York, 1991

light is enormous on the scale of the atomic distances, suggesting that light scattering should not be an appropriate tool for the characterization of the local properties of sub-nanometer thickness layers.

Previous studies of the Raman spectra of samples made of ultra-thin Ge and Si layers revealed that changes in the growth conditions, such as the variation of the growth temperature between 250 C and 600 C and the Ge layer thickness between 4 and 17 Å, produced large changes in the Raman spectra arising from these layers.[4,7,8] The changes included shifts in the Raman frequencies of the different bands comprising our spectra which were consistent with the formation of a thin, homogeneous Ge layer at the lower growth temperatures. At the higher growth temperatures, the observed Raman frequency shifts showed that the Ge layers were actually alloys with significant interdiffusion between the Ge layers and the Si-cap layers.[4] The Raman spectra also displayed large modifications in the relative intensities of the different Raman lines arising from the three types of local bonding in these samples. At high growth temperatures and for Ge layers less than 4 Å thick, the ratios of the intensities of the Ge-Ge phonon Raman scattering near 300 cm^{-1}, as compared to those of the Ge-Si phonon scattering above 400 cm^{-1}, I_{Ge}/I_{Ge-Si}, were similar to the ratios observed for Ge_xSi_{1-x} alloys when x was near 0.5. However, when the growth temperatures were held below 400 C and the Ge layer thicknesses were above 7 Å, the Raman spectra showed strong Ge-Ge phonon scattering and much weaker Ge-Si scattering. This is similar to what would be expected if there was a continuous Ge layer with the Ge-Si bonds only occurring at the interfaces between the Ge layer and the Si substrate and cap layer. These results showed that growth temperatures < 400 C were needed to produce ultra-thin homogeneous layers of Ge and Si with good crystallinity using MBE.[4]

The ratios of the layer and interface Raman scattering intensities were, however, considerably greater than expected assuming that the Ge layer and the Ge-Si modes had comparable Raman scattering cross sections. In fact, the growth temperature and layer thickness dependences observed in the Raman spectra obtained from Ge layers as thin as 7 Å are associated with two distinct changes of the properties of these samples. The changes in the layer thicknesses alter the relative number of bonds of the different types as previously suggested[4]. In addition, there are also thickness dependent changes in the electronic structure of the layers. These can produce large resonant enhancements of the Ge layer signals for excitation energies near 2.2 eV. A principle result of this paper is that Raman spectra characteristic of well defined Ge layers, and the resonant enhancement of the Ge Raman scattering cross sections by the $E_1 - E_1 + \Delta_1$ gaps of Ge, appear almost simultaneously as the Ge layer thickness increases beyond about 4 atomic layers. The existence of E_1-gap-like structures in the Ge layer dielectric function has been independently verified through direct measurements of the dielectric response of these layers using spectroscopic ellipsometry. These results raise interesting questions about the localized character of both the Raman active optical phonons in the Si-Ge system and the E_1 gaps in Si-Ge layers and how together they contribute to the sensitivity of Raman spectroscopy as a local probe in Si-Ge layers. Our demonstration of how resonant Raman effects can modify the interpretation of the relative intensities of Raman lines in Si-Ge layer systems when the layer thicknesses are at the nanometer level is complemented by related results which have been obtained on ordered $Ge_{.5}Si_{.5}$ alloys. As in the case of the thicker quantum wells discussed above, we find that the ordering of the Ge and Si atoms in alternating pairs of Ge and Si planes normal to the (111) axes produces changes in the intensities of the different Raman lines. These cannot be completely described only by the changes in the number of bonds of each type due to ordering. The electronic structure effects can modify bond counting effects in the Raman spectra even at the two atomic layer level.

EXPERIMENTAL DETAILS

The samples studied in this paper were all grown by molecular beam epitaxy. The growth temperatures were generally kept low, ranging between 300 and 425 C except where noted otherwise. The growth chamber pressures, methods of monitoring the incident fluxes, deposition rates (0.3±0.1 Å/sec for Ge epitaxy and 3 Å/sec for Si epitaxy) and other significant parameters affecting the growth of these layers have been discussed in detail elsewhere.[4] The samples studied in this paper have been characterized by a number of different structural tools, including: medium energy ion scattering, Rutherford backscattering and electron microscopy. Experimental results obtained from the different probes produced quantitatively consistent results for the Ge layer crystallinity and thicknesses. An electron micrograph of a 17 Å layer of Ge grown on Si(100) and covered by about 50 Å of a silicon cap layer is shown in Figure 1. The Ge layer appears as a continuous, dark band in this picture. Detailed examination of this and other images shows that the layer is uniformly strained with no visible sign of any dislocations. The micrographs suggest that the top and bottom surfaces of the layer are flat and parallel on the scale of a fraction of the 17 Å layer thickness. There is clearly no evidence of large island growth at the > 200 Å scale as has been observed in other papers on the epitaxial growth of Ge layers at higher temperatures.[9] It should be noted that the strain limit as calculated from equilibrium theory for Ge grown epitaxially on Si is about 10 Å so that the film shown in Figure 1 cannot be in thermal equilibrium.[10] Attempts to grow considerably thicker Ge layers on Si(100) resulted in observable departures from epitaxial, layer by layer growth. Our use of relatively thin, individual strained layers, rather than thick strained superlattice samples where the signals will be

Figure 1. An electron micrograph of a 17 Å Ge layer grown by MBE at about 300 C on Si(100) and covered by a 50 Å Si cap layer. The Ge layer is the dark line on the diagonal of the picture. The Si substrate and buffer are to the left of the dark band indicating the Ge layer while the Si cap is the light band to its right.

stronger due to the larger number of Ge layers and Ge-Si interfaces, arises from concerns about the properties of thicker, strained films. By using Ge films whose total thicknesses are below 20 Å, Figure 1 shows that we can avoid problems associated with the effects of inhomogeneous strains, Stranski-Krastanov type growth, etc. Roughness at the interface can give the appearance of increased interfacial mixing with respect to layer formation and complicate the analysis of our results. The simplicity of the single layer system also provides for greater confidence in the decomposition of the ellipsometric data to obtain the Ge layer dielectric functions.

Raman spectra were excited using a variety of laser lines from Ar ion, dye and Ti-saphire lasers over a spectral range between 1.7 and 3.5 eV. The scattered light was analyzed by a triple monochromator and detected by either an imaging photomultiplier or a charge coupled device imager. The ellipsometric results described in this paper were obtained using a rotating polarizer spectroscopic ellipsometer of the Strobie-Rao-Dignam design.[11] This design includes an additional fixed polarizer ahead of the rotating polarizer before the sample. The additional polarizer completely removes all effects due to d.c. errors in the measurements. The system has been used for measurements over a spectral range between 1.34 and 5.4 eV and for angles of incidence near 70°. The detailed procedures for the reduction of the experimentally measured spectroscopic values of the ellipsometric parameters Φ and Δ to the derived layer dielectric functions have been described in detail elsewhere.[6,11] To summarize, the reduction of the experimental data to derive the layer dielectric response was done using a two step process. In the first step, the published bulk dielectric functions were used to obtain spectroscopic values for the thicknesses of the individual layers in the structure shown in Figure 1. The Ge layer thickness, the Si cap layer thickness and the thickness of the oxide layer on the Si cap were the free parameters that were varied to obtain a best fit over the spectral range of the experiment to the measured ellipsometric parameters for the dielectric response of the structure shown in Figure 1. The best fit value for the Ge layer thickness was then used with the bulk dielectric functions of Si and the surface oxide layer to derive a dielectric function for the Ge layer from the experimentally obtained values of the ellipsometric parameters. Since there are only two experimental parameters Φ and Δ, it was possible to derive the Ge layer $\varepsilon_1(\omega) + \varepsilon_2(\omega)$ unambiguously. Freeouf et al. found that 20% changes in the Ge layer thickness produced only changes in the amplitude of the derived Ge layer dielectric response with no significant changes in its spectral structure.[6] In fact, the spectroscopically obtained thicknesses for the Ge layers were all within 15% of the Ge layer thicknesses obtained by glancing angle Rutherford back-scattering experiments using 1 MeV He atoms on these samples. An important question was whether the procedure described above could actually converge to generate the Ge layer response function or whether it generated a value for the Ge layer thickness which produced a bulk Ge-like response function for the Ge layer. We shall see, in fact, that Ge layer dielectric functions were obtained which differed substantially from that of bulk Ge. Although the purpose of this paper is to show how the spectroscopic ellipsometry results allow us to understand the behavior of the Raman spectra as a function of the Ge layer thickness and composition, it should be noted that the agreement between the dielectric functions generated by the ellipsometric studies, and the resonant Raman spectra of these layers, provides independent confirmation of the accuracy of the ellipsometric determination of the Ge layer dielectric response. The chief limitation to the ellipsometric determination of the Ge layer dielectric function was the small thicknesses of the Ge layers. The Ge layer contributions to the total dielectric response of the Ge-Si composite system could be small. This problem became very severe for photon energies above 2.8 eV where the strong E_1 contributions to the dielectric response of Si appear. The large substrate and Si cap contributions to the dielectric response introduced monotonically varying backgrounds to the derived Ge layer dielectric functions for photon energies below the onset of the Si E_1 gap.

The resonant Raman spectrum of bulk Ge for excitation energies in the visible, like most of the other optical properties of Ge, is dominated by transitions across the E_1 and $E_1 + \Delta_1$ gaps.[12] These produce a single broad peak in the Raman scattering intensity at 2.24 eV with a full width at half maximum of about 150 meV. Experimental studies of resonant Raman scattering from semiconductors such as Ge, Si and GaAs, have shown that in all of these materials, the dominant contribution to the intensity of the Raman scattering for photon energies either below or near the E_1 gaps arises from the E_1 gaps.[13] While lower lying gaps such as E_o produce a weak enhancement of the bulk Raman scattering for excitation energies near E_o, the E_1 gaps can produce order of magnitude increases in the Raman scattering intensities under resonant excitation and are responsible for most of the experimentally observed Raman signals for photon energies below these gaps.[13] The first order Raman scattering observed in the back scattering geometry from a semiconductor with the zincblende structure involves a Γ_{25} symmetry phonon. The magnitude of the Raman tensor for light scattering by this phonon can be described by a single parameter $d(\omega)$ where the scattering efficiency is proportional to $\omega^4 d^2(\omega)$ and ω is the frequency of the light. Cerdeira et al.[12] showed that the experimental results for the dispersion of this Raman parameter for Ge could be described by an expression

$$d(\omega) = (\frac{1}{(4\pi)})((\frac{1}{(2\sqrt{3})})d_{1,0}^5(\frac{(d\varepsilon)}{(d\omega)}) + 2 \frac{\sqrt{2}}{\sqrt{3}} d_{3,0}^5 \frac{(\varepsilon^+ - \varepsilon^-)}{(\Delta_1)}) \frac{(n_o + 1/Q/n_o)}{(a_o)} \quad (1)$$

where a_o is the lattice constant, ε^+ and ε^- are respective contributions of the E_1 and $E_1 + \Delta_1$ gaps to the dielectric response, and the d's are the two and three band contributions, respectively, to the deformation potentials. Δ_1 is the splitting of the heavy and light hole valence bands at the L point. The matrix element couples the $n_o + 1$ and n_o optical phonon states through the phonon displacement operator Q. Cerdeira et al.[12] showed that for Ge, the two-band terms were negligible while $d_{3,0}^5$ is about 40 eV. It has been shown that $d_{3,0}^5$ has about the same value in Ge, Si and GaAs.[13] At room temperature, $E_1 = 2.12$ eV, while $E_1 + \Delta_1 = 2.32$ eV and the $(\varepsilon^+ - \varepsilon^-)/\Delta_1$ term in equation 1 produces the experimentally observed peak in the resonant Raman scattering at 2.24 eV. Near this peak in the resonance for the Raman scattering from Ge, the Raman tensor component of Ge has a value more than an order of magnitude larger (640 Å^2) than its value in similar materials such as GaAs and Si (60 Å^2) at photon energies far from the comparable resonances for GaAs and Si.[13] Since the intensity of the Raman scattering depends on the square of the Raman tensor, this means that for Ge or Si layers which are thinner than the optical penetration depths but are thick enough for the $E_1 - E_1 + \Delta_1$ gaps to be well defined, the Ge layer Raman scattering near 2.2 eV can be more than two orders of magnitude stronger than the Si Raman scattering.

The properties of the E_1 gaps in bulk group IV and III-V semiconductors have been widely studied. The dependence of the transition energies of Ge on external strain was studied by Chandrasekhar and Pollak,[14] who determined that the splitting of the E_1 transitions was augmented by applied strains. In fact, extension of their results to the 4% lattice deformations associated with the epitaxial growth of Ge on Si(100) would result in splittings of the order of 1 eV, in contrast to the 200 meV value for Δ_1 in the unstrained bulk. Exciton effects have also been associated with these transitions and effective exciton radii of the order of 15-20 Å have been deduced from electroreflectance measurements.[14] The E_1 transitions in bulk Si occur at energies near 3.4 eV and calculations of the band offsets for Ge layers grown on Si substrates suggest that the Si conduction band at the L point is slightly above the Ge L point conduction band minimum.

The Ge L point heavy hole valence band minimum is well above the Si L point valence band minimum.[15] This raises the possibility that the Ge E_1 transitions can be localized in a Ge-Si structure. Little is known about the changes expected in the energies of these levels due to confinement effects in ultra-thin structures. Garriga et al.[16] have seen both a relative insensitivity to well size for these transitions in GaAs-AlAs superlattices as well as a splitting of the E_1 and $E_1 + \Delta_1$ gaps for small superlattice periodicities which they associated with zone folding effects in the band structure. While there is an enormous literature on the behavior of the E_1 gaps in bulk Ge,[17] there is only a small amount of experimental literature on the effects of nanometer size structures on the E_1 gaps in Group IV and III-V materials.

The Raman spectrum of the Si-Ge strained layer system consists of three lines near 300, 400 and 500 cm^{-1}. It has been shown theoretically[18,19] for a multilayer Si-Ge system, that the highest energy line is associated with a vibrational mode which is completely localized on the Si layers with negligible amplitude in the Ge layers. The structure near 400 cm^{-1} is commonly referred to as the Si-Ge line by analogy with the similar structure observed in the random alloys. For the case of ordered layers, the vibrational amplitude of a longitudinal optical phonon at these energies penetrates far into the Si layers with some amplitude on the Ge atoms at the Si-Ge interfaces so that it is not, rigorously speaking, an interface mode. Its intensity will be very sensitive to disorder at the interface both due to the rapid increase in the number of Ge-Si bonds with increasing disorder and the effects of interface disorder on the wavefunctions of these modes.[18] The lowest energy mode has been described as quasi-localized with a small vibrational amplitude in the Si layers but with most of its amplitude in the Ge layers. The differing spatial characters of these modes mean that the Ge vibrations will couple strongly to the Ge E_1 gap, the Ge-Si Raman line will couple weakly to this gap and and Si vibrations should be relatively insensitive to its presence. If a well defined Ge layer supports a Ge-like E_1 gap, then the Ge mode Raman scattering will show a large enhancement when this gap is resonantly excited. This enhancement will have a much smaller effect on the Ge-Si scattering due to its small spatial extent in the Ge layer. This is in contrast to the behavior in the homogeneous random alloy where all three phonons will show similar resonant enhancements at the alloy E_1 gaps.

EXPERIMENTAL RESULTS

In Figure 2, we show the Raman spectra of an 11 Å layer of Ge in Si(100) for several excitation energies between 1.9 and 3.5 eV. The spectra in Figure 2 have been normalized against the intensities of the first order, symmetry-allowed Raman scattering from the Si substrate and cap layer. It is clear that the intensity of the Ge-Ge scattering, I_{Ge}, near 300 cm^{-1} is strongly dependent on the excitation energy. Although the first order Raman scattering cross section for the Si substrate and cap layer changes by almost two orders of magnitude over this frequency range, the ratio of the intensity of the Ge-Si scattering near 415 cm^{-1}, I_{Ge-Si}, with respect to I_{Si}, does not show any large change. On an absolute scale, while I_{Ge} shows a decrease with increasing photon energy between 2.5 and 3.5 eV, I_{Ge-Si} shows a significant increase, comparable to that of the scattering from the Si substrate and cap layer. As a result, under 3.5 eV photon excitation, I_{Ge} can be weaker than I_{Ge-Si} arising from the shoulder near 415 cm^{-1}. The Raman spectrum then qualitatively resembles that obtained from a dilute alloy layer of Ge_xSi_{1-x} where $x < .5$. As seen in Figure 2, this can occur in a sample whose Raman spectrum obtained under lower photon energy excitation is consistent with a physical structure consisting of a well defined Ge layer where most of the Ge atoms are bonded to other Ge atoms and only the Ge atoms at the Ge-Si interfaces are bonded to Si atoms.[4] Since the frequencies of the Ge-Ge and Ge-Si phonons in Figure 2 are inconsistent with the phonon frequencies of a $Ge_{1-x}Si_x$ random alloy and are consistent with the expected phonon frequencies for a strained Ge layer,[8] questions can be raised as to why their relative intensities for photon energies near 3.5 eV appear to be consistent with Raman scattering from an Si rich alloy rather than a well defined Ge layer. In previous studies,[4] the strength of I_{Ge} in Figure 2, as compared to I_{Ge-Si} at excitation energies below 2.8 eV, has been used to identify the presence of a Ge layer bounded by relatively abrupt interfaces with the Si substrate and cap. However, Figure 2 shows that the ratio of the intensities of these two lines reverses around 3.5 eV. As discussed in our introduction, we shall show that this behavior reflects the presence of the Ge E_1 transitions near 2.2 eV and the Si E_1 gap at energies above 3 eV.

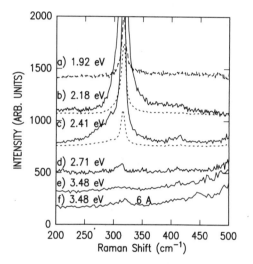

Figure 2. The Raman spectra of an 11 Å layer of Ge in Si(100) excited using a number of different photon energies between 1.92 and 3.5 eV. Also shown is the Raman spectrum of a 6 Å Ge layer excited at 3.5 eV.

In Figure 3, we show the dependence on the excitation photon energy of the Raman scattering efficiency for the near 300 cm^{-1} Ge phonon measured on single 4, 7 and 17 Å Ge layers grown in Si(100) at about 300 C. These results are normalized against the Ge layer thickness so that if the scattering cross sections were independent of Ge layer thickness, the three curves would coincide. This is clearly not the case in Figure 3. The resonant Raman response of the 17 Å sample (dot-dashed line) is similar in shape to that of bulk Ge (solid line) while the 7 Å Ge layer (dashed line) shows both a loss of scattering efficiency and a shift of its peak to higher photon energies. The 4 Å layer (dotted line) shows no resonant enhancement over this range of photon energies. Figures 4a and b show $\varepsilon_1(\omega) + i\varepsilon_2(\omega)$ obtained from spectroscopic ellipsometry for these 4, 7 and 17 Å Ge layers. Also shown in Figure 4 is the bulk dielectric response for Ge. From Figure 4, we see that the strong resonances in the Raman scattering cross sections

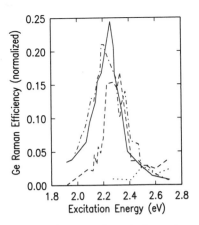

Figure 3. The excitation wavelength dependence of I_{Ge} for 4 (dotted curve), 7 (dashed curve) and 17 (dot dashed curve) Å layers of Ge in Si(100). The curves are normalized against the intensity of the Si Raman scattering and the Ge layer thickness. The solid curve shows the shape of the bulk $E_1 - E_1 + \Delta_1$ first order Raman scattering resonance in Ge.

of the 7 and 17 Å Ge layers are matched by the appearance of bulk Ge-like E_1 transitions in $\varepsilon_1(\omega) + i\varepsilon_2(\omega)$ for these samples. There is no experimental evidence for such a transition in the 4 Å film. The oscillator strength for the bulk Ge-like transition for the 17 Å Ge layer is about 40% that of the bulk transition. The integrated strength of the optical transition at 2.32 eV observed in ε_2 for the 7 Å Ge layer is about 30% of the strength of the $E_1 - E_1 + \Delta_1$ transitions in the 17 Å Ge layer. The decrease in the strength of this transition with decreasing Ge layer thickness is qualitatively consistent with the decrease in the strength of the Ge scattering under resonant excitation as seen in Figure 3. In Figure 5, the excitation energy dependence of I_{Ge-Si} is shown. The coupling of the Ge-Si modes to the strong Ge layer resonances shown in Figure 4 is considerably weaker than the coupling of the Ge modes in Figure 3. For the 7 and 17 Å Ge layers, the coupling of the Ge-Si modes to the Ge layer E_1 transition is at least an order of magnitude weaker than the coupling of the Ge modes. This is qualitatively consistent with the fact the Ge-Si mode amplitudes are largely located on the Si atoms with only the Ge atoms at the Ge-Si interface being involved in the displacement pattern of the modes. The results in Figures 3 to 5 show that the Ge dielectric response assumes many of its dominant bulk characteristics in this energy range for Ge layer thicknesses near 10 Å.

This behavior is in marked contrast to that shown by the different Raman modes of the homogeneous alloy system. In the homogeneous alloys, all three modes show almost the same dependence of the intensity on the excitation energy.[20] The Raman frequencies for the Ge-Ge and Ge-Si modes in a 4 Å thick Ge layer are consistent with expectations for a strained alloy with a $Ge_{.5}Si_{.5}$ composition. The observation that the ratio of the intensities of these lines in the 4 Å Ge layer is independent of excitation energy between 1.8 and 2.8 eV is also consistent with the presence of an alloy-like layer rather than a well defined Ge layer. The E_1 gap of the alloy system $Ge_{1-x}Si_x$ is a strong function of the alloy composition moving to higher photon energies from 2.12eV for $x = 0$ to 2.8 eV for $x = 0.5$.[21] The effects of strain on this transition as extrapolated from

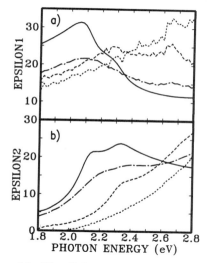

Figure 4a and b. The dielectric functions $\varepsilon_1(\omega) + \varepsilon_2(\omega)$ for 4 (dotted curve), 7 (dashed curve) and 17 (dot-dashed curve) Å epitaxial Ge layers in Si(100) for $1.8 < \hbar\omega < 2.8$ eV. The solid curve is the bulk Ge dielectric response.

the results of Chandrasekhar and Pollak[14] suggest that the E_1 gap in both the alloy and pure Ge will shift down in energy. However, the down shift in the alloy will be smaller than in pure Ge because of the smaller lattice mismatch with a bulk Si substrate. As a result, any Ge-Si-like E_1 resonance will be well above 2.5 eV and outside our spectral range. The experimental evidence that there is a localized electronic transition in the Ge layers carries with it the consequence that there should be a confinement shift associated with the localization of the transition to the Ge layer. This appears to be what is observed as the layer thickness is reduced from 17 to 7 Å and both the peaks in the resonant Raman profiles and $\varepsilon_2(\omega)$, shift to higher energies. The disappearance of this transition at these photon energies for thinner Ge layers presumably reflects the change in the character of these transitions from Ge-like to alloy-like. The coincidence of the resonant Raman spectra, and the experimentally derived $\varepsilon_1(\omega) + \varepsilon_2(\omega)$, provide strong evidence for the presence of Ge-like E_1 transitions in Ge layers as thin as 7 Å. The failure to observe a strong enhancement at these energies in I_{Ge-Si} shows that the electronic transition responsible for the resonance in I_{Ge} is localized on the Ge layers.

The appearance of a bulk-like E_1 gap in the electronic structure of the 17 Å Ge layer and its absence in the 4 Å Ge layer, means the intensity of the Ge layer Raman scattering for excitation energies near the E_1 will show a non-linear dependence on Ge layer thickness. At a photon energy of 2.22 eV, the intensity of the Ge layer scattering changes by about a factor of 5 as the Ge layer thickness changes from 17 Å to 7 Å and by about a factor of > 15 as the layer thickness changes from 7 Å to 4 Å. As seen from Figures 3 and 4, much of this increase must be attributed to the appearance of a bulk Ge-like optical transition near 2.2 eV as the Ge layer thickness approaches and exceeds

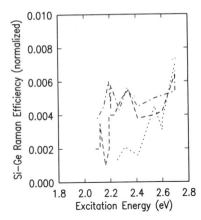

Figure 5. The excitation photon energy dependence of I_{Ge-Si} for 4 (dotted curve), 7 (dashed curve) and 17 (dot-dashed curve) Å Ge layers grown in Si(100). The results are normalized against the Si Raman scattering intensity.

Table 1. The parameters describing the E_1 like electronic transitions in ultra-thin Ge layers grown epitaxially in Si(100).

GE THICKNESS (Å)

	4	7	17	bulk
E_1 (eV)	-	2.32	2.13	2.12
$E_1 + \Delta_1$	-	-	2.33	2.32
Γ_1 (eV)	-	.127	.130	.075
Γ_Δ	-	-	.089	.075
Φ(radians)	-	3.99	4.03	3.93
ratio of E_1 amplitude to $E_1 + \Delta$ amplitude			4.2	2

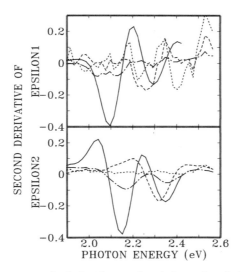

Figure 6. The second derivatives of $\varepsilon_1(\omega)$ and $\varepsilon_2(\omega)$ obtained by spectroscopic ellipsometry for 17 (dot-dashed curve), 7 (dashed curve) and 4 (dotted curve) Å Ge layers in Si(100). The solid curves are the second derivatives of the bulk Ge response.

10 Å. The failure to observe a similar sizable resonant enhancement of the Ge-Si Raman intensities shows that the large ratios of the Ge-Ge scattering intensities to the Ge-Si scattering intensities reported previously,[4] reflect both the increase in the number of Ge-Ge bonds relative to the number of Ge-Si bonds as the Ge layer thickness increases and the appearance of a new optical transition near 2.2 eV which couples strongly only to the Ge-Ge phonons. The change in I_{Ge}/I_{Ge-Si} from a value $>> 1$ for excitation energies below 2.5 eV to a value < 1 for an excitation energy of 3.4 eV, is due to the presence of the Si E_1 gap. Because the Ge-Si mode has a large amplitude in the Si layer while the Ge mode is largely confined to the Ge layers, the Ge-Si modes will show a large enhancement of their scattering intensities at the Si E_1 resonance in comparison to the Ge modes.

Our results on the behavior of the gaps near 2.2 eV in ultra-thin Ge layers grown epitaxially on Si (100) are consistent with modulation spectroscopy studies of the $E_1 - E_1 + \Delta_1$ gap in bulk Ge. In Table 1, we show the spectroscopic results obtained from ellipsometry for the E_1 gap in the Ge layers discussed above and for bulk Ge. The different energies presented in Table 1 can be derived from the experimental results, especially if suitable derivatives are taken of the data in Figures 4 to remove the effects of the broad, slowly varying background arising from the large Si substrate response at higher photon energies. The values for Γ_1 and Γ_Δ are the widths of the two transitions, while Φ is the phase shift in the dielectric response arising from the exciton effects in these transitions. Figure 6 gives the second derivative of the bulk dielectric function of Ge and of the experimentally derived dielectric functions for the 7 and 4 Å Ge layers in Si(100). Also shown is the second derivative of the bulk which is spectroscopically similar to the second derivative of the 17Å Ge layer dielectric response. The single peak seen in the second derivative of the 7 Å layer ε_2, resembles the 17 Å layer result if E_1 was shifted to 2.32 eV and Δ_1 reduced considerably. The similarities of the transitions observed by spectroscopic ellipsometry on these thin layer samples to the bulk $E_1 - E_1 + \Delta_1$ gaps show they can have an effect on the resonant Raman scattering similar to that of the bulk E_1 gaps.

Recently, Legoues et al.[22] found that unstrained Si_5Ge_5 alloys display spontaneous ordering if they are grown at low temperatures. This ordering involves the presence of alternating sequences of pairs of Ge and pairs of Si planes in an AABBAABB..... stacking normal to the (111) crystallographic axes. The ordered structure proposed by Legoues et al. produces significant changes in the number of nearest neighbor atoms of the different types as compared to the random alloy. In the random alloy, the fraction of Ge-Ge, Ge-Si and Si-Si nearest neighbors is 0.25:0.5:0.25. In the ordered phase of Legoues et al., this ratio changes to 0.375:0.25:0.375. In the random alloys, the relative intensities of the first and second order Raman lines for the different vibrational modes closely follow the relative fractions of the different nearest neighbor bonds. Figure 7a shows the second order Raman spectrum of a Ge_5Si_5 random alloy sample. The spectrum in Figure 7b was obtained by taking the difference of the Raman scattering obtained from an ordered Ge_5Si_5 alloy with respect to the Raman spectrum of the random alloy. The second order spectrum is shown since it produces a reasonably accurate representation of the phonon density of states in Ge_xSi_{1-x} alloys. It is less sensitive than the first order Raman spectrum to long range ordering effects, which could modify any comparison of Raman spectra obtained from random and ordered alloys.[23] The three broad peaks between 550 and 1000 cm^{-1} in the random alloy spectrum in Figure 7 correspond to the excitation of 2 Ge-Ge, 2 Ge-Si and 2 Si-Si vibrations. The intensity of the difference spectrum has been doubled so that if the Raman intensities were dictated solely by the numbers of bonds of each type, the increases in the Ge-Ge and Si-Si scattering, and the decrease in the intensity of the Ge-Si scattering, would be equal to the strengths of these Raman lines in the random alloy. We find in Figure 7 that the changes in the intensities of the Ge-Ge and Ge-Si scattering are about a factor of three smaller than expected, suggesting that the ordering is incomplete. More significantly, we find that the Si-Si scattering actually decreases in intensity where it should increase. Spectroscopic ellipsometry of these samples shows no large changes in the dielectric response on ordering. The E_1 alloy transition shows a small 60 meV shift to low energies. While this shift is consistent with a small (2-3%) increase in the Ge concentration from 50%, such a small change in composition cannot explain the decrease in the intensities of the Ge-Si and Si-Si peaks.

The ordered phase of Legoues et al. produces a significant change in the bonding of the Si atoms, reducing the average number of nearest neighbor Ge atoms for each Si atom by 50%. As the Si atoms segregate into Si layers, for excitation energies in the visible, the efficiency of the Raman scattering from the Si-like vibrations can be expected to decrease. This is because the Raman cross sections in the visible for bulk Si are considerably smaller than for either Ge or Ge_xSi_{1-x} alloys since the silicon E_1 gaps are at much higher energies (3.4 eV). If the electronic structure of the random alloy is spatially uniform, then the Si-Si bonds can modulate the alloy $E_1 - E_1 + \Delta_1$ gaps to produce Raman scattering. As the alloy orders, the Si-Si vibrations become strongly localized on the Si layers and couple less strongly to the alloy-like E_1 gaps. While the number of Si-Si bonds can increase, the shift to higher energies of the Si layer E_1 transitions can produce a decrease in the Raman cross sections in the visible, resulting in the observed decrease of the Si-Si scattering in Figure 7.

CONCLUSION

Our results show that electronic structure changes associated with the presence of homogeneous Ge layers in Ge-Si structures can produce changes in the Raman scattering intensities of the different Raman active modes. These can be comparable to or larger than the changes associated with increases in the number of Ge-Ge bonds as compared to the case of the random alloy. The electronic structure derived changes are associated with the development of the Ge layer E_1 gaps for Ge layer thicknesses less than 10 Å. This conclusion raises some interesting questions concerning the properties of the E_1 transitions in Ge. The observation of bulk-like Ge $E_1 - E_1 + \Delta_1$ transitions in a fully strained Ge layer only 17 Å thick is surprising. The splitting between the two states should be considerably enhanced by the 4% lattice mismatch between the Si substrate and Ge. As mentioned earlier, extrapolation of the results of Chandrasekhar and Pollak for the strain induced splitting of these states to that of the epitaxial growth of Ge on Si, would produce a splitting of almost 1 eV. There can also be significant shifts in the transition energies due to confinement effects if the E_1 transitions are treated in

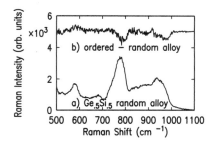

Figure 7. a) The second order Raman spectrum of disordered $Ge_{.5}Si_{.5}$. b) The change in the second order Raman spectrum of the alloy when it shows the ordered phase. (b) is normalized as described in the text.

the effective mass approximation. However, Garriga et al., in ellipsometric studies of $(GaAs)_m(AlAs)_n$ superlattices, have observed only small confinement shifts and also the presence of additional E_1 -like transitions. They attributed the latter to zone folding effects on the band structure.[16] Our results suggest that theoretical studies of the E_1 transitions in ultra-thin Ge quantum wells should be pursued to determine why these transitions appear to be so bulk-like and apparently insensitive to strain and confinement effects in these ultra-thin layers.

ACKNOWLEDGEMENT

We thank F. K. Legoues for electron microscopy studies of the samples discussed in this paper and V. P. Kesan for the ordered alloy samples.

REFERENCES

1. S. S. Iyer, G. L. Patton, J. M. Stork, B. S. Meyerson, and D. L. Harame, IEEE Trans. Elec. Dev. ED 36 ,2043 (1989).
2. G. Abstreiter, K. Eberl, E. Freiss, W. Wegscheider and R. Zachai, J. Cryst. Growth 95 , 431 (1989).
3. E. A. Montie, G. F. A. can de Walle, D. J. Gravesteijn, A. A. van Gorkum, and C. W. T. Bulle-Lieuwma, Appl. Phys. Lett. 56 , 340 (1990).

4. S. S. Iyer, J. C. Tsang, M. Copel, P. Pukite and R. Tromp, Appl. Phys. Lett. 54 , 219 (1989) and S. S. Iyer, Silicon Molecular Beam Epitaxy, in "Si Epitaxy," ed. by G. J. Baliga, Academic Press, Orlando (1986), p. 91.

5. J. C. Tsang and S. S. Iyer, IEEE J. Quantum Elec. 25 , 1008 (1989).

6. J. L. Freeouf, J. C. Tsang, F. K. Legoues and S. S. Iyer, Phys. Rev. Lett. 64 , 315 (1990).

7. J. C. Tsang, S. S. Iyer, J. A. Calise and B. A. Ek, Phys. Rev. B40 , 5886 (1989).

8. J. C. Tsang, S. S. Iyer, P. Pukite and M. Copel, Phys. Rev. B39 , 13545 (1989).

9. K. Sakamoto, T. Sakamoto, S. Nagai, G. Hashiguchi, K. Kuniyoshi and Y. Bando, Jpn. J. Appl. Phys. 26 , 666 (1987).

10. R. People and J. C. Bean, Appl. Phys. Lett. 47 , 322 (1985).

11. J. L. Freeouf, Appl. Phys. Lett. 53 , 2426 (1988).

12. F. Cerdeira, W. Dreybrodt and M. Cardona, Solid State Commun. 10 , 591 (1972).

13. M. Cardona in "Light Scattering in Solids II," ed. by M. Cardona and G. Guntherodt, Topics Appl. Phys., Vol. 50 ,Springer, Berlin, Heidelberg (1983) p. 19.

14. M. Chandrasekhar and F. H. Pollak, Phys. Rev. B15 , 2127 (1977).

15. C. van de Walle and R. M. Martin, Phys. Rev. B34 , 5621 (1986).

16. M. Barriga, M. Cardona, N. E. Christensen, F. Lautenschlager, T. Isu and K. Ploog, Phys. Rev. B36 , 3254 (1987).

17. L. Vina, P. Logothetides and M. Cardona, Phys. Rev. B30 , 1919 (1984).

18. A. Fasolino, E. Molinari and J. C. Mann, Phys. Rev. B39 , 3923 (1989).

19. M. I. Alonso, M. Cardona, and G. Kanellis Solid. State Commun. 69 ,479 (1989).

20. F. Cerdeira, M. I. Alonso, D. Niles, M. Garriga, M. Cardona, E. Kasper and H. Kibbel, Phys. Rev. B40 , 1361 (1989).

21. J. Humlicek, M. Garriga, M. I. Alonso and M. Cardona, J. Appl. Phys. 65 , 2827 (1989).

22. F. K. Legoues, V. J. Kesan and S. S. Iyer, Phys. Rev. Lett. 64 , 40 (1990).

23. J. S. Lannin, Phys. Rev. B16 , 1510 (1977).

STRAIN CHARACTERIZATION OF SEMICONDUCTOR
STRUCTURES AND SUPERLATTICES

E. Anastassakis

Physics Department
National Technical University
Athens 157 73, Greece

INTRODUCTION

The effects of strains on the Raman-active optical phonons of crystalline materials are reviewed, with emphasis on Si and zincblende (ZB) or wurtzite (W) cubic crystals, their strained epilayers (EL) and superlattices (SL). In the presence of strains the symmetry of the crystal is altered and the phonon degeneracies may be lifted with an upward or downward shift of their frequencies. To a first approximation the shifts are linear in the strain components; the corresponding rates lead to the so-called **phonon deformation potentials (PDP)**. The latter allow the calculation of built-in strains from any observed frequency shifts. The dynamical secular equation (DSE) associated with the phonon provides the necessary phenomenological basis for converting frequency shifts to strains and vice versa. The same problem, i.e., the DSE of triply-degenerate $q \approx 0$ optical phonons of cubic crystals, is treated here in an arbitrary system of axis $x'_1 x'_2 x'_3$ relative to the system $x_1 x_2 x_3$ of the crystallographic axes $<100>$. This generalization is dictated by the need of handling phonon shifts and splittings in EL and SL which are grown along an arbitrary direction x'_3. For this purpose, it is necessary to have precise knowledge of the strain field, namely, the full array of its non-zero tensor components. This is a problem of elasticity theory and will be addressed independently. Strained polycrystalline films will also be treated. For this purpose, the PDP of a polycrystal are expressed in terms of the PDP of the corresponding single crystal, by use of averaging procedures. Examples will be given of recent determinations of reliable PDP values in materials of current interest, and also selective cases from the literature, where strain characterization through Raman Scattering (RS) is successfully demonstrated. In what follows, Greek (Latin) tensor indices stand for single cartesian (suppressed) indices running from 1(1) to 3(6), according to the usual code of crystal physics[1].

STRAINS AND PHONON FREQUENCIES - PDP

In the presence of a symmetric strain ε the triple degeneracy of the $q \approx 0$ optical phonon of frequency ω_0 is lifted, due to anisotropic changes in the lattice constants. [Antisymmetric strains are commented on in the last section.] In general, each of the new eigenvectors of the strain-split phonon exhibits its own frequency Ω (eigenvalue) which, for ε in the elastic regime, varies linearly with the components $\varepsilon_{\kappa\varrho}$. The new frequencies $\Omega(\varepsilon_{\kappa\varrho})$ can be measured directly from the Raman spectra. Observation of a shift $\Delta\Omega = \Omega - \omega_0$ may therefore be attributed to tensile or compressive strains present in the scattering volume. If these strains are inhomogeneous, the shifts $\Delta\Omega$ assume a range of values. As a result the observed spectrum becomes broader and asymmetric, in general. In short, the Raman spectra obtained under various physical or environmental conditions can lead to information about strains and their spatial distribution. A full microscopic treatment of the problem is rather involved. However, the following[2,3] general considerations have proved themselves sufficient for most applications in materials characterization in the last two decades.

We start with a phenomenological definition of the normal coordinate u_j associated with a phonon j and an appropriate crystal potential energy ϕ. Within the harmonic approximation and with no strains present, the mode frequency ω_j is associated with a diagonal effective force constant K_j^0 (in general, a second-rank tensor), according to the definition

$$K_{j,\alpha\beta}^0 = \left(\frac{\partial^2 \phi}{\partial u_{j\alpha}\partial u_{j\beta}}\right) = K_{j,\beta\alpha}^0 \equiv \delta_{\alpha\beta} K_j^0 = \omega_j^2 \delta_{\alpha\beta}, \tag{1}$$

where $\delta_{\alpha\beta}$ is the Kronecker delta. In the presence of a symmetric strain, K_j^0 and ω_j can be expanded to terms linear in $\varepsilon_{\kappa\varrho}$,

$$K_j \cong K_j^0 + \Delta K_j \tag{2}$$

$$\Omega_j \cong \omega_j + \Delta\Omega_j, \tag{3}$$

where

$$\Delta K_{j,\alpha\beta} = \left(\frac{\partial K_{j,\alpha\beta}}{\partial \varepsilon_{\kappa\varrho}}\right) \varepsilon_{\kappa\varrho} \equiv K_{j,\alpha\beta\kappa\varrho} \varepsilon_{\kappa\varrho}. \tag{4}$$

Summation is implied over repeated indices. The new mode coefficient $K_{j,\alpha\beta\kappa\varrho}$ is a fourth-rank tensor, symmetric in interchanging the two symmetric pairs $\alpha\beta$ and $\kappa\varrho$. The effect of strain is to mix components (of the originally degenerate mode j) with different indices. This means that the tensor ΔK_j is not necessarily diagonal when referred to $x_1 x_2 x_3$. The DSE now becomes[2]

$$\left| \Delta K_{j,\alpha\beta} - \lambda_j \delta_{\alpha\beta} \right| = 0, \tag{5}$$

where

$$\lambda_j = \Omega_j^2 - \omega_j^2 \cong 2\omega_j \Delta\Omega_j. \tag{6}$$

Diagonalization of Eqn. (5) yields the new eigenvectors and eigenfrequencies of the phonon j; in order to do this, it is necessary first to establish the exact tensor form of ΔK_j, i.e., to find the non-zero components of $K_{j,\alpha\beta\kappa\varrho}$ for the specific phonon j. This is done by use of group theory. The criterion for $K_{j,\alpha\beta\kappa\varrho}$ to have non-zero components, by crystal and phonon symmetry arguments, lies in the reduction

$$[\Gamma(j)\times\Gamma(j)]_s\times\Gamma(\varepsilon) = n_1\Gamma_1 + ... \tag{7}$$

where $\Gamma(j)$ is the mode irreducible representation of the crystal point group. $\Gamma(\varepsilon)$ is the reducible representation for the strain tensor and Γ_1 is the symmetric irreducible representation of the point group. The index S indicates that only the symmetrized part of the Kronecker product should be considered. If $n_1 = 0$, all components of $K_{j,\alpha\beta\kappa\rho}$ are identically zero by symmetry. If $n_1 \neq 0$, the tensor is allowed to exhibit as many as n_1 (non-zero) independent components. Thus, in the case of triply-degenerate phonons (ω_0) from O_h or T_d structures, the reduction of Eqn. (7) yields $n_1 = 3$, hence there are three components K_{11}, K_{12}, K_{44} in suppressed notation, relative to $x_1x_2x_3$. The index j has been omitted for simplicity. Often, these three components are designated by p, q, r, respectively. They are defined as **PDP**. They connect the strain components with the changes induced on the phonon frequency (squared) in the same way the elastic constants connect strain components with stresses (or vice versa).

In practice, it is more convenient to work with PDP which are normalized to ω_0^2. Such dimensionless parameters usually take values from 0 to $\pm$ 5 for the various phonons of different materials and are easier to use. We give these definitions below for diamond-type structures since we will frequently use this notation

$$\tilde{K}_{11} \equiv K_{11}/\omega_0^2 = p/\omega_0^2$$

$$\tilde{K}_{12} \equiv K_{12}/\omega_0^2 = q/\omega_0^2 \tag{8}$$

$$\tilde{K}_{44} \equiv K_{44}/\omega_0^2 = r/\omega_0^2.$$

In view of Eqn. (4), the DSE of Eqn. (5) takes the following form, in $x_1x_2x_3$,

$$\begin{vmatrix} K_{11}\varepsilon_1 + K_{12}(\varepsilon_2 + \varepsilon_3) - \lambda & K_{44}\varepsilon_6 & K_{44}\varepsilon_5 \\ K_{44}\varepsilon_6 & K_{11}\varepsilon_2 + K_{12}(\varepsilon_3 + \varepsilon_1) - \lambda & K_{44}\varepsilon_4 \\ K_{44}\varepsilon_5 & K_{44}\varepsilon_4 & K_{11}\varepsilon_3 + K_{12}(\varepsilon_1 + \varepsilon_2) - \lambda \end{vmatrix} = 0. \tag{9}$$

If the strain tensor is known in component form, Eqn. (9) is readily diagonalized. In terms of the eigenvalues λ_s the new frequencies are

$$\Omega_s = \omega_0 + \Delta\Omega_s \cong \omega_0 + \lambda_s/2\omega_0, \tag{10}$$

where s = 1, 2, 3. The eigenvectors obtained from Eqn. (9) correspond to the polarizations of the perturbed phonons expressed in $x_1x_2x_3$. From an experimental point of view, the new phonon components can be best seen in RS experiments, since by proper use of polarization selection rules it is possible to observe each eigenvector separately.

It becomes clear now that in order to obtain residual strains from the Raman spectra (i.e., from shifts $\Delta\Omega_j$), it is necessary to know the values of PDP for one of the modes j. A control experiment is therefore required for each material whereby the shifts $\Delta\Omega_j$ are measured as a function of an externally applied stress. Such control experiments have been done for a number of materials and will be discussed in a later section.

It should be emphasized that in the DSE of Eqn. (9), all strain and PDP components ε_i and K_{ij} refer to $x_1 x_2 x_3$. It often happens, however, that the component array of ε is known as ε', i.e., relative to a rotated system $x_1' x_2' x_3'$. There are two ways to proceed then, (i) first obtain ε from ε' by rotation, and then use Eqn. (9), and (ii) obtain K_{1m} from K_{ij} by rotation and then apply Eqn. (5) in $x_1' x_2' x_3'$. [The secular equation has the same form as Eqn. (5) in all systems of axes.] Usually, the phonon eigenvectors are simply related, if not identical, to the primed system itself; this makes the second approach more advantageous, since then it is easier to probe individual phonon eigenvectors by performing backward RS experiments along any of the axes $x_1' x_2' x_3'$. This situation arises in SL or EL grown along x_3'. The mismatch strain tensor in this case has the simplest form in the primed system; it often happens, in fact, that the new eigenvectors of the Raman-active phonons are along x_3' and on the plane normal to x_3'. Clearly, the former (latter) are of longitudinal (transverse) polarization, the phonon wavevector $\mathbf{q}$ being along x_3' for backward scattering along x_3'.

Another point that should be stressed is that the entire formalism above is based on the effects of **strains** on the phonon frequencies. The strains may be either induced (by external stresses σ_i) or built-in strains (due to growth conditions or modification of the material). The latter case is most frequently encountered in strained EL or SL, where the strains are produced by lattice mismatch or by the different thermal expansion coefficients of the two material layers involved. In either case, it is essential that the non-zero strain components be known before any diagonalization of Eqn. (5) is undertaken.

In the next two sections we will review cases of both types of strains. For simplicity, we first consider the triply-degenerate phonon of Si and then the TO-LO split phonon of ZB and W structures.

STRAINS PRODUCED BY APPLIED STRESSES

We consider strains produced by an externally applied stress.

Uniaxial Stress along [001], [111]

A uniaxial stress σ applied along [001] or [111] in any cubic crystal produces symmetric strains which, in $x_1 x_2 x_3$, are given by

$$\varepsilon_{\kappa\varrho} = \sigma \begin{pmatrix} S_{12} & \cdot & \cdot \\ & \ddots & \cdot \\ & & S_{11} \end{pmatrix}_{\|[001]} , \quad \frac{\sigma}{3} \begin{pmatrix} S_{11}+2S_{12} & S_{44}/2 & \\ & \ddots & \\ & & \end{pmatrix}_{\|[111]} , \quad (11)$$

where S_{ij} are the elastic compliances, dots stand for zero and connected entries are equal. When either of these two strains is introduced in Eqn. (9), the triple degeneracy splits into a singlet ($\Omega_3 = \Omega_s$) with eigenvector parallel to the stress axis, and a doublet ($\Omega_1 = \Omega_2 = \Omega_d$) with mutually perpendicular eigenvectors normal to the stress. Part of the shift ($\Delta\omega_h$) of each component is due to the hydrostatic content of the applied stress. The total (observed) shifts of the singlet and doublet components can be expressed in terms of $\Delta\omega_h$ (not directly observable) and the splitting $\Delta\Omega$ between singlet and doublet as follows,

$$\Delta\Omega_s = \Omega_s - \omega_0 = \Delta\omega_h + \frac{2}{3}\Delta\Omega$$

$$\Delta\Omega_d = \Omega_d - \omega_0 = \Delta\omega_h - \frac{1}{3}\Delta\Omega, \tag{12}$$

where

$$\Delta\omega_h = (\Delta\Omega_s + 2\Delta\Omega_d)/3 \tag{13a}$$

$$= \sigma\omega_0 (\tilde{K}_{11} + 2\tilde{K}_{12})(S_{11} + 2S_{12})/6, \tag{13b}$$

and

$$\Delta\Omega \equiv \Omega_s - \Omega_d = \begin{cases} \sigma\omega_0 (\tilde{K}_{11} - \tilde{K}_{12})(S_{11} - S_{12})/2 \equiv \Delta\Omega^{(001)}, & \sigma\|[001] \\ \\ \sigma\omega_0 \tilde{K}_{44} S_{44}/2 \equiv \Delta\Omega^{(111)}. & \sigma\|[111] \end{cases} \tag{13c}$$

These two stress directions are most suitable for control experiments; the latter consist in determing the slopes $(\Delta\Omega_s/\sigma)$ and $(\Delta\Omega_d/\sigma)$ for the two directions of σ. The values of $\tilde{K}_{11}$, $\tilde{K}_{12}$, $\tilde{K}_{44}$ are then obtained from Eqns. (13a,b) and (13c). Furthermore, the mode Gruneisen parameter is written as

$$\gamma = -\left(\frac{\partial\ln\omega_0}{\partial\ln V}\right)_T = -\frac{B}{\omega_0}\left(\frac{\Delta\omega_h}{\sigma/3}\right) = -(\tilde{K}_{11} + 2\tilde{K}_{12})/6, \tag{14}$$

where B is the bulk modulus and $\sigma/3$ the effective hydrostatic content of these particular stresses. It is reminded that compressive stress and pressure should be taken as negative.

Uniaxial Stress along [110]

Complete removal of the triple degeneracy is obtained by a uniaxial stress $\sigma\|[110]$. The corresponding strain tensor in $x_1 x_2 x_3$ is

$$\varepsilon_{\kappa\varrho} = \frac{\sigma}{2}\begin{pmatrix} S_{11} + S_{12} & S_{44}/2 & . \\ & \bullet & \\ & & 2S_{12} \end{pmatrix}_{\|[110]} \tag{15}$$

Diagonalization of Eqn. (9) leads to three single components at frequencies Ω_1, Ω_2, Ω_3 such that

$$\Delta\Omega_1 = \Omega_1 - \omega_0 = \Delta\omega_h + \frac{1}{6}\Delta\Omega^{(001)} - \frac{1}{2}\Delta\Omega^{(111)}$$

$$\Delta\Omega_2 = \Omega_2 - \omega_0 = \Delta\omega_h + \frac{1}{6}\Delta\Omega^{(001)} + \frac{1}{2}\Delta\Omega^{(111)} \tag{16a}$$

$$\Delta\Omega_3 = \Omega_3 - \omega_0 = \Delta\omega_h - \frac{1}{3}\Delta\Omega^{(001)},$$

where $\Delta\omega_h$, $\Delta\Omega^{(001)}$, $\Delta\omega^{(111)}$ are given in Eqns. (13b,c). Now Eqn. (13a) becomes

$$\Delta\omega_h = (\Delta\Omega_1 + \Delta\Omega_2 + \Delta\Omega_3)/3. \tag{16b}$$

The three eigenvectors (phonon polarizations) are along [110], [110] and [001], respectively. This stress configuration offers no new information; it is often used for a consistency check.

Purely Bisotropic Stress

Suppose an isotropic planar stress σ is applied normal to the [001] or [111] direction of any cubic crystal. This situation arises, for instance, when a planar force is distributed radially along the periphery of a wafer. Also, it is believed to occur between growing grains of polycrystalline layers. The usual term found in the literature for this case is **biaxial stress**. However, according to Ref. 1, biaxial stress requires that $\sigma_3 = 0$ and $\sigma_1 \neq \sigma_2 (\neq 0)$ which is not the present case, in as much as here $\sigma_3 = 0$ but $\sigma_1 = \sigma_2$. The term **purely bisotropic stress** is introduced to describe the present case. The more general configuration where $\sigma_3 \neq 0$ and $\sigma_1 = \sigma_2$ is termed **bisotropic stress**. Analogous terms are suggested for strains.

In the $x_1 x_2 x_3$ system, the strain produced by the purely bisotropic stress σ is

$$\varepsilon_{\kappa\varrho} = \sigma \begin{pmatrix} S_{11}+S_{12} & \cdot & \cdot \\ & \cdot & \\ & & 2S_{12} \end{pmatrix}_{\perp [001]} , \quad \frac{\sigma}{3}\begin{pmatrix} 2(S_{11}+2S_{12}) & -S_{44}/2 & \\ & \cdot & \\ & & \cdot \end{pmatrix}_{\perp [111]} , \tag{17}$$

i.e., we have a bisotropic and a general strain, respectively. Diagonalization of Eqn. (9) reveals partial splitting of the phonon to singlet and doublet components polarized perpendicular and parallel to the plane, respectively. Equations (12) and (14) are still valid with

$$\Delta\omega_h = (\Delta\Omega_s + 2\Delta\Omega_d)/3 = \sigma\omega_0(\tilde{K}_{11}+2\tilde{K}_{12})(S_{11}+2S_{12})/3 \tag{18a}$$

$$\Delta\Omega \equiv \Omega_s - \Omega_d = \begin{cases} -\sigma\omega_0(\tilde{K}_{11}-\tilde{K}_{12})(S_{11}-S_{12})/2 \equiv -\Delta\Omega^{(001)}, & \sigma \| [001] \\ \\ -\sigma\omega_0\tilde{K}_{44}S_{44}/2 \equiv -\Delta\Omega^{(111)}. & \sigma \perp [111] \end{cases} \tag{18b}$$

Notice that the splitting $\Delta\Omega$ is opposite in sign of that for the corresponding uniaxial stress cases.

Uniaxial Strain

Uniaxial strains along [001] or [111] have the following matrix forms, in $x_1 x_2 x_3$,

$$\varepsilon_{\kappa\varrho} = \varepsilon \begin{pmatrix} \cdot & \cdot & \cdot \\ \cdot & \cdot & \cdot \\ \cdot & \cdot & 1 \end{pmatrix}_{\|[001]} , \quad \frac{\varepsilon}{3}\begin{pmatrix} 1 & 1 & 1 \\ \cdot & 1 & 1 \\ \cdot & \cdot & 1 \end{pmatrix}_{\|[111]} . \tag{19}$$

Equation (9) gives, for $\varepsilon \| [001]$

$$\Delta\Omega_s = \tilde{K}_{11}\omega_0\varepsilon/2 \tag{20a}$$

$$\Delta\Omega_d = \tilde{K}_{12}\omega_0\varepsilon/2 \tag{20b}$$

$$\Delta\Omega_h = (\Delta\Omega_s + 2\Delta\Omega_d)/3 = \omega_0(\tilde{K}_{11}+2\tilde{K}_{12})\varepsilon/6 = -\gamma\omega_0\varepsilon \tag{20c}$$

and, for $\varepsilon \| [111]$

$$\Delta\Omega_s = \omega_0(\tilde{K}_{11} + 2\tilde{K}_{12} + 4\tilde{K}_{44})\varepsilon/6 \tag{21a}$$

$$\Delta\Omega_d = \omega_0(\tilde{K}_{11} + 2\tilde{K}_{12} - 2\tilde{K}_{44})\varepsilon/6 \tag{21b}$$

$$\Delta\omega_h = (\Delta\Omega_s + 2\Delta\Omega_d)/3 = \omega_0(\tilde{K}_{11} + 2\tilde{K}_{12})\varepsilon/6 = -\gamma\omega_0\varepsilon . \tag{21c}$$

Situations involving uniaxial strains occur in ion-implanted materials and during pulsed laser annealing.

Zincblende-type crystals

The triply-degenerate $q\approx0$ phonon in ZB materials is split by the phonon electric field into a transverse (TO, ω_T) and a longitudinal (LO, ω_L) component. The effective force constants of the TO, LO phonons are different, in general, and so are the corresponding PDP. Thus, we expect to have two independent sets of PDP, i.e., $\tilde{K}_{ij}^T$ and $\tilde{K}_{ij}^L$, ij=11, 12, 44. The results obtained in this section are valid independently for the TO and LO phonons, and therefore all related parameters should be characterized by an additional index T or L, respectively.

LATTICE MISMATCH STRAINS - HIGH SYMMETRY DIRECTIONS OF GROWTH

The most interesting cases of built-in strains occur in lattice mismatched epilayers or superlattices of two different materials 1 and 2 (lattice constants a_1 and a_2) which are grown along x_3' with direction cosines l_3, m_3, n_3. Assuming that the mismatch strain is entirely absorbed by material 1, then $\varepsilon_\parallel = (a_2 - a_1)/a_1 = \Delta a/a_1$ represents the maximum in-plane isotropic strain in material 1, i.e., $\varepsilon_1 = \varepsilon_2 \equiv \varepsilon_\parallel \neq \varepsilon_3 \equiv \varepsilon_\perp$. The in-plane axes x_1', x_2' are chosen to form, together with x_3', a right-handed orthogonal system $x_1' x_2' x_3'$. To simplify the form of the DSE we assume in this section that the symmetry is high enough to ensure that, except for $\varepsilon_\parallel$ and $\varepsilon_\perp$, all other strain components are zero (bisotropic strain). The non-zero stress components allowed in this case are σ_1' and σ_2' ($\neq\sigma_1'$, in general) i.e., we have a biaxial stress configuration ($\sigma_3'=0$). If, in addition, $\sigma_1' = \sigma_2'$ for all choices of x_1', x_2', then we have a purely bisotropic stress configuration. Knowledge of the stresses is not really necessary for handling the DSE, but they are derived here only for completeness. The most general case of arbitrary directions of $x_1' x_2' x_3'$ will be treated in the following section.

Setting $\sigma_3'=0$, after having expressed the elastic stiffnesses in $x_1' x_2' x_3'$, yields the following general results for the tetragonal distortion $\Delta\varepsilon = \varepsilon_\parallel - \varepsilon_\perp$ and the in-plane stresses,

$$\Delta\varepsilon \equiv \frac{(C_{11} + 2C_{12})\varepsilon_\parallel}{(C_{12} + 2C_{44}) + C(l_3^4 + \text{c.p.})} \tag{22a}$$

$$\sigma_{\kappa\kappa}' = \left\{ 2C_{44} + C\,[\,(l_3^4 + \text{c.p.}) - (l_3^2 l_\kappa^2 + \text{c.p.})\,] \right\} \Delta\varepsilon, \tag{22b}$$

where $C \equiv C_{11} - C_{12} - 2C_{44}$. By c.p. we designate the cyclic permutation $l \rightarrow m \rightarrow n \rightarrow l$

179

of the direction cosines l_i, m_i, n_i corresponding to x'_i. It is clear from (22b) that the condition for having purely bisotropic stress $(\sigma_{11} = \sigma_{22})$ is

$$l_3^2 (l_1^2 - l_2^2) + \text{c.p.} = 0 .\tag{22c}$$

Equations (22) continue to hold even if $\varepsilon_{\parallel} < \Delta a/a_1$, due to dislocations, etc. It is also assumed that the strain field is homogeneous over the entire volume of the strained layer. We apply Eqns. (22) to the following three geometries most frequently encountered in EL and SL technology:

Growth along $[l_3 m_3 n_3] = [001]$

With $[l_1 m_1 n_1] = [100]$, $[l_2 m_2 n_2] = [010]$, Eqns. (22) give the well-known results

$$\varepsilon_{\perp} = -\frac{2C_{12}}{C_{11}} \varepsilon_{\parallel}\tag{23a}$$

$$\sigma_1 = \sigma_2 = (C_{11} - C_{12})\Delta\varepsilon,\tag{23b}$$

in agreement with (22c). Equation (9) is directly applicable and gives

$$\Delta\Omega_s = \omega_0 (2\tilde{K}_{12}\varepsilon_{\parallel} + \tilde{K}_{11}\varepsilon_{\perp})/2\tag{23c}$$

$$\Delta\Omega_d = \omega_0 \left[(\tilde{K}_{11} + \tilde{K}_{12})\varepsilon_{\parallel} + \tilde{K}_{12}\varepsilon_{\perp} \right]/2\tag{23d}$$

$$\Delta\omega_h = (\Delta\Omega_s + 2\Delta\Omega_d)/3 = \omega_0(\tilde{K}_{11} + 2\tilde{K}_{12})(2\varepsilon_{\parallel} + \varepsilon_{\perp})/6.\tag{23e}$$

Growth along $[l_3 m_3 n_3] = [111]/\sqrt{3}$

With $[l_1 m_1 n_1] = [11\bar{2}]/\sqrt{6}$ and $[l_2 m_2 n_2] = [\bar{1}10]/\sqrt{2}$ we find

$$\varepsilon_{\perp} = -2 \left(\frac{C_{11} + 2C_{12} - 2C_{44}}{C_{11} + 2C_{12} + 4C_{44}} \right) \varepsilon_{\parallel}\tag{24a}$$

$$\sigma'_1 = \sigma'_2 = 2C_{44}\Delta\varepsilon.\tag{24b}$$

The strain tensor is expressed in $x'_1 x'_2 x'_3$; we choose to remain in this system and rewrite Eqns. (4, 5) with $K_{\alpha\beta\kappa\varrho} = K_{ij}$ replaced by K'_{ij}. Eqn. (5) then becomes

$$\begin{vmatrix} (K'_{11} + K'_{12})\varepsilon_{\parallel} + K'_{13}\varepsilon_{\perp} - \lambda & -K'_{36}\Delta\varepsilon & -K'_{35}\Delta\varepsilon \\ -K'_{36}\Delta\varepsilon & (K'_{21} + K'_{22})\varepsilon_{\parallel} + K'_{23}\varepsilon_{\perp} - \lambda & -K'_{34}\Delta\varepsilon \\ -K'_{35}\Delta\varepsilon & -K'_{34}\Delta\varepsilon & (K'_{31} + K'_{32})\varepsilon_{\parallel} + K'_{33}\varepsilon_{\perp} - \lambda \end{vmatrix} = 0\tag{25}$$

where, for this choice of axes,

$$K'_{11} = K'_{22} = \tfrac{1}{2} (K_{11} + K_{12} + 2K_{44})$$

$$K'_{33} = \frac{1}{3}(K_{11} + 2K_{12} + 4K_{44})$$

$$K'_{12} = K'_{21} = \frac{1}{6}(K_{11} + 5K_{12} - 2K_{44})$$ (26)

$$K'_{13} = K'_{31} = K'_{32} = \frac{1}{3}(K_{11} + 2K_{12} - 2K_{44})$$

$$K'_{36} = K'_{35} = K'_{34} = 0,$$

and likewise for $\tilde{K}'_{ij}$. In deriving the transformed components of Eqns. (26) we have taken into account that K_{ij} is identical, symmetry-wise, to the fourth-rank tensor C_{ij} and that $K'_{16} + K'_{26} + K'_{36} = 0$ (Ref. 4). Eqn. (25) gives,

$$\Delta\Omega_s = \omega_0 \left[2\tilde{K}'_{13}\varepsilon_{\parallel} + \tilde{K}'_{33}\varepsilon_{\perp} \right]/2$$ (27a)

$$\Delta\Omega_d = \omega_0 \left[(\tilde{K}'_{11} + \tilde{K}'_{12})\varepsilon_{\parallel} + \tilde{K}'_{13}\varepsilon_{\perp} \right]/2$$ (27b)

$$\Delta\omega_h = (\Delta\Omega_s + 2\Delta\Omega_d)/3 = \omega_0(\tilde{K}_{11} + 2\tilde{K}_{12})(2\varepsilon_{\parallel} + \varepsilon_{\perp})/6.$$ (27c)

The eigenvector of the s(d) component is $\parallel(\perp)$ to x'_3.

Growth along $[l_3 m_3 n_3] = [110]/\sqrt{2}$

We choose $[l_1 m_1 n_1] = [001]$ and $[l_2 m_2 n_2] = [1\bar{1}0]/\sqrt{2}$. The condition of Eqn. (22c) predicts $\sigma'_1 \neq \sigma'_2$. The calculation yields

$$\varepsilon_{\perp} = -\left(\frac{C_{11} + 3C_{12} - 2C_{44}}{C_{11} + C_{12} + 2C_{44}} \right) \varepsilon_{\parallel}$$ (28a)

$$\sigma'_1 = \left(\frac{C_{11} - C_{12} + 2C_{44}}{C_{11} + C_{12} + 2C_{44}} \right)(C_{11} + 2C_{12})\varepsilon_{\parallel}$$ (28b)

$$\sigma'_2 = \left(\frac{4C_{44}}{C_{11} + C_{12} + 2C_{44}} \right)(C_{11} + 2C_{12})\varepsilon_{\parallel} .$$ (28c)

The triple degeneracy is completely removed, as can be seen from (25), with K_{ij} expressed in $x'_1 x'_2 x'_3$. The calculation is lengthy but straightforward.

LATTICE MISMATCH STRAINS - ARBITRARY DIRECTION OF GROWTH

For an arbitrary direction of growth of EL or SL the component array of strains/stresses is not so simple. Such lower symmetry directions of growth are increasingly attracting interest because of the new physical and technical aspects behind their higher anisotropy. A complete solution is presented here[5].

We assume, as before, that $\varepsilon_{\parallel}$ is known from independent arguments. The basic equations are,

$$\sigma'_i = C'_{ij}\, \varepsilon'_j \tag{29}$$

$$C'_{ij} = C_{ij} + C(T_{ij} - T^o_{ij}) \tag{30a}$$

$$C = C_{11} - C_{12} - 2C_{44} \tag{30b}$$

$$T_{ij} = T_{\lambda\mu\nu\varrho} = l_\lambda l_\mu l_\nu l_\varrho + m_\lambda m_\mu m_\nu m_\varrho + n_\lambda n_\mu n_\nu n_\varrho, \tag{30c}$$

where $T^o_{ij} = 1$ for $i = j \le 3$, and zero otherwise. Furthermore, the boundary conditions in this generalized case are,

$$\varepsilon'_1 = \varepsilon'_2 = \varepsilon_{\parallel}, \; \varepsilon'_6 = \sigma'_3 = \sigma'_4 = \sigma'_5 = 0 \tag{31a}$$

$$\varepsilon'_3 \equiv \varepsilon_{\perp}, \; \varepsilon'_4, \; \varepsilon'_5, \; \sigma'_1, \; \sigma'_2, \; \sigma'_6 \neq 0 \,. \tag{31b}$$

Equations (30a,c) have the advantage that they allow the invariant properties of C'_{ij} and T_{ij} to be fully exploited, and this simplifies matters. The invariant properties are[4],

$$\sum_j^3 T_{ij} = \sum_j^3 T_{ji} = 1 \; (=0) \qquad i = 1,2,3 \; (=4,5,6) \tag{32a}$$

$$\sum_j^3 C'_{ij} = \sum_j^3 C'_{ji} = C_{11} + 2C_{12} \; (=0) \tag{32b}$$

$$T_{12} - T_{66} = 0, \qquad T_{14} - T_{56} = 0, \quad \text{c.p.} \tag{32c}$$

$$C'_{12} - C'_{66} = C_{12} - C_{44} = \text{c.p.}, \qquad C'_{36} - C'_{45} = \text{c.p.} = 0, \tag{32d}$$

where c.p. refers to the cyclic permutation $1 \to 2 \to 3 \to 1$ or $4 \to 5 \to 6 \to 4$. From (29)-(32) the following general results are obtained[5],

$$\Delta\varepsilon = 3B \left[C^2_{44} + CC_{44}(1-T_{33}) + C^2(T_{31}T_{32}-T^2_{36}) \right] \varepsilon_{\parallel}/\Delta \tag{33a}$$

$$= 3B \left[C^2_{44} + CC_{44}(1-T_{33}) + 3C^2(l_3 m_3 n_3)^2 \right] \varepsilon_{\parallel}/\Delta \tag{33b}$$

$$\varepsilon_{\perp} = -\left[2C_{12}C^2_{44} + C\,(C_{11}+3C_{12})(1-T_{33})C_{44}/2 \right.$$
$$\left. + C^2(2C_{11}+4C_{12}-C_{44})(l_3 m_3 n_3)^2 \right] \varepsilon_{\parallel}/\Delta \tag{34}$$

$$\varepsilon'_4 = 3BC \left[C_{44}T_{34} + C(T_{31}T_{34}-T_{35}T_{36}) \right] \varepsilon_{\parallel}/\Delta \tag{35a}$$

$$\varepsilon'_5 = 3BC \left[C_{44}T_{35} + C(T_{32}T_{35}-T_{36}T_{34}) \right] \varepsilon_{\parallel}/\Delta \tag{35b}$$

$$\sigma'_1 = 3B\varepsilon_{\parallel} - (C_{12}+CT_{31})\Delta\varepsilon + C(T_{14}\varepsilon'_4 + T_{15}\varepsilon'_5) \tag{36a}$$

$$\sigma_2' = 3B\varepsilon_\| - (C_{12} + CT_{32})\Delta\varepsilon + C(T_{24}\varepsilon_4' + T_{25}\varepsilon_5') \tag{36b}$$

$$\sigma_6' = -CT_{36}\Delta\varepsilon + C(T_{25}\varepsilon_4' + T_{14}\varepsilon_5') , \tag{36c}$$

where

$$\Delta = C^3 \begin{vmatrix} T_{33} - 1 + C_{11}/C & T_{34} & T_{35} \\ T_{34} & T_{32} + C_{44}/C & T_{36} \\ T_{35} & T_{36} & T_{31} + C_{44}/C \end{vmatrix} \tag{37a}$$

$$= C_{11}C_{44}^2 + C(C_{11} + C_{12})(1 - T_{33})C_{44}/2 + C^2(C_{11} + 2C_{12} + C_{44})(l_3 m_3 n_3)^2 \tag{37b}$$

In addition, the elastic free energy density is found to be

$$U(\varepsilon_\|) = \frac{1}{2}\sum_{i=1}^{6}\varepsilon_i'\sigma_i' = \frac{1}{2}\varepsilon_\|(\sigma_1' + \sigma_2') = \frac{3}{2}B\varepsilon_\|(3\varepsilon_\| - \Delta\varepsilon)$$

$$= \frac{3}{2}B\varepsilon_\|(2\varepsilon_\| + \varepsilon_\perp) \tag{38a}$$

$$= \frac{3}{2}B\left[\frac{4C_{44}(C_{11} - C_{12}) + C(C_{11} - C_{12})(1 - T_{33}) + 6C^2(l_3 m_3 n_3)^2}{2C_{11}C_{44} + C(C_{11} + C_{12})(1 - T_{33}) + 2C^2(1 + 3B/C_{44})(l_3 m_3 n_3)^2}\right]\varepsilon_\|^2. \tag{38b}$$

The results of Eqns. (33)-(36) describe the complete array of strain/stress components subject to the conditions of Eqns. (31). The necessary parameters for their calculation are the components C_{ij}, $\varepsilon_\|$ and certain components of T_{ij}, which are easily obtained from Eqn. (30c). Depending on the choice of axes it may, in fact, happen that many of the T_{ij} components are zero and this simplifies the computation significantly. In principle, the above formalism could also be based on $\varepsilon_i' = S_{ij}\sigma_j'$ instead of Eqn. (29). Due to the conditions of Eqns. (31), however, the computation becomes more involved and the results are necessarily left formulated in terms of S_{ij}. Often in the literature, the off-diagonal strain/stress components are ignored. The solution presented here is most general and easily applicable. It is even possible to predict the non-zero components for the specific system of axes, simply by examining the various components of T_{ij}. Furthermore, conditions can be established for having particular strain/stress components equal to zero. For instance, in order to have $\varepsilon_4' = \varepsilon_5' = \sigma_6' = 0$, it is necessary and sufficient that $T_{34} = T_{35} = T_{36} = 0$, according to Eqns. (35a,b) and (36c). It is easy then to show that the earlier results of Eqns. (22), for which the conditions $\varepsilon_4' = \varepsilon_5' = \sigma_6' = 0$ were assumed to be valid, can be derived from the most general expressions of Eqns. (33a) and (36a,b), simply by setting $T_{34} = T_{35} = T_{36} = 0$.

Having established the complete component array of the strain tensor in the system $x_1' x_2' x_3'$, the DSE of Eqn. (5) can be readily used with

$$\Delta K_{\alpha\beta}' = \Delta K_i' = K_{ij}'\varepsilon_j' , \qquad\qquad i = 1\text{-}6$$

$$= (K_{11}+2K_{12})\varepsilon_{\parallel} - K'_{i3}\Delta\varepsilon + (K'_{i4}\varepsilon'_4 + K'_{i5}\varepsilon'_5), \qquad i=1\text{-}3 \qquad (39a)$$

$$= \qquad - K'_{i3}\Delta\varepsilon + (K'_{i4}\varepsilon'_{i4} + K'_{i5}\varepsilon'_5). \qquad i=4\text{-}6 \qquad (39b)$$

Notice that K'_{ij} transforms under rotation in the same way C'_{ij} does, i.e.[4],

$$K'_{ij} = K_{ij} + K(T_{ij}-T^0_{ij}),$$

with $K = K_{11}-K_{12}-2K_{44}$. Formally, this completes the discussion on strain-split phonons in **EL and SL grown along an arbitrary direction**. Some typical directions are treated next, to show the effectiveness of the general approach presented here in determining the complete strain/stress field.

<u>Growth along $[l_3m_3n_3] = [111]/\sqrt{3}$</u>

With $[l_1m_1n_1] = [11\bar{2}]/\sqrt{6}$ and $[l_2m_2n_2] = [\bar{1}10]/\sqrt{2}$, we easily find

$$T_{13}= T_{23}= T_{33}=1/3, \quad T_{25} = -T_{15} = \sqrt{2}/6, \quad T_{14}= T_{24}= T_{34}= T_{35}= T_{36}= 0$$

and the results of Eqns. (24a,b) are reproduced. The same results hold for any other system which is produced by rotating x'_1, x'_2, about x'_3 by an arbitrary angle θ. Hence the stress is purely bisotropic ($\sigma'_1=\sigma'_2$, for all choices of x'_1, x'_2, and $\sigma'_3=0$) while the strain is bisotropic ($\varepsilon'_1=\varepsilon'_2$, $\varepsilon'_3\neq0$). The same comment applies when $x'_1x'_2x'_3$ coincide with the $<100>$ axes. The results in this case are given by Eqns. (23a,b).

<u>Growth along $[\bar{1}10]/\sqrt{2}$</u>

With $[l_1m_1n_1]= [111]/\sqrt{3}$ and $[l_2m_2n_2]= [11\bar{2}]/\sqrt{6}$ we have

$$T_{13}= 2T_{23}= 1/3, \quad T_{33}= 1/2, \quad T_{36}= \sqrt{2}/6, \quad T_{14}= T_{24}= T_{15}= T_{25}= T_{34} =T_{35}= 0$$

$$\Delta\varepsilon = \varepsilon_{\parallel}-\varepsilon_{\perp} = \frac{6B\varepsilon_{\parallel}}{C_{11}+C_{12}+2C_{44}}, \quad \varepsilon'_4 = \varepsilon'_5= 0 \qquad (41a)$$

$$\sigma'_1 = (C_{11}-C_{12}+10C_{44})\Delta\varepsilon/6 \qquad (41b)$$

$$\sigma'_2 = (C_{11}-C_{12}+4C_{44})\Delta\varepsilon/3 \qquad (41c)$$

$$\sigma'_6 = -\sqrt{2}\,C\Delta\varepsilon/6. \qquad (41d)$$

These results are to be contrasted to Eqns. (28), i.e., the results for growth along $[110]/\sqrt{2}$ with x'_1 and x'_2 along $[001]$ and $[1\bar{1}0]/\sqrt{2}$, respectively. $\Delta\varepsilon$, ε'_4, ε'_5 remain the same but the stress components are different.

Application of Eqns. (41) to Si with $C_{11}= 166$, $C_{12}= 64$, $C_{44}= 80$ in GPa at 300K, gives $\Delta\varepsilon = 1.51\varepsilon_{\parallel}$, $\varepsilon_{\perp}= -0.51\varepsilon_{\parallel}$, $\sigma'_1= 227\varepsilon_{\parallel}$, $\sigma'_2= 212\varepsilon_{\parallel}$, $\sigma'_6= 21\varepsilon_{\parallel}$, in GPa. On the other hand, Eqns. (28) yield $\sigma'_1= 198\varepsilon_{\parallel}$ and $\sigma'_2= 242\varepsilon_{\parallel}$. A typical value for $\varepsilon_{\parallel}$

in the Si layer of a Si/Ge short period superlattice is 0.04 [9]. Hence, $\sigma'_{1,2} \sim 8$ to 10 GPa and $\sigma'_6 \sim \sigma'_{1,2}/10$.

Growth along [11$\bar{2}$]/$\sqrt{6}$

We take $[l_1 m_1 n_1] = [\bar{1}10]/\sqrt{2}$, $[l_2 m_2 n_2] = [111]/\sqrt{3}$, then

$$2T_{13} = T_{23} = 1/3, \quad T_{33} = 1/2, \quad T_{14} = -T_{34} = \sqrt{2}/6, \quad T_{24} = T_{15} = T_{25} = T_{35} = T_{36} = 0$$

$$\Delta\varepsilon = \frac{18 B\varepsilon_\parallel (C_{11} - C_{12} + C_{44})}{3(C_{11} + C_{12} + 2C_{44})(C_{11} - C_{12} + C_{44}) - C^2} \tag{42a}$$

$$\varepsilon'_4 = -\frac{\sqrt{2}\, C\Delta\varepsilon}{2(C_{11} - C_{12} + C_{44})}, \quad \varepsilon'_5 = \sigma'_6 = 0 \tag{42b}$$

$$\sigma'_1 = 6B\varepsilon_\parallel - 2(C_{11} + 2C_{12} + C_{44})\Delta\varepsilon/3 \tag{42c}$$

$$\sigma'_2 = 3B\varepsilon_\parallel - (C_{11} + 2C_{12} - 2C_{44})\Delta\varepsilon/3. \tag{42d}$$

Application of Eqns. (42) to Si as before gives: $\Delta\varepsilon = 1.53\varepsilon_\parallel$, $\varepsilon_\perp = -0.53\varepsilon_\parallel$, $\varepsilon'_4 = 0.34\varepsilon_\parallel$, $\sigma'_1 = 207\varepsilon_\parallel$, $\sigma'_2 = 226\varepsilon_\parallel$, in GPa. Notice that ε'_4 is an appreciable fraction of $\varepsilon_\parallel$, and as such it should be taken into consideration in relevant applications.

POLYCRYSTALLINE MEDIA

Next we consider media of randomly distributed polycrystallites. For convenience we take again Si as the basis for the discussion. The first question is to express the elastic (C^*_{ij}) and lattice dynamical (K^*_{ij}) parameters of poly-Si in terms of those for bulk Si using an appropriate averaging procedure. Then, solving the DSE is straightforward.

Elastic Stiffness

The so-called Voigt-Reuss-Hill (VRH) average is adopted here. According to this, the poly-Si elastic stiffness components are given by

$$C^*_{ij} = \frac{1}{2}(C^V_{ij} + C^R_{ij}), \tag{43}$$

where $C^{V(R)}_{ij}$ are the Voigt (Reuss) averages given by [10],

$$C^V_{11} = C_{11} - 2C/5, \quad C^V_{12} = C_{12} + C/5 \tag{44a}$$

$$C^V_{44} = (C^V_{11} - C^V_{12})/2 = C_{44} + C/5 \tag{44b}$$

$$C^R_{11} = C^V_{11} - 2C^2/5C', \quad C^R_{12} = C^V_{12} + C^2/5C' \tag{45a}$$

$$C^R_{44} = (C^R_{11} - C^R_{12})/2 = C^V_{44} - 3C^2/10C' \tag{45b}$$

185

with $C' = 3(C_{11}-C_{12})/2+2C_{44}$. The PDP of poly-Si are obtained likewise, i.e.,

$$K^*_{ij} = \tfrac{1}{2} (K^V_{ij}+K^R_{ij}), \tag{46}$$

where K^V_{ij} obeys similar expressions to Eqn. (44), with $K = K_{11}-K_{12}-2K_{44}$. The Reuss average is more complicated to obtain, i.e.,

$$K^R_{11} = L^R_{11}C^R_{11}+2L^R_{12}C^R_{12} \tag{47a}$$

$$K^R_{12} = L^R_{11}C^R_{12}+L^R_{12}(C^R_{11}+C^R_{12}) \tag{47b}$$

$$K^R_{44} = (K^R_{11}-K^R_{12})/2 = L^R_{44}C^R_{44}. \tag{47c}$$

L^R_{ij} is the Reuss average of a new tensor L_{ij}, i.e.,

$$L^R_{11} = L_{11}-2L/5, \qquad L^R_{12} = L_{12}+L/5 \tag{48a}$$

$$L^R_{44} = L^R_{11}-L^R_{12} = L_{44}+2L/5, \tag{48b}$$

where $L = L_{11}-L_{12}-L_{44}$, and L_{ij} is defined from

$$L_{ij} = K_{im}S_{mj} \equiv \omega^2_o \tilde{L}_{ij}. \tag{49}$$

Thus, starting from Eqn. (49) backwards, the PDP for poly-Si can be calculated in terms of the three values of $\tilde{K}_{ij}$ (Table I), C_{ij} (168, 65, 80 GPa, 80K) of bulk Si. The results are: $\tilde{L}_{ij}=(-4.4, \ -8.8, \ -8.75)\times10^{-3}$ GPa^{-1}, $\tilde{L}^*_{ij}=(-10.2, \ -6.0, \ -4.15)\times10^{-3}$ GPa^{-1} and $\tilde{K}^*_{ij}=(-2.55, \ -2.0, \ -0.28)$.

Strained Polycrystalline Systems

In a bisotropically strained polycrystalline layer with $\varepsilon^*_{\parallel}$ known from independent sources, the normal strain $\varepsilon^*_{\perp}$ and stresses are the same as for a crystalline layer grown along [001]. The results are identical to those in Eqns. (23a,b) except that all parameters should be replaced by their starred counterparts. The DSE of Eqn. (9) is valid with only diagonal terms, and with K_{ij} replaced by K^*_{ij}. The phonon frequency splittings are those in Eqs. (23c,d,e). If, on the other hand, a strain is applied externally on a bulk poly-crystalline system, the general formalism based on Eqn. Eqn. (9) and K^*_{ij} should be followed. The definition of axes here is arbitrary.

CONTROL EXPERIMENTS

Following the early measurements of RS on Si under stress twenty years ago[11], a number of materials were subsequently studied by various workers, including diamond- and ZB-type structures, fluorides and some uniaxial crystals. A tabulation of the results up to 1977 can be found in Ref. 3. The interest on PDP has been revived in recent years in connection with strain characterization of material systems through Raman spectroscopy. A number of crystals, mainly III-V, ZB structures have been examined or re-examined through control experiments, and accurate values for the PDP are now available for use in

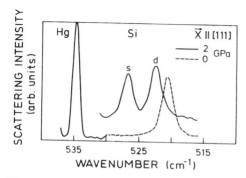

Fig. 1. Raman spectrum of Si under 2 GPa along [111] at 110 K. Nd:YAG laser. Hg lamp for calibration.

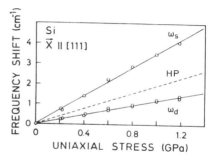

Fig. 2. Singlet (doublet) shifts $\Delta\Omega_s$ ($\Delta\Omega_d$) of Si versus X∥[111] at 110 K. Nd:YAG laser. HP: hydrostatic shift calculated from Eqn. (13a).

diagnostic work. The RS experimental procedure has been described in detail in the literature and will not be repeated here.

The first control experiment was performed on Si [11] using a He-Ne laser for excitation (632.8nm). Later it was repeated with the 647.1 nm line of a Kr[+] laser[12]. Both these experiments suffer from the fact that the penetration of light is limited to layers near the surface. This means that any variation (relaxation) of the applied stress near the surface will affect the values of PDP. The remedy to this is to use laser lines in the transparent region of the crystal and perform 90° scattering. This procedure has been followed for a number of III-V crystals and more recently for Si [13], using the 1.064 nm line of the Nd:YAG CW laser. The measurements on Si were performed at 110 K, in order to quench the indirect gap luminescence, which at 300 K is too strong to allow observation of any Raman signals. A typical spectrum and data are shown in Figs. 1 and 2. These newer values for Si. and other semiconductors are compiled in Table I and are recommended for use in applications, as more accurate and free of stress relaxation effects.

The measurements of the two sets of PDP in ZB-type crystals requires the full power of polarization selection rules in order to discriminate the stress-split components of the LO and TO phonons. The limitations of stress relaxation near the surface still exist in the case of opaqueness. Moreover, backward scattering prohibits the measurement of the singlet component of the LO phonon. The first PDP data on GaAs, GaSb, InAs, ZnSe [14] and InSb [20] were obtained in the region of opaqueness (with the exception of ZnSe) and were not complete since the PDP of the LO components could not and/or were not measured. An interesting scattering geometry, involving oblique-incident forward scattering in the transparent region was first applied to GaP [21] and more recently to GaAs, In-hardened GaAs, AlSb, and InP (Table I).

APPLICATIONS

The phenomenological approach reviewed here has been used routinely in strain characterization problems in the last two decades. Some selective references are made here with emphasis on work published after 1984. Work published prior to 1984 is reviewed in Ref. 22.

Table I. Dimensionless Phonon Deformation Potentials and Gruneisen parameters for diamond- and zincblende-type crystals. Values in parenthesis refer to LO phonons. Frequencies in cm^{-1}. V (S) : Volume (Surface) scattering.

Material	ω_0	$\tilde{K}_{11}$	$\tilde{K}_{12}$	$\tilde{K}_{44}$	γ	T(K)	Ref.	V(S)
Si	521	-1.85	-2.3	-0.7	1.07	110	13	V
Ge	300	-1.45	-1.95	-1.1	0.89	300	14	S
Diamond	1332	-2.9	-1.9	-1.2	1.12	300	15	V
GaAs	269 (292)	-2.4 (-1.7)	-2.7 (-2.4)	-0.9 (-0.55)	1.29 (1.09)	300	16	V
GaAs:In	268 (291)	-2.35 (-1.55)	-2.65 (-2.4)	-0.55 (-0.15)	1.27 (1.05)	300	17	V
InP	304 (351)	-2.5 (-1.6)	-3.2 (-2.8)	-0.5 (-0.2)	1.48 (1.19)	300	18	V
AlSb	317 (338)	-2.1 (-1.6)	-2.6 (-2.6)	-0.7 (-0.3)	1.21 (1.15)	300	19	V
GaSb	231 (241)	-1.9 (...)	-2.35 (...)	-1.1 (...)	1.10 (...)	300	14	V
InAs	218 (243)	-0.95 (...)	-2.1 (...)	-0.75 (...)	0.85 (...)	300	14	V
ZnSe	205 (250)	-2.75 (...)	-4.0 (...)	-0.45 (...)	1.80 (...)	300	14	V
InSb	180 (191)	-2.45 (...)	-3.3 (...)	-0.6 (...)	1.50 (...)	77	20	S
GaP	366 (403)	-1.35 (-1.45)	-1.95 (-2.5)	-0.6 (-0.5)	0.87 (1.07)	300	21	V

Heterostructures and Superlattices of Diamond-type Materials

The classical example of a strained heterostructure is the Si-on-Sapphire (SOS) film[23]. It exhibits a bisotropic strain due to lattice mismatch between Si and Sapphire, and the difference in their thermal expansion coefficients. Using Eqns. (12,18) and the most recent PDP values of Si from Table I, we find for backward scattering along [001],

$$\Delta\Omega_s = \Omega_s - \omega_o = \omega_o [\tilde{K}_{11}S_{12} + \tilde{K}_{12}(S_{11} + S_{12})] \, \sigma \qquad (50a)$$

$$\sigma_{\parallel}(Kb) = -2.2 \, \Delta\Omega_s(cm^{-1}). \qquad (50b)$$

Accordingly, the observed shift of +3.5 cm^{-1} in Ref. 23 is interpreted as a (compressive) stress of ~7.7 Kb. It is emphasized that Eqn. (50b) is valid only for Si and only for backward scattering along [001].

Extensive Raman diagnostic work reported on Si heterostructures includes: conditions for growing stress-free films using laser annealing[24] or combination of implantation and laser or thermal annealing[25-28], topography (i.e., spatial distribution) of such strains[29,30], quality control of Si/CaF$_2$[31], Si/GaP[32], Si/SiO$_2$[33] heterostructures, and others[34,35].

In addition to strains on SOS films, residual strains induced by lattice defects and/or implantation treatment of bulk or poly-Si has been the subject of continuous investigations[36,37]. CW and pulsed-laser annealing of Si samples of various preparations have been shown to induce tensile strains, as inferred from the downward shift of the Raman spectra[38,39].

The most interesting cases are provided by strained SL. We take, as an example, the Si/Ge$_x$Si$_{1-x}$ superlattice, a most promising taylored structure for high performance optoelectronic devices. By proper choice of excitation frequency it is possible to determine the level of strains, and their distribution among the layers, as a function of thickness[40,41]. Similar situations have been confirmed in strained Si/Ge superlattices[9,42,43]. The work in Ref. 9, in particular, refers to short period Si/Ge SL grown along the "low-symmetry" direction [110]. The frequency splitting is found in accordance with Eqn. (28a). The two TO split-phonon components are separated by use of polarization selection rules. No RS by the LO component is allowed for backward geometry along [110].

Spatial distribution of strains in Si heterostructures can be probed and mapped by use of Raman microprobing techniques. Classical examples refer to studies of Si formations (stripes, islands, etc.) under built-in strain[44-48]. Such observations are particularly useful since they can be made during intermediate phases of crystal growth, thus assisting in the identification of the processes responsible for the strain field. Other types of heterostructures have been studied likewise, such as Ge and GaAs-Ge films grown on SiO$_2$-coated Si substrates[49], and Si-on-quartz films[50]. A comprehensive review on the possibilities of Raman microprobing is given in Ref. 51.

Heterostructures and Superlattices of Zincblende-type Materials

The application of Raman probing and microprobing in the analysis of strains has been applied extensively to materials with ZB structure. Again, the problems considered refer to: quality of structures after treatment (implantation, laser or heat annealing, etc.), residual strains, and microprobing applications for heterostructures and superlattices. The difficulty here is that the PDP values for the LO phonons are not the same as those for the TO phonons; information on all six parameters $\widetilde{K}_{ij}^T$ and $\widetilde{K}_{ij}^L$ has become available for a few materials only very recently (Table I). Often the difference between $\widetilde{K}_{ij}^L$ and $\widetilde{K}_{ij}^T$ is not recognized in the literature, or it is simply overlooked. Very frequently the values of $\widetilde{K}_{ij}^T$ are used in place of $\widetilde{K}_{ij}^L$. Moreover, when neither set of PDP is known, the PDP of similar materials are used. Both practices lead only to approximate results for the strains. Some characteristic examples of strain analysis in ZB structures are given next. Unless specified, backward scattering along [001] by LO phonons is involved in the presence of a bisotropic strain. Thus, the results of previous sections are directly applicable with an index L added to all phonon related parameters.

The strains induced by mechanical and/or chemical treatment of surfaces of bulk materials (GaAs, InAs, GaSb, InP) have been monitored by RS[52,53]. Encapsulation of GaAs, or its alloys, by Si$_3$N$_4$ has been shown to produce strains[54,55] which in turn induce lattice disorder and subsequent increase of forbidden scattering by TA or LA phonons[55]. Residual strains in GaAs induced by laser pyrolytic etching in a CCl$_4$ gas atmosphere were studied with Raman microprobing[56].

Numerous combinations of materials in thin film/substrate heterostructures have been investigated with RS in order to establish the existence of misfit strains[57,65] and their dependence on layer thickness (e.g., ZnSe/GaAs, Refs. 57,65), topography (GaAs/Si, Ref. 58; $In_xGa_{1-x}As/InP$, Ref. 59), dislocations ($In_xGa_{1-x}P/GaAs$, Ref. 60; 3C-SiC/Si, Ref. 61), preparation details (GaP/GaAs and GaAs/GaP, Ref. 62; ZnSe/GaAs, Ref. 63; 3C-SiC/Si, Ref. 64), substrate orientation (3C-SiC/Si, Ref. 64), and alloy concentration ($Ga_{1-x}In_xAs/GaAs$, Ref. 65; $Ga_{1-x}In_xAs/InP$ or GaAs, Ref. 66; $Ga_{1-x}In_xP/GaAs$, Ref. 67).

Strain diagnostics in GaAs-based superlattices is rather limited. The existence of misfit strains and their dependence on superlattice parameters has been studied in $GaAs/In_xGa_{1-x}As$[68,69], in $GaAs/In_xAl_{1-x}As$[70,71] and monolayer GaP/InP [72]. The well-known combination $GaAs/Al_xGa_{1-x}As$ is nearly lattice-matched and does not present an interesting case for strain characterization. On the contrary, the strained superlattices GaSb/AlSb exhibit sufficient lattice mismatch (~0.65% at 300 K) to manifest itself through Raman spectra[73-75]. It is interesting that by appropriate choice of AlSb (GaSb) for a buffer layer, the bisotropic stress can be made to confine itself almost entirely within the GaSb (AlSb) component of the superlattice[74].

Very pronounced shifts of phonon frequencies induced by misfit strains appear in II-VI superlattices where the lattice mismatch varies typically from 5% to 13%. Related studies have shown strong dependence of the TO and LO phonon frequencies on the ratio of layer thicknesses in ZnTe/ZnSe[76-79], ZnTe/ZnS[80], ZnTe/CdTe[81], and in ZnSe/ZnS[82]. In fact, this ratio can be linked quantitatively to the in-plane strains and lead to speculations for PDP values of the bulk materials[77,80,82]. Other studies on II-VI superlattices include CdTe/ZnTe[83] and $ZnSe/ZnS_xSe_{1-x}$[84,85].

Implantation-induced and Laser-induced Strains

Implantation of crystals by energetic ions has been extensively used in recent years for modifying the physical properties of materials. RS is an effective technique for studying the lattice damage (defects, strains) caused by implantation. The presence of lattice imperfections is manifested by the asymmetric broadening of the Raman bands. Fitting such irregular bands to the so-called **spatial correlation model (SCM)** can give information about the average size of microcrystallites[86,87]. Creation of amorphous or heavily damaged layers under the surface, results in residual strains which are believed to be uniaxial, in the direction normal to the surface [88,89].

When a laser beam is sharply focused on the surface of an absorbing material, inhomogeneous temperature (T) and thermoelastic fields are produced. The problem of calculating the T distribution on the surface of semiconductors has been treated in detail in recent years[90] assuming Gaussian (CW) laser beam profiles. The thermoelastic response of the material has been considered only recently, the material being treated as elastically isotropic (poly-Si, Ref. 91; crystalline Si, Ref. 92). In Fig. 3(a) we have the results in a three dimensional presentation of the T distribution on the surface of polycrystalline Si, due to 1.5 W of the 514.5 nm Ar^+ line focused on a spot of 4 μm diameter[91]. Figure 3(b) shows the corresponding thermoelastic deformation of the surface, i.e., the displacement (U_z) of atoms along z (positive z axis downwards) caused by the net thermal expansion. This represents the solution of the following differential equation for the displacement $\mathbf{U}$ of an elastic semi-infinite isotropic medium subject to an external heating force,

$$(C_{12}+C_{44})\nabla(\nabla\cdot\mathbf{U})+C_{44}(\nabla\cdot\nabla)\mathbf{U} = \nabla(B\alpha\Delta T),\tag{51}$$

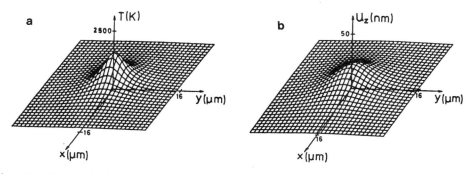

Fig. 3. Theoretical calculation of temperature distribution due to 1.5 W of 514.5 nm focused on the surface of polycrystalline Si (a), and surface deformation (b).

where α is the linear thermal expansion coefficient and $\Delta T = T - T_0$ is the local temperature rise over the background temperature T_0. The values of B, α, and C_{ij} used in Eqns. (47) are those appropriate for polycrystalline Si [10]. For crystalline Si, the form of Eqns. (47) becomes more complicated and depends on the orientation of the surface relative to the crystallographic axes. In the simplest case of the (100) surface the term $C \sum_{i=1}^{3} \hat{x}_i (\partial^2 U_i / \partial^2 x_i)$ should be added to the left-hand side of Eqns. (47). Such extended analysis has not been reported yet. Having obtained U, it is a matter of straightforward calculation to obtain the components of strains ε as a function of the cylindrical coordinates r and z.

Both the temperature and strain inhomogeneities are expected to affect the bandshape of the Raman spectrum. The scattering volume consists of many elementary scatterers at different T and ε. The net scattering intensity will be represented by an asymmetrically broadened and shifted band. The stronger the inhomogeneities the more evident the asymmetries of the bandshapes. The effects of such T distributions have already been observed in Raman spectra[93]. Thermoelastic strain effects on Raman spectra have not been reported yet. Approximate calculations based on a coordinate-dependent DSE, [Eqn. (9)], predict splitting of the Si phonon into singlet-doublet components at the center of the beam, while away from the center the degeneracy is completely removed[92,94]. The concomitant frequency shifts, however, are rather small (≤ 1 cm^{-1} for bulk Si, 2-3 cm^{-1} for Si-on-SiO$_2$ films) compared to the thermal shifts (~ 40 cm^{-1}) and this will make their observation rather difficult. On the other hand, even such small shifts may lead to miscalculation of T by 30-120 K, and this is rather serious in related applications of laser processing[92].

CLOSING REMARKS

The purpose of this contribution was to demonstrate the applicability of Raman spectroscopy as a non-destructive, contactless diagnostic tool for observing and analyzing strains. Prerequisite for this approach is the knowledge of the phonon deformation potentials which can be obtained from control experiments of RS under uniaxial stress in regions of transparency. An alternative spectroscopic procedure has been applied to GaAs[95,96] and InP[18], that is, piezoreflectance in the restrahlen region (far IR). This technique too is sensitive to strain relaxations near the surface since the penetration of the IR radiation in the restrahlen region is small and rather nonuniform. On the other hand, the selection rules allow the measurement of the complete set of PDP for the TO and LO components. The piezoreflectance technique is preferred to piezo-Raman in the region of opaqueness, but not so when compared to piezo-Raman in the region of transparency.

X-ray diffraction is the classical way for strain characterization. Comparative measurements performed recently on the same heterostructure of InAs/GaAs with X-ray diffraction and RS by the same researchers[97], have demonstrated the superiority of the latter technique, provided accurate PDP values are available. [The main advantages of RS appear to be the possibility for microprobing and the fine in-depth resolution which is made possible by the large variety of laser lines; on the other hand, the strain sensitivity of the X-ray method (10^{-7}) is two to three orders of magnitude higher than that of RS.]

From a theoretical point of view, the experimental values of PDP can be used to obtain information on harmonic and anharmonic force constants as well as lattice-dynamical and third-order elastic constants[15,98,13]. Existing PDP values of crystalline materials have been shown to lead to the PDP of the corresponding polycrystalline material[10]. Strain characterization of poly-Si[99] (SOI) pressure sensors is important in predicting their performance characteristics[99].

The entire subject of strain characterization through Raman spectroscopy has been developed so far on the assumption of symmetric strains. Antisymmetric strains are also known to exist in crystals with lower-than-cubic symmetry. They can be generated by shear acoustic waves and are, therefore, of dynamical character. It has been shown by Nelson and Lax that such strains are responsible for antisymmetric contributions to the photoelastic tensor[100]. These contributions can be written as a linear combination of components of the strain-independent dielectric impermeability. The same approach has been applied to the gyration tensor[101]. In view of these facts, it would be reasonable to expect that antisymmetric strains (i.e., local rotations on a unit cell scale) introduce antisymmetric contributions to the PDP of phonons of non-cubic crystals, i.e., components of the form

$$K_{j,(\alpha\beta)[\kappa\varrho]} = K_{j,(\alpha\beta)\kappa\varrho} = -K_{j,(\alpha\beta)\varrho\kappa} . \qquad (52)$$

By analogy to the dielectric impermeability and gyration tensor, these terms could be expressed as a linear combination of the strain-independent components $K^0_{j,\alpha\beta}$. According to Eqn. (1), the latter are strictly diagonal ($=\omega_j^2$) and are defined in a subspace of the crystal point group with dimensionality equal to the degree of degeneracy of the corresponding mode j (i.e., 2 or 1). No such phenomena have been investigated so far, either theoretically or experimentally. Symmetry restrictions, however, are expected to limit severely the possibilities of observing such effects (see for instance, Table I of Ref. 102).

Judging from the way the PDP components have been used in the literature (see remarks at the beginning of the subsection on ZB crystals, previous section), it appears necessary to emphasize that the sets of PDP for the TO and LO phonons in ZB structures are different in general, and that indiscriminate use of $\widetilde{K}^T_{ij}$ in problems involving backward scattering by LO phonons can only lead to approximate results. Physically speaking, the difference K^L_{ij}-K^T_{ij} reflects the strain dependence of the optical dielectric constant ε_∞, and of the transverse effective charge e^*_T through their strain derivatives $k^e_{ij} = (\partial\varepsilon_\infty/\partial\varepsilon)_{ij}$ (elasto-optical coefficients) and $\widetilde{M}_{ij} = (\partial\ln e^*_T/\partial\varepsilon)_{ij}$, both fourth-rank tensors. The exact relationship is[102]

$$K_{ij}^L - K_{ij}^T = \omega_T^2 (2\tilde{M}_{ij} - k_{ij}^e/\varepsilon_\infty - 1) \Delta\varepsilon/\varepsilon_\infty, \tag{53}$$

where $\Delta\varepsilon = \varepsilon_0 - \varepsilon_\infty$, ε_0 being the static value of the dielectric constant. [The -1 term in (53) is absent for ij = 44, 55, 66.]

Although, at present, complete sets of PDP values are available only for a handful of ZB materials (Table I), the possibilities for supplementing the list with ZnSe, ZnS, ZnTe, AlP, AlAs, CdTe, and HgTe seem realistic. For these materials there exist laser lines below their fundamental gaps, and thus Raman measurements in the region of transparency are possible, in principle. On the other hand, the situation is not so promising for small gap semiconductors · like Ge, GaSb, InSb, and InAs for which no laser lines are available at present, suitable for Raman measurements in the region of transparency. Additional difficulties arise for some materials, like AlAs or 3C-SiC which cannot be grown in forms appropriate for control experiments. An alternative approach would be to grow unstrained thin films of such materials on wafers and then measure Raman shifts while the wafer is mechanically stressed. This can be done either by a central point force applied normal to the wafer[103,104] or by a hydrostatic pressure, inside a gas pressure cell, the wafer serving as one of its windows. In either case, a careful strain calibration of the apparatus is necessary before it is used for measuring phonon deformation potentials.

ACKNOWLEDGEMENTS

Partial support was provided by the General Secretariat for Research and Technology, Greece, and by the Max Plank Institute for Solid State Research, Stuttgart. The lasting collaboration with M. Cardona is acknowledged with appreciation.

REFERENCES

1.   J.F. Nye, "Physical Properties of Crystals", Clarendon Press, Oxford (1964)
2.   E. Anastassakis and E. Burstein, J. Phys. Chem. Solids 32: 563 (1971).
3.   E.M. Anastassakis, "Dynamical Properties of Solids", Vol. 4, G.K. Horton and A.A. Maradudin, Eds., North-Holland (1980) p. 157.
4.   E. Anastassakis and E. Liarokapis, phys. stat. sol. (b) 69: 137 (1988).
5.   E. Anastassakis, J. Appl. Phys. 68: 4561 (1990).
6.   R.W. Vook and F. Witt, J. Appl. Phys. 36: 2169 (1965).
7.   F. Witt and R.W. Vook, J. Appl. Phys. 39: 2773 (1969).
8.   M. Murakami and T. Yogi, J. Appl. Phys. 57: 211 (1985).
9.   E. Friess, H. Brugger, K. Eberl, G. Krotz, and G. Abstreiter, Solid State Commun. 69: 899 (1989).
10. E. Anastassakis and E. Liarokapis, J. Appl. Phys. 62: 3346 (1987).
11. E. Anastassakis, A. Pinczuk, E. Burstein, F.H. Pollak, and M. Cardona, Solid State Commun. 8: 133 (1970).
12. M. Chandrasekhar, J.B. Renucci, and M. Cardona, Phys. Rev. B17: 1623 (1978)
13. E. Anastassakis, A. Cantarero, and M. Cardona, Phys. Rev. B41: 7529 (1990).
14. F. Cerdeira, C.J. Buchenauer, F.H. Pollak, and M. Cardona, Phys. Rev. B5: 580 (1972).
15. M.H. Grimsditch, E. Anastassakis, and M. Cardona, Phys. Rev. B18: 901 (1978).
16. P. Wickboldt, E. Anastassakis, R. Sauer, and M. Cardona, Phys. Rev. B35: 1362 (1987).
17. E. Anastassakis and M. Cardona, Solid State Commun. 64: 543 (1987).
18. E. Anastassakis, Y.S. Raptis, M. Hunnerman, W. Richter, and M. Cardona, Phys. Rev. B38: 7702 (1988).
19. E. Anastassakis and M. Cardona, Solid State Commun. 63: 893 (1987).
20. E. Anastassakis, F.H. Pollak, and G.W. Rubloff, "Proc. 11th Int. Conf. on

Physics of Semiconductors", Polish Scientific Publishers, Warsaw (1972) p. 1188.

21. I. Balslev, phys. stat. sol. (b) 61: 207 (1974).
22. E. Anastassakis, "Physical Problems in Microelectronics" J. Kassabov, Ed., World Scientific, Singapore (1985) p. 128.
23. T. Englert, G. Abstreiter, and J. Pontcharra, Solid State Electr. 23: 31 (1980).
24. G.A. Sai-Halasz, F.F. Fang, T.O. Sedwick, and A. Segmuller, Appl. Phys. Lett. 36: 419 (1980).
25. K. Yamazaki, M. Yamada, K. Yamamoto, and K. Abe, Japn. J. Appl. Phys. 20: L299, L371 (1981).
26. Y. Ohmura, T. Inoue, and T. Yoshii, J. Appl. Phys. 54: 6779 (1983).
27. F. Moser and R. Beserman, J. Appl. Phys. 54: 1033 (1983).
28. Y. Kobayashi, M. Nakamura, and T. Suzuki, Appl. Phys. Lett. 40: 1040 (1982).
29. M. Nakamura, Y. Kobayashi, and K. Usami, Jpn. J. Appl. Phys. 23: 687 (1984).
30. K. Yamazaki, M. Yamada, K. Yamamoto, and K. Abe, Jpn. J. Appl. Phys. 23: 681, (1984).
31. M.B. Stern, T. R. Harrison, V.D. Archer, P.F. Liao, and J.C. Bean, Solid State Commun. 51: 221 (1984).
32. P.M.J. Maree, R.I.J. Olthof, J.W.M. Frenken, J.F. van der Veen, C.W.T. Bulle-Lieuwma, M.P.A. Vieger, and P.C. Zalm, J. Appl. Phys. 58: 3097 (1985)
33. D.J. Olego, H. Baumgart, and G.K. Celler, Appl. Phys. Lett. 52: 483 (1988).
34. J.C. Tsang and S.S. Iyer, IEEE J. Quant. Elect. QE-25: 1008 (1989).
35. J. Gonzalez-Hernandez and R. Tsu, Solid State Commun. 69: 637 (1989).
36. S. Nakashima, S. Oitma, A. Mitsuishi, T. Nishimura, T.Fukumoto, and Y. Akasaka, Solid State Commun. 40: 765 (1981).
37. J. Takahashi and T. Makino, J. Appl. Phys. 63: 87 (1987).
38. R.J. Nemanish and D. Haneman, Appl. Phys. Lett. 40: 785 (1982).
39. H.S. Tan, M.H. Kuok, S.C. Ng, C.K. Ong, and S.H. Tang, J. Appl. Phys. 55: 1116 (1984).
40. F. Cerdeira, A. Pinczuk, J.C. Bean, B. Batlogg, and B.A. Wilson, Appl. Phys. Lett. 45: 1138 (1984).
41. G. Abstreiter, H. Brugger, T. Wolf, H. Jorke, and H.J. Herzog, Phys. Rev. Lett. 54: 2441 (1985).
42. E. Kasper, H. Kibbel, H. Jorke, H. Brugger, E. Friess, and G. Abstreiter, Phys. Rev. B38: 3599 (1988).
43. S.J. Chang, C.F. Huang, M.A. Kallel, K.L. Wang, R.C. Bowman, Jr., and P.M. Adams, Appl. Phys. Lett. 53: 1835 (1988).
44. S.A. Lyon, R.J. Nemanich, N.M. Johnson, and D.K. Biegelsen, Appl. Phys. Lett. 40: 316 (1982).
45. S.R.J. Brueck, B. Tsaur, J. Fan, D. Murphy, T. Deutsch, and D. Silversmith, Appl. Phys. Lett. 40: 895 (1982).
46. P. Zorabedian, F. Adar, Appl. Phys. Lett. 43: 177 (1983).
47. S. Nakashima, Y. Inoue, M. Miyauchi, and A. Mitsuishi, J. Appl. Phys. 54: 2611 (1983).
48. K. Yamazaki, R.K. Uotani, K. Nambu, M. Yamada, K. Yamamoto, and K. Abe, Jpn. J. Appl. Phys. 23: L403 (1984).
49. T. Nishioka, Y. Shinoda, and Y. Ohmachi, J. Appl. Phys. 57: 276 (1985).
50. Y.M. Cheong, H.L. Marcus, and F. Adar, J. Mater. Res. 2: 902 (1987).
51. S. Nakashima and M. Hango, IEEE J. Quant. Elect. QE-25: 965 (1989).
52. D.J. Evans and S. Ushioda, Phys. Rev. B9 :638 (1974).
53. H. Shen and F.H. Pollak, Appl. Phys. Lett. 45: 692 (1984).
54. T. Nakamura, A. Ushirokawa, and T. Katoda, Appl. Phys. Lett. 38: 13 (1981).
55. T. Kamijoh, A. Hashimoto, H. Takano, and M. Sakuta, Appl. Phys. Lett. 44: 1084 (1984); J. Appl. Phys. 55: 3756 (1984).
56. M. Takai, H. Nakai, S. Nakashima, T. Minamisono, K. Gamo, and S. Namba, Jpn. J. Appl. Phys. 24: L755 (1985).
57. S. Nakashima, A. Fujii, K. Mizoguchi, A. Mitsuishi, and K. Yoneda, Jpn. J. Appl. Phys. 27: 1327 (1988).

58. Y. Huang, P.Y. Yu, M. Charasse, Y. Lo, and S. Wang, Appl. Phys. Lett. 20: 192 (1987).
59. S. Emura, S. Gonda, Y. Matsui, and H. Hayashi, Phys. Rev. B38: 3280 (1988).
60. T. Kato, T. Matsumoto, M. Hosoki, and T. Ishida, Jpn. J. Appl. Phys. 26: L1597 (1987).
61. Z.C. Feng, W.J. Choyke, and J.A. Powell, J. Appl. Phys. 64: 6827 (1988).
62. T. Nomura, Y. Maeda, M. Miyao, M. Hagino, and K. Ishikawa, Jpn. J. Appl. Phys. 26: 908 (1987).
63. T. Matsumoto, T. Kato, M. Hosoki, and T. Ishida, Japn. J. Appl. Phys. 26: L576 (1987).
64. H. Mukaida, H. Okumura, J.H. Lee, H. Daimon, E. Sakuma, S. Misawa, K. Endo, and S. Yoshida, J. Appl. Phys. 62: 254 (1987).
65. G. Burns, C.R. Wie, F.H. Dacol, G.D. Pettit, and G.M. Woodall, Appl. Phys. Lett. 51: 1919 (1987).
66. M.J.L.S. Haines, B.C. Cavenett, and S.T. Daven, Appl. Phys. Lett. 55: 849 (1989).
67. R.M. Abdelouhab, R. Braunstein, K. Barner, M.A. Rao, and H. Kroemer, J. Appl. Phys. 66: 787 (1989).
68. M. Nakayama, K. Kubota, H. Kato, and S. Sano, Solid State Commun. 51: 343 (1984).
69. F. Iikawa, F. Cerdeira, C. Vazquez-Lopez, P. Motisuke, M.A. Sacilotti, A. P. Roth, and R.A. Masut, Solid State Commun. 68: 211 (1988).
70. M. Nakayama, K. Kubota, H. Kato, T. Kanata, S. Chika, and N. Sano, J. Appl. Phys. 58: 4342 (1985).
71. M. Nakayama, K. Kubota, H. Kato, S. Chika, and N. Sano, Appl. Phys. Lett. 48: 281 (1986).
72. R.M. Abdelouhab, R. Braunstein, M.A. Rao, and H. Kroemer, Phys. Rev. B39: 5857 (1989).
73. B. Jusserand, P. Voisin, M. Voos, L.L. Chang, E.E. Mendez, and L. Esaki, Appl. Phys. Lett. 46: 678 (1985).
74. G.P. Schwartz, G.J. Gualtieri, W.A. Sunder, L.A. Farrow, D.E. Aspnes, and A.A. Studna, J. Vac. Sci. Technol. A5, 1500 (1988); also Phys. Rev. B36: 4868 (1987).
75. P.V. Santos, A.K. Sood, M. Cardona, K. Ploog, Y. Ohmori, and H. Okamoto, Phys. Rev. B37: 6381 (1988).
76. M. Kobayashi, M. Konagai, K. Takahashi, and K. Urabe, J. Appl. Phys. 61: 1015 (1987).
77. S. Nakashima, Y. Nakakura, H. Fujiyasu, and K. Mochizuki, Appl. Phys. Lett. 48: 236 (1986).
78. S. Nakashima, A. Wada, H. Fujiyasu, M. Aoki, and H. Yang, J. Appl. Phys. 62: 2009 (1987).
79. Y.H. Wu, H. Yang, A. Ishida, H. Fujiyasu, S. Nakashima, and K. Tahara, Appl. Phys. Lett. 54: 239 (1989).
80. L.H. Shon, K. Inoue, and K. Murase, Solid State Commun. 62: 621 (1987).
81. M.K. Jackson, R.H. Miles, T.C. McGill, and J.P. Faurie, Appl. Phys. Lett. 55: 786 (1989).
82. L.H. Shon, K. Inoue, O. Matsuda, K. Murase, T. Yokagawa, and M. Ogura, Solid State Commun. 67: 779 (1988).
83. J. Menendez, A. Pinczuk, J.P. Valladares, R.D. Feldman, and R.F. Austin, Appl. Phys. Lett. 50: 1101 (1987).
84. D.J. Olego, K. Shahzad, J. Petruzzello, and D. Cammack, Phys. Rev. B36: 7674 (1987).
85. D.J. Olego, K. Shahzad, D.A. Cammack, and H. Cornelissen, Phys. Rev. B38: 5554 (1988).
86. H. Richter, Z.P. Wang, and L. Ley, Solid State Commun. 39: 625 (1981).
87. K.K. Tiong, P.M. Amirtharaj, F.H. Pollak, and D.E. Aspnes, Appl. Phys. Lett. 44: 122 (1984).
88. E. Anastassakis and J. Tatarkiewicz, Appl. Phys. Lett. 50: 245 (1987).
89. E. Anastassakis, K.K. Biskupska, W. Graeff, E. Liarokapis, J. Tatarkiewicz, and K. Wieteska, "Proc. 19th Intern. Conf. Physics of Semiconductors", Ed.

W. Zawadzki, Vol. 1, Institute of Physics, Polish Academy of Sciences Warsaw (1989), p. 753.

90. E. Liarokapis and Y.S. Raptis, J. Appl. Phys. 57: 5123 (1985) and references therein.

91. E. Liarokapis and E. Anastassakis, Physica Scripta 38: 84 (1988); also, J. Appl. Phys. 63: 2615 (1988).

92. L.P. Welsh, J.A. Tuchman, and I.P. Herman, J. Appl. Phys. 64: 6274 (1988).

93. J. Raptis, E. Liarokapis, and E. Anastassakis, Appl. Phys. Lett. 44: 125 (1984).

94. Y.S. Raptis, PhD thesis 1988, National Technical University, Athens, Greece (unpublished).

95. B.A. Weinstein and M. Cardona, Phys. Rev. B5: 3120 (1972).

96. M. Hunermann, W. Richter, J. Saalmuller, and E. Anastassakis, Phys. Rev. B 35: 5381 (1986).

97. A.C. Diebold, S.W. Steinhauser, and R.P. Mariella, Jr., J. Vac. Sci. Technol. B7: 365 (1989).

98. C.S.G. Cousins, L. Gerward, J.S. Olsen, B. Selsmark, B.J. Sheldon, and G.Webster, Semicond. Sci. Technol. 4: 333 (1989).

99. J. Suski, V. Mosser, and J. Goss, Sensors and Actuators 17: 405 (1989).

100. E. Anastassakis, J. Phys. C: Solid State Phys. 16: 3329 (1983).

101. D.F. Nelson and P.D. Lazay, Phys. Rev. Lett. 25: 1187 (1970).

102. E. Anastassakis, Phys. Rev. B12: 5934 (1973).

103. I.I. Novak, V.V. Baptizmanskii, and L.V. Zhoga, Opt. Spectrosc. 43: 145 (1977).

104. V.V. Baptizmanskii, I.I. Novak, and Yu.F. Titoves, Sov. Phys. Solid State 21: 1915 (1979).

CHARACTERIZATION OF STRAIN AND EPITAXIAL QUALITY IN Si/Ge

HETEROSTRUCTURES

D. J. Lockwood and J.-M. Baribeau

Division of Physics
National Research Council
Ottawa K1A 0R6, Canada

ABSTRACT

Raman scattering and x-ray diffraction techniques have proved to be most useful for characterizing epitaxial semiconductor material. X-ray methods such as double crystal diffractometry can be used to determine the strain partition, layer thickness, alloy composition and epitaxial quality in Si/Ge heterostructures, and some of this information can also be obtained from Raman measurements of the acoustic and optic phonons. The relative merits of the two techniques are evaluated from the application of both methods in determining the strain relaxation in $Si_{0.8}Ge_{0.2}$ epilayers and $Si/Si_{1-x}Ge_x$ superlattices on Si and the epitaxial quality of $(Si_mGe_n)_p$ atomic layer superlattices on (100) Si and Ge substrates and on $Si_{0.5}Ge_{0.5}$ buffer epilayers. From the observed linear relationship between the Raman frequency of the Si-Si vibration line and the x-ray relative Bragg angle in the alloy epilayers, we deduce a Raman shift with stress of 0.5 cm^{-1}/kbar, which is about 20% higher than the pure Si value. Because the optic mode frequencies vary linearly with strain, Raman measurements may be used, for example, to estimate the activation energy of nucleation/propagation of strain relieving misfit dislocations. This activation energy for the alloy epilayer is deduced to be 1.2 eV.

INTRODUCTION

Raman scattering and x-ray diffraction are useful techniques for characterizing epitaxial layers. They can provide information on layer thickness, alloy composition, strain, epitaxial quality, and interface sharpness in single and multilayer structures.[1,2] We have applied both techniques in investigating strain relaxation in $Si/Si_{1-x}Ge_x$ superlattices and epilayers, and epitaxial quality in $(Si_mGe_n)_p$ atomic layer superlattices (ALS) grown on Si and the Ge substrates and on $Si_{1-x}Ge_x$ buffer epilayers. Representative results from these studies are presented in the following sections and the applicability and relative merits of the two techniques are evaluated.

EXPERIMENTAL METHODS

The various $Si/Si_{1-x}Ge_x$ heterostructures were grown in a VG Semicon V80 MBE system on 100 mm (100) Si and 50 mm (100) Ge Czochralski wafers. The thick

Light Scattering in Semiconductor Structures and Superlattices
Edited by D.J. Lockwood and J.F. Young, Plenum Press, New York, 1991

197

Si/Si$_{1-x}$Ge$_x$ multilayers were grown at a temperature of 500 °C using a deposition rate of ~ 0.5 nms^{-1}. The short-period superlattices were grown typically at 350 °C and at 0.05 nms^{-1}. Details on the substrate preparation and growth methodology are given elsewhere.[3]

The Raman scattering measurements were performed in the quasi-backscattering geometry described elsewhere.[4] The spectra were generally excited with 300 mW of 457.9 nm argon laser light, analyzed with a Spex 14018 double monochromator, and detected with a cooled RCA 31034A photomultiplier. The polarization of the incident laser light was contained within the scattering plane, while the unanalyzed scattered light was directed through a polarization scrambler into the spectrometer. The sample was contained in a helium gas atmosphere to eliminate Raman scattering from air. The Raman laser beam illuminates a slit shaped area of approximately 0.1 × 1 mm^2 on the sample surface.

X-ray double crystal diffractometry (DCD) was used to obtain structural information and the strain partition of the various heterostructures. Rocking curves about the (400) Bragg reflection were recorded with a BEDE 150 diffractometer in a (+,-) non-dispersive geometry using a Si(100) first-crystal monochromator and Cu K$_\alpha$ radiation. In this set-up the x-ray beam spot size at the sample crystal is estimated to be ~ 0.1 mm^2. In the present work the Bragg angles are measured relative to the Si substrate (400) reflection at ~ 34.58°. Experimental rocking curves were simulated using a kinematical formalism[5] to obtain the physical dimensions of the heterostructures and evolution of the strain upon thermal processing. Conventional (400) and glancing incidence θ-2θ measurements were also performed using a Philips PW1820 vertical goniometer.

ANNEALING INDUCED RELAXATION IN SUPERLATTICES

The parameters of two Si/Si$_{1-x}$Ge$_x$ superlattices grown on Si(100) are given in Table 1. These samples demonstrate the two kinds of behavior found experimentally on annealing a large number of superlattices of various compositions. Sample 567 with a low x value is relatively stable on annealing to high temperatures, while the other sample exhibits a high degree of metastability on annealing. Annealing treatments were performed on air-exposed pieces of the samples in high vacuum (approximately 10^{-7} Pa) for times ranging from 30 min to 4 h at temperatures from 600 to 900 °C.

Raman Spectroscopy

The effect of annealing sample 349 on its Raman spectrum is shown in Figs. 1 and 2. The frequencies and intensities of longitudinal acoustic (LA) phonons in the as-grown sample are given in Table 2. The phonon frequencies and intensities can be calculated[4] using the Rytov[6] and photoelastic coupling[7] models, respectively. The folded LA mode frequencies are given approximately by $\omega_m = \langle v \rangle (q_m \pm q)$, where $q_m = 2\pi m/t$ and m = 0, 1, 2...is the folding index. The su-

Table 1.   Structural parameters in Si/Si$_{1-x}$Ge$_x$ superlattices: layer thickness t (nm) and alloy composition x.

Sample	Number of periods	t$_{Si}$		t$_{SiGe}$		x	
		X-ray	Raman	X-ray	Raman	X-ray	Raman
349	40	10.9±0.5	11.1±0.1	7.3±0.5	7.6±0.1	0.25±0.02	0.25±0.01
567	20	32.0±0.5	-	10.4±0.5	-	0.10±0.02	-

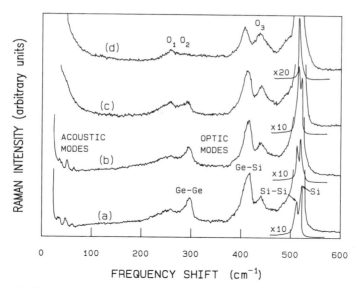

Fig. 1. Raman spectra of sample 349 (a) as-grown, (b) annealed for 30 min at 600 °C, (c) 30 min at 800 °C, and (d) 4 h at 900 °C. The spectral resolution was 3.1 cm^{-1}.

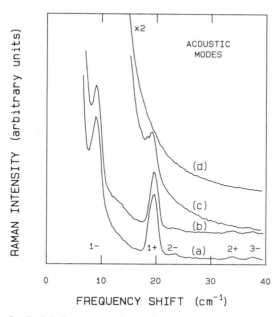

Fig. 2. Folded acoustic phonons in the Raman spectrum of sample 349 under the same annealing conditions as in Fig. 1. The phonon doublets are labeled by folding index m (see Table 2). The spectral resolution was 1.6 cm^{-1}.

Table 2. Experimental and calculated Raman peak frequencies and intensities for folded acoustic phonons in sample 349 taking t_{Si} = 11.1 nm, t_{SiGe} = 7.6 nm, x = 0.25, and $<v>$ = 8028 m/s. The calculated intensity is normalized to agree with the mean experimental intensity for the m = 1 doublet.

m	ω_{expt} (cm^{-1})	ω_{calc} (cm^{-1})	I_{expt} (counts/s)	I_{calc}
1–	9.0	9.0	860	830
1+	19.6	19.6	800	830
2–	23.4	23.3	48	70
2+	34.0	33.9	41	70
3–	37.8	37.7	37	41
3+	48.4	48.3	39	41
4–	52.0	52.0	22	48
4+	62.4	62.6	14	48

perlattice period t = t_{Si} + t_{SiGe}, where t_{Si} and t_{SiGe} are the Si and $Si_{1-x}Ge_x$ layer thicknesses, respectively, $<v>$ is the average sound velocity of the superlattice,[6,7] and q is the light-scattering wave vector transfer. The Raman intensities are given by $I_m \alpha\ m^{-2} \sin^2 (m\pi t_{Si}/t)$.[7] The Raman results given in Table 2 were accurately modeled by theory taking t = 18.7 nm and x = 0.25, and using the pure Si value of 4.595 for the refractive index of the superlattice at 457.9 nm. The values for t and x are highly constrained by the frequency data. The intensity data is more "noisy" and thus the value for t_{Si} = 11.1 ± 0.1 nm is less certain.

At higher frequencies the optical-mode Raman spectrum exhibits first-order features characteristic of the silicon and alloy layers (see Fig. 1). The spectrum of a Si-$Si_{1-x}Ge_x$ superlattice exhibits three main peaks attributed to scattering from longitudinal optical (LO) phonons corresponding largely to the vibrations of the Ge—Ge, Ge—Si, Si—Si bonds in the alloy layers,[8] and a fourth peak due to the optical lattice vibration in the Si layers. The line positions are sensitive to the strain within the layers and the measurement of the line shifts provides a direct indication of strain relaxation in the superlattice.[9] The weaker optic phonon lines near 255 cm^{-1} (0_1) and 435 cm^{-1} (0_3) have been attributed to a particular Si-Ge ordering within the alloy layer.[9,10] A third line (0_2) due to this ordering[10] is expected near 295 cm^{-1}, but this line is often obscured by the stronger Ge-Ge peak.

Table 3. Raman optical-mode frequencies (in cm^{-1}) of the as-grown and annealed sample 349 (see Fig. 1).

Vibrational mode	As-grown	600 °C anneal for 30 min	800 °C anneal for 30 min	900 °C anneal for 4 h
Alloy layer mode 0_1	254.0	254.9	254.3	254.8
Alloy layer Ge-Ge mode	292.2	291.5	288.8	Not measurable
Alloy layer Ge-Si mode	411.8	411.7	406.4	401.9
Alloy layer mode 0_3	436.7	437.0	434.5	432.3
Alloy layer Si-Si mode	510.6	510.6	511.2	512.6
Si layer mode	519.1	519.0	517.1	Not measurable

For the as-grown sample, the Ge-Ge, Ge-Si, and Si-Si alloy vibrational modes (see Table 3) are shifted up in frequency from the corresponding unstrained bulk alloy positions[8] for x = 0.25 by an amount consistent with the strain ε_{SiGe} in the alloy layers.[9] The intense peak at 519.1 cm^{-1} is due to the lattice vibration in the Si layers. Its frequency is slightly lower than the bulk Si value of 520.1 cm^{-1} indicating a relatively small stress of 1.0 cm$^{-1} \times$ 2.5 kbar/cm^{-1} (see Ref. 11) corresponding to a perpendicular strain value ε_{Si} of 0.14 $\times$ 10^{-2} in the Si layers of the superlattice. This is in good agreement with the value of 0.13$\times$10^{-2} for ε_{Si} deduced from the x-ray diffraction measurements (see Table 4).

On annealing, these Raman peaks generally shift in frequency as shown in Table 3. The three alloy layer Ge-Ge, Ge-Si, and Si-Si vibrational modes should shift down in frequency together when the strain is removed.[9] However the Si-Si mode frequency shifts upwards on annealing at temperatures of 800 °C and higher indicating the average alloy concentration in the layer is also changing. The Si layer mode shifts down in frequency after annealing at the higher temperatures, because of the diffusion of Ge into the Si layers. The sensitivity of the Si-Si mode in the alloy layers and the corresponding mode in the Si layers to the annealing treatment is shown by the reversal in intensity of the two peaks (see Fig. 1). After the 900 °C anneal the optic phonon spectrum has the appearance of a bulk alloy spectrum with x $\cong$ 0.1, i.e., the superlattice has become a single alloy layer. A full analysis of the optic phonon spectrum can only come from a study of the effects of annealing on a single strained alloy layer, because only then can frequency shifts due to strain be separated from frequency shifts due to interdiffusion between the Si and alloy layers.

The frequencies of the modes near 255 and 435 cm^{-1} are remarkably insensitive to annealing, which is consistent with their assignment to a particular Si-Ge ordered state,[10] and it is notable that the extent of this ordering increases on annealing, as evidenced by the relative increase in intensity of these modes. Thus, the annealing allows more atoms to diffuse into lattice positions corresponding to this Si-Ge ordering and they prefer to stay there.

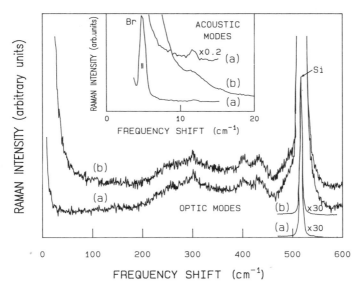

Fig. 3. Raman spectra of sample 567 (a) as-grown and (b) after annealing for 30 min at 800 °C. The low-frequency region is shown in the inset. For this measurement a Kr laser exciting line (λ = 468 nm) of 170-mW power was used.

The high peak intensities of the LA-phonon Raman lines see in Figs. 1 and 2 indicate sharp interfaces (plus or minus one atomic spacing). The acoustic phonon scattering vanishes from the Raman spectrum after annealing at the higher temperatures. The two most intense lines at 9.0 and 19.6 cm^{-1} in Fig. 2 comprise the m = 1 doublet separated by twice the Brillouin frequency $\langle v \rangle q = 5.3$ cm^{-1}. The m = 2 doublet and the lower-frequency component of the m = 3 doublet are also visible with a lower intensity. These lines do not shift in frequency with annealing. Their intensities are scarcely affected by annealing at temperatures up to 700 °C, but after the 800 °C anneal their intensities are considerably reduced (by about one-half) and they then disappear after the 4-h 900 °C anneal. The intensities of the acoustic modes are a sensitive indicator of interface quality[7,9,12] and these results show that the interface sharpness is unaffected by annealing at temperatures up to 700 °C, but thereafter the interfaces are degraded due to atomic diffusion between layers. The results obtained after annealing at 900 °C indicate a complete absence of well-defined interfaces between the original layers.

The corresponding Raman spectra for sample 567 are displayed in Fig. 3. The absence of any significant shifts in the frequencies of the LO modes (the optic mode spectra are virtually identical) indicates no measurable thermally induced relaxation. Some loss of interface sharpness is revealed by the damping of the LA folded modes after annealing. This is shown for the upper component of the m = 1 doublet (at 11.9 cm^{-1}) in the inset to Fig. 3, where the intense peak at 5.2 cm^{-1} is the Brillouin line. The optic phonon for the Si layers occurs at 520.1 cm^{-1} for the as-grown sample and at 519.8 cm^{-1} for the annealed sample. This indicates that the Si layers are unstrained and are truly lattice matched to the substrate in the as-grown sample and, again, that the annealing scarcely affects the superlattice structure.

X-ray Diffraction

Figure 4 displays a series of (400) x-ray rocking curves from sample 349 as-grown and after various annealing treatments. Before annealing, the rocking curve exhibits sharp satellites which account for the superlattice periodicity [Fig. 4(a)]. An excellent fit of the experimental data is obtained from the kinematical intensity simulation.[13] In particular, the width of the satellite reflections is well reproduced indicating good homogeneity in thickness and composition. A progressive broadening and damping of the superlattice reflections are observed upon annealing. Also, the zero-order reflection is seen to shift toward the substrate peak indicating progressive strain relaxation in the sample.

Table 4. The effect of annealing on the strain in samples 349 and 567, as deduced by double-crystal x-ray diffraction; ε_{Si} and ε_{SiGe} are the strain values in the growth direction measured with respect to the Si substrate lattice constant.

Sample	Heat treatment	ε_{Si} ($\times 10^{-2}$)	ε_{SiGe} ($\times 10^{-2}$)	$\Delta\theta$ (arcsec)
349	as-grown	0.13	1.93	0
	30 min at 600 °C	0.05	1.85	210
	30 min at 700 °C	0	1.7	290
	30 min at 800 °C	0	1.60	600
	4 h at 900 °C	0	0.42	1050
567	as-grown	0.01	0.66	0
	30 min at 800 °C	0	0.66	0

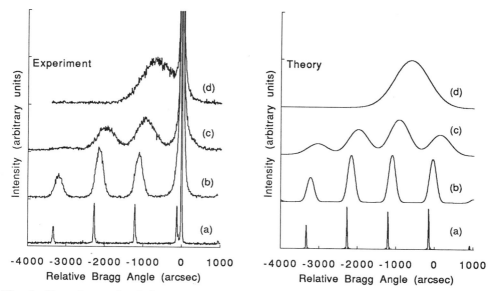

Fig. 4. Experimental (left panel) and theoretical (right panel) double-crystal (400) rocking curves from sample 349 (a) as-grown, (b) annealed 30 min at 600 °C, (c) 30 min at 800 °C, and (d) 4 h at 900 °C.

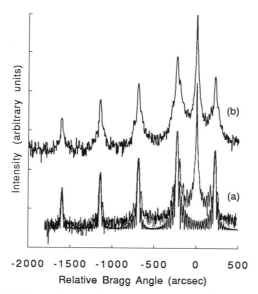

Fig. 5. Experimental and theoretical (lower trace) double crystal (400) rocking curves from sample 567 (a) as-grown and (b) after annealing 30 min at 800 °C.

Simulated theoretical profiles for the results obtained after the different annealing treatments are also shown in Fig. 4. Very good agreement is obtained using smaller values for the strain and introducing a spectral broadening $\Delta\theta$ (see Table 4). Notice that upon annealing the high-order satellites are weaker in the experiment than what is predicted from theory. This is an indication of a loss of interface sharpness due to interdiffusion and is most clearly seen in the sample annealed at 900 °C [Fig. 4(d)]. In that case an excellent fit was obtained assuming that the superlattice structure has been destroyed by annealing and consequently that the sample may be approximated by a fully relaxed homogeneous alloy layer of average composition x = 0.10. The considerable broadening of the satellites observed upon annealing probably originates from the combined effects of threading dislocations in the epilayer and strain variation across the x-ray sampling area. The residual perpendicular strain values decrease as the annealing temperature is raised, as shown in Table 4. (Here the strain is given with respect to the bulk $Si_{1-x}Ge_x$ lattice constant and x is obtained from classical elasticity theory, assuming that Veguard's law is valid.)

A different behavior was observed for sample 567, which has a smaller built-in strain (i.e., lower Ge concentration), as its structural properties appeared to be less affected by thermal annealing. Figure 5 shows the (400) x-ray rocking curves of sample 567 before and after annealing for 30 min at 800 °C. The simulated curve displayed in Fig. 5 is also in very good agreement with the measured spectrum of the as-grown sample. Contrary to sample 349, no significant broadening or shift of the main satellites are observed upon annealing. With double-crystal diffractometry, a relative change in lattice constant of the order of 1×10^{-4} can be measured which, in the case of a partially relaxed heterostructure, corresponds to having an interfacial array of misfit dislocations of average spacing of ~ 5 µm. The absence of a peak shift indicates that this structure has retained most of its strain upon thermal treatment. The spectra of Fig. 5 exhibit subsidiary oscillations frequently seen in double crystal diffraction experiments. These are known as Pendelossung fringes and originate from an interference phenomenon involving the wave fields diffracted below and above the epilayer/substrate interface. When compared to the experimental profile, the simulated curve displays some discrepancy for these secondary peaks and this may indicate a small inhomogeneity in thickness or composition in the superlattice.[14] The damping of the secondary peaks observed upon heating may be attributed to a smearing of the interfaces. It may possibly also indicate a loss of interfacial integrity due to the onset of relaxation. Complementary measurements using techniques having higher strain resolution have confirmed the metastability of sample 567 upon annealing.[15]

Comparison of Techniques

For the as-grown sample 349, the x-ray and Raman determinations of the superlattice parameters are in remarkably good agreement (see Table 1). The as-grown superlattices are commensurate with the Si substrate and the lattice mismatch is mostly accommodated by the tetragonal distortion of the alloy layers. However, in the case of sample 349, both techniques discovered a slight strain in the Si layers and agreed very well on the magnitude of that strain. The origin of that strain is unclear, but it may arise from the difference in thermal expansion coefficients of Si and $Si_{1-x}Ge_x$ or from the presence of a small fraction of Ge in the Si layers due to segregation phenomena.

For the annealed samples, the Raman results are entirely consistent with the x-ray measurements in that both techniques show that the interface sharpness is considerably degraded in sample 349 by annealing at temperatures above 700 °C. However, it is difficult to provide a quantitative comparison between the two techniques. Unlike the Raman intensities, which are a direct but not absolute measure of interface quality, it is only a difference between theory and experiment for satellite intensities that appears in the x-ray study. That is, the loss of satellite intensity must be reproduced by an appropriate modeling of the material distribution at the interface before the x-ray results can be used directly to measure interface sharpness. Both techniques agree that the superlattice structure has been destroyed by the 4-h 900 °C anneal, which hasbeen confirmed by transmission electron microscopy (TEM) measure-ments,[16] and that the sample approximates a relaxed homogeneous alloy layer with $x \cong 0.1$. The Raman and x-ray techniques show strain relaxation occurs in sample 349 on annealing, but no quantitative comparison between the techniques can be made because of atomic interdiffusion at the higher annealing temperatures. For sample 567, the apparent stability (within the strain resolution of the techniques) of the super-lattice even after annealing at 800 °C is directly confirmed by both techniques.

ANNEALING INDUCED RELAXATION IN EPILAYERS

The problem of annealing induced interdiffusion between Si and $Si_{1-x}Ge_x$ layers that complicated the analysis of strain relaxation from the Raman spectrum of superlattices can largely be overcome by studying single *thick* alloy layers. In this case the effects of interdiffusion between the single Si-substrate/alloy-layer interface are negligible, especially when using rapid thermal annealing (RTA).

Two samples were investigated in this study: one with $t_{SiGe} = 160$ nm and $x = 0.19$ (sample 738) and the other with $t_{SiGe} = 170$ nm and $x = 0.175$ (sample 869). The annealing treatments were performed in a Heatpulse 410 rapid thermal annealer, under nitrogen flow, for 3 min over a temperature range of 550–850 °C. The annealing was performed on individual ~ 1 cm^2 samples taken near the center of the wafer. The same sample, at each anneal temperature, was investigated by both experimental techniques to avoid possible variations in heat treatment.

Raman Spectroscopy

Raman spectra showing features due to scattering from longitudinal optic phonons are given in Fig. 6. Two strong peaks are observed near 514 and 520 cm^{-1} in the as-grown sample and arise from lattice vibrations in the $Si_{0.82}Ge_{0.18}$ epilayer and Si substrate, respectively. Weaker features are located near 256, 288, 408, and 436 cm^{-1} that all arise from vibrational modes in the alloy epilayer. As before, the peaks labeled Ge-Ge, Ge-Si, and Si-Si in Fig. 6 are largely associated with vibrations involving the corresponding $Si_{0.82}Ge_{0.18}$ alloy bonds.[8] The two remaining weaker peaks are again attributed to vibrational modes of a particular kind of Si-Ge ordering in the alloy.[10] On annealing, the Ge-Ge, Ge-Si, and Si-Si peaks shift to lower frequency, consistent with strain relaxation.[9] For example, after the 850 °C anneal these peaks are located near 287, 405, and 510 cm^{-1}, respectively (see Fig. 6). The two other weaker peaks have also decreased slightly in frequency to 255 and 433 cm^{-1}. Thus, the shift in peak frequency between the as-grown and 850 °C anneal cases is small, and is most readily discerned for the Si-Si line (see inset to Fig. 6). The Si-Si line width increases slightly (by 0.2 cm^{-1} overall) with increasing anneal temperature up to 850 °C but its integrated intensity remains constant. Similar Raman results were obtained from sample 738.

In the course of the Raman measurements several physical changes in surface of sample 869 became apparent at the higher annealing temperatures. The as-grown 550 and 600 °C annealed samples all appeared smooth (i.e., low surface scatter) in the laser beam, which was close to grazing incidence on the sample. However, for samples annealed above 600 °C the smooth spread of the incident laser beam along the sample <110> directions was broken up by the appearance of dark lines. These lines were parallel to <110> directions and to varying extents, depending on the sample, had a greater density in say the <110> direction compared with the <1$\bar{1}$0> direction (as seen in SEM images[17] of the surface topography).   At the same time weaker subsidiary streaks of laser light appeared with equal intensity above and below the main streak due to the incident laser beam. When the sample was rotated so that the incident light was along <100> directions, the dark lines and subsidiary streaks disappeared.   The subsidiary streak effect was reproduced artificially by a reflection-type blazed diffraction grating when the ruled grooves were perpendicular to the incident laser beam.   This indicates that the subsidiary streaks observed for the annealed epilayers result from diffraction of light by surface ledges running along <110> directions, and the random pattern of parallel dark lines in the incident beam position is probably due to shadowing by these ledges.

The sample-averaged peak frequency of the Raman Si-Si line, ω_{Si-Si}, as well as the shift in frequency, $\Delta \omega_{Si-Si}$, with anneal temperature are given in Table 5. The errors given in the table represent the accuracy of measuring the relative position and shift of the Si-Si line in each sample (the absolute instrumental error in measuring any line position is $\pm$ 0.3 cm^{-1}).   It is not possible to directly relate the Raman frequency shift to a misfit density, as in the case of the DCD peak shift.   In principle, however, the absolute frequencies of the principal alloy lines could be used to obtain the stress in the $Si_{0.82}Ge_{0.18}$ alloy epilayer, but the necessary constants derived from bulk alloy spectra have yet to be determined.   A study of the uniaxial pressure dependence of the Raman spectrum of bulk Si has shown that the 520 cm^{-1} line shifts by 0.40 cm^{-1}/kbar.[11]   The corresponding constant for the Si-Si line in the alloy should be close to this value.   The Si-Si line occurs at 508 $\pm$ 1 cm^{-1} in bulk (i.e., fully relaxed) $Si_{0.82}Ge_{0.18}$ (Refs. 8, 18 and 19) and thus the stress in the alloy epilayer is $\sim$ 15 kbar in the as-grown sample and is still $\sim$ 5 kbar in the sample annealed at 850 °C.

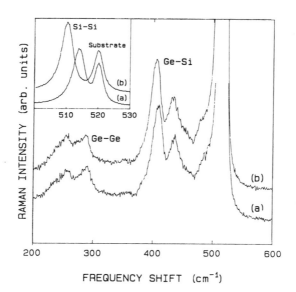

Fig. 6. Raman spectra for sample 869 (a) as grown and (b) after annealing at 850 °C. The inset gives details of the epilayer Si-Si mode near 515 cm^{-1}.

Table 5. Misfit dislocation densities ($\rho_{\parallel}$) as a function of anneal temperature as deduced by x-ray diffraction (DCD). An asterisk denotes a dislocation density below the detection limit of DCD at the indicated temperature. The DCD calculated vertical strain shift ($\Delta\varepsilon_{\perp}$) and measured relative Bragg angle ($\Delta\theta_B$) and the Raman frequency (ω_{Si-Si}) and frequency shift ($\Delta\omega_{Si-Si}$) of the epilayer Si-Si Raman line are also presented.

Annealing temperature (°C)	DCD $\rho_{\parallel}$ (μm^{-1})	DCD $-\Delta\varepsilon_{\perp}$ ($\times 10^{-3}$)	DCD $-\Delta\theta_B$ (arcsec)	Raman ω_{Si-Si} (cm^{-1})	Raman $\Delta\omega_{Si-Si}$ (cm^{-1})
500	*	*	1844 ± 10	514.0 ± 0.1	0
550	*	*	1849 ± 10	513.8 ± 0.1	0.2
600	0.6	0.09	1839 ± 10	513.8 ± 0.1	0.2
650	0.9	0.13	1834 ± 10	513.6 ± 0.1	0.4
700	4.5	0.67	1757 ± 20	512.8 ± 0.2	1.2
750	12	1.81	1600 ± 50	511.5 ± 0.3	2.5
800	19	2.86	1450 ± 50	510.7 ± 0.2	3.3
850	21	3.23	1400 ± 50	510.2 ± 0.1	3.8

X-ray Diffraction

The DCD (400) rocking curves from the $Si_{0.82}Ge_{0.18}$ samples annealed at the indicated temperatures are shown in Fig. 7. In the as-grown specimen the $Si_{1-x}Ge_x$ alloy epilayer gives rise to a well defined Bragg reflection that corresponds to a vertical strain, $\varepsilon_{\perp}$, of $(5.66 \pm 0.05) \times 10^{-3}$ and a Ge fraction, x, of 0.175 ± 0.005. The displacement of the $Si_{0.82}Ge_{0.18}$ alloy reflection on annealing indicates a change in the lattice constant in the growth direction due to a strain reduction within the alloy layer. A value for epilayer thickness of (0.17 ± 0.01) μm is deduced from the spacing between the Pendelossung thickness fringes seen in the DCD rocking curve.[5] The observation of these features generally indicates that the underlying heterointerface is sharply defined. Furthermore, the fact that the alloy reflection peak width is similar to the fringe spacing indicates that there are no substantial long range (~ 0.1 μm) depth compositional fluctuations ($< 1\%$) within the sample. No significant change in the alloy peak is seen in the samples annealed below 650 °C. Some damping of the secondary fringes is, however, evident at 600 °C, which suggests that, as for the superlattice case, some loss of interfacial integrity, possibly due to a low density misfit dislocation array, has already occurred at this anneal temperature. The $Si_{0.82}Ge_{0.18}$ epilayer reflection is seen to progressively shift with annealing temperature in the 650–850 °C range indicating that relaxation is enhanced at high temperature.

The DCD peak shift is also accompanied by a significant peak broadening generally attributed to the formation of structural imperfections in the relaxed structure. The broadening of the alloy Bragg reflection may possibly originate from the nonuniform relaxation of the epilayer over the x-ray sampling area (~ 0.5 mm^2) or from the presence of threading dislocations within the crystal. In the latter case the crystal may be modeled as a mosaic of small slightly misoriented crystallites. For randomly distributed dislocations, the threading dislocation density, ρ_d, is related to the x-ray broadening, $\delta\theta$, by the expression $\rho_d = (\delta\theta)^2/9b^2$ where b is the relevant Burgers vector. This expression yields values for ρ_d of about 10^8 and 10^9 cm^{-2} in the samples annealed at 700 and 850 °C, respectively. These values are clearly an overestimate of ρ_d since no evidence of threading dislocations, at these density levels, was found by cross-sectional TEM

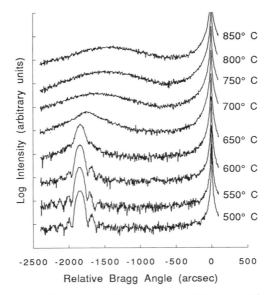

Fig. 7. X-ray diffraction (400) rocking curves from sample 869 as-grown and after annealing at the indicated temperatures up to 850 °C.

in the samples.[17] This suggests that nonuniform relaxation is a more likely explanation of the alloy reflection broadening. The broadening of the (400) x-ray peak may also be related to the presence of ridges at the sample surface. One can estimate the degree of lattice misorientation that is required to produce the significant broadening observed in the x-ray alloy line. For example, a 1 μm slab (typical of the surface ledge spacing in the 700 °C anneal sample[17]) tilted by the presence of one 60° type misfit dislocation ($b_\perp \sim 0.3$ nm) would lead to a local tilt $\sim \pm 60$ arcsecs, which is comparable to the observed broadening of ~ 200 arcsecs.

The physical data and deduced misfit dislocations densities, $\rho_{\parallel}$, obtained by the DCD technique are given as a function of anneal temperature in Table 5. In order to deduce a misfit dislocation density from the x-ray diffraction data one must use a model linking misfit density to the measured strain shift in the epilayer.[20] Using classical elasticity theory (tetragonal strain model) the measured variation in vertical strain (i.e., along the growth direction), $\Delta\varepsilon_\perp$, in the $Si_{1-x}Ge_x$ epilayer can be related to a misfit dislocation density, $\rho_{\parallel}$, as follows:

$$\Delta\varepsilon_\perp = -\left(\frac{2\nu}{1-\nu}\right)\Delta\varepsilon_{\parallel} = -\left(\frac{2\nu}{1-\nu}\right)\rho_{\parallel}\, b_{\parallel} = -1.5 \times 10^{-4}\, \rho_{\parallel} \;[\mu m]\,, \tag{1}$$

where $\Delta\varepsilon_{\parallel}$ is the variation in the in-plane strain from the fully strained condition $\varepsilon_{\parallel} = (a_{SiGe}-a_{Si})/a_{SiGe}$, ν is the $Si_{1-x}Ge_x$ Poisson ratio ($\nu_{Si} = 0.273 \cong \nu_{Ge}$), $\rho_{\parallel}$ is the average misfit density (in units of μm^{-1}) and $b_{\parallel}$ is the in-plane slip per dislocation (in nm). Here it has been assumed that the dislocations are of the 1/2<110> 60° type for the $Si_{0.82}Ge_{0.18}$ sample[21] so that $b_{\parallel} = 0.19$ nm. The introduction of one interface misfit per μm would lead to a vertical strain variation of 1.5×10^{-4}.

The Raman data for $\omega_{Si\text{-}Si}$ are compared with the DCD relative Bragg angle results in Fig. 8. The error bars on the Raman frequencies are the minimum error of $\pm$ 0.1 cm^{-1} for most anneal temperatures. However, in the case of the samples annealed at 700, 750 and 800 °C the error bars are larger because the different Raman meaurements across the sample area gave frequencies which differed by more than the normal measurement error of $\pm$ 0.1 cm^{-1}. The effect was most noticeable for the 750 °C annealed sample. For all three samples, the Raman mapping of the sample area revealed nonuniform epilayer relaxation, as indicated by the $\omega_{Si\text{-}Si}$ frequency, on a large (mm) scale. The nonuniform relaxation occurs at annealing temperatures where the rate of change in $\omega_{Si\text{-}Si}$ with regard to anneal temperature is greatest.

Figure 8 shows that the Raman and x-ray results are in close correspondence with regard to annealing effects. At first, there is little change in relative Bragg angle or $\omega_{Si\text{-}Si}$ with annealing temperature. The Raman results are sensitive enough to show that some relaxation has occurred at 550 °C but more definitive evidence of the relaxation induced shift is only apparent at 650 °C (see Table 5). At anneal temperatures above 650 °C there is a sharp drop in the relative Bragg angle and $\omega_{Si\text{-}Si}$ consistent with the large increase in the dislocation line density observed by TEM and charge collection microscopy (EBIC).[17] The results indicate that full relaxation is not obtained by annealing at 850 °C and suggest that the relaxation shift will continue to occur at a reduced rate at anneal temperatures above 850 °C. The greatest differences between the DCD and Raman data occur at the anneal temperatures between 600 and 800 °C for both samples. It is now apparent from the above comments concerning sample nonuniformity that these differences are due to the sampling of different regions of the epilayer in the x-ray and Raman measurements.

The close correspondence between the DCD and Raman results provides encouragement for a closer examination of the relationship between the two sets of

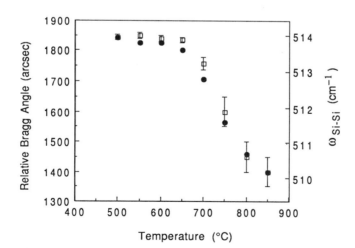

Fig. 8. Raman Si-Si mode frequency ($\omega_{Si\text{-}Si}$) [squares] and DCD relative (400) Bragg angle ($\Delta\theta_B$) [circles] as a function of annealing temperature for sample 869. Similar results were obtained for sample 738.

results. The shift in the frequency of the Si-Si longintudinal optic phonon peak is primarily due to a change in the average lattice constant in the epilayer. A reduction in frequency indicates a larger lattice constant than in the as-grown material and hence a smaller biaxial compressive stress. The shift in frequency $\delta\omega = \omega_{Si-Si} - (\omega_{Si-Si})_0$ [denoted as ω and ω_0 hereafter] is related to the strain by the relationship[22]

$$\delta\omega = \frac{p}{2\omega_0}\varepsilon_\perp + \frac{q}{\omega_0}\varepsilon_\parallel = \frac{1}{\omega_0}\left(\frac{p\nu}{\nu-1} + q\right)\varepsilon_\parallel = b\varepsilon_\parallel = b(S_{11} + S_{12})\tau, \qquad (2)$$

where p and q are phenomenological material parameters[11], S_{11} and S_{12} are the alloy compliances, and τ is the built-in interfacial biaxial stress. Equation (2) can be rewritten in terms of the measured physical parameters ω and $\Delta\theta_B$ to yield

$$\omega - \omega_0 = \frac{1}{2\omega_0}\left(p + q\frac{\nu-1}{\nu}\right)[\Delta\theta_B - (\Delta\theta_B)_0]\cot\theta_B, \qquad (3)$$

where $(\Delta\theta_B)_0$ is the relative Bragg angle for the fully relaxed alloy layer $((\Delta\theta_B)_0 \cong 1060$ arcsec for x = 0.175). Equation (3) shows that the Raman frequency of the Si-Si vibration line varies linearly (assuming that p and q are not function of strain) with the strain (relative Bragg angle). An estimate of ω_0 can be obtained by plotting ω vs $\Delta\theta_B$, as shown in Fig. 9. This plot confirms the linear relationship that exists between the Raman peak shift and the strain. Extrapolating the best linear fit of Eq. (3) yields the bulk alloy Si-Si frequency (ω_0), values for the expression $\{p + q (\nu-1)/\nu\}$, and parameter b given in Table 6. This table shows that the values obtained for $\{p + q (\nu-1)/\nu\}$ and b are reasonably close to the respective pure Si values.[11,22] The extrapolated value for ω_0 for both samples is also in good agreement with the literature value[8,18,19] of 508 ± 1 cm^{-1} for x = 0.2. Our b values are also in good agreement with the value of 930 ± 90 cm^{-1} reported by Halliwell et al.[23] from a study of as-grown $Si_{1-x}Ge_x$ epilayers with $0.09 < x < 0.20$. For the Si-Ge Raman peak, Cerdeira et al.[24] found a much lower value of b (455 cm^{-1}) for $x \leq 0.65$.

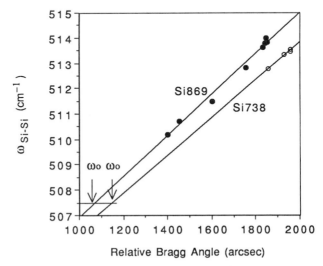

Fig. 9. The Si-Si Raman frequency shift plotted as a function of the relative epilayer Bragg angle for samples 869 and 738.

Table 6. Parameters obtained from a fit of Eq. (3) to the plots of ω_{Si-Si} versus $\Delta\theta_B$ for samples 738 and 869.

Sample	x	ω_0 (cm^{-1})	$p+q(\nu-1)/\nu$ ($\times 10^5$ cm^{-2})	b (cm^{-1})
869	0.175	507.4	11.6	859
738	0.19	507.5	10.6	784
Si	0	520.1	9.7	717
Alloy	0.2	508		

From Eq. (2), and by linear interpolation between the Si and Ge compliances using

$$S_{11}(x) = [7.7(1-x) + 9.79x] \times 10^{-4} \text{ /kbar} \tag{4}$$

$$S_{12}(x) = -[2.1(1-x) + 2.67x] \times 10^{-4} \text{ /kbar} \tag{5}$$

we find the stress coefficients for the Si-Si line given in Table 7. The small difference between the stress coefficients for samples 869 and 738 indicates they could be in error by as much as 10%. The parameters determined for sample 869 are more accurate than those found for sample 738, as the sample 869 data extend over a much wider $\Delta\theta_B$ range (see Fig. 9). Nevertheless, they are accurate enough to indicate that the x = 0.2 coefficient is approximately 20% higher than the pure Si value.[11,22]

Because of the linear relationship between the Raman shift and the strain, the optical phonon frequency data can be used to determine material properties associated with strain. For example, the plot of frequency versus the inverse of the annealing temperature shown in Fig. 10 yields an activation energy of 1.2 ± 0.3 eV for the strain relaxation. This number is in good agreement with the value of 1.5 ± 0.3 eV determined from direct measurements of the misfit dislocation density.[17] The physical interpretation of the 1.2 eV activation energy is not straightforward. It does not represent the activation energy for a single process such as dislocation glide in the epilayer, but rather the net result of the various dislocation nucleation and propagation processes that have taken place during the 3 min anneals. This value is comparable to the activation energies for dislocation glide in Si (2.2 eV) and Ge (1.6 eV)[25,26] which suggests that this may be the rate limiting process in the relaxation phenomenon.

Table 7. Stress coefficient (cm^{-1}/kbar) for the Si-Si line in Si$_{1-x}$Ge$_x$.

Sample	x	Stress coefficient
869	0.18	0.51
738	0.19	0.46
Si	0	0.40

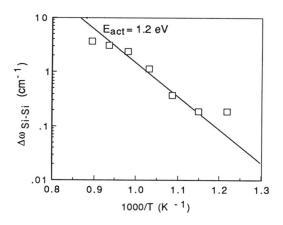

Fig. 10. Frequency of the Si-Si line as a function of the reciprocal anneal temperature for sample 869. The straight line has a slope corresponding to an activation energy of 1.2 eV.

ATOMIC LAYER SUPERLATTICES

Recently, Si/Ge ALSs have attracted considerable attention, because of the possible direct character of their band structure and resulting possibilities for the design of novel Si-based optoelectronic devices.[27] These ALSs are of the form $(Si_mGe_n)_p$ comprising p repeats of m monolayers of Si and n monolayers of Ge grown on Si or Ge substrates or $Si_{1-x}Ge_x$ buffers. As for thick layer structures, Raman spectroscopy and x-ray diffraction have proved to be versatile non-destructive techniques for investigating the dynamical, electronic, and structural properties of these artificial semiconductor crystals.[28] Due to a gain in sensitivity, glancing incidence and (400) θ-2θ x-ray measurements were performed on these thin structures. A comparison is made here of the information provided by the two techniques for two different cases.

Case 1 – Same Nominal ALSs

We consider a nominal $(Si_8Ge_{12})_8$ ALS grown on a 300 nm Ge buffer on Si(100). The (400) and glancing incidence x-ray results from sample 861 and sample 868 are shown in Fig. 11. The (400) curves exhibit diffracted peaks from the Ge buffer and Si substrate at $\sim$ 66° and $\sim$ 69°, respectively. The ALS is revealed by a weak peak at $\sim$ 67.5° which corresponds to the average perpendicular lattice constant in the periodic structure. The glancing incidence curves display broad satellite reflections arising from the superperiodicity of the ALS. The oscillations also seen in the traces are due to reflection of x-rays at the bottom interface. Analysis of the x-ray diffraction data in Fig. 11 yields the sample parameter values listed in Table 8. The two samples have similar periods but slightly different average m and n values. The relative crystalline perfection can be estimated qualitatively from visual inspection of the curves of Fig. 11. The weakness of the ALS diffracted peak in the (400) scan and the damping of the oscillations observed in sample 861 suggest that it is of poorer quality than sample 868.

The Raman spectra of the two samples are given in Fig. 12. The peaks near 300 and 500 cm^{-1} in both samples arise from vibrational modes confined largely to the Ge and Si layers, respectively.[29] The peak near 400 cm^{-1} can be either a lower frequency Si-layer mode or a vibrational mode associated with blurred (or

Table 8. Structural parameters of the $(Si_8Ge_{12})_8$ ALSs grown on a Ge buffer on Si.

Sample	m	n	Period (nm)
861	6.9	19.6	3.68
868	7.2	17.3	3.39

disordered) interfaces.[29,30] A prominent peak due to a folded acoustic mode is visible near 50 cm^{-1}. The frequencies of these modes for samples 861 and 868 are given in Table 9. The intensities of the folded acoustic modes are a sensitive indicator of interface quality.[9,12,29] The higher intensities and narrower linewidths for the folded modes in sample 868 compared with sample 861 indicate a superior interface quality. The lower frequency of the unresolved m = 1 doublet near 50 cm^{-1} in sample 861 versus sample 868 is consistent with the larger period found by x-ray diffraction.

The x-ray and Raman techniques have both shown that sample 861 is of inferior quality to sample 868. Although it is not possible to obtain quantitative information about the differences between the two ALSs from these two techniques, sufficient information is obtained to decide which of the two growths is superior.

Case 2 – Different Nominal ALSs

In this study, three $(Si_m Ge_n)_{100}$ ALSs with different (m,n) combinations were grown on a 200 nm $Si_{0.4}Ge_{0.6}$ buffer layer. The (400) and glancing incidence x-ray results are shown in Fig. 13. To enhance the features due to the epitaxial layer, the diffraction profile from a clean (100)Si substrate was

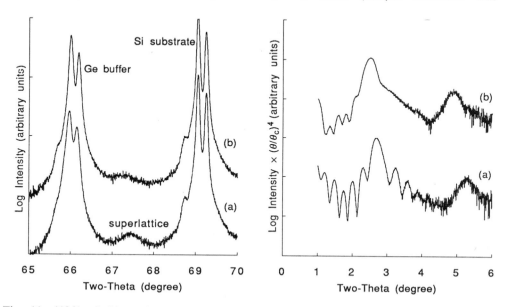

Fig. 11. (400) (left) and glancing incidence (right) θ-2θ x-ray diffraction from samples 868 (a) and 861 (b). In the glancing incidence data, the intensity is multiplied by a factor $(\theta/\theta_c)^4$ to emphasize details at high angle. θ_c is the angle of total external reflection of Si for Cu K_α x-rays ($\sim 0.22°$).

Table 9. Frequencies (in cm^{-1}) of peaks in the Raman spectra of samples 861 and 868.

Sample	Folded mode	Ge layer	Interface	Si layer
861	49.6	301.4	396.2	490.1
868	52.7	302.4	397.6	489.3

subtracted from the (400) data. All curves exhibit a strong peak near 67.2° originating from the Si-Ge buffer layer. For sample 879, a second peak near 68.2° is also seen and is believed to be due to the buffer. It would result from a sudden composition change during growth. The ALS diffraction peak is seen between 68 and 69° and shifts to higher angles as the m/n ratio increases. (In sample 879, the ALS peak is probably masked by the buffer layer peak at 68.2°.) The glancing incidence profile (Fig. 13) of each sample exhibits the first superlattice satellite peak, which allowed us to obtain the structural parameters listed in Table 10. The relative weakness and large breadth of the satellite indicate that these three samples have poor crystallinity. Also, the presence of intensity fluctuations away from the satellite that are well above the instrumental noise suggest significant variations in their periodicity.

The Raman spectra of these samples are shown in Fig. 14. As for the other ALSs considered in case 1, three optical phonon peaks are observed in each sample at frequencies near 300 cm^{-1} (Ge layers), 400 cm^{-1} (lower frequency Si mode/blurred interface), and 500 cm^{-1} (Si layers), respectively. Broad and generally weak folded-acoustic modes are visible at frequencies below 250 cm^{-1} in all three samples. Sample 879 (Si$_6$Ge$_4$) exhibits a relatively sharper and more intense line near 154 cm^{-1} compared with the 95 and 160 cm^{-1} lines in sample 880 (Si$_7$Ge$_3$) and the 105 and 205 cm^{-1} lines in sample 881 (Si$_8$Ge$_2$). This indicates that the interfaces of sample 879 are of better quality than the other samples. However, the large widths of the folded acoustic modes and the optic modes indi-

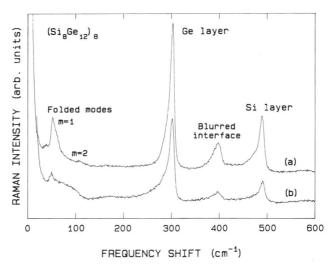

Fig. 12. Raman spectra of (a) sample 868 and (b) sample 861.

Table 10. Parameters of $(Si_mGe_n)_{100}$ ALSs grown on 200 nm of $Si_{0.4}Ge_{0.6}$ on Si.

| Sample | Nominal | | X-ray | | Period |
	m	n	m	n	(nm)
879	6	4	6.2	4.5	14.8
880	7	3	7.5	4.4	16.3
881	8	2	5.4	2.8	11.7

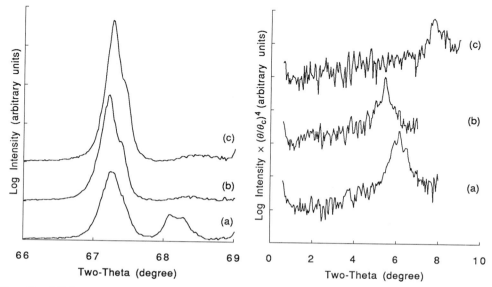

Fig. 13. (400) with substrate signal subtracted (left) and glancing incidence (right) θ-2θ x-ray diffraction from samples (a) 879, (b) 880, and (c) 881.

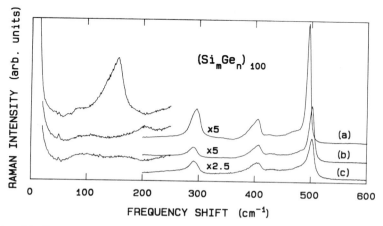

Fig. 14. Raman spectrum of a (a) Si_6Ge_4, (b) Si_8Ge_2, and (c) Si_7Ge_3 100-period ALS grown on a buffer of 200 nm of $Si_{0.4}Ge_{0.6}$ on (100) Si.

cate poor sample quality overall (e.g., variations in layer widths along the growth direction). The frequency of the Si-layer optical mode shifts up from 495 cm^{-1} in sample 879 to 501 cm^{-1} in the other samples. However, the Ge-layer vibration shifts down in frequency from 295 cm^{-1} in sample 879 to 290 cm^{-1} in the others. This greater (lower) strain in the Si (Ge) layers of sample 879 compared with samples 880 and 881 is consistent with the closer lattice match of the sample 879 ALS structure to the Si$_{0.4}$Ge$_{0.6}$ buffer layer.

For these three samples, x-ray diffraction and Raman scattering have both shown that the actual ALS structures are far from the ideal ones. The samples have relatively poor interfaces, their crystallinity is suspect, and there are significant fluctuations in their layer thicknesses/periodicities along the growth direction.

CONCLUSIONS

This joint x-ray and Raman study of a wide variety of thick and ultrathin Si/Ge heterostructures has shown that the two techniques provide similar, but also complementary information about the physical properties of these structures. X-ray diffraction provides direct information on strain partition, layer thickness and alloy composition. Raman spectroscopy of optic phonons can also be used to obtain measurements of strain, but less directly than in the case of the x-ray technique. For superlattices, analysis of the Raman scattering from folded acoustic phonons yields the layer thicknesses and alloy composition directly, and also provides qualitative information on interface quality. Both techniques are very useful for quickly determining in a non-destructive manner the overall epitaxial quality of a given heterostructure.

The complementary nature of the two techniques has allowed physical properties such as a dislocation activation energy and stress coefficients to be derived from the Raman spectra of Si$_{1-x}$Ge$_x$ alloy epilayers. This serves to emphasize the value of applying diffraction and light scattering techniques together when characterizing samples.

ACKNOWLEDGEMENT

We thank H.J. Labbé for technical assistance with the Raman measurements.

REFERENCES

1. See, for example, M. Cardona and G. Güntherodt, eds., "Light Scattering in Solids V", Springer, Berlin (1989).
2. A. Guinier, "X-Ray Diffraction", Freeman, San Francisco, 1963.
3. J.-M. Baribeau, T.E. Jackman, P. Maigné, D.C. Houghton, and M.W. Denhoff, J. Vac. Sci. Technol. A 5, 1898 (1987).
4. D.J. Lockwood, M.W.C. Dharma-wardana, J.-M. Baribeau, and D.C. Houghton, Phys. Rev. B35, 2243 (1987).
5. J.-M. Baribeau, Appl. Phys. Lett. 52, 105 (1988).
6. S.M. Rytov, Akust. Zh. 2, 71 (1956).
7. C. Colvard, T.A. Gant, M.V. Klein, R. Merlin, R. Fischer, H. Morkoç, and A.C. Gossard, Phys. Rev. B31, 2080 (1985).
8. W.J. Brya, Solid State Commun. 12, 253 (1973).
9. D.C. Houghton, D.J. Lockwood, M.W.C. Dharma-wardana, E.W. Fenton, J.-M. Baribeau, and M.W. Denhoff, J. Cryst. Growth 81, 434 (1987).
10. D.J. Lockwood, K. Rajan, E.W. Fenton, J.-M. Baribeau, and M.W. Denhoff, Solid State Commun. 61, 465 (1987).
11. M. Chandrasekhar, J.B. Renucci, and M. Cardona, Phys. Rev. B17, 1623 (1978).

12. H. Brugger, E. Friess, G. Abstreiter, E. Kasper, and H. Kibbel, Semicond. Sci. Technol. $\underline{3}$, 1166 (1988).

13. J.-M. Baribeau, Song Kechang, and K. Munro, Appl. Phys. Lett. $\underline{54}$, 1781 (1989).

14. S.J. Barnett, G.T. Brown, D.C. Houghton, and J.-M. Baribeau, Appl. Phys. Lett. $\underline{54}$, 1781 (1989).

15. D.C. Houghton, J.-M. Baribeau, Song Kechang, and D.D. Perovic, MRS Symp. Proc. $\underline{130}$, 159 (1989).

16. D.J. Lockwood, J.-M. Baribeau, and P.Y. Timbrell, J. Appl. Phys. $\underline{65}$, 3049 (1989); Can. J. Phys. $\underline{67}$, 351 (1989).

17. P.Y. Timbrell, J.-M. Baribeau, D.J. Lockwood, and J. McCaffrey, J. Appl. Phys. $\underline{67}$, 6292 (1990); J. Electron. Mater. $\underline{19}$, 657 (1990).

18. M.A. Renucci, J.B. Renucci, and M. Cardona, in: "Light Scattering in Solids", M. Balkanski, ed., Flammarion, Paris (1971), p. 326.

19. M.I. Alonso and K. Winer, Phys. Rev. B$\underline{39}$, 10056 (1989).

20. S.N.G. Chu, A.T. Macrander, K.E. Strege, and W.D. Johnson, Jr., J. Appl. Phys. $\underline{57}$, 249 (1985).

21. R. Hull and J.C. Bean, J. Vac. Sci. Technol. A$\underline{7}$, 2580 (1989).

22. Th. Englert, G. Abstreiter, and J. Pontcharra, Solid State Electron. $\underline{23}$, 31 (1980).

23. M.A.G. Halliwell, M.H. Lyons, S.T. Davey, M. Hockly, C.G. Tuppen, and C.J. Gibbings, Semicond. Sci. Tech. $\underline{4}$, 10 (1989).

24. F. Cerdeira, A. Pinczuk, J.C. Bean, B. Batlogg, and B.A. Wilson, Appl. Phys. Lett. $\underline{45}$, 1138 (1984).

25. H. Alexander and P. Haasen, Solid State Physics $\underline{22}$, 27 (1968).

26. H. Alexander, Dislocations in Solids $\underline{7}$, 113 (1986).

27. See, for example, G. Fasol, A. Fasolino, and P. Lugli, eds., "Spectroscopy of Semiconductor Microstructures", Plenum, New York (1989).

28. See other papers in this book for recent examples of the application of Raman spectroscopy to ALS characterization.

29. M.W.C. Dharma-wardana, G.C. Aers, D.J. Lockwood, and J.-M. Baribeau, Phys. Rev. B$\underline{41}$, 5319 (1990).

30. D.J. Lockwood, in: "Condensed Systems of Low Dimensionality", J.L. Beeby, P.K. Bhattacharya, P.Ch. Gravelle, F. Koch, and D.J. Lockwood, eds., Plenum, New York (1991).

RAMAN SCATTERING CHARACTERIZATION OF STRAIN IN (001) AND (111)

GaSb/AlSb SINGLE QUANTUM WELLS AND SUPERLATTICES AND IN

METASTABLE Ge_xSn_{1-x} ALLOYS

G. P. Schwartz

AT&T Bell Laboratories
Murray Hill, NJ 07974

INTRODUCTION

There is growing interest in strained-layer structures which employ metastable alloys stabilized by coherent epitaxy or involve growth orientations other than (100) for which novel piezoelectric effects are possible. In order to realize the full potential of these structures it is necessary to characterize the strain retained following growth. Although there are several techniques commonly utilized for this purpose, not all are appropriate for very thin layers designed to employ a single quantum well. Raman scattering measurements of optical phonon shifts provide a very basic tool for such characterization, and data can often be collected even in the very thin film limit.

This paper will consider three topic areas related to Raman scattering measurements of strain. The first concerns a comparison of measured and calculated optic phonon shifts in strained-layer GaSb/AlSb superlattices and single quantum wells grown on (001) and (111)B-oriented substrates. The second topic treats strain relaxation measurements as a function of layer thickness. The data will be discussed in the context of establishing an equilibrium critical layer thickness for the GaSb/AlSb system. The last topic addresses recent measurements of the optic modes found in metastable diamond structure Ge_xSn_{1-x} alloys grown on InSb substrates and their relation to phase segregation with the introduction of misfit dislocations.

(100) AND (111)-ORIENTED GaSb/AlSb

Most superlattices and quantum well devices fabricated using molecular beam epitaxy (MBE) have utilized the (001) growth orientation. Recent studies and calculations suggest that some degree of superlattice bandstructure tailoring is possible for other orientations. For example, novel piezoelectric fields[1] are expected to manifest in strained-layer superlattices grown on (111) planes, and the valence bands for nonzero momentum differ significantly for the (001) and (111) quantization directions.[2] Despite the additional degrees of freedom associated with growth on generalized (hkl) planes, only a nominal number of studies have examined such growth. The basic problem encountered by MBE is that growth on surfaces

Light Scattering in Semiconductor Structures and Superlattices
Edited by D.J. Lockwood and J.F. Young, Plenum Press, New York, 1991

219

other than (100) tends to be faceted. In some systems such as GaAs, specific substrate tilt angles have been found such that growth on (111)[3] and (110)[4] planes proceeds without faceting. Equivalent studies for GaSb have yet to be conducted, although it is known that growth on nominally oriented (nonvicinal) (111)B planes is accompanied by faceted growth in which the surface facet density increases as the layer thickness increases.

The GaSb/AlSb heterojunction is known to be Type I, with both holes and electrons confined in the GaSb wells.[5] For growth on GaSb buffer layers the GaSb wells would be unstrained and the AlSb barriers under compression. Of somewhat more interest is the case where superlattices are grown on a fully relaxed AlSb buffer, leaving the GaSb wells under tension and the barriers unstrained. The layer thickness necessary to achieve a fully relaxed AlSb buffer, however, is typically 5-10 times the critical layer thickness. Since the facet density depends on layer thickness for growth on nonvicinal (111) planes, the criteria of full buffer relaxation and a relatively smooth initial growth surface cannot both be simultaneously achieved. There remains a basic issue concerning the magnitude of the strain retained in GaSb single and multiple well structures growth on (111) AlSb buffer layers. It is the purpose of this section to examine this issue.

Estimates of the frequency shifts of the optic phonons are given by the solution of the dynamical secular equation for zincblende materials under strain.[6] The longitudinal and transverse (L,T) optic modes split into singlet (S) and doublet (D) components which are distinguished by whether the relative displacement amplitude is parallel (S) or perpendicular (D) to the tetragonal strain axis. For biaxially stressed films the tetragonal strain axis and the normal to the surface plane coincide for nonvicinal surfaces. The singlet and doublet shifts (Ω_S, Ω_D) relative to the unstrained frequency ω_0 can be expressed as[7]

$$\left[\frac{\Omega_S - \omega_0}{\omega_0}\right]^{L,T} = 2\left[\frac{\Omega_H}{\omega_0}\right] - \frac{2}{3}\left[\frac{\Omega}{\omega_0}\right], \quad \text{and}$$

$$\left[\frac{\Omega_D - \omega_0}{\omega_0}\right]^{L,T} = 2\left[\frac{\Omega_H}{\omega_0}\right] + \frac{1}{3}\left[\frac{\Omega}{\omega_0}\right], \quad \text{where}$$

$$\left[\frac{\Omega_H}{\omega_0}\right] = \left[\frac{p + 2q}{6\omega_0^2}\right](S_{11} + 2S_{12})\sigma$$

$$\left[\frac{\Omega}{\omega_0}\right] = \left[\frac{p - q}{2\omega_0^2}\right](S_{11} - S_{12})\sigma \quad (001)$$

$$= \left[\frac{rS_{44}}{2\omega_0^2}\right]\sigma \quad (111)$$

For the zincblende structure, p, q, and r are the three nonzero values of the phonon deformation potential, S_{ij} represent the elastic compliances, and σ is the in-plane stress. The L and T character of p, q, r, and ω_0 have been suppressed in the notation but are implicit in calculating the optic phonon shifts. Since the in-plane strains ε_{xx} and ε_{yy} are equal to the fractional lattice mismatch for coherent epitaxy, it is often convenient to recast the equations in terms of strain using the stress-strain relations of Table 1.

Table 1 Stress-strain Relations for the Biaxial Stress Model

	(001)	(111)
$\dfrac{\varepsilon_{xx}}{\sigma}$	$S_{11} + S_{12}$	$S_{11} + S_{12} + S_0/3$
$\dfrac{\varepsilon_{yy}}{\sigma}$	$S_{11} + S_{12}$	$S_{11} + S_{12} + S_0/3$
$\dfrac{\varepsilon_{zz}}{\sigma}$	$2S_{12}$	$2S_{12} - 2S_0/3$

The anisotropy index $S_0 \equiv S_{44}/2 - S_{11} + S_{12}$ reflects the fact that zincblende materials are not elastically isotropic even though the structure is cubic. For coherent epitaxy $\varepsilon_{xx} = \varepsilon_{yy} = f$, which for GaSb/AlSb has a magnitude of 6.5×10^{-3} at room temperature and $\sim 8.3 \times 10^{-3}$ at 77K.

Table 2 presents the physical constants necessary to evaluate the optic phonon shifts. The phonon deformation potentials were taken from references 8 and 9. A complete set of separate longitudinal and transverse values in not currently available for GaSb. The strain parameters shown are for coherent epitaxy at room temperature.

Table 2 Physical Constants Used to Evaluate Ω_S and Ω_D for GaSb and AlSb

Parameter	GaSb	AlSb
$(p+2q)/6\omega_0^2$	-1.10	-1.21 (T); -1.18 (L)
$(p-q)/2\omega_0^2$	$+0.22$	$+0.27$ (T); $+0.48$ (L)
r/ω_0^2	-1.08	-0.71 (T); -0.34 (L)
S_{11}^a	1.583	1.665
S_{12}	-0.496	-0.552
S_{44}	2.315	2.402
S_0	-0.922	-1.015
$\varepsilon_{xx}, \varepsilon_{yy}$	$+6.5 \times 10^{-3}$	-6.5×10^{-3}

$^aS_{ij}$ units 10^{12} dynes/cm^2

The calculated singlet and doublet frequency shifts are presented in Table 3. The particular mode which is being examined is determined by the selection rules for Raman scattering. On the (001) surface the scattering configuration employed was $Z(XY)\bar{Z}$ where $X=[100]$, $Y=[010]$, and $Z=[001]$. Only the LO mode is allowed, and the longitudinal mode displacement is parallel to the tetragonal strain axis for backscattering from (001). For the (111) orientation, the TO mode is selected by $Z'(X'Y')\bar{Z}$, with $X'=[1\bar{1}0]$, $Y'=[11\bar{2}]$, and $Z'=[111]$, with eigenvector displacements perpendicular to the strain axis.

Table 3 Calculated Values (cm^{-1}) for $\Delta\Omega_S$ and $\Delta\Omega_D$ for Strained Layer Growth on the (001) and (111) Planes

	GaSb		AlSb	
	(001)	(111)	(001)	(111)
$\Delta\Omega_S$	−2.3	−0.9	+3.3 (T) +4.0 (L)	+2.1 (T) +3.0 (L)
$\Delta\Omega_D$	−1.6	−3.2	+2.2 (T) +1.9 (L)	+4.3 (T) +4.1 (L)

Although confinement can also contribute to shifts in the optic mode frequencies, none of the layer widths in any of the samples to be discussed are less than 50Å. For this case the confinement shifts are less than the measurement uncertainity.

Table 4 presents representative data for measured GaSb optic mode shifts for samples grown on 2000Å AlSb buffers over GaSb bulk substrates. The spectra were recorded at room temperature using 5145Å excitation in the $Z(XY)\bar{Z}$ and $Z'(X'Y')\bar{Z}'$ configurations respectively on (001) and (111) faces. The unstrained TO and LO frequencies measured were 225.4 and 234.6 cm^{-1}. All spectra were taken in a cell evacuated to ~10^{-7} torr. For multiple quantum well samples the total superlattice never exceeded 1200Å and the barrier widths were 75-100Å.

The following trends appear in the data. For growth of multiple well structures on the (001) orientation, the observed shift in ω_{LO} is reasonably close to the calculated shift

Table 4 Measured GaSb LO and TO Strain Shifts in Single and Multiple Quantum Wells Grown on 2000Å AlSb Buffer Layers

(001) MQW	d_{GaSb} (Å)	$\Delta\omega_{LO}$ (cm^{-1})
	50	−2.2 ± 0.2
	75	−2.0 ± 0.3
	100	−1.6 ± 0.5
(001) SQW		
	50	−1.1 to −1.6
(111) MQW	d_{GaSb} (Å)	$\Delta\omega_{TO}$ (cm^{-1})
	50	−2.4 ± 0.3
	75	−2.3 ± 0.3
	100	−1.6 ± 0.5
(111) SQW		
	60	−1.7 ± 0.4
	80	−1.5 ± 0.4
	110	−1.2 ± 0.4

(-2.3 cm^{-1}) for coherent epitaxy for well widths below 75Å. Above 100Å some strain relief is apparent. We will return this point in the following section concerning the critical layer thickness. Samples grown with a single quantum well consistently exhibit less strain retention. The relatively poor quality of the first well to grow on (001) substrates is believed to be due to the high defect density initially present on the surface of the AlSb buffer. These defects could originate from defects in the substrate or misfit dislocations at the buffer/substrate interface which thread to the surface predominantly along (111) planes. The precise manner in which such defects could lower the "apparent" strain is not well understood. Two possible mechanisms include local strain relaxation around dislocation boundaries and partially faceted growth in the vicinity of defects. Growth of multiple periods is known to both smooth out rough interfaces and reduce the threading dislocation density, so that multiple quantum well samples appear to be well characterized as fully strained. Another consideration is that the first well might grow on a buffer which is not fully relaxed, thereby importing less strain than anticipated based on unstrained lattice constants. In this case however, it is difficult to reconcile the observation that multiple well structures on (001) appear to be fully strained, since coherent epitaxy on a partially relaxed buffer would not yield the full lattice-mismatch strain.

For scattering from the (111) face, even the multiple well structures only exhibit ~75% of the calculated strain. Although some of the discrepancy may be associated with uncertainties in the values of the phonon deformation potentials, most is believed to be due to the enhanced facet roughness of the AlSb buffer when grown on (111) planes. The first well to grow is consequently of poor quality, and growth of subsequent wells is inadequate to suppress formation of non (111) planes in superlattices limited to ~10 periods.

CRITICAL LAYER THICKNESS MEASUREMENTS

One of the basic parameters associated with strained-layer epitaxy is the equilibrium critical layer thickness beyond which the strain is relieved via the introduction of misfit dislocations. Although it is possible to achieve relatively thick strained layers under conditions of metastable, low temperature growth, the critical thickness for a given strain estimated from elastic theory[10-12] subject to thermodynamic equilibrium is much less. For example, the estimated equilibrium critical thickness of the GaAs/AlSb system is of order 60-180Å. This issue becomes critical when trying to grow and characterize fully strained superlattices and single quantum wells.

A series of single layer and superlattice samples were grown on (001) GaSb and the optic mode shift of the AlSb LO phonon examined. The data are presented in Fig. 1. The largest observed shifts in $\Delta\omega_{LO}$ were ~$+2.2$ cm^{-1} at room temperatures for superlattice samples with an AlSb layer width of 50-75Å. This shift is smaller than the estimated shift ($+3.3 - 4$ cm^{-1}) for coherent epitaxy presented in Table 3, and suggests that all samples are partially relaxed. However, three of the superlattice samples with nominal AlSb layer widths of 50, 75 ans 100Å were also examined by x-ray diffraction using a high resolution four reflection monochromator.[13] The tetragonal lattice constant along [001] was consistent with coherent epitaxy assuming that only the AlSb layers were strained. A comparison of numerically simulated and measured (004) and (002) reflections of the full width at half maximum of superlattice satellite lines also gave no evidence for strain relaxation. The disparity between the Raman scattering and x-ray diffraction results is believed to be due to strong optical absorption at 5145Å in GaSb which makes the former measurement more sensitive to the strain in the layers nearest the superlattice/air interface. Any relaxation associated with surface oxidation would bias the Raman measurements toward lower strain relative to a bulk measure such as x-ray diffraction.

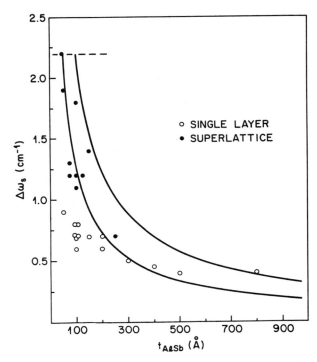

Fig. 1 Raman shift of the AlSb LO phonon versus layer thickness.

The single layer data exhibit even more strain relaxation. These samples were capped with 25Å of GaSb, and it appears that a cap of that thickness may be insufficient to retard the oxidation of AlSb. Rutherford backscattering measurements on single AlSb layers capped by relatively thick (~250Å) amorphous Be indicate a critical thickness on the order of 100-170Å.[13] Strain relaxation develops slowly, and nearly 25% of the coherent strain is retained in films 500Å thick. Fully relaxed films typically require a growth thickness of 1000-2000Å. The solid curves are from model calculations for strain relief using the theory of Matthews and Blakeslee[12] with assumed critical thicknesses of 50 and 100Å. A 100Å critical thickness represents an upper bound based on the Raman data, with the caveat that those measurements are potentially biased toward lower values as discussed previously.

METASTABLE Ge_xSn_{1-x} ALLOYS

Bandstructure calculations[14] for crystalline Ge_xSn_{1-x} alloys condensed in the diamond structure indicate that this material should exhibit semimetallic or semiconductor properties depending on the composition. Of particular interest was the prediction of a tunable direct gap of 0-500 meV for $0.4 < x < .8$. The diamond structure is a metastable phase for this alloy; under equilibrium conditions, neither component exhibits any appreciable solid solubility in the other and phase segregation occurs.

The problem of stabilizing the diamond structure can be overcome by growing on a zincblende structure if the lattice mismatch is sufficiently small that growth proceeds via coherent epitaxy. The spectre of metastability becomes an important issue if misfit dislocations are generated at some point during growth. The question of phase segregation for growth with and without misfit dislocations is addressed qualitatively in this section.

Samples of nominal thickness 1200Å were grown on (001) InSb at room temperature with germanium concentrations of 0 (α-Sn), 8, 10, and 13 atomic percent.[15] X-ray diffraction indicated that the first three samples were coherently strained, whereas the 13 at.% film was partially relaxed. The low growth temperature permitted films to be grown with thicknesses which considerably exceed equilibrium critical thickness estimates.

Raman spectra for the four samples are shown in Fig. 2. The data were processed with a 10 point filter because peak count rates were only 3-10 cps with ~100 mW of 5145Å excitation. Typical data acquisition times were 15-20 hours/spectrum.

A quantitative description of the unstrained optic mode frequencies versus concentration for diamond structure $Ge_x Sn_{1-x}$ alloys is not presently available. The following discussion assumes that the system exhibits three mode behavior similar to the IV-IV alloy $Ge_x Si_{1-x}$.

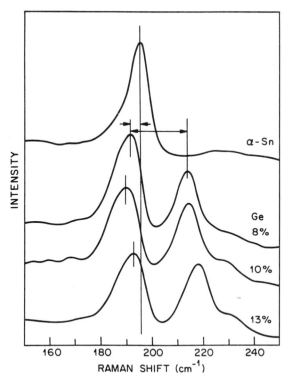

Fig. 2   Raman spectra of $Ge_x Sn_{1-x}$ alloys.

For $Ge_x Sn_{1-x}$ with x > 0, the low frequency peak would be assigned to Sn-Sn vibrations and the higher frequency peak to Ge-Sn vibrations. At these concentrations we were unable to observe a Ge-Ge mode which would lie at still higher frequencies. The separation of the two peaks (x>0) is a crude measure of concentration, although it is not expected to be linear over any extensive concentration range. Both peaks will be shifted by the residual strain in the film. Relative to the InSb substrate, the estimated in-plane strain for coherent epitaxy is +0.15% (α-Sn), −0.89% (8% Ge), −1.15% (10% Ge), and −1.54% (13% Ge). Figure 3 is a plot of the Sn-Sn mode frequency versus in-plane strain. The Sn-Sn mode is expected to decrease slightly for a system which exhibits three mode behavior as x increases, so a total frequency downshift $\Delta\omega$ plotted in Fig. 3 contains contributions both from concentration dependence and tensile strain which both act to increase $\Delta\omega$ to larger negative values. The strain component is expected to be the dominant contributor over a narrow concentration range so the drop in the magnitude of $\Delta\omega$ for the 13% alloy confirms the x-ray diffraction result that this film is partially relaxed.

The measured separation between the Sn-Sn and Ge-Sn modes was 22.4, 24.3, and 25.2 cm^{-1} for the 8, 10, and 13 at.% Ge alloys respectively. Of particular interest is the fact that the 13% sample still exhibits alloy-like modes indicative of the fact that strain relief has not precipitated a total phase separation. The x-ray diffraction studies did detect β-Sn, but we were unable to identify this phase with Raman scattering. The "apparent" concentration of the alloy component of the 13% film can be estimated if one assumes a roughly linear dependence of mode splitting on concentration over a limited concentration range. The splittings for the 8 and 10% samples provide a value of ~0.95 cm^{-1}/at.% Ge, so the

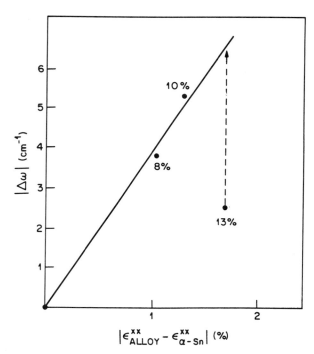

Fig. 3 Sn-Sn mode shift versus in-plane strain assuming coherent epitaxy.

"apparent" alloy phase in the 13% sample is closer to 11%. It is unlikely, however, that the alloy phase composition is homogeneous for conditions where coherent epitaxy is lost during some stage of the growth. Nevertheless, phase segregation is sufficiently hindered that some residual diamond structure $Ge_x Sn_{1-x}$ component remains even when $\sim 2/3$ of the stabilizing strain is relaxed.

For $x \sim 0.4$, $Ge_x Sn_{1-x}$ alloys are expected to switch from being semimetals to direct gap semiconductors. Growth of (Ge,Sn) on fully relaxed (Al,Ga)Sb buffer layers is currently being examined. For the lattice constant range between AlSb and GaSb, lattice match to (Ge,Sn) is possible for $.42 < x < .47$. The estimated direct gap of (Ge,Sn) in this regime spans the tunable range 9-20 microns. Because of the metastable character of the diamond structure, it remains to be determined whether such films can be processed into detector structures.

ACKNOWLEDGEMENTS

The work on (Ge,Sn) alloys is an ongoing collaboration with E. A. Fitzgerald. The x-ray data came from the laboratories of R. Kortan and W. Lowe. G. J. Gualtieri collaborated with the crystal growth of all GaSb/AlSb samples.

REFERENCES

1. D. L. Smith, Solid State Commun. *57*, 919 (1986).

2. M. Huong, Y. C. Chang, and W. I. Wang, J. Appl. Phys. *64*, 4609 (1988).

3. T. Hayakawa, M. Kondo, T. Suyama, K. Takahasi, S. Yamamoto, and T. Hijikata, Jpn. J. Appl. Phys. *26*, L302 (1987).

4. L. T. P. Allen, E. R. Weber, J. Washburn, and Y. C. Pao, Appl. Phys. Lett. *51*, 670 (1987).

5. G. J. Gualtieri, G. P. Schwartz, R. G. Nuzzo, and W. A. Sunder, Appl. Phys. Lett. *49*, 1037 (1986).

6. E. M. Anastassakis, "Dynamical Properties of Solids,," Chap. 3, ed. Horton & Maradudin, North Holland, (1980).

7. B. Jusserand, P. Voisin, M. Voos, L. L. Chang, E. E. Mendez, and L. Esaki, Appl. Phys. Lett. *46*, 678 (1985).

8. F. Cerdeira, C. J. Buchenauer, F. Pollak, and M. Cardona, Phys. Rev. B*5*, 580 (1972).

9. E. Anastassakis and M. Cardona, Solid State Commun. *63*, 898 (1987).

10. F. C. Frank and J. H. van der Merwe, Proc. Roy. Soc. A*198*, 205 (1949).

11. J. W. Matthews, S. Mader and T. B. Light, J. Appl. Phys. *41*, 3800 (1970).

12. J. W. Matthews and A. F. Blakeslee, J. Crystal Growth *29*, 273 (1975).

13. H. J. Gossmann, B. A. Davidson, G. J. Gualtieri, G. P. Schwartz, A. T. Macrander, S. E. Slusky, M. H. Grabow, and W. A. Sunder, J. Appl. Phys. *66*, 1687 (1989).

14. D. W. Jenkins and J. D. Dow, Phys. Rev. B *36*, 7994 (1987).

15. E. A. Fitzgerald, A. R. Kortan and W. P. Lowe, private communication.

THE RAMAN LINE SHAPE OF SEMICONDUCTOR NANOCRYSTALS

Philippe M. Fauchet

Laboratory for Laser Energetics and
Department of Electrical Engineering
University of Rochester
Rochester, NY 14623-1299

ABSTRACT

Most properties of micro/nanocrystals are different from those of bulk crystals. Because of their size, it is often difficult to characterize these objects. In recent years, microcrystalline silicon and other types of semiconductor microcrystals have been investigated by Raman spectroscopy. Analysis of the optic phonon line shape has been used to deduce the object size. In this review, our current understanding of the optic phonon line shape is presented with emphasis on objects of ~10 nm size. In the simplest model, the peak position, width and symmetry of the optic phonon lines are modified in a way that is a direct measure of size for spherical crystals. Although the model has been especially successful for spherical silicon grains, it can also be applied to other semiconductors and generalized to non-spherical shapes. A discussion of several significant factors that may complicate the interpretation of the spectrum is given, with emphasis on stress/strain, electromagnetic resonances for isolated grains, absorption effects in mixed phase films, surface or interface vibrations, and crystallinity of the film.

INTRODUCTION

In modern semiconductor technology, ultrasmall objects are of increasing importance. The precise characterization of micro/nanocrystalline semiconductors is, however, difficult. For example, the grain size is not easily measured with conventional techniques, the uniformity and crystallinity of the grains are difficult to assess, and the grain boundaries may play a dominant role. Raman scattering has been among the most successful nondestructive diagnostic tools for bulk and reduced dimensionality semiconductors.[1] In the past few years, semiconductor nanocrystals have also been

Light Scattering in Semiconductor Structures and Superlattices
Edited by D.J. Lockwood and J.F. Young, Plenum Press, New York, 1991

characterized by Raman scattering. We have previously reviewed how the first-order optic phonon line shape can be used to measure grain size, mostly in microcrystalline silicon.[2]

In this paper, we focus on semiconductor nanocrystals in the ~10 nm size range. A very simple but quantitative model, which predicts the evolution of the first order optic phonon line shape as a function of the diameter of the spherical nanocrystal, is first presented and compared to data. Given its simplicity, the model works surprisingly well for silicon. We generalize it to other semiconductors and other shapes. Preliminary results on the second-order optic phonon line shape can however not be explained with the same model. The predictions of the simple theory can be dramatically altered by many factors. We consider several important effects: homogeneous and heterogeneous stress/strain; electromagnetic resonances and surface modes; composite films; crystallinity of the grains. Because many of these factors are under active investigation, we will discuss the results of many different groups but attempt a synthesis only when a consensus is emerging.

A SIMPLE MODEL FOR SPHERES

Introduction

Consider a hypothetical material having the optic phonon dispersion shown in Fig. 1. The momentum of a photon is extremely small on the scale of the Brillouin zone, and thus we can say that, in an infinite crystal, no momentum is exchanged during Raman scattering, and a wave vector selection rule applies. Therefore, only phonons with negligible wave vectors may contribute to Raman scattering, and the Raman shifts will be ω_1 and ω_2. Suppose now that the size of the crystal becomes finite, typically of the order of 10 nm. We assume further that the dispersion curves are unaffected and that no new modes are created by the size reduction. The phonon can no longer be described by a plane wave; it is now a wave packet whose spatial dimensions are comparable to the crystallite size. An uncertainty in wave vector is thus introduced, which will be larger for smaller grain sizes because the wave packet becomes more localized in real space. Since the phonon dispersion curves are in general not flat, new frequencies will be introduced in the Raman line when contributions from phonons having significantly different wave vectors start playing a role, as shown in Fig. 1. For branch 1 of our hypothetical material, ω is an increasing function of wave vector $\mathbf{q}$, and thus the Raman line shift increases, whereas for branch 2, ω is a decreasing function of $\mathbf{q}$, and thus the Raman line shift decreases. These ideas were initially put forth by Nemanich et al.[3] and Richter et al.[4] The Raman line not only shifts but also broadens and becomes asymmetric for small crystallites. The critical size at which these changes become observable is a function of the dispersion curve of the material and of the intrinsic phonon lifetime. For Si at room temperature, the critical diameter is about 20 nm.

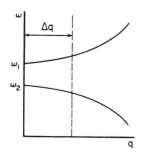 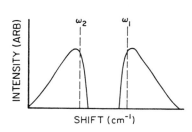

Fig. 1   Optical phonon dispersion curves of a hypothetical material and Raman spectrum for a small crystallite made of that material.

Model

We now review in detail the model put forth by Richter et al.[4] and follow the discussion in our review.[2] Consider the wave function of a phonon of wave vector q_0 in an infinite crystal:

$$\Phi(q_0, r) = u(q_0, r)e^{iq_0 \cdot r} \tag{1}$$

where $u(q_0, r)$ has the periodicity of the lattice. Let us restrict this phonon to a sphere of diameter L. The phonon confinement changes the phonon wave function to

$$\psi(q_0, r) = W(r, L)\Phi(q_0, r) = \psi'(q_0, r)u(q_0, r) \tag{2}$$

where $W(r, L)$ is the phonon weighting function such that

$$|\psi(q_0, r)|^2 = |W(r, L)|^2. \tag{3}$$

It is clear that $W(r, L)$ describes the phonon confinement, although it is not a known function a priori. To calculate the Raman spectrum, we first expand $\psi'(q_0, r)$ in a Fourier series:

$$\psi'(q_0, r) = \int d^3q C(q_0, q)e^{iq \cdot r} \tag{4}$$

with Fourier coefficients $C(q_0, q)$ obtained from

$$C(q_0, q) = \frac{1}{(2\pi)^3} \int d^3r \psi'(q_0, r)e^{-iq \cdot r}. \tag{5}$$

The microcrystal phonon wave function is a superposition of eigenfunctions with q vectors centered at q_0. Thus, the first-order Raman spectrum $I(\omega)$ is given by

$$I(\omega) \;=\; \int \frac{d^3q\,|C(0,\mathbf{q})|^2}{\left[\omega - \omega(\mathbf{q})\right]^2 + \left(\Gamma_o/2\right)^2} \tag{6}$$

where $\omega(\mathbf{q})$ is the phonon dispersion curve; Γ_o is the natural line width (inversely proportional to the intrinsic phonon lifetime); $\mathbf{q}_o = 0$, which is appropriate for first-order Raman scattering; and the integration must be performed over the entire Brillouin zone.

So far, we have not discussed the function $W(\mathbf{r},L)$. In order to carry out numerical calculations, one has to write a specific phonon confinement function. Ideally, the choice of $W(\mathbf{r},L)$ should be based on physical considerations. For example, Nemanich et al.,[3] who considered orthorhombic crystallites with dimensions $L_i(i = x,y,z)$, used for $W(\mathbf{r},L)$

$$W(\mathbf{r},L) \;\cong\; \prod_{i=x,y,z} \frac{\sin\!\left(qL_i/2\right)}{q/2} \tag{7}$$

in analogy to the well-known x-ray scattering formula. However, they noted that because they sampled microcrystallites with a finite size distribution, it was appropriate to replace their $W(\mathbf{r},L)$ with a Gaussian function.

We have investigated the influence of the choice of $W(\mathbf{r},L)$ on the quality of the fit between theory and experiments[5] for the case of silicon spherical microcrystallites. Three types of functions were considered, which are given below together with their Fourier coefficients:

$$W(\mathbf{r},L) \;=\; \frac{\sin(2\pi r/L)}{2\pi r/L} \qquad\qquad |C(0,\mathbf{q})|^2 \;\cong\; \frac{\sin^2(qL/2)}{\left(4\pi^2 - q^2L^2\right)^2} \tag{8a}$$

$$W(\mathbf{r},L) \;=\; \exp\!\left(-4\pi^2 r/L\right) \qquad\qquad |C(0,\mathbf{q})|^2 \;\cong\; \frac{1}{\left(16\pi^4 - q^2L^2\right)^4} \tag{8b}$$

$$W(\mathbf{r},L) \;=\; \exp\!\left(-8\pi^2 r^2/L^2\right) \qquad\qquad |C(0,\mathbf{q})|^2 \;\cong\; \exp\!\left(-q^2L^2/16\pi^2\right) . \tag{8c}$$

The first confinement function, Eq. (8a), was chosen by analogy with the ground state of an electron confined to a hard sphere. The amplitude of the phonon is zero at the boundary. The second confinement function, Eq. (8b), which can be written as $\exp(-\alpha r)$, was chosen by analogy with a wave in a lossy medium. The third confinement function, Eq. (8c), which can be written as $\exp(-\alpha r^2/L^2)$, was also proposed by Richter et al.[4] The best confinement function is given by Eq. (8c) for which it is possible to select a value for α that fits the data well.[5]

Results

The integration of Eq. (6) must be performed over the entire Brillouin zone. The phonon dispersion curves are anisotropic. However, as long as the diameter L remains large enough, significant contributions to the integral come from a relatively small region at the center of the Brillouin zone. In that region, the anisotropy is small, and the integration can be performed on an average dispersion curve and for a spherical Brillouin zone. The Raman line shape thus calculated differs from that of single-crystal Si, because ω decreases as $|\mathbf{q}|$ increases. Three convenient parameters are used to describe quantitatively the Raman line shape: $\Delta\omega$, the central frequency shift with respect to the peak of crystalline Si; Γ, the full width at half maximum (FWHM); and Γ_a/Γ_b, the ratio of the half width at half maximum (HWHM) on the low-energy side to the HWHM on the high-energy side. Figure 2 shows the calculated Raman line shape for spherical Si crystallites of different sizes (assuming T = 300 K). Figure 3 quantifies these changes and indicates the general agreement with data from various sources, in which L was measured by x-ray scattering[4,6] or transmission electron microscopy (TEM).[7]

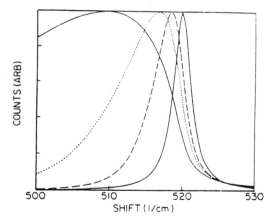

Fig. 2   Calculated evolution of the optic phonon Raman line for spherical silicon crystallites of various sizes (L = 3 nm, 6 nm, 10 nm, and single crystal).

GENERALIZATION OF THE SIMPLE MODEL

Other Semiconductors

The magnitude of the effect of microcrystallinity on different materials depends on the optic phonon dispersion relation in the region centered at the Γ point. As an example let us compare silicon and germanium. We assume a spherical Brillouin zone with an isotropic dispersion curve given by:

$$\omega(q) = A + B \cos(\pi q) . \tag{9}$$

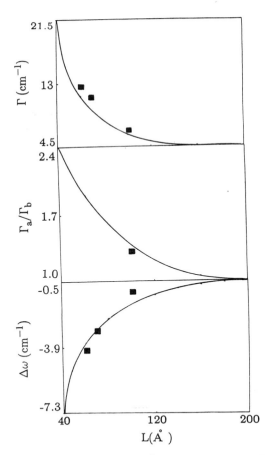

Fig. 3   Calculated shift $\Delta\omega$, asymmetry Γ_a/Γ_b, and line width Γ as a function of the diameter L of silicon crystallites. The data points are from Refs. 4, 6, and 7.

Variations in B lead to different Raman line shapes for the same crystallite size. Figure 4 is a comparison of the Raman spectra of silicon and germanium for microcrystalline spheres.[8] To obtain the same changes in germanium, the size must be smaller than that of the same structure made of silicon. The Raman line shape parameters scale simply with the ratio of the amplitudes $B_{Ge}/B_{Si} \sim 0.5$, modified slightly by the smaller lattice constant of silicon which leads to an observed ratio of ~ 0.57. It has been shown very recently that this simple analysis applies also to diamond.[9]

Generalization to other semiconductors, such as GaAs alloys or II-VI materials is possible. In fact, Parayanthal and Pollak[10] have treated the effect of alloy disorder in $Al_xGa_{1-x}As$ and $In_xGa_{1-x}As$ using the same model based on relaxation of the wave vector selection rule, and found agreement between theory and experiments. Tiong et al.[11] and Nakamura and Katoda[12] have applied the model to the case of ion implantation in GaAs.

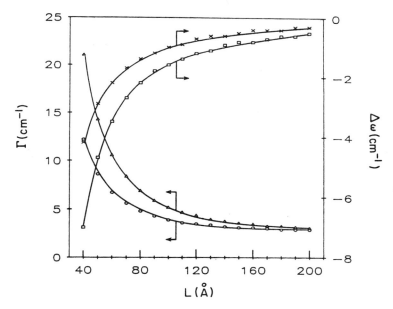

Fig. 4   Calculated relationship for shift and line width as a function of diameter for silicon
and germanium crystallites: Si shift ($\Box$) and line width (Δ), and Ge shift (x) and
line width (o).

Both groups were able to measure the average size of the undamaged regions after ion
implantation, as a function of the fluence, energy, and type of ion. However, Burns et al.[13]
have pointed out that the changes in the Raman spectrum of ion-bombarded GaAs may be
due to the combined effects of strain and a change in the ionic plasma frequency caused by
the formation of defects rather than phonon confinement.

The original work of Rossetti et al.,[14] and subsequent investigations of the Raman
spectrum of II-VI semiconductor nanocrystals, have not been focused on the phonon line
shape. No comparison is available with the present theory. In all heteropolar materials, the
long-range electrostatic interaction complicates the interpretation of the spectra, as pointed
out be Kanellis et al.[15] in a series of papers on thin ionic slabs of GaAs.

Other Shapes

As discussed in Refs. 2 and 5, we have applied a similar phonon confinement model
to silicon cylinders and slabs, and found agreement with the little data available. Of greater
interest is the ability of Raman spectroscopy to distinguish between objects of different

shapes containing the same number of atoms. We find that the peak shift is a function of the number of atoms but is quite insensitive to the shape, whereas the low wave number tail, which controls the broadening and asymmetry, is sensitive to the shape. We take the expression of Eq. (9), with A = 490 cm^{-1} and B = 30 cm^{-1}. The Raman line shape of silicon is calculated for three different geometries: sphere, rod, and disk. Here, the rod is a cylinder of length equal to four times the diameter, and the disk is a cylinder of diameter equal to four times the length. In Fig. 5, we show the calculated Raman spectra for grains of each shape containing 2000 atoms.[8] Whereas the peak position is virtually identical, the low wave number tail is very different. The increasing asymmetry from spheres to disks and rods is rooted in the fact that one and two dimensions are much smaller respectively, which leads to an increased contribution from lower frequency phonons. In Fig. 6, we show the evolution of two parameters describing the line shape as a function of size. These results suggest that size and shape can be determined simultaneously. However, for nonspherical objects having an aspect ratio not too different from one, all line shapes may be in practice undistinguishable from that of a sphere. To date, we are not aware of any experiments that can be used to test the result of Figs. 5 or 6.

BEYOND THE SIMPLE MODEL

Second-order Raman Scattering

Two-phonon spectra are more complex than first order spectra; they involve contributions from phonons throughout the Brillouin zone and peaks in the spectrum result

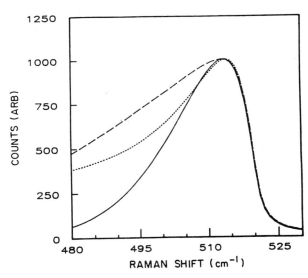

Fig. 5   Calculated Raman spectra for silicon crystallites containing 2000 atoms and having spherical (solid line), disklike (dashed line) and rodlike (dotted line) shapes.

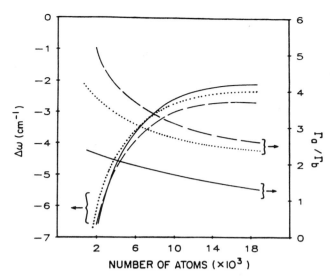

Fig. 6   Calculated relationship for shift and asymmetry as a function of number of atoms for the three shapes of Fig. 5, spheres (—), rods (·····), and disks (– –).

from critical points in the phonon density of states modified by appropriate momentum and symmetry selection rules. The dominant contribution to the second order spectrum is from phonons near the edge of the Brillouin zone where the phonon dispersion relation is relatively flat. Early work implied that the primary effect of finite crystal size was to relax selection rules.[3] As the crystal size decreased the peak widths increased and new peaks appeared in the spectrum corresponding to relaxation of the momentum and symmetry selection rules, respectively.

Although first and second order spectra are both altered by finite crystal size effects, the interpretation of the two phonon spectra is more complicated. As an example, we consider the second order Raman spectrum of microcrystalline Si in the spectral region from $550 \ cm^{-1}$ to $1100 \ cm^{-1}$.[16] Figure 7 shows the evolution of the first and second order Raman spectra of microcrystalline Si with decreasing crystal size. In materials with the diamond structure, the second order optic phonon spectrum is dominated by overtones, e.g., 2TO(X) and 2LO(Γ), while many combinations, e.g., TO(L)+LO(L), are symmetry forbidden.[17,18] The main peak from $900 \ cm^{-1}$ to $1000 \ cm^{-1}$ in the second order spectrum of crystalline Si is due to several overtones: 2TO(X), 2O(Q), 2O(S_I), 2TO(W), and 2TO(L). The dominant peaks in both the second order spectra and the first order spectra exhibit a down shift and broadening as the crystal size decreases. The effect is more pronounced in the two-phonon spectra because the changes imposed by microcrystallinity

237

affect each phonon which leads to a larger absolute shift and broadening. Of particular interest is the comparison between the spectra for the two samples with identical grain size. In this case the first order spectra of the microcrystals are virtually identical but the second order spectra are radically different.

We cannot apply the phonon confinement model to the second order spectra. Since the phonon energy is at a minimum at the zone edge, relaxing momentum conservation should shift the two phonon peaks to higher energy and not to lower energy as observed. The short wavelength phonons involved in the two-phonon spectra may not experience significant confinement, perhaps because the amorphous matrix that surrounds the microcrystal has many vibrational modes that are at the same frequency as the zone edge optical phonons. This is in contrast to the zone-center optic phonons which have energies 4–5 meV above the dominant vibrations of the amorphous material. In addition to the down shift and broadening of the second order spectrum, there is relaxation of the symmetry selection rules that forbid combinations of zone edge phonons. This is clearly seen in the spectra of crystalline and microcrystalline Si: the crystalline spectrum contains weak scattering from TO+TA combinations at ~630 cm^{-1} but as the crystal size decreases, the magnitude of the scattering in this spectral region generally increases.

At present, we do not have a model that explains the second order spectra. Several of the considerations discussed in the rest of this section for first-order spectra may also apply. It appears clear that the second order spectrum is at least as dramatically altered by size effects as the first order spectrum. More work is necessary to explain these promising results.

Stress/Strain

Stress/strain is another factor of importance, especially for thin films. For example, silicon deposited on SiO_2 is under planar tensile stress. It is possible to vary the grain size over a wide range by laser irradiation of Si on SiO_2. However, the optic phonon line shape of nanocrystalline Si is considerably affected by stress.[19] The Raman line simply shifts with uniform applied stress. For common semiconductors, a line shift of 1 cm^{-1} corresponds to $\gtrsim 10^9$ dynes/cm^2 of stress or $\gtrsim 0.1\%$ strain. If the stress is nonuniform, or heterogeneous, the line also broadens.

For nanocrystals in a thin film, stress is usually not uniform. Therefore, the peak shifts, the line broadens, and the asymmetry changes. If the data still fit on all the curves of Fig. 3, the effect of stress is negligible. If the data do not fit the $\Delta\omega$ versus L curve only, the stress is uniform and can be measured. Otherwise, it is essentially impossible to separate the effects of size and stress. Also note the work of Veprek et al.[20] who prepared films containing Si crystallites (and no amorphous tissue—see below) and were able to measure the compressive stress within the grain.

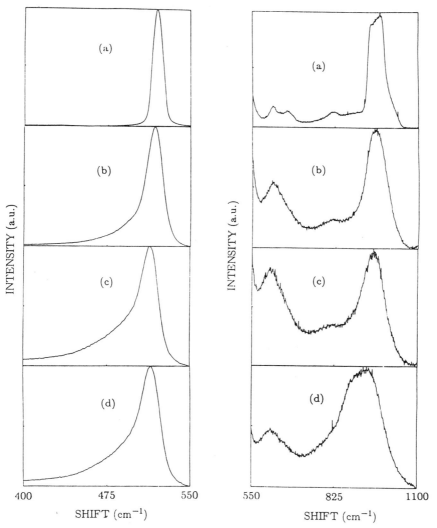

Fig. 7 (left) First order Raman spectra for three silicon crystal sizes (a) wafer, (b) 10 nm, (c) and (d) 5 nm. (right) Second order Raman spectra for the same samples. As the size decreases, the dominant second order peak at 970 cm^{-1} changes strikingly and the symmetry forbidden combination line at 630 cm^{-1} also changes.

Electromagnetic Resonances and Surface Modes

Consider objects that are either isolated or embedded in a matrix with a very different index of refraction. There will be two major consequences: electromagnetic resonances may be observed due to local field effects, and contributions to the Raman spectrum from surface modes may appear. Several groups have prepared isolated crystalline grains of Si or Ge.[21,22,23] As shown in Fig. 8, for a diameter <8 nm, the spectrum becomes entirely amorphous-like with a broad (>50 cm^{-1}) TO phonon line centered at 480 cm^{-1}. An explanation has been proposed by Hayashi *et al.*[24] Assume that each isolated microcrystallite is made of a crystalline core of radius a, surrounded by a shell which is a disordered layer of thickness (b-a), as shown in the inset of Fig. 9. The Raman spectrum is thus made of two components, one coming from the core and the other from the shell. The amorphous-like component would be produced by vibrations in the shell, and the crystalline component would be produced by core vibrations. The amorphous components, which completely dominate for 2a < 10 nm, are enhanced by two mechanisms. First, the vibrational amplitudes in the shell, as calculated by Hama and Matsubara[25] are larger than in the core. Second, the electric field is much larger in the shell than in the core. To show this, Hayashi *et al.* described the dielectric function of the shell as given by a mixture of air and Si to which they applied a Maxwell-Garnet formula. They found that the electric field in the boundary layer depends on the azimuthal angle Θ, measured from the direction of the incident beam, and that it was about ten times larger than the electric field in the core at $\Theta = 0°$. The ratio R of the Raman intensity in the shell to that in the core is given by

$$ R = \frac{V_{shell}}{V_{core}} \left| \frac{E_{shell}}{E_{core}} \right|^4 \frac{\sigma_{shell}}{\sigma_{core}} \tag{10} $$

which must be integrated over the azimuthal angle. For equal cross sections σ_{core} and σ_{shell}, the enhancement of the electric fields dominates the volume ratio and $R \cong 10$.

In this model, the observed spectrum results from the combined effect of electromagnetic resonances and surface modes. As we have already discussed in Refs. 2 and 16, the amplitude of the vibration at grain boundaries is most likely much smaller than at free boundaries and electromagnetic resonances of the type discussed above disappear in films. This may explain why the amorphous-like spectrum observed in most films made of Si nanocrystals is very small.[2] However, there is not adequate description of surface/interface vibrational modes in nanocrystals. Veprek[20] has also discussed possible pitfalls in interpreting the amorphous-like components in Si films. From these considerations, we conclude that when electromagnetic resonances and surface modes play a role, the correct interpretation of the spectra is very difficult and the simple model given earlier may not apply.

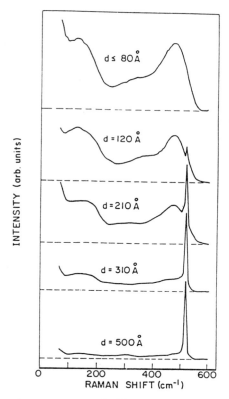

Fig. 8   Raman spectra of gas-evaporated silicon particles of average diameter d. For
d < 8 nm, the observed spectrum is identical to that of amorphous silicon. After
Ref. 24.

Composite Films

Films containing nanograins often contain a volume fraction that is amorphous or at
least strongly disordered. In polycrystalline silicon films, the intergrain material often
appears as a weak shoulder close to 495 wave numbers.[2,19,26] This interpretation has
been disputed[20,27] but we have shown in Ref. 2 it is the most likely. When the grain size
decreases below 10 nm, the volume fraction of the intergrain material increases and the low
wave number tail of the Raman line starts overlapping the 495 cm⁻¹ line. The grain size can
only be determined from a careful line shape analysis, which becomes especially difficult
below 5 nm (see Fig. 2). In films made of other semiconductors, the problem is similar,
since the amorphous Raman spectrum can be modeled by the spectrum of nanocrystals in the
limit of grain size comparable to the interatomic spacing.

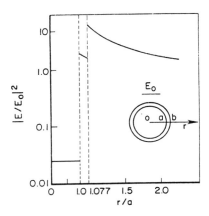

Fig. 9   Calculated light intensity distribution using the core/shell model for isolated silicon spherical crystallites. After Ref. 24.

Very recently, Nemanich *et al.*[28] have investigated Raman scattering in microcrystalline films containing phases with different optical absorption constants. They examined carbon films containing graphite and diamond regions, and silicon films containing microcrystalline and amorphous regions. In each case, the optical absorption differs widely between the two phases. They mostly considered larger grain size than those of interest here. Using a very simple model based on the attenuation of light in each grain, they showed that the ratio of the peak intensities for the two phases was not only a function of the volume fraction of the phases but also of the grain sizes. Figure 10 illustrates this point for a mixed phase Si film. The analysis becomes somewhat suspect when $\alpha L \lesssim 1$ and each grain becomes uniformly illuminated. In addition, if the optical constants of the two phases differ strongly, a more realistic theory that takes into account the redistribution of the electric field is required. The treatment of the previous section becomes relevant. As the interest in composite films is growing (for example, SiC or SiGe[29]), this area of research will expand.

Crystallinity of the Grains

So far, we have assumed that the grains are crystallographically perfect. Actual grains can be perfect or be made of several nanocrystals or contain many types of defects. It has been reported that the size inferred from Raman scattering could be in some cases much smaller than the size obtained from transmission electron microscopy (TEM) measurement.[30] In the same study, the electrical conductivity was correlated to the Raman size, not the TEM size. Some crystallographic imperfections probably decouple vibrations from one region of the grain to another, thereby confining the phonons to a subgrain region.

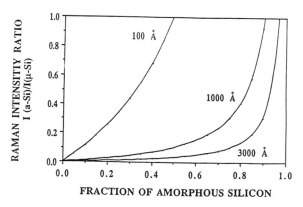

Fig. 10 Calculated ratio of the Raman intensity from the amorphous and crystalline regions of microcrystalline silicon as a function of the volume fraction of the amorphous regions. The three different domain sizes are indicated. After Ref. 28.

Thus, the major obstacle to transport would not come from the grain boundaries. This interpretation is plausible, although to our knowledge, there is no real proof for it. For example, it would be desirable to identify the crystallographic imperfections that confine phonons to a region of a grain.

CONCLUSION

Raman scattering is a very useful diagnostic tool for microcrystalline semiconductors. The first order Raman spectrum is used to measure grain size with considerable success. We have pointed out several factors which, if overlooked, may lead to erroneous conclusions. Some of these factors are not understood and cannot be quantified presently. More experimental and theoretical work is needed, for example, to model the second order spectrum that has been shown to be so sensitive to size effects. Progress also relies on the ability of the material scientists to grow or fabricate well-controlled samples. Eventually, Raman scattering could also become a useful diagnostic tool for monitoring *in-situ* the growth of nanocrystals.

ACKNOWLEDGMENT

The author gratefully acknowledges the work of Ian Campbell and support from the National Science Foundation (ECS-8657263), Coherent, Newport Research Corporation and Solarex through the Presidential Young Investigator program and from the Alfred P. Sloan Research Foundation.

REFERENCES

1.   See for example this volume.

2.   P. M. Fauchet and I. H. Campbell, Crit. Rev. Solid State Mater. Science 14, S79 (1988).

3.   R. J. Nemanich and S. A. Solin, Phys. Rev. B 20, 392 (1979); R. J. Nemanich, S. A. Solin, and R. M. Martin, Phys. Rev. B 23, 6348 (1981).

4.   H. Richter, Z. P. Wang, and L. Ley, Solid State Commun. 39, 625 (1981).

5.   I. H. Campbell and P. M. Fauchet, Solid State Commun. 58, 739 (1986).

6.   Z. Iqbal and S. Veprek, J. Phys. C 15, 377 (1982).

7.   B. Goldstein, C. R. Dickson, I. H. Campbell, and P. M. Fauchet, Appl. Phys. Lett. 53, 2672 (1988).

8.   I. H. Campbell and P. M. Fauchet, Proc. 18th Int. Conf. Physics of Semiconductors, O. Engstrom, Ed., World Scientific, Singapore, 1357 (1987).

9.   Y. M. LeGrice and R. J. Nemanich, to appear in Mat. Res. Soc. Symp. Proc. (1990).

10.   P. Parayanthal and F. H. Pollak, Phys. Rev. Lett. 52, 1822 (1984).

11.   K. K. Tiong, P. M. Amirtharaj, F. H. Pollak, and D. E. Aspnes, Appl. Phys. Lett. 44, 122 (1984).

12.   T. Nakamura and T. Katoda, Jpn. J. Appl. Phys. 23, L552 (1984).

13.   G. Burns, F. H. Dacol, C. R. Wie, E. Burstein, and M. Cardona, Solid State Commun. 62, 449 (9187).

14.   R. Rossetti, S. Nakahara, and L. E. Brus, J. Chem. Phys. 79, 1086 (1983).

15.   G. Kanellis, J. F. Morhange, and M. Balkanski, Phys. Rev. B 28, 3390 (1983); Phys. Rev. B 28, 3398 (1983); Phys. Rev. B 28, 3406 (1983).

16.   P. M. Fauchet and I. H. Campbell, Mat. Res. Soc. Symp. Proc. 164, 259 (1990).

17.   P. A. Temple and C. E. Hathaway, Phys. Rev. B 7, 3685 (1973).

18.   J. L. Birman, Phys. Rev. 131, 1489 (1963).

19.   P. M. Fauchet, I. H. Campbell, and F. Adar, Appl. Phys. Lett. 47, 479 (1985); I. H. Campbell P. M. Fauchet, and F. Adar, Mat. Res. Soc. Symp. Proc. 53, 311 (1986).

20.   S. Veprek, F. A. Sarott, and Z. Iqbal, Phys. Rev. B 36, 3344 (1987); S. Veprek, Mat. Res. Soc. Symp. Proc. 164, 39 (1990).

21.   S. Hayashi, M. Ito, and H. Kanamori, Solid State Comm. 44, 75 (1982); S. Hayashi and H. Abe, Jpn. J. Appl. Phys. 23, L824 (1984).

22.   T. Okada, T. Iwaki, K. Yamamoto, H. Kasahara, and K. Abe, Solid State Commun. 49, 809 (1984); T. Okada, T. Iwaki, H. Kasahara, and K. Yamamoto, Jpn. J. Appl. Phys. 24, 161 (1985); J. Phys. Soc. Jpn. 54, 1173 (1985).

23.   T. Kanata, H. Murai, and K. Kubota, J. Appl. Phys. 61, 969 (1987).

24.   S. Hayashi, Jpn. J. Appl. Phys. 23, 665 (1984); S. Hayashi and K. Yamamoto, Superlattices and Microstructures 2, 581 (1986).

25. T. Hama and T. Matsubara, Prog. Theor. Phys. <u>59</u>, 1407 (1978).

26. P. M. Fauchet, Scanning Electron Microsc. <u>2</u>, 425 (1986).

27. J. F. Morhange, G. Kanellis, and M. Balkanski, Solid State Commun. <u>31</u>, 805 (1979).

28. R. J. Nemanich, E. C. Buehler, Y. M. LeGrice, R. E. Shroder, G. N. Parsons, C. Wang, G. Lucovsky, and J. B. Boyce, Mat. Res. Soc. Symp. Proc. <u>164</u>, 265 (1990).

29. P. M. Fauchet and I. H. Campbell, SPIE Proc. <u>1055</u>, 204 (1989).

30. J. Gonzalez-Hernandez, G. H. Azarbayajani, R. Tsu, and F. H. Pollak, Appl. Phys. Lett. <u>47</u>, 1350 (1985).

RAMAN SCATTERING OF III-V AND II-VI

SEMICONDUCTOR MICROSTRUCTURES

M. Watt, A.P .Smart, M.A. Foad, C.D.W. Wilkinson, H.E.G. Arnot
and C.M. Sotomayor Torres

Nanoelectronics Research Centre, Department of Electronics and
Electrical Engineering, Glasgow University, Glasgow, G12 8QQ, U K

INTRODUCTION

Electronic and optoelectronic devices exploit size quantization effects when they are made sufficiently small that the de Broglie wavelength of the charge carriers becomes comparable to the size of the device (sub-micron). Recent advances in fabrication techniques including electron beam lithography have enabled the physical realization of such devices. The main impetus of developments so far has been with III-V GaAs-based semiconductors. For optoelectronic applications, these materials are limited by energy gap constraints primarily to the infra-red wavelengths suitable for fibre-optic communications. In order to extend the technology of semiconductor laser diodes and detectors to the visible regions of the spectrum, we must turn to II-VI materials. These present a new challenge with regard to both growth and fabrication, particularly with regard to the opportunities for growth over III-V or Silicon substrates which have the inherent potential for large scale integration with present technology. Here we report some of our work on the characterization of fabrication processes by Raman scattering.

Raman scattering is an excellent tool with which to study fabricated structures. Unlike many other methods, it is contactless and non-destructive; by changing the excitation wavelength we can vary the penetration depth and look closely at the layer nearest to the surface; spatial measurements are limited only by the laser spot size, thus permitting uniformity to be measured on a scale of some tens of microns. For these reasons, we use Raman scattering to complement our other characterization techniques of photoluminescence and Schottky diode measurements.

EXPERIMENTAL

We report Raman scattering studies of the reactive ion etch (RIE) effect on nominally-undoped MOCVD-grown ZnSe. Our previous work on the effect of RIE on bulk GaAs involved the etchant species SiCl4 [1]. However for ZnSe, we have investigated a variety of etchant species including SiCl4, Argon with CHF3 and CH4/H2. Other work in this area has been published by Clausen, et al. [2] who obtained good quality etch profiles with BCl3/Ar as the RIE species. The RIE machines used for this work are an Electrotech 340 and a Plasmatechnology RIE80. RIE is one of the

Light Scattering in Semiconductor Structures and Superlattices
Edited by D.J. Lockwood and J.F. Young, Plenum Press, New York, 1991

247

major pattern transfer techniques presently employed in VLSI technology and typically involves ions with energies of some hundreds of electron volts. The process involves both chemical and physical effects leading to desirable profile characteristics such as anisotropic etching, vertical sidewalls and etch uniformity. A full account of the etching techniques employed here will be published shortly [3].

The Raman spectra were recorded at room temperature in the nearly-backscattering geometry with a spot focus on the sample using various lines from an Argon ion laser as the excitation source in order to vary the resonance conditions. The scattered light was dispersed in a Jobin-Yvon U1000 double spectrometer and detected using standard photon counting techniques.

EFFECT OF RIE ON RAMAN SPECTRA

Theory

In this work we are using Raman scattering to characterize the effect of RIE on the surface of a crystalline semiconductor. As discussed above, the etch process involves both chemical and physical elements, each of which will have a different effect on the sample. We suspect that the main cause of damage to the crystal will be the physical etching, i.e., the ion bombardment as chemical wet-etching can leave very clean and ordered surfaces and is a usual prerequisite for most fabrication and growth processes.

The energies of the etchant ions are of the order of some hundreds of electron volts. Therefore at least one order of magnitude smaller than the energies used in ion implantation techniques. This helps to keep the damage to a minimum and consequently we find no evidence of ion implantation which we would expect to manifest itself as additional features in the Raman spectra arising from local impurity modes.

Our previous work on GaAs samples [1] showed that the two major effects of dry-etch damage in the Raman spectra are shifts and asymmetric broadening of the LO phonon peak together with the emergence of the TO phonon (symmetry-forbidden in first-order in our geometry). Two models have been developed to explain these effects but neither offers a complete analysis as discussed below. In addition, work on ion implantation damage in GaAs demonstrates that 2LO phonon scattering is much more sensitive to damage than 1LO scattering [3]. We comment on this later.

The model developed by Richter, et al. [5] and later used by Tiong, et al. [6] to interpret data on ion-bombarded and implanted GaAs, we shall call the phonon confinement model. Defects in the crystal structure are assumed to limit the phonon coherence length in real space, i.e., the extent of crystallinity as perceived by the phonon. In turn, this invalidates the infinite crystal approximation which relaxes the $q=0$ first-order selection rule thereby allowing more of the phonon dispersion to contribute to the first-order scattering and resulting in a shift to lower energies (softening) and asymmetric broadening of the LO phonon peak.

The behaviour of the LO phonon is well described within this model, but the behaviour of the TO phonon cannot be explained. The arguments brought forward by Tiong, et al. [6] to support the phonon confinement model rely on a comparatively flat phonon dispersion for the TO phonon as compared to that of the LO. Figure 1 shows that this is not the case unless the whole of the Brillouin zone is considered, corresponding to correlation lengths of less than 20Å and an amorphous sample structure.

The effective charge model of Burns, et al. [9] accounts for the shifts of both the

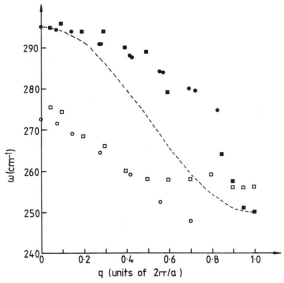

Fig. 1. Phonon dispersion data for GaAs. The circles are Raman scattering data by
Sood , et al. [7] and the squares are the neutron scattering data of Dolling,
et al. [8]. The full symbols are longitudinal modes while the open symbols are
transverse modes. The dashed line is the cosine approximation to the LO
dispersion used by Tiong, et al. [6].

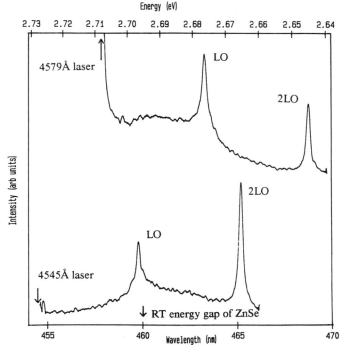

Fig. 2. Raman spectra excited by 4579Å (upper trace) and 4545Å (lower trace)
radiation. The spectra were recorded at room temperature with a spectral
resolution of 1cm^{-1} in the z(x,x&y)z configuration. The two sharp features are
the LO and the 2LO phonons of ZnSe while the broad structure at around
460nm we identify as the PL emission from the sample.

TO and the LO phonons but neglects the asymmetry of the LO phonon peak. Both the TO and the LO phonons are assumed to have moved as a result of tetragonal distortion and strain induced in the crystal parallel to the direction of ion implantation. The residual extra shift of the LO phonon is explained in terms of a change in the effective charge of the material which alters the TO-LO splitting. The $4 cm^{-1}$ shift of the LO phonon was accounted for by estimating a defect density of 10% comprising both vacancies and interstitial atoms. As was pointed out by Anastassakis, et al. [10], this defect density would be consistent with a spatial correlation length of some tens to hundreds of Ångstroms in terms of the phonon confinement model, thus suggesting that a dual treatment is needed to preserve consistency.

Results and Discussion

Figure 2 shows the Raman spectrum obtained from the as-grown ZnSe sample with two different excitation wavelengths: 4545 and 4579Å. As the excitation energy varies with respect to the fundamental energy gap (which we expect at about 4600Å at room temperature), we observe photoluminescence emission as a background in the spectra. The identification of the broad structure as photoluminescence (PL) is based on the observation that it does not shift in wavelength as the laser line is changed and it cannot therefore be a Raman feature. When we use 4880Å excitation, which lies below the bandgap, we observe no PL background as expected.

The second observation we can make from Figure 2 is that the relative intensities of the LO phonon and the 2LO phonon change with excitation wavelength. This is a well-documented phenomenon [see, e.g., Leite, et al. [11] and Scott, et al. [12]] which arises because of the different dependence of the 1LO and 2LO Raman cross sections on the resonance conditions.

In Figure 3, we plot the Raman spectra obtained from the control (unetched) and three samples etched with various etch parameters using CH_4/H_2 as the RIE gas. The excitation in this case was at 4579Å thus the PL emission is visible as a background to the Raman spectra. The PL signal is stronger for the etched samples than for the control material and we suggest two interpretations of this result. The first is based on the fact that secondary electron micrographs of the samples before and after etching show a general qualitative improvement in surface smoothness after etching. The improved surface after etching could possibly result in an enhancement of the luminescence intensity as a result of the removal of non-radiative surface defects by the etch process. Residual lumps of material on the surface after etching can be shown to be a material defect in that clusters of non-stoichiometric material is present just below the sample surface before etching and is merely uncovered by the etch process. We can therefore neglect the contribution of such clusters to the spectra as they are present both before and after etching. The second possible interpretation of the increase in the PL intensity after etching is related to hydrogen passivation of the ZnSe. It has been shown by Pearton, et al. [13] that hydrogen in a crystalline semiconductor can passivate defects and therefore reduce non-radiative recombination sites and lead to an increase in the intensity of the observed PL signal. Although most of the work reported was for III-V and group IV semiconductors, extension to CdTe and ZnTe suggested that the principle may be more widely applicable and this could be tested by performing experiments on annealed samples.

The Raman spectra recorded with the 4880Å line from the control and same three etched samples are shown in Figure 4. Two additional features arising from the GaAs substrate are present in the spectra at 268 and $290 cm^{-1}$. We identify these as the GaAs TO and LO phonon peaks respectively. No difference in the spectra from the etched and the unetched samples could be observed, which agrees with our observations of the resonantly-excited spectra. Because some resonant enhancement of the 2LO phonon is still evident at this wavelength, we cannot use the technique of Wagner, et al. [4] to probe the damage to the crystal.

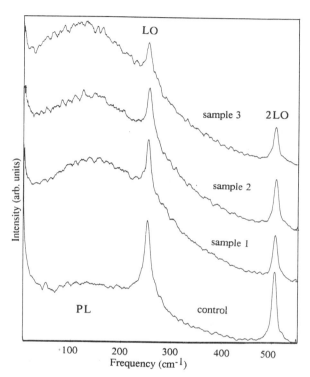

Fig. 3. Raman spectra recorded with 4579Å excitation from the control sample (lowest trace) and three samples etched with different etch parameters. The PL emission which appears in these spectra at around $150cm^{-1}$ shows an increase in intensity with respect to the LO phonon in the etched samples. The spectra are displaced vertically for clarity. The spectral resolution is $1cm^{-1}$ and the spectra were recorded at room temperature in the z(x,x&y)z geometry.

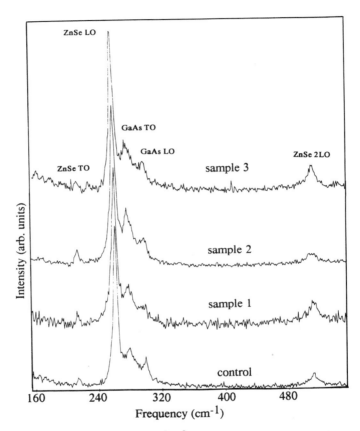

Fig. 4. Raman spectra recorded with 4880Å excitation from the control sample (lowest trace) and three samples etched with different etch parameters. Some resonant effects are still evident by the strength of the 2LO phonon in these spectra. The spectra are displaced vertically for clarity. The spectral resolution is 1cm⁻¹ and the spectra were recorded at room temperature in the z(x,x&y)z geometry.

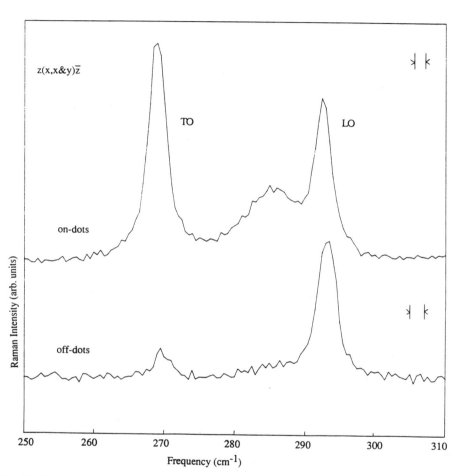

Fig. 5. Raman spectra from one of the **quantum** cylinder samples. The upper trace
shows the on-pattern spectrum while the lower trace shows the off-pattern
spectrum. The spectra are displaced vertically for clarity. The spectral resolution
is 2cm^{-1} and the spectra were recorded at room temperature in the z(x,x&y)z
geometry using 4880Å excitation.

RAMAN SCATTERING FROM QUANTUM CYLINDERS OF GALLIUM ARSENIDE

Raman scattering from microstructures is of interest because the enhanced contribution to the scattering from the surfaces of the sample yields information about those surfaces. We have studied the Raman spectra from arrays of GaAs cylinders with diameters in the range 60-200nm and heights ranging from 140 to 570nm. The cylinders were fabricated using a modified Phillips PSEM 500 scanning electron microscope to produce the mask patterns in a negative resist on GaAs. The samples were then reactive ion etched with $SiCl_4$ as discussed above, leaving an array 200µm by 200µm of free-standing cylinders. Each sample had both a patterned area and an unpatterned area, both of which had been etched, thus allowing direct comparison of the effect of the quantum cylinders on the Raman spectra independently of any variations in the GaAs starting material or etch conditions.

The Raman spectra from the cylinder arrays showed not only a TO and LO phonon, but also a surface phonon as shown in Figure 5. (A full account of the surface phonon investigations has been published separately [14].) The behaviour of this phonon was well-described in terms of the dielectric continuum model developed by Ruppin, et al. [15] and good agreement was obtained between experimental and theoretical values for the surface phonon frequency change with cylinder diameter. Surface phonon character was demonstrated unambiguously by showing a shift of the surface phonon frequency as the dielectric constant of the surrounding medium was changed.

CONCLUSIONS

We have reported Raman scattering experiments of reactive ion etched ZnSe in which we used various excitation wavelengths in order to change the resonance conditions. We have observed no effect of the RIE processing on the Raman spectra from our samples but attribute this to enhanced scattering due to the near resonance of the excitation light with the fundamental energy gap of the sample. Preliminary luminescence data recorded at room temperature suggest that hydrogen passivation of defects occurring during the etch process may play a role in the enhanced luminescence intensities observed from the etched samples.

ACKNOWLEDGEMENTS

We are grateful to M Razeghi for providing us with the ZnSe sample and we acknowledge the financial support of the SERC.

REFERENCES

1.   M. Watt, C.M. Sotomayor Torres, R. Cheung, C.D.W. Wilkinson, H.E.G. Arnot, and S.P. Beaumont, J. Mod. Opt. 35, 365 (1988).
2.   E.M. Clausen, H.G. Craighead, M.C. Tamargo, J.L. deMiguel and L.M.Schiavone, Appl. Phys. Lett. 53, 690 (1988).
3.   M.A. Foad, to be published.
4.   J. Wagner and Ch. Hoffman, Appl. Phys. Lett. 50, 682 (1987).
5.   H. Richter, Z.P. Wang and L. Ley, Solid State Comm. 39, 625 (1981).
6.   K.K. Tiong, P.M. Amirtharaj, F.H. Pollak and D.E. Aspnes, Appl. Phys. Lett. 44, 122 (1984).
7.   A.K. Sood, J. Menendez, M. Cardona and K. Ploog, Phys. Rev. Lett. 54, 2111 (1985).
8.   G. Dolling, and J.L.T. Waugh, in "Lattice Dynamics", R.F. Wallis, ed., Pergamon, London (1965) p. 19.
9.   G. Burns, F.H. Dacol, C.R. Wie, E. Burstein and M. Cardona, Solid State Comm. 62, 449 (1987).

10. E. Anastassakis, K.M. Biskupska, W. Graeff, E. Liarokapis, J. Tatarkiewicz, and K. Wieteska, in "Proc. of the 19th Int. Conf. on Physics and Semiconductors", W. Zawadzki, ed., Polish Academy of Sciences, Warsaw (1988) p. 753.

11. R.C.C. Leite, T.C. Damen and J.F. Scott, in "Light Scattering in Solids", G.B. Wright, ed., Springer-Verlag, New York (1969) p. 359.

12. J.F. Scott, R.C.C. Leite and T.C. Damen, Phys. Rev. 188, 1285 (1969).

13. S.J. Pearton, J.W. Corbett and T.S. Shi, Appl. Phys. A 43, 153 (1987).

14. M. Watt, C.M. Sotomayor Torres, H.E.G. Arnot and S.P. Beaumont, Semicond. Sci. Technol. 5, 285 (1990).

15. R. Ruppin and R. Englman, Rep. Prog. Phys. 33, 149 (1970).

TOWARDS TWO DIMENSIONAL MICRO-RAMAN ANALYSIS OF SEMICONDUCTOR

MATERIALS AND DEVICES

B. Wakefield and W.J. Rothwell

British Telecom Research Laboratories
Martlesham Heath
Ipswich IP5 7RE, UK.

ABSTRACT

Raman scattering spectroscopy has already proved to be a valuable technique for the non-destructive analysis of strain, composition, periods and layer thicknesses in semiconductor superlattices and epitaxial layers. There is now the need to extend the technique to electronic and optoelectronic device structures, and indeed to be able to make such measurements on individual devices. This latter requirement demands the capability of analysing with a spatial resolution of the order of one micrometer. Micro-Raman scattering techniques have recently become available, and these are capable of yielding the necessary spatial resolution.

For device related applications the requirement is often not simply for point analyses, but for areal distributions of material properties. In particular, measurements on cross sections through device structures may be required. At the British Telecom Research Laboratories we have been working towards the development of a Micro-Raman scattering system that will be capable of producing such two dimensional information on a microscopic scale.

The progress achieved to date will be reported, and some preliminary results reporting the use of the system to map out the strain around dislocations in a semiconductor will be presented. Also included will be an account of some of the problems encountered, and an assessment made of their likely effect on the realisation of the goal.

Light Scattering in Semiconductor Structures and Superlattices
Edited by D.J. Lockwood and J.F. Young, Plenum Press, New York, 1991

257

RAMAN SPECTROSCOPY FOR CHARACTERIZATION OF LAYERED

SEMICONDUCTOR MATERIALS AND DEVICES

Hans Brugger

Daimler Benz AG
Research Center
D-7900 ULM, FRG

INTRODUCTION

Novel semiconductor heterojunctions, superlattices and quantum well structures exhibit a large number of new effects and offer the possibility for future device applications with reduced dimensionality. Considerable advances in the growth and processing of materials have allowed the production of high quality microstructures. Powerful techniques are necessary for the characterization of the large number of new materials and devices. Optical methods provide a non-contact and non-destructive way for the investigation of important electronic, optical and structural parameters. Inelastic light scattering has been developed as a versatile and convenient analytical tool in this field. Recently, the combination of Raman spectrometers with optical microscopes made this technique even more attractive for microanalysis.[1]

In this article the ability of inelastic light scattering for characterization of layered materials for device applications and devices themselves is reported. Only the strong phonon excitations as a local probe for microstructures are discussed. Fig. 1 shows schematically the Raman backscattering technique which is widely used on opaque materials. The different blocks list the various parameters obtained from phonon Raman scattering.

MICRO RAMAN SPECTROSCOPY

High spatial resolution is obtained by combining a Raman spectrometer with a conventional optical microscope in order to focus the Raman laser beam onto the sample surface and to collect the scattered light. The achieved spot size on the focal plane is diffraction limited and can be varied by use of different lenses and/or working distances. Values below $1 \mu m$ are achieved with usually used excitation lines from argon and krypton ion lasers.

Light Scattering in Semiconductor Structures and Superlattices
Edited by D.J. Lockwood and J.F. Young, Plenum Press, New York, 1991

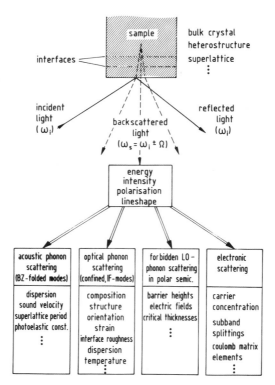

Fig. 1. Schematics of Raman backscattering in semiconductors.

Two-dimensional Raman images are obtained from line scans with the specimen (device, microstructure) mounted on a high precision (0.1μm) x,y-translation stage. Positioning and aligning are achieved by observing the structure with a video camera under high magnification. We report about measurements at room temperature and under air ambient. This allows the devices to be investigated during standard operation.

Temperature Mapping on Laser Mirrors

GaAs related power lasers are used for applications in optical communication systems, printers and optical disk memory systems. The maximum output power is limited by catastrophic optical damage or output power saturation due to local heating on laser mirrors. Knowing the local operating temperature distribution on the facets is very useful to explore degradation processes, to see the influence of different coatings and to assess the impact of thermal crosstalk in laser arrays. Several methods have been utilized for laser temperature measurements.[2,3,4,5]

Temperature mapping experiments on facets of ridge-waveguided Al(x)Ga(1-x)As/GaAs single quantum well (SQW) graded-index separate-confinement heterostructure (GRIN-SCH) laser diodes (LD) with high accuracy are discussed. A cross section of the cleaved facet region with Al-profile is shown in the inset of Fig. 2. The shaded ellipsoidal area shows schematically the near-field spot of the emerging laser beam.

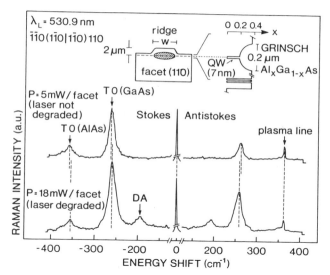

Fig. 2. Typical AS and S phonon spectra of a cleaved AlGaAs
laser facet during device operation with a ridge width w=3μm.
The inset shows a cross section with Al-profile. The plasma
line at 363 cm^{-1} is useful as an energetic reference.

Antistokes (AS) and Stokes (S) phonon Raman spectra from the
center region (inside near-field spot) are also shown in
Fig. 2. The spatial resolution was about 0.9μm. The LD device
was operating during measurement at a low optical power level
(upper curves) and at a high power level and after heavy
degradation (lower curves). The signals show the symmetry
allowed TO-phonons in the Al(x)Ga(1-x)As material. The
asymmetric shape originates from contributions of material
with different x. The high power operation leads to a strong
facet heating. The variation of the intensities, the shift of
the phonon energies and the line broadening give information
about the local temperature. We also observed an additional,
disorder-activated (DA) phonon structure at about 195 cm^{-1} in
the heavily degraded and uncoated LD. Its existence
demonstrates a strong crystal damage at facets of cleaved
degraded LD.[5]

We used the I_S/I_{AS} intensity ratio of the integrated
signal (area under the curve) of the GaAs-like TO-phonon to
estimate the local temperature employing the following
expression:[6]

$$I_S/I_{AS} \approx C \cdot K \cdot [(\omega_L - \Omega)/(\omega_L + \Omega)]^3 \cdot \exp(\hbar\Omega/kT) \qquad (1)$$

where ω_L and kT are the exciting laser energy and the crystal
temperature, respectively. For the AlGaAs phonon energy we
used $\Omega = 32.6$ meV. All the prefactors of the exponential
expression were on the order of 0.9. K is a correction factor
due to the difference of reflectivity and absorption of the
material at $\omega_L \pm \Omega$ (values are taken from ref. 7). C is an
experimental correction factor due to different response of
the detector and gratings at $\omega_L \pm \Omega$. The temperature

dependence of the Raman susceptibility has been neglected. The influence of local heating effects due to the probing Raman laser beam were measured on facets with the LD off as a function of intensity. Typical temperature rises are of the order of 10 K for a power density of 100 kW/cm^2.

Fig. 3 shows the results of a series of Raman measurements on lasers with 8μm wide ridges and facets with different technologies/mirror coatings. The temperature rise ΔT relative to the reference temperature at threshold current (P = 0) is plotted as a function of optical output power of LD at the facet. Cleaved, uncoated and etched mirrors show a strong nonlinear ΔT(P) dependence with typical values of ΔT > 100 K for P > 1 MW/cm^2. The simultaneous measurement of the power/current P(I) dependence of the LD shows a continuous increase of the threshold current at high ΔT. Once degradation occurred, an increase of ΔT is noted as a function of time at contant P. ΔT up to 1000 K has been measured on highly degraded cleaved facets. In contrast LD with λ/2-thick Al$_2$O$_3$ passivation and λ/4-thick Al$_2$O$_3$ anti-reflection (AR) layers[8] can withstand up to five times the power densities of uncoated ones without significant degradation and moderate heating. The dielectric coatings are transparent for the probing Raman laser light.

Measured lateral temperature profiles at cleaved front facets are shown in Fig. 4. The Raman beam was focused to about 0.9μm in diameter. The curves were obtained from line

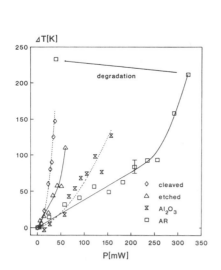

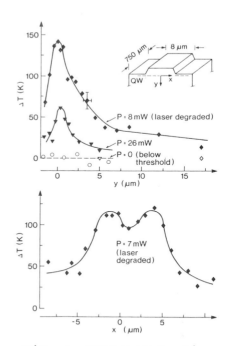

Fig. 3. Temperature rise on laser facets as a function of optical output power for different uncoated and coated mirrors.

Fig. 4. Temperature line scans at a cleaved laser mirror. Center of near-field spot: x = y = 0.

scans starting at the edge of the top contact ($y = -2\mu m$) towards the substrate. As expected, no temperature variation is observed within the experimental error below the lasing threshold. In the lasing mode, however, very localized hot spot regions of about $5\mu m$ extension are observed around the active layer with a pronounced temperature rise ΔT for the degraded LD even at low operating levels. This demonstrates the correlation between degradation and facet heating. Horizontal line scanning along the active layer was performed in x-direction of the same degraded LD as shown in the lower part of Fig. 4. The temperature distribution exhibits two bumps inside the waveguide region. Near-field measurements of the same LD under the same P/I-conditions revealed a two-mode pattern. It seems possible that the temperature profile images the LD optical density distribution on the facet.

The active region lies about $2\mu m$ beneath the ridge surface. Therefore the operating temperature distribution inside the active region along the ridge is not directly accessible to Raman spectroscopy because of the surface sensitivity of this technique, which is typically $< 0.1\mu m$. In this case the intrinsic electroluminescence (EL) signal of the quantum well in the active layer on open-contact samples has been detected perpendicular to the long cavity axis.[5] A continuous red-shift of the EL from resonator bulk towards the facets is observed, indicating hot facet regions and a cold cavity. The measurements have shown together with the Raman results, that considerable heating effects occur only at surfaces of cleaved facets and they are correlated with degradation effects on AlGaAs/GaAs SQW GRIN-SCH LD.

Stress Mapping on LOCOS Structures

Strain often occurs in semiconductors during the device fabrication process. Especially the local oxidation of silicon (LOCOS) at high temperatures results in intrinsic strain due to the different thermal expansion coefficients of Si and SiO_2. Faster degradation in metal-oxid-silicon (MOS) and bipolar devices is expected. The stress distribution in silicon with LOCOS has been measured by Brunner and Abstreiter[9] and Kobayashi et al.,[10] who detected the LO-phonon shift of Si adjacent to and beneath the oxide.

Fig.5 shows a line scan perpendicular to the transparent oxide stripe with $4\mu m$ width and $0.45\mu m$ thickness. The phonon shift is negative for the Si material under the oxide stripe and positive adjacent to the oxide edges. The results can be understood as a local compensation of uniaxial tensile and compressive stress in Si, respectively, along the [110] direction and negligible macroscopic strain elsewhere. The blue excitation line allows a surface sensitive sampling with high spatial resolution.

MATERIAL CHARACTERIZATION OF STRAINED LAYER SUPERLATTICES

The characterization of materials for semiconductor applications by inelastic light scattering has been significantly enhanced in the last 20 years. It goes beyond

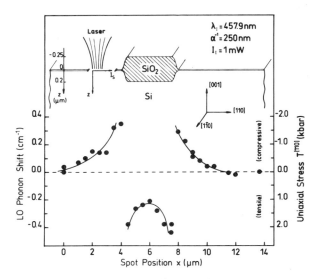

Fig. 5. Line scan of the Si LO-phonon across a LOCOS
structure. From ref. 9.

the scope of this article to cover all the exciting recent
developments. To demonstrate the usefulness of phonon Raman
spectroscopy as a microscopic probe for various properties we
selected a few examples on the field of Si/Ge. This system is
very promising for future device applications.[11]

Composition, Strain, Period Length

In mixed crystals like $Al(x)Ga(1-x)As$ the phonon spectra
show a two mode behaviour (see Fig. 2). From the energetic
positions of the GaAs-like and AlAs-like optical modes a
direct determination of the composition is possible.[12]
$Si(x)Ge(1-x)$ alloys exhibit three dominant modes. Selected
spectra from thick alloy films grown on GaAs substrates are
shown in Fig. 6. The frequencies of the Ge-Ge, Si-Ge and Si-
Si modes and their intensity ratios are fingerprints of
composition.[13] This can be used to determine the Si and Ge
content in a non-destructive and non-contact way. The
selection rules can be used to get information on the
orientation of the films. Phonon frequency, linewidth and
line shape also reflect in a direct way, the crystallinity of
the grown film. The dashed line in Fig. 6 (x = 1) is obtained
from a monocrystalline Si wafer. The full line spectrum is
obtained from a $0.2 \mu m$ thick Si grown on the lattice
mismatched GaAs substrate. The broad and asymmetric lineshape
reflects the high amount of disorder in the polycrystalline
film.

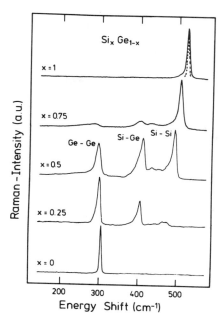

Fig. 6. Raman spectra of thick crystalline bulk Si(x)Ge(1-x) alloys grown on GaAs substrates.

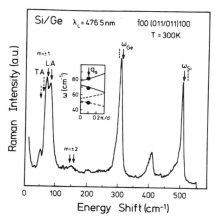

Fig. 7. Room temperature Raman spectrum of a symmetrically strained Si/Ge superlattice. Inset: Continuum calculation of the folded acoustic modes with experimental points.

There is considerable interest in the properties and
device applications of strained superlattices. They are
promising because of the possibility of tailoring the
electronic and optical properties and band-gap conversion
from indirect-gap materials to direct-gap superlattices. A
typical Raman spectrum of a new type of elastically strained
Si/Ge superlattice with strain symmetrization is shown in
Fig. 7.[14] The sample consists of a periodic sequence of
12 monolayers (ML) Si and 8 ML Ge with total thickness of
200 nm on top of an appropriate SiGe alloy buffer layer. The
difference in lattice constant between Si and Ge is about 4%.
The thickness ratio d_{Si}/d_{Ge} corresponds exactly to the
inverse strain ratio $\varepsilon_{Ge}/\varepsilon_{Si}$ yielding to a negligible average
strain of the superlattice:

$$\varepsilon_{AV} \approx (1/d) \cdot (d_S \cdot \varepsilon_{Si} + d_{Ge} \cdot \varepsilon_{Ge}) \approx 0 \tag{2}$$

Hence the total thickness of the superlattice is not limited
by a critical value.

Two strong and sharp modes at 306 cm^{-1} and 506 cm^{-1}
appear in Fig. 7 indicating good material quality. The
phonons are shifted in energy due to the biaxial strain in
the layers. This has been used to determine quantitatively
stress distribution.[15,16] It follows:

$$\Delta\omega(\text{cm}^{-1}) = -a \cdot \mathcal{T}(\text{kbar}) \tag{3}$$

with a = 0.40 (Si) and 0.35 (Ge). For comparison, the
energetic positions of lattice vibrations in unstrained bulk
Ge and Si are also marked by dashed lines. The arrows
indicate the expected phonon frequencies of stressed Ge
(-34 kbar, biaxial compressive) and Si (30 kbar, biaxial
tensile) in this structure as expected from the growth
parameters. The small negative shifts of the phonon modes
relative to the expected ones is due to confinement effects
of the optical vibrations in the very thin layers.[17] There is

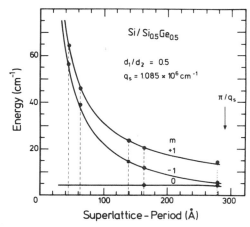

Fig. 8. Calculated frequencies for the first three eigenmodes
of folded LA phonons and measured frequencies (dots).

a weak Si-Ge mode around 400 cm^{-1} indicating good material quality. This mode is mainly attributed to interface roughness in the samples.[18] Intermixing strongly enhances the intensity of this mode.[15]

Another widely used method involving Raman scattering for superlattice studies is the measurement of zone folding effects on the phonon dispersion.[19] In Fig. 7 a strong folded acoustic doublet (m = ±1) appears in the spectrum. The next higher mode is also seen. It provides evidence of a periodic layer sequence with sharp interfaces. The dispersion curves of these modes have been calculated using a layered elastic continuum model and are also drawn in the inset of Fig. 7. The calculated values agree quite well with the experimental frequencies. The energetic positions are very sensitive to the period length $d = d_{Si} + d_{Ge}$ rather than to the thickness ratio d_{Si}/d_{Ge}. This method allows an accurate determination of period lengths as shown in Fig. 8 for a series of $Si/Si_{0.5}Ge_{0.5}$ strained layer superlattices.[20]

Annealing Study

The optical phonon spectra contain information about the local properties of the individual layers such as composition and strain distribution. The energies of the acoustic modes are sensitive to the superlattice period. The intensities of the phonon doublets scale with the square of the Fourier-components of the superlattice photoelastic constant which is determined by the composition profile. Hence information about the interface abruptness is obtained. Raman scattering has been used as a powerful tool for the investigation of interdiffusion effects after annealing treatments of samples.[15,21,22]

Fig. 9 shows Raman spectra of the optical phonons after annealing of a short period superlattice sample with 12 ML Si and 8 ML Ge. The temperature treatment was performed in a quartz tube furnace under vacuum conditions. The sample was heated by irradiation for 15 min at a temperature T_A. The arrows and the dashed lines mark the phonon energies before the temperature treatment (from Fig. 7.). For T_A > 600 C the Si-Ge vibration gains in intensity and the Ge-mode shifts to lower frequencies, but the Si energy is not affected up to $T_A \approx$ 800 C. A fourth mode originates on the low energy side of the Si phonon. At T_A = 900 C the observed spectrum is typical for a SiGe-alloy with Si content between 0.5 and 0.6, close to the expected value of 0.6 (12 ML Si + 8 ML Ge) for a complete homogeneous alloy. The Si substrate phonon at 520 cm^{-1} clearly appears in the annealed sample indicating a change in the electronic band structure and higher transparency for the used laser light. Similar effects have been seen on identical superlattices but with shorter period lengths.[14] In Fig. 10 all the measured phonon frequencies are summarized and plotted as a function of T_A. The Si-Ge mode shows a striking behaviour. In the temperature range between 750 C and 800 C the energy passes a maximum value typical for a concentration around $x \approx$ 0.5.[13] From these results a simple picture of the interdiffusion processes is obtained. For T_A < 600 C, no change in the spectrum, and consequently in

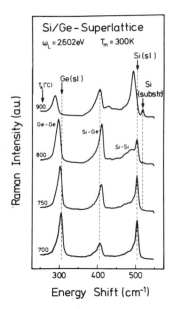

Fig. 9. Raman spectra of Si/Ge superlattices after each increment of annealing.

the concentration profile of the sample, is observed. For $T_A > 600$ C, the formation of an alloy layer with increasing Si content occurs by diffusion of Si into the Ge layers. However, there are still rather clean Si core layers left in the sample as seen from the sharp Si phonon line at 506 cm^{-1}. At high temperatures a complete intermixing of both SiGe layers and Si cores is observed. The average strain in the superlattice is not changed with annealing.

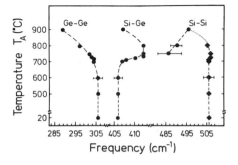

Fig. 10. Phonon frequencies after annealing treatment.

Interdiffusion alters the concentration profile inside the periodic structure but not the period length itself. This was confirmed by the observation of no shift in energy of the folded acoustic phonon modes.[15] However, the intensities inside the first doublet and the intensity ratios between the first and the second order doublets provide comprehensive information about a change in the inner structure of the supercell.[23] The intensity drop of the second order doublet was even faster in comparison with the first one. This also reflects a change in concentration profile during annealing. From the Raman data, rough values of diffusion coefficients in short period Si/Ge superlattices are obtainable.[15]

IN SITU MEASUREMENTS DURING GROWTH

The most important features in modern electronic and optoelectronic solid state devices are interfaces between two different materials such as semiconductors, metals and oxides. Heterojunctions are characterized by band-edge discontinuities and barrier heights which are a measure of the Fermi-level position inside the gaps on both sides of the interface.

Lattice Matched Heterostructures

It has been shown that forbidden LO-phonon Raman scattering in the polar semiconductor GaAs can be used as a sensitive technique to study the formation of barrier heights at surfaces or interfaces.[24,25] We report about Raman

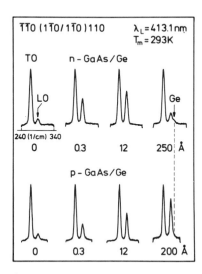

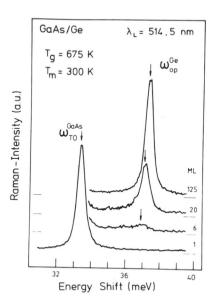

Fig. 11. Allowed TO- and forbidden LO-phonon scattering in GaAs with different Ge overlayers.

Fig. 12. Phonon spectra of thin crystalline GaAs/Ge heterojunctions.

measurements, which were performed during the MBE-growth of
Ge on n- and p-type (110) GaAs substrates. A conventional
Raman setup was combined with a vacuum growth chamber in
order to monitor in situ the phonon behaviour of the growing
heterostructure.

Fig. 11 shows two series of Raman spectra obtained from
n- and p-type GaAs covered with various crystalline
overlayers of Ge which were grown under the same conditions.
Only the TO-phonon is symmetry allowed in this backscattering
geometry. Under resonance conditions with GaAs, however, a
strong enhancement of LO-phonon intensity I_{LO} is observed due
to higher order symmetry breaking mechanisms. In our case,
the presence of the macroscopic surface electric field E in
GaAs (band bending effects) influences the phonon intensity
ratio via $I_{LO}/I_{TO} \sim E^2 \sim V_b$.[24] Therefore I_{LO}/I_{TO} is a local
probe of the barrier height V_b in GaAs at the surface or
heterointerface. The interface formation is followed from
submonolayer coverages of Ge up to overlayers of about 25 nm.
The information depth is limited by the absorption
coefficient of the Raman light in the Ge film. The barrier
height on n-GaAs passes a maximum at about 1.2 nm Ge, but
shows a continuous increase for p-type substrates as clearly
seen in Fig. 11.

To reduce photovoltaic effects[28] due to the probing
laser beam, a low excitation power density is used. Typical

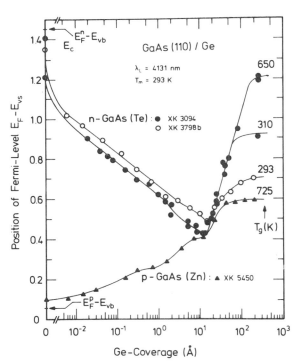

Fig. 13. Position of Fermi-level in GaAs at the interface of
the heterojunction as a function of Ge coverage.

values are of the order of 0.1kW/cm^2 (laser wavelength
413.1nm) at the GaAs surface. The measurements were performed
at constant sample temperature T_m = 293 K. The MBE growth for
crystalline overlayers is done at elevated temperatures T_g in
the range of 300 C to 400 C.

By just switching the excitation wavelength to Ge-
resonance (green light), the optical Ge-phonon is greatly
enhanced and allows the quality of the growing Ge-film to be
investigated in the same run as shown in Fig. 12. The GaAs
LO-phonon is completely suppressed.

Forbidden LO-phonon scattering primarily yields
arbitrary numbers of phonon intensity ratios which depend on
experimental conditions. The I_{LO}/I_{TO} values for each series
have been evaluated in terms of barrier heights in [eV] by
use of calibration points. It allows the quantitative
determination of the Fermi-level (E_F-E_{vs}) relative to the
valence band edge at the surface of GaAs.[26] This is shown in
Fig. 13. At about 6 ML (1.2 nm) Ge thickness, the Fermi-level
positions on n- and p-type samples are nearly the same. This
strengthens the assumption that the Fermi-level is fixed at
the Ge surface rather than at the GaAs/Ge interface. From the
present data, maximum values for the valence band offset can
be deduced: $E_v \leq 0.5$ eV.[26]

Lattice-Mismatched Heterostructures

The method of forbidden LO-phonon scattering has
also been applied to strained layer systems. Fig. 14 shows
measured I_{LO}/I_{TO} values from a strained Si(0.5)Ge(0.5)
overlayer on n-GaAs. In the low coverage region the behaviour
is similar to the GaAs/Ge system (Fig. 11). Above a critical
thickness, however, the strained film relaxes by misfit
dislocations close to the interface. This causes a Fermi-
level pinning and consequently an abrupt change in barrier
height, which is directly reflected in the phonon behaviour.
The method has been used to determine critical thicknesses
for various Si(x)Ge(1-x) overlayers with different lattice
mismatch. This is shown in Fig. 15 together with results from
low energy diffraction experiments[27] and theoretical curves.

In the same run, the nature of the growing Si$_{0.5}$Ge$_{0.5}$
film is monitored with allowed phonon scattering as a
function of film thickness as shown in Fig. 16. At
thicknesses of only a few monolayers, the characteristic
three-mode behaviour of SiGe is observed. The signals gain in
intensity with increasing film thickness, concomitant with a
decreasing intensity of the GaAs TO-mode due to light
absorption. For thin films, however, the positions of
Si$_{0.5}$Ge$_{0.5}$ phonons are shifted downwards compared with the
expected bulk modes due to the tensile biaxial strain
(lattice mismatch 2%). Above the critical thickness the peaks
move to higher energies, approaching the bulk values.
Measurements on various Si(x)Ge(1-x) films with different x
on (110) GaAs substrates have shown that the Si-Ge mode
shifts linearly with strain:

$$\Delta\omega (\text{cm}^{-1}) \approx -36 \ x \approx -8.8 \ \mathcal{E} \ (\%) \tag{4}$$

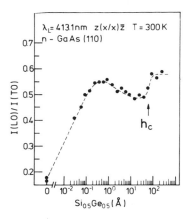

Fig. 14. Normalized intensity
of forbidden LO-phonon scattering in GaAs for various
coverages of Si(0.5)Ge(0.5).

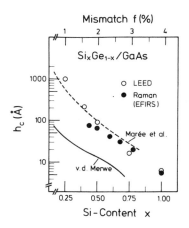

Fig. 15. Critical thickness
of strained SiGe overlayers
on GaAs substrates.

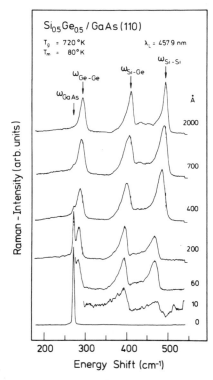

Fig. 16. Raman spectra of $Si_{0.5}Ge_{0.5}$ on GaAs substrates.

272

ACKNOWLEDGEMENTS

The fruitful collaboration with S. Beeck, P.W. Epperlein and E. Friess is gratefully acknowledged. I would like to thank G. Abstreiter, V. Graf, E. Kasper and A. Oosenbrug for supporting the work.

REFERENCES

1. see for example: "Special Issue on Laser Diagnostics of Semiconductors", eds. P.M. Fauchet and J.C. Tsang, IEEE J. Quantum Electronics $\underline{25}$, 963 (1989).
2. J.C. Dyment, Y.C. Cheng and A.J. SpringThorpe, J. Appl. Phys. $\underline{46}$, 1739 (1975).
3. T.L. Paoli, IEEE J. Quantum Electron. $\underline{QE-11}$, 498 (1975).
4. S. Todoroki, J. Appl. Phys. $\underline{60}$, 61 (1986).
5. H. Brugger and P.W. Epperlein, Appl. Phys. Lett. $\underline{56}$, 1049 (1990).
6. A. Compaan and H.J. Trodahl, Phys. Rev $\underline{B29}$, 793 (1984).
7. D.E. Aspnes, S.M. Kelso, R.A. Logan and R. Bhat, J. Appl. Phys. $\underline{60}$, 754 (1986).
8. H. Brugger, P.W. Epperlein, S. Beeck and G. Abstreiter, Proc. 16th Int. Symp. on GaAs and Rel. Comp. 1989, Karuizawa, Japan, eds. T. Ikoma and H. Watanabe, IOP Publishing Ltd., Inst. Phys. Conf. Ser. $\underline{106}$, 771 (1990).
9. K. Brunner and G. Abstreiter, Appl. Surf. Sci. $\underline{39}$, 116 (1989).
10. K. Kobayashi, Y.Inoue, T. Nishimura, T. Nishioka, H. Arima, M Hirayama and T. Matsukawa, Ext. Abstr., 19th Conf. Solid State Devices Mater., Tokyo, Japan, 323 (1987).
11. E. Kasper, $\underline{in}$ "Heterostructures on Si: One Step Further with Silicon", eds. Y. Nissim, K. Ploog and E. Rosencher, Cargèse (May 1988), and references therein.
12. G. Abstreiter, E. Bauser, A. Fischer and K. Ploog, Appl. Phys. $\underline{16}$, 345 (1978).
13. M.A. Renucci, J.B. Renucci and M. Cardona, in Proc. 2nd Int. Conf. on Light Scattering in Solids, ed. M. Balkanski, Paris, Flammerion, 326 (1971).
14. E. Kasper, H. Kibbel, H. Jorke, H. Brugger, E. Friess and G. Abstreiter, Phys. Rev. $\underline{B38}$, 3599 (1988).
15. H. Brugger, E. Friess, G. Abstreiter, E. Kasper and H. Kibbel, Semicond. Sci. Technol. $\underline{3}$, 1166 (1988).
16. G. Absteiter, H. Brugger, T. Wolf, H. Jorke and H.J. Herzog, Phys. Rev. Lett. $\underline{54}$, 2441 (1985); Surf. Sci. $\underline{174}$, 640 (1986).
17. R. Zachai, E. Friess, G. Abstreiter, E. Kasper and H. Kibbel, Proc. of the 19th Int. Conf. on Physics of Semicond., Warshaw (1988).
18. M.I. Alonso, F. Cerdeira, D. Niles, M. Cardona, E. Kasper and H. Kibbel, J. Appl. Phys. $\underline{66}$ (1989), 5645.
19. "Light Scattering in Solids V", eds. M. Cardona and G. Güntherodt, Vol. $\underline{66}$, Springer, Berlin (1989).
20. H. Brugger, G. Abstreiter, J. Jorke, H.J. Herzog and E. Kasper, Phys. Rev. $\underline{B33}$, 5928 (1986); Superlatt. Microstruct. $\underline{2}$, 451 (1986).
21. D. Levi, S.L. Zhang, M.V. Klein and H. Morkoc, Phys. Rev. $\underline{B36}$, 8032 (1987).

22. D.J. Lockwood, J.M. Baribeau and P.Y. Timbrell,
    J. Appl. Phys. <u>65</u>, 3049 (1989).
23. B. Jusserand, D. Paquet, F. Mollot, F. Alexandre and
    G. LeRoux, Phys. Rev. <u>B35</u>, 2808 (1987).
24. F. Schäffler and G. Abstreiter,
    Phys. Rev. <u>B34</u>, 4017 (1986).
25. H. Brugger, F. Schäffler and G. Abstreiter,
    Phys. Rev. Lett. <u>52</u>, 141 (1984).
26. H. Brugger and G. Abstreiter, to be published.
27. K. Eberl, G. Krötz, T. Wolf, F. Schäffler and
    G. Abstreiter, Semicond. Sci. and Technol. <u>2</u>, 561 (1987).
28. M.H. Hecht, Phys. Rev. <u>B41</u> 7918 , (1990).

RAMAN SPECTROSCOPY OF DOPANT IMPURITIES AND DEFECTS IN GaAs LAYERS

Joachim Wagner

Fraunhofer-Institut für Angewandte Festkörperphysik
Tullastrasse 72, D-7800 Freiburg, Federal Republic of Germany

INTRODUCTION

Raman spectroscopy has found widespread use for the investigation of intrinsic phonon modes and electronic excitations of free carriers in bulk semiconductors as well as in semiconductor heterostructures [1-3]. Raman scattering by impurities and defects [4, 5], in contrast, has been used so far only in a limited number of examples. This also contrasts to other techniques, such as photoluminescence and absorption spectroscopy, which are used routinely for the investigation of defects and impurities.

This is, at least partially, due to the weakness of Raman signals from defects and impurities because of their small concentrations as compared to the density of host lattice atoms. In the last few years, the advent of optical multichannel detectors allowed to increase the sensitivity of the Raman technique considerably [6], which opens new possibilities for the study of defects and impurities by Raman scattering. This is, in particular, true for the recent work on Raman spectroscopy of dopant impurities in planar (δ-) doped GaAs [7] which would have been hardly possible using conventional single channel detection.

In Raman spectroscopy, impurities and defects can be observed either via scattering by internal electronic excitations [4, 5] or via scattering by localized vibrational modes [2, 8-11]. Both types of scattering probe directly individual impurities and defects. The intensity of the scattered light per unit solid angle $\partial S/\partial\Omega$ can be written as [12]

$$\partial S/\partial\Omega \sim \sigma\, n\, V \tag{1}$$

where σ denotes the scattering cross section per impurity, n is the impurity concentration, and V is the scattering volume. If the sample is opaque for the light used to excite the Raman spectrum, V is proportional to the probing depth $1/(2\alpha)$ where α denotes the absorption coefficient. Using light for which the material is fully transparent, V is for backscattering geometry proportional to the thickness of the sample. Eq. 1 shows that, for a given scattering cross section per impurity and a given scattering volume, the measured Raman intensity is directly proportional to the impurity concentration. There-

Light Scattering in Semiconductor Structures and Superlattices
Edited by D.J. Lockwood and J.F. Young, Plenum Press, New York, 1991

fore, it is straight forward to calibrate the Raman technique and to make it a versatile tool for quantitative materials characterization. This is in contrast to photoluminescence spectroscopy, where the measured signal intensity is not only determined by the impurity concentration but also by the carrier life time.

From Eq. 1 it is also evident that there are two ways to enhance the Raman signal from scattering by impurities and defects. One possibility is to use incident light for which the material is transparent. This leads to a fairly large scattering volume limited only by the thickness of the sample, which is typically in the range 0.5 - 5 mm. This approach has been used successfully to study impurities in bulk semiconductors such as Si [4, 5] and GaAs [4, 5, 13, 14]. In the case of GaAs, a detection limit for Raman scattering by electronic excitations of residual shallow acceptors of $\leq 5 \times 10^{14}$ cm^{-3} has been reported [14] which demonstrates the sensitivity of this technique.

Alternatively, the incident photon energy can be chosen to match an electronic interband transition which is well above the fundamental band gap of the material. For this kind of optical excitation, the probing depth $1/(2\alpha)$ is, for e.g., GaAs in the range of 10 - 100 nm [15, 16]. The scattering cross section, on the other hand, can be enhanced significantly by this approach as it has been demonstrated for scattering by electronic excitations of acceptors incorporated in quantum well heterostructures [3, 17-19] and for scattering by impurity induced local vibrational modes (LVM) in GaAs [11]. In the latter case optical excitation in resonance with the E_1 band gap has been used, leading to a probing depth of only 10 nm [15, 16]. This makes the above approach particularly well-suited to study, e.g., dopant impurities in thin highly doped epitaxial layers,[11, 20, 21] including δ-doped structures [7].

Defects introduced in large concentrations by ion implantation or reactive ion beam etching can also affect Raman scattering by intrinsic phonon modes. The first-order phonon Raman spectrum gets modified by the relaxation of the constraints imposed by momentum conservation as studied, e.g., in detail for ion implanted GaAs [22, 23]. Resonant dipole forbidden first-order and allowed second-order scattering by longitudinal optical (LO) phonons is sensitive to ion beam induced defects via the broadening of the corresponding resonances in the Raman efficiency [23-25].

In the following we concentrate on GaAs and discuss Raman spectroscopy of dopant impurities and defects via vibronic excitations. Firstly, we focus on the effect of ion beam damage on resonant Raman scattering by intrinsic LO phonons both in ion implanted and in reactive ion etched GaAs. Secondly, recent work on the Raman spectroscopy of impurity induced LVM in highly-doped GaAs is reviewed. It is further demonstrated how this spectroscopic technique can be applied to the analysis of the incorporation and spread of dopant atoms in δ-doped GaAs structures.

EFFECT OF ION BEAM INDUCED DEFECTS ON RAMAN SCATTERING BY INTRINSIC PHONON MODES

Fig. 1 displays the effect of ion implantation on the first-order deformation potential scattering [22, 23]. A series of Raman spectra is shown for material implanted with ^{29}Si at an ion energy of 100 keV with doses ranging from 1×10^{13} to 1×10^{16} cm^{-2}. With increasing implantation dose, the LO(Γ) phonon line shifts towards lower energies and broadens. These effects have been explained by Tiong et al. [26] by a "spatial correlation

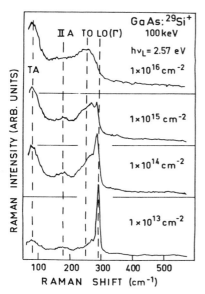

Fig. 1 Room-temperature Raman spectra of $^{29}Si^+$-implanted GaAs. Implantation doses are indicated in the figure. The spectra were excited at 2.57 eV with the incident light polarized along a (110) crystallographic direction and the scattered light not analyzed for its polarization. Spectral resolution was 6 cm⁻¹.

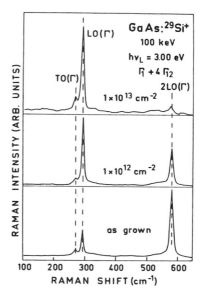

Fig. 2 Room-temperature Raman spectra of as-grown and $^{29}Si^+$-implanted GaAs. Implantation doses are given in the figure. The spectra were excited at 3.00 eV and recorded in the x(y,y)x̄ scattering geometry with a spectral resolution of 5 cm⁻¹.

model". This model is based on a phonon confinement concept [27] with the phonon localization length being as small as ~ 50 Å for the highest damage level for which a LO phonon peak is resolved. As can be seen from Fig. 1 for doses exceeding 1×10^{15} cm⁻², the LO phonon line disappears and three broad bands are observed which arise from disorder activated first-order scattering in amorphized GaAs. These bands develop gradually for doses in the range 10^{13} - 10^{15} cm⁻², indicating that in this range the material consists of damaged crystalline, as well as amorphized, regions.

The effect of ion implantation induced damage on dipole-forbidden but defect induced 1LO and resonant 2LO phonon scattering is shown in Fig. 2, which displays a sequence of Raman spectra excited at 3.00 eV in resonance with the E_1 band gap energy [28]. In as-grown GaAs, the defect induced 1LO phonon line intensity is weak and the Raman spectrum is dominated by 2LO phonon scattering. Upon ion implantation the defect induced 1LO phonon line increases and the 2LO phonon peak decreases in intensity. Already for a moderate implantation dose of 1×10^{13} cm⁻², where first-order deformation potential scattering is only modified slightly (Fig. 1), resonant defect induced 1LO and 2LO phonon scattering shows a drastic change with the relative intensities of both phonon peaks inverted.

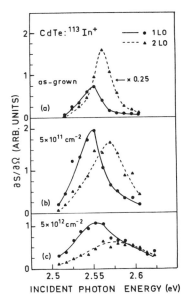

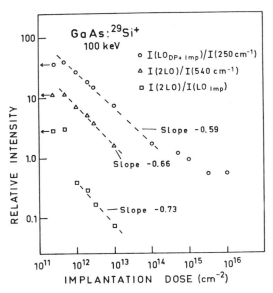

Fig. 3 Measured scattering efficiencies $\partial S/\partial \Omega$ for first-order (1LO) and second-order (2LO) Raman scattering in (a) as-grown and (b) and (c) ^{113}In$^+$-implanted CdTe. The full (1LO) and dashed (2LO) lines are drawn to guide the eye. In (a) the 2LO resonance curve displayed has been reduced by a factor of 4.

Fig. 4 Relative intensities of allowed first-order scattering $I(LO_{DP+Imp})/I(250$ cm^{-1}), allowed second-order scattering $I(2LO)/I(540$ cm^{-1}), and allowed second-order normalized to symmetry-forbidden but defect induced first-order scattering $I(2LO)/I(LO_{Imp})$ versus implantation dose. The arrows at the vertical scale indicate the corresponding relative intensities in as-grown GaAs.

This strong effect of implantation damage on the 1LO and 2LO phonon scattering intensities for excitation in resonance with, e.g., the E_1 band gap of GaAs, is caused by a damage induced broadening of the corresponding resonance in the scattering efficiency [24, 25]. This has been demonstrated explicitly for ^{113}In implanted CdTe [25], as shown in Fig. 3. Here the $E_0 + \Delta_0$ gap resonance in the 2LO phonon scattering efficiency broadens and decreases in height continuously with increasing implantation dose (Fig. 3a-c). The resonance for 1LO phonon scattering also broadens with increasing dose, but its height first increases for low doses because this scattering process is defect induced [28] (Fig. 3a and b). For higher implantation doses the increase in defect concentration is overcompensated by the broadening of the resonance leading eventually also to a decrease in the 1LO phonon scattering intensity (Fig. 3c).

The damage induced broadening of the above resonances for 1LO and 2LO phonon scattering is understood as follows [23]. It has been shown for GaAs that structures in the spectrum of the dielectric function ϵ, which are related to interband transitions such as across the E_1 band gap, get broadened and smeared out upon ion implantation [29]. The resonances in the efficiency for defect induced 1LO and 2LO phonon scattering can be approximated by $|\partial^2\chi/\partial\omega^2|^2$, where χ denotes the electric susceptibility and ω the incident photon energy [28]. Thus any change in ϵ, which is related to χ by $\epsilon = 1 + 4\pi\chi$, strongly affects the resonances for scattering by LO phonons.

The effect of ion implantation damage on various scattering processes by intrinsic phonon modes in GaAs is summarized in Fig. 4 [24]. It shows the intensity of resonant 2LO phonon scattering normalized to that of dipole forbidden but defect induced 1LO scattering $(I(2LO)/I(LO_{Imp}))$ plotted versus implantation dose. This ratio decreases strongly with increasing dose, which has been exploited to study the lateral homogeneity of ion implantation in GaAs wafers [24]. For comparison also, the 2LO scattering intensity normalized to the less resonant second-order scattering by two transverse optical (TO) phonons $(I(2LO)/I(540\ cm^{-1}))$, as well as the scattering intensity of first-order scattering by LO phonons normalized to the TO phonon band in amorphized GaAs $(I(LO_{DP+Imp})/I(250\ cm^{-1}))$, are plotted in Fig. 4 [24].

The sensitivity of resonant LO phonon scattering to ion beam induced damage in polar semiconductors in general, and in GaAs in particular, can be used to assess the damage induced by reactive ion etching (RIE) processes. This is demonstrated in Fig. 5 where the intensity of resonant 2LO phonon scattering, normalized to that of dipole forbidden but defect induced 1LO scattering $(I(2LO)/I(LO_{Imp}))$, is plotted versus the bias voltage applied during the RIE process. The material processed was n-type GaAs with a free carrier concentration of $2.7 \times 10^{17}\ cm^{-3}$ which was exposed to a CHF_3 plasma. For bias voltages exceeding 300 V, the ratio $I(2LO)/I(LO_{Imp})$ decreases, indicating a degradation of the material within the probing depth of 10 nm underneath the surface. The

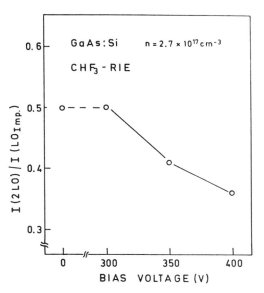

Fig. 5   Relative intensity of allowed 2LO scattering normalized to symmetry-forbidden but defect induced 1LO scattering in n-type GaAs plotted versus bias voltage applied during reactive ion etching.

degradation of the material detected by Raman spectroscopy correlates with a degradation of the electrical properties of Schottky diodes fabricated on the same RIE processed substrates [30]. The above results clearly show that resonant Raman scattering by intrinsic LO phonon modes provides a useful tool to analyze residual defects of RIE processing steps. This is of particular interest for the lateral processing of semiconductor heterostructures to fabricate, e.g., quantum wires and quantum dots [31].

RAMAN SCATTERING FROM LOCALIZED VIBRATIONAL MODES

Dopant impurities which are significantly lighter than the host lattice atoms, give rise to localized vibrational modes (LVM). These modes, which are observed in GaAs doped, e.g., with Si, Be, or C, are higher in frequency than the intrinsic host lattice phonon modes [32]. For a given dopant atom, the frequency and fine structure of the LVM indicate the lattice site occupied by the atom [32]. Infrared absorption spectroscopy is the commonly used technique to study LVM in bulk material and in relatively thick (> 1 μm) epitaxial layers [32]. Recently Raman spectroscopy has been used to investigate impurity induced LVM in heavily doped GaAs layers [7, 10, 11, 20, 21, 33-36].

In the following subsection, Raman scattering by LVM is discussed for homogeneously doped GaAs layers, whereas in the subsection following it emphasis is laid on the application of this Raman technique to the study of dopant incorporation in δ-doped GaAs structures.

Homogeneously Doped Layers

Fig. 6 displays the low-temperature Raman spectrum of a heavily doped GaAs MBE layer. The upper spectrum, which was excited at 3.00 eV in resonance with the E_1 band gap, shows the LVM of ^{28}Si on a Ga site (Si_{Ga}) at 384 cm^{-1} [32] superimposed on the intrinsic second-order phonon spectrum [28]. For comparison, the second-order phonon spectrum recorded under identical conditions from an undoped reference sample is displayed in the lower half of Fig. 6.

The analysis of the incorporation of Si into GaAs is of particular interest because Si can be incorporated on a Ga site (Si_{Ga}), where it acts as a donor, as well as on an As site (Si_{As}) acting as an acceptor [32]. In addition, electrically inactive Si_{Ga} - Si_{As} next neighbouring pairs can be formed, as well as complexes with, presumably, intrinsic defects labelled Si-X and Si-Y [32]. Si-X is believed to act as an acceptor and Si-Y might also be an acceptor or electrically inactive [32, 33, 37, 38]. Consequently, the electrical properties of heavily Si doped GaAs depend strongly on the balance between these different impurities and impurity complexes as shown in Fig. 7. There, a sequence of low temperature Raman spectra is displayed, which were recorded from MBE grown GaAs layers doped with Si at different concentrations. The total Si concentration [Si] measured by secondary ion mass spectroscopy (SIMS) increases monotonically from 4.7 x 10^{18} cm^{-3} to 4 x 10^{19} cm^{-3} (Fig. 7a-c). For a total Si concentration of 4.7 x 10^{18} and 1.2 x 10^{19} cm^{-3} (Fig. 7a and b), the free carrier concentration measured by Raman spectroscopy of the coupled plasmon-phonon modes [2] increases from 3.8 x 10^{18} to 8.6 x 10^{18} cm^{-3}. When the total Si concentration is further increased to 4 x 10^{19} cm^{-3} (Fig. 7c), the free carrier concentration drops to 5.4 x 10^{18} cm^{-3}. These differences in the electrical activation of the

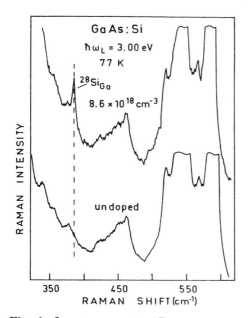

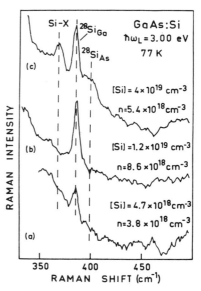

Fig. 6 Low-temperature Raman spectra of Si doped GaAs grown by MBE (top) and of an undoped reference sample (bottom). The spectra were excited at 3.00 eV and recorded with a spectral resolution of 5 cm^{-1}.

Fig. 7 Low-temperature Raman spectra of MBE grown GaAs:Si with different dopant concentrations [Si] given in the figure. The free carrier concentration n is also given. For all spectra, which were excited at 3.00 eV and recorded with a spectral resolution of 5 cm^{-1}, the second-order phonon spectrum has been subtracted.

dopant is explained by the Si site occupancy in the different samples. For the two lower Si concentrations, only the LVM of $^{28}Si_{Ga}$ is observed acting as a donor (Fig. 7a and b). For the highest Si concentration, the $^{28}Si_{Ga}$ donor gets compensated by the acceptors $^{28}Si_{As}$ and Si-X, which produce LVM at 399 and 370 cm^{-1}, respectively [32, 33].

The incorporation of Si dopant atoms into different lattice sites depends very much on the details of the growth conditions used for the epitaxy of the GaAs layer [33, 37, 39]. This applies similarly to the incorporation of Si into ion implanted and thermally annealed material [40]. It is illustrated in Fig. 8, which shows a sequence of LVM Raman spectra taken from a series of samples all implanted with 100 keV ^{29}Si at a dose of 1 x 10^{16} cm^{-2}. Then the samples were subjected to different annealing processes. Wafers A and D were rapid thermal annealed with the sample surface protected by a SiO$_2$ (wafer A) or a Si$_3$N$_4$ (wafer D) capping layer. Wafers B and C were furnace annealed under arsine overpressure (wafer B) or with the surface protected by the proximity technique (wafer C). It is evident that the Si site distribution within the probing depth of 10 nm underneath the surface depends strongly on the details of the annealing process. The relative concentrations of Si$_{Ga}$ and Si$_{As}$ vary considerably and the formation of the Si-X complex is observed exclusively for annealing under a SiO$_2$ protective cap [40].

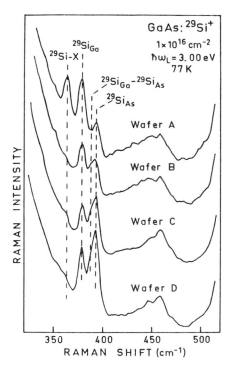

Fig. 8 Low-temperature Raman spectra of GaAs implanted with 1×10^{16} ^{29}Si/cm^2 and annealed under various conditions. Wafer A was capped with SiO_2 followed by rapid thermal annealing. Wafers B and C were furnace annealed with the surface protected by arsine overpressure (B) or the proximity technique (C). Wafer D capped with Si_3N_4 was processed by rapid thermal annealing. The Raman spectra excited at 3.00 eV were recorded with a spectral resolution of 5 cm^{-1}.

Another interesting point is the dependence of the Si LVM Raman spectrum on the incident photon energy. Fig. 9a shows a spectrum excited at 3.00 eV in resonance with the E_1 band gap, whereas the spectrum displayed in Fig. 9b was excited below that resonance at 2.71 eV. The LVM of Si_{As} and Si-X are observed for both incident photon energies. The LVM of Si_{Ga}, in contrast, is the dominant peak for excitation at 3.00 eV but absent in the spectrum excited at 2.71 eV. The difference in behaviour for the Si_{Ga} LVM on one hand, and the Si_{As} and Si-X LVM on the other, might be either due to the different lattice sites occupied or to the different electrical activity as a donor and an acceptor, respectively [21].

It is interesting to compare the above findings with the resonance behaviour of Raman scattering by the $^9Be_{Ga}$ LVM. Be is an acceptor occupying a Ga site. The corresponding LVM Raman spectra are displayed in Fig. 10. Both for excitation in resonance at 3.00 eV (Fig. 10a) and below resonance at 2.71 eV (Fig. 10b), the LVM of $^9Be_{Ga}$ is observed at 482 cm^{-1}, which indicates a resonance behaviour for Be_{Ga} different to the one for Si_{Ga} but similar to the Si_{As} and Si-X acceptors [21]. This may lead to the conclusion that the difference in resonance behaviour for Si_{Ga} and Si_{As} is due to the different electrical activity. However, more work is necessary in order to understand the underlying scattering mechanisms.

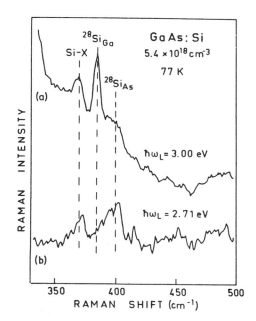

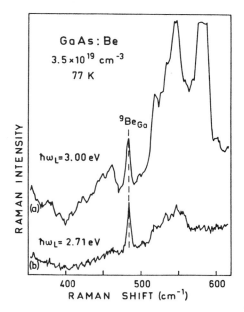

Fig. 9 Low-temperature Raman spectra of MBE grown GaAs:Si excited at different photon energies indicated in the figure. Free carrier concentration is also given. Spectral resolution was set to 5 cm^{-1}.

Fig. 10 Low-temperature Raman spectra of MBE grown GaAs:Be excited at different photon energies indicated in the figure. The nominal dopant concentration is also given. Spectral resolution was 5 cm^{-1}.

Recently, doping of GaAs with carbon has received considerable interest for the fabrication of thin heavily doped layers in, e.g., heterostructure bipolar transistors [41-43]. Carbon in GaAs gives rise to a LVM with a frequency of 582 cm^{-1} for $^{12}C_{As}$ [32]. The Raman spectrum of a heavily carbon doped layer grown by metalorganic molecular beam epitaxy (MOMBE) is displayed in Fig. 11a [44]. Fig. 11b shows the corresponding spectrum of an undoped reference sample and the difference spectrum [(a) - (b)] is plotted in Fig. 11c. The LVM of $^{12}C_{As}$ is clearly resolved in the Raman spectrum. All the spectra were excited at 2.71 eV, because for excitation at 3.00 eV in resonance with the E_1 gap, the resonantly enhanced scattering by 2LO phonons strongly overlaps with the $^{12}C_{As}$ LVM.

Raman spectroscopy by LVM has also been made a quantitative technique. The cross section per impurity $\partial S/\partial \Omega$ has been determined for scattering by the $^{28}Si_{Ga}$ and the $^{9}Be_{Ga}$ LVM to $\partial S/\partial \Omega = (3.2 \pm 0.6) \times 10^{-24}$ Sr^{-1}cm^2 and $\partial S/\partial \Omega = (1.1 \pm 0.6) \times 10^{-24}$ Sr^{-1}cm^2, respectively [20, 45]. These cross sections are considerably larger than the cross section per host lattice atom of $\partial S/\partial \Omega = 9 \times 10^{-26}$ Sr^{-1}cm^2 for dipole allowed intrinsic first-order scattering by LO phonons [46]. The above values refer to low temperatures (77 K) and excitation at 3.00 eV.

Because it is difficult to measure accurately absolute scattering intensities, also calibration factors have been derived using the strength of intrinsic second-order phonon scattering as a "built-in" reference [33, 45]. Based on this internal reference, a linear relationship between the $^{9}Be_{Ga}$ LVM intensity and the Be_{Ga} concentration has been found

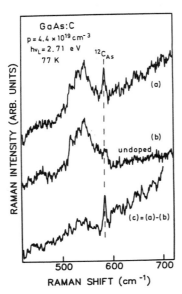

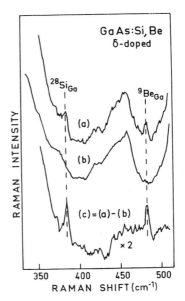

Fig. 11 Low-temperature Raman spectrum of (a) carbon doped GaAs grown by MOMBE and of (b) an undoped reference sample. The difference spectrum [(a) - (b)] is displayed in (c). The spectra were excited at 2.71 eV and recorded with a spectral resolution of 5 cm^{-1}.

Fig. 12 Low-temperature Raman spectra of (a) a Si and Be δ-doped GaAs sawtooth superlattice and of (b) an undoped reference sample. The spectrum of the superlattice with the intrinsic second-order phonon spectrum subtracted is displayed in (c).

for concentrations ranging from 3×10^{18} up to 1×10^{20} cm^{-3} [45]. This proves the validity of the above concept and the derived calibration factors. The detection limit for both Si_{Ga} and Be_{Ga} is $\approx 2 - 3 \times 10^{18}$ cm^{-3}.

Dopant Incorporation in δ-Doped Structures

The applicability of Raman scattering by LVM as a quantitative technique, combined with its sensitivity to even very thin doped layers due to the small probing depth of 10 nm for optical excitation in resonance with the E_1 band gap,[15] allows us to study the incorporation of dopants in δ-doped GaAs [7]. This is demonstrated in Fig. 12 for a Si and Be δ-doped GaAs sawtooth superlattice [47]. This superlattice consists of a first layer 5.3 nm underneath the sample surface δ-doped with 5×10^{12} cm^{-2} Be followed by an alternating sequence of Si and Be δ-doped layers with a spacing of 10.5 nm in-between and a doping concentration of 1×10^{13} cm^{-2} each. Due to the small probing depth of the Raman experiment for excitation at 3.00 eV in resonance with the E_1 band gap, essentially only the topmost two layers doped with Be and Si, respectively, are probed. The spectrum (a) shows the $^{28}Si_{Ga}$ LVM at 384 cm^{-1} as well as the $^{9}Be_{Ga}$ LVM at 482 cm^{-1} superimposed on a background of intrinsic second-order phonon scattering [28]. Fig. 12b displays the second-order phonon spectrum of an undoped reference sample and Fig. 12c shows the difference spectrum [(a) - (b)] after subtraction of the second-order phonon spectrum. Both the $^{28}Si_{Ga}$ and the $^{9}Be_{Ga}$ LVM are well resolved in Fig. 12c which demonstrates the sensitivity of the Raman technique to even a single δ-doped layer.

The δ-doping technique has recently found considerable interest both for fundamental studies and for device applications [47]. An important question related to this doping technique is the incorporation and the spread of the dopant atoms along the growth direction. A number of experimental techniques has been applied to characterize δ-doping, such as capacitance-voltage (C-V) profiling [48], magnetotransport measurements [49-51], secondary-ion mass spectroscopy (SIMS) [52-54], and Raman scattering by LVM [7]. In the following, a discussion of how Raman spectroscopy of LVM provides information on the depth distribution of those dopant atoms incorporated on lattice sites will be given.

The basic idea is to grow a series of GaAs layers containing a single δ-doping spike of, e.g., Si at a nominal depth z_0 underneath the surface with all the samples identical except for the different depths z_0. The measured normalized Si_{Ga} LVM intensity $I(Si_{Ga})/I(540$ cm$^{-1})$, where the intrinsic second-order phonon scattering strength at 540 cm^{-1} is used as a reference, can be written for each sample with a given Si depth profile $[Si_{Ga}(z-z_0)]$ as

$$I(Si_{Ga})/I(540 \text{ cm}^{-1}) = k \int_{o}^{\infty} [Si_{Ga}(z-z_0)] \, e^{-2\alpha z} \, dz. \qquad (2)$$

Here k is a calibration factor and $e^{-2\alpha z}$ describes the decay of the Raman sensitivity versus z, which is the coordinate normal to the surface. α is the absorption coefficient with $1/(2\alpha) = 10$ nm for excitation at 3.00 eV and 77 K [15]. With the above mentioned series of samples at hand, $I(Si_{Ga})/I(540$ cm$^{-1})$ can be measured for a variety of depths z_0 and the actual dopant depth profile is, in principle, obtained by solving Eq. 2 for $[Si_{Ga}(z)]$.

In Fig. 13, the normalized Si_{Ga} LVM intensity $I(Si_{Ga})/I(540$ cm$^{-1})$ is plotted versus the nominal depth z_0 for a series of samples grown by MBE at a substrate temperature of 580°C with a nominal dopant density of 8 x 10^{12} cm^{-2}. The LVM intensity is found to increase up to a depth of $z_0 \approx 20$ nm followed by a rapid decrease for larger values of z_0. These experimental data can neither be explained by a very narrow Gaussian depth distribution with a full width at half maximum (FWHM) of 2 nm (dotted curve in Fig. 13), nor by a strongly broadened symmetric distribution with a FWHM of 23 nm (dashed curve in Fig. 13). The doping profiles, which have been assumed for the calculation [Eq. 2] of the $I(Si_{Ga})/I(540$ cm$^{-1})$ versus z_0 curves, plotted in the main part of Fig. 13, are displayed in the inset.

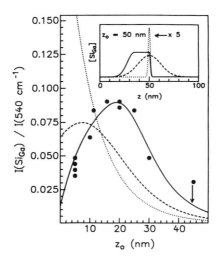

Fig. 13 Normalized $^{28}Si_{Ga}$ LVM scattering intensity $I(Si_{Ga})/I(540\ cm^{-1})$ versus nominal depth z_0 of the doping spike. The samples were grown by MBE at a substrate temperature of 580°C with an intended doping level of 8 x 10^{12} Si/cm^2. The inset displays different depth profiles of the Si concentration $[Si_{Ga}]$ used to calculate [Eq. (2)] the corresponding $I(Si_{Ga})/I(540\ cm^{-1})$ versus z_0 curves which are displayed in the main part of the figure.

To fit the experimental data, one has to postulate that, after the interruption of growth and deposition of the dopant atoms, these atoms are incorporated in the newly grown material at a given 3-dimensional dopant density [7, 55]. Dopant atoms, which are not incorporated in the material when finishing the growth of the layer, deposited after the growth interruption just sit on the surface and are not detected by the Raman experiment [7]. This leads to a strongly asymmetric depth profile with a constant Si_{Ga} concentration over a certain distance in growth direction and an abrupt decrease towards the substrate (solid curve in the inset of Fig. 13). Assuming a spread of the dopant atoms along the growth direction of $\approx$ 20 nm, a good quantitative agreement with the experimental data is obtained [7], as shown by the solid curve in the main part of Fig. 13.

The actual spread of the dopant atoms in δ-doped structures depends very much on the growth conditions, such as substrate temperature, and on the 2-dimensional dopant density [49-54]. A reduction in the 2-dimensional dopant density and a lowering of the substrate temperature are both expected to reduce the actual width of the doping spike. This is illustrated by the LVM data of Fig. 14 which were obtained from a series of Si δ-doped samples grown at a nominal substrate temperature of 500°C with a doping level of 2.8 x 10^{12} cm^{-2}. The resulting doping profile is much narrower with the measured width partially determined by the experimental depth resolution, which is of

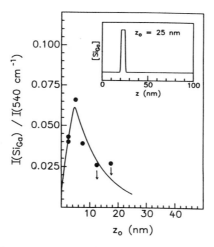

Fig. 14 Same as Fig. 13 for a sequence of samples grown at a nominal substrate temperature of 500°C with an intended doping level of 2.8×10^{12} cm^{-2}.

the order of $1/(2\alpha)$. Therefore the actual width of the doping profile can only be estimated to ≤ 5 nm.

In n-type δ-doped structures space charge effects induce a potential well in which quantized electron subbands are formed [47]. The actual shape of this potential well depends on the spread of the dopant atoms. Ideal δ-doping with the dopant atoms confined to one atomic plane leads to a V-shaped potential well whereas a large spread of the dopant atoms results in a parabolic well [49, 50]. The shape of this well, in turn, determines the energy spacings of the electron subbands. Therefore Raman spectroscopy of intersubband transitions provides complementary information on the actual width of doping spikes in nominally δ-doped GaAs. For the present samples grown at a substrate temperature of 580 °C with a doping density of 8×10^{12} cm^{-2} spin-density excitation energies of 24.8 and 36.7 meV were found for a nominal depth of $z_0 = 30$ nm [57]. Self-consistent electronic subband calculations, using the actual spread of the dopant atoms measured by the above described LVM Raman experiments (see Fig. 13) as an input parameter, yield spin-density excitation energies of 25.1 and 36.5 meV for transitions between the second and third and the third and fourth subband, respectively. This is in good agreement with the experimental values [57]. It demonstrates that a consistent picture is obtained for the spread of dopant atoms in Si δ-doped GaAs from the combination of LVM and subband Raman spectroscopy.

CONCLUSIONS

The effect of impurities and defects on the Raman spectroscopy of vibronic excitations has been discussed. Resonant Raman scattering by intrinsic LO phonon modes has been shown to be sensitive to ion beam induced damage which allows to assess, e.g., residual defects in reactive ion etched semiconductor structures. Dopant atoms significantly lighter than the host lattice atoms are accessible by Raman spectroscopy via light scattering by localized vibrational modes. This technique has a detection limit of $\leq 2 - 3 \times 10^{18}$ cm^{-3} corresponding to an area density of $\leq 2 - 3 \times 10^{12}$ cm^{-2} for Si and Be in GaAs. This opens the possibility to study dopant impurities in quasi-two-dimensional systems such as, e.g., planar (δ-) doped GaAs structures. For such structures the incorporation and spread of the dopant atoms has been analyzed and correlated with the energy spacings of the electron subbands formed in the space charge induced potential well.

ACKNOWLEDGMENTS

The author wants to thank M. Ramsteiner (IAF Freiburg), R. Murray, and R.C. Newman (both with the Imperial College London) for their cooperation in the course of this work as well as P. Koidl and H.S. Rupprecht for many helpful discussions and continuous support of the work at the IAF. Thanks are further due to F. Eisen (IAF Freiburg), K. Köhler (IAF Freiburg), W. Pletschen (IAF Freiburg), K. Ploog (MPI/FKF Stuttgart), and M. Weyers (RWTH Aachen) for providing samples, as well as to A. Maier for help with the preparation of this manuscript.

REFERENCES

1.   M. Cardona, in: "Light Scattering in Solids II", eds. M. Cardona and G. Güntherodt, Springer, Berlin (1982) p. 49 and references therein.
2.   G. Abstreiter, A. Pinczuk, and M. Cardona, in: "Light Scattering in Solids IV", eds. M. Cardona and G. Güntherodt, Springer, Berlin (1984) p. 5 and references therein.
3.   A. Pinczuk and G. Abstreiter, in: "Light Scattering in Solids V", eds. M. Cardona and G. Güntherodt, Springer, Berlin (1989) p. 153 and references therein.
4.   G.B. Wright and A. Mooradian, in: "Proceedings of the 9th Int. Conf. on the Physics of Semiconductors", Nauka, Leningrad (1969) p. 1067; A. Mooradian, in: "Laser Handbook", Vol. 2, eds. F.T. Arecchi and E.O. Schulz-Dubois, North Holland, Amsterdam (1972) p. 1409.
5.   M.V. Klein, in: "Light Scattering in Solids I", ed. M. Cardona, Springer, Berlin (1975) p. 148.
6.   J. Tsang, in: "Light Scattering in Solids V", eds. M. Cardona and G. Günterhordt, Springer, Berlin (1989) p. 233 and references therein.
7.   J. Wagner, M. Ramsteiner, W. Stolz, M. Hauser, and K. Ploog, Appl. Phys. Lett. 55, 978 (1989).
8.   G. Contreras, M. Cardona, and A. Axmann, Solid State Commun. 53, 861 (1985).
9.   G. Contreras, M. Cardona, and A. Compaan, Solid State Commun. 53, 857 (1985).
10. T. Nakamura and T. Katoda, J. Appl. Phys. 57, 1084 (1985).
11. M. Ramsteiner, J. Wagner, H. Ennen, and M. Maier, Phys. Rev. B 38, 10669 (1988) and references therein.
12. K. Jain, S. Lai, and M.V. Klein, Phys. Rev. B 13, 5448 (1976).

13. K. Wan and R. Bray, Phys. Rev. B 32, 5265 (1985).
14. J. Wagner, M. Ramsteiner, H. Seelewind, and J. Clark, J. Appl. Phys. 64, 802 (1988); J. Wagner, Physica Scripta T29, 167 (1989).
15. M. Cardona and G. Harbeke, J. Appl. Phys. 34, 813 (1962).
16. D.E. Aspnes and A.A. Studna, Phys. Rev. B 27, 985 (1983).
17. D. Gammon, R. Merlin, W.T. Masselink, and H. Morkoc, Phys. Rev. B 33, 2919 (1986).
18. K.T. Tsen, J. Klem, and H. Morkoc, Solid State Commun. 59, 537 (1986).
19. P.O. Holtz, M. Sundaram, R. Simes, J.L. Merz, A.C. Gossard, and J.H. English, Phys. Rev. B 39, 13293 (1989).
20. J. Wagner, M. Ramsteiner, R. Murray, and R.C. Newman, in: "Materials Science Forum", Vols. 38-41, Trans Tech, Switzerland (1989), p. 815.
21. J. Wagner and M. Ramsteiner, IEEE J. Quantum Electron. 25, 993 (1989).
22. T. Nakamura and T. Katoda, J. Appl. Phys. 53, 5870 (1982) and references therein.
23. J. Wagner and C.R. Fritzsche, J. Appl. Phys. 64, 808 (1988) and references therein.
24. J. Wagner, Appl. Phys. Lett. 52, 1158 (1988).
25. A. Lusson, J. Wagner, and M. Ramsteiner, Appl. Phys. Lett. 54, 1787 (1989).
26. K.K. Tiong, P.M. Amirtharaj, F.H. Pollak, and D.E. Aspnes, Appl. Phys. Lett. 44, 122 (1984).
27. H. Richter, Z.P. Wang, and L. Ley, Solid State Commun. 39, 625 (1981).
28. R. Trommer and M. Cardona, Phys. Rev. B 17, 1865 (1978).
29. D.E. Aspnes, S.M. Kelso, C.G. Olson, and D.W. Lynch, Phys. Rev. Lett. 48, 1863 (1982).
30. J. Wagner, W. Pletschen, and G. Kaufel, to be published.
31. For a recent review, see: A. Forchel, H. Leier, B.E. Maile, and R. Germann, in: "Advances in Solid State Physics", Vol. 28, Vieweg, Braunschweig (1988) p.99.
32. R.C. Newman, Mat. Res. Soc. Symp. Proc. Vol. 46, 459 (1985) and references therein.
33. R. Murray, R.C. Newman, M.J.L. Sangster, R.B. Beall, J.J. Harris, P.J. Wright, J. Wagner, and M. Ramsteiner, J. Appl. Phys. 66, 2589 (1989).
34. M. Holtz, R. Zallen, A.E. Geissberger, and R.A. Sadler, J. Appl. Phys. 59, 1946 (1986).
35. T. Kamijoh, A. Hashimoto, H. Takano, and M. Sakuta, J. Appl. Phys. 59, 2382 (1986).
36. R. Ashokan, K.P. Jain, H.S. Mavi, and M. Balkanski, J. Appl. Phys. 60, 1985 (1986).
37. J. Maguire, R. Murray, R.C. Newman, R.B. Beall, and J.J. Harris, Appl. Phys. Lett. 50, 516 (1987).
38. H. Ono and R.C. Newman, J. Appl. Phys. 66, 141 (1989).
39. J.H. Neave, P.J. Dobson, J.J. Harris, P. Dawson, and B.A. Joyce, Appl. Phys. A 43, 195 (1983).
40. J. Wagner, H. Seelewind, and W. Jantz, J. Appl. Phys. 67, 1779 (1990).
41. T.F. Kuech, M.A. Tischler, P.-J. Wang, G. Scilla, R. Potemski, and F. Cardone, Appl. Phys. Lett. 53, 1317 (1988).
42. K. Saito, E. Tokumitsu, T. Akatsuka, M. Miyauchi, T. Yamada, M. Konagai, and K. Takahashi, J. Appl. Phys. 64, 3975 (1988).
43. C.R. Abernathy, S.J. Pearton, R. Caruso, F. Ren, and J. Kovalchik, Appl. Phys. Lett. 55, 1750 (1989).
44. J. Wagner and M. Weyers, to be published.

45. J. Wagner, M. Maier, R. Murray, R.C. Newman, R.B. Beall, and J.J. Harris, J. Appl. Phys., to appear.

46. M.H. Grimsditch, D. Olego, and M. Cardona, Phys. Rev. B 20, 1758 (1979).

47. For a recent review, see: K. Ploog, M. Hauser, and A. Fischer, Appl. Phys. A 45, 233 (1988).

48. E.F. Schubert, J.B. Stark, B. Ullrich, and J.E. Cunningham, Appl. Phys. Lett. 52, 1508 (1988).

49. F. Koch and A. Zrenner, Mater. Sci. Eng. B1, 221 (1989) and references therein.

50. A. Zrenner, F. Koch, R.J. Williams, R.A. Stradling, K. Ploog, and G. Weimann, Semicond. Sci. Technol. 3, 1203 (1988).

51. M. Santos, T. Sajoto, A. Zrenner, and M. Shayegan, Appl. Phys. Lett. 53, 2504 (1988).

52. R.B. Beall, J.B. Clegg, and J.J. Harris, Semicond. Sci. Technol. 3, 612 (1988).

53. J.B. Clegg and R.B. Beall, Surf. Interface Anal. 14, 307 (1989).

54. A.-M. Lanzillotto, M. Santos, and M. Shayegan, Appl. Phys. Lett. 55, 1445 (1989).

55. A. Zrenner, F. Koch, and K. Ploog, Inst. Phys. Conf. Ser. 91, 171 (1988).

56. G. Abstreiter, R. Merlin, and A. Pinczuk, IEEE J. Quantum Electron. 22, 1771 (1986).

57. J. Wagner, M. Ramsteiner, D. Richards, G. Fasol, and K. Ploog, Appl. Phys. Lett., January 14 (1991).

RAMAN MICROPROBE STUDY OF SEMICONDUCTORS

S. Nakashima

Department of Applied Physics
Osaka University
Suita, Osaka 565, Japan

INTRODUCTION

Recent advances in technologies and fabrication of highly integrated semiconductor devices have increased the need for technique of non-destructive characterization and microanalysis. Raman microprobe spectroscopy has considerable promise as a diagnostic tool in the field of semiconductors[1].

While the ultimate resolution of the Raman microprobe can never approach that of an electron microscope (EM), it has important

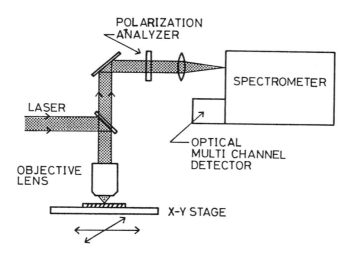

Fig.1 Schematic diagram of Raman microscope.

Light Scattering in Semiconductor Structures and Superlattices
Edited by D.J. Lockwood and J.F. Young, Plenum Press, New York, 1991

291

advantages in other aspects. For example, specimens must be kept in a vacuum and special preparation for the specimen is required for the EM measurement. In contrast, the Raman microprobe does not require a vacuum and special preparation. By using various excitation lines with different penetration depths, it is possible to perform depth profiling of physical quantities. More recent development of optical multichannel detectors (OMD) enabled us to obtain Raman spectra of ultra thin films of semiconductors[2]. Furthermore, the combination of the Raman microscope and OMD resulted in the quantitative characterization of semiconductors because accurate determination of the Raman frequency and band shape is possible by this combination.

The translation of specimens or the scanning of laser beams provides us two-dimensional Raman images. The scanning Raman microscope is useful for obtaining not only the spatial inhomogeneity of physical quantities but also physical processes.

The application of Raman microprobe to semiconductors has been reviewed recently[1]. In this paper more emphasis will be placed on the Raman mapping from which we get information about the transformation of semiconductor structures: crystal growth of thin films, recovery of crystallinity in implanted semiconductors and thermal conversion of SiC polytypes. Characterization of composite structures such as p-n junction and heterojunction is discussed.

Raman microprobe measurement at low temperature was made on GaP diodes with graded impurity distribution. Raman parameters were obtained as a function of the impurity concentration. The analysis of the Raman parameters has allowed us to examine the electron-LO phonon interaction in GaP.

RAMAN MICROSCOPE

A Raman microscope consists of an optical microscope and a spectrometer with a photomultiplier or an OMD. The Raman measurements of a local area are made by (1) point illumination, (2) line illumination and (3) global illumination methods[3].

In the method (1), laser light is focused on a sample surface by an objective lens and the scattered light is collected by the same objective lens as shown in Fig.1. A minimum spot size which is limited by diffraction determines the spatial resolution. Although Raman spectra are obtained with a spatial resolution comparable to the wavelength of a laser used, focused laser beams often cause local heating or sample degradation. Accordingly, this method limits the use of the microprobe to robust samples. Raman images are obtained by translating a sample stage or by scanning the light probe two-dimensionally and adjusting the optical image of scattered light so as to be focused onto a slit.

In the line illumination method (2), one-dimensional spectral profile is obtained if one uses a charge coupled devices (CCD)[4]. Scanning the line beam provides two-dimensional intensity images. Two-dimensional Raman spectral profile is obtained by using the CCD and scanning a line image on the slit.

The global illumination (3) is used for easily damaged materials such as biomolecules. Since in this case the image of the scattered light is introduced into a spectrometer through the opening of a slit and the spatial resolution is determined by the slit width, the spatial resolution is not high compared with the method (1).

Another important advantage of the Raman microprobe is that no adjustment of optical elements is needed when changing samples. The shift of the focused image on the entrance slit arising from the adjustment causes apparent change in Raman frequency. The rate of the frequency change due to the shift of the image is comparable to the reciprocal linear dispersion of the spectrometer.

The error in Raman frequency measurements arising from mechanical scan of the spectrometer can be avoided when we use an OMD. Accuracy of frequency determination better than 0.1 cm^{-1} is obtained when one uses the Raman microprobe together with the OMD and compares the Raman frequency of a specimen tested with that of a reference crystal. This procedure is analogous to the difference Raman spectroscopy, for which the spectra of different samples are traced in the same scan and the differential spectra are analyzed assuming a Lorentzian or Gaussian line shape for Raman band[5]. More recently, the Hadamard transform Raman microprobe has been proposed by Treado and Morris[6]. In this system, loosely focused light is used and the image of the scattered light is collimated to a Hadamard mask which is translated in X and Y directions. The spatially encoded image is focused onto a spectrometer and detected. Two-dimensional Raman images are obtained by transforming the Raman signals obtained at each mask position. This method has the advantage of saving time and of preventing specimens from the damage arising from local heating because loosely focused laser light is used. It was reported that the spatial resolution of Hadamard transform images was close to the diffraction limited resolution[6].

DETERMINATION OF LOCAL CRYSTALLOGRAPHIC ORIENTATIONS

Raman scattering in crystals is a two photon process. The Raman intensity depends on polarization directions of incident and scattered light relative to the crystal axes.

This anisotropy is significant even for cubic crystals. Hence, the Raman intensity measurement for different polarization directions enables us to determine the crystal axes of specimens with unknown orientations[7,8]. The polarization measurement, together with the Raman microprobe, makes it possible to determine the orientations of local area of crystals. This technique is useful for determining the crystal orientation of small semiconductor crystallites because it is a non-destructive technique and the spatial resolution ($\sim 1 \mu m$) is better than that of conventional x-ray diffraction analyses.

Here we consider a diamond type crystal such as silicon. For a back scattering geometry shown in Fig.2, the wave vectors of the incident and scattered light are given by

$$k_i = (-\sin\theta\cos\phi, -\sin\theta\sin\phi, -\cos\theta)$$

$$k_s = -k_i$$

(1)

where θ is the angle between k_s and $\langle 001 \rangle$ axis of the crystal, ϕ is the angle between the $\langle 100 \rangle$ axis and the projection of k_s onto (001)

surface. Let ψ_0 be the angle between the projection from $\langle 001 \rangle$ axis on X-Y plane and X-axis in the laboratory system and $e_i(e_s)$ and the X-axis make angle $\psi(\psi')$. The unit polarization vectors of the incident and scattered light are expressed by

$$e_i = (\cos\theta\cos\phi\cos(\psi+\psi_0) - \sin\phi\sin(\psi+\psi_0),$$

$$\cos\theta\sin\phi\cos(\psi+\psi_0)+\cos\phi\sin(\psi+\psi_0),$$

$$-\sin\theta\cos(\psi+\psi_0)) \cdot \qquad (2)$$

$e_s(e_s{}^x, e_s{}^y, e_s{}^z)$ is given by replacing ψ by ψ'. The Raman scattering intensity is given by the following equation.

$$S = \eta\Sigma(e_i \cdot R_j \cdot P_s)^2 \qquad (3)$$

The prefactor η is a constant of proportionality related to the standard Raman cross section. Substituting Eq.(2) into Eq.(3) and arranging in terms of angle ψ', we get[8]

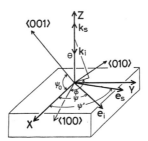

Fig.2   Configuration of Raman polarization measurements with the back scattering geometry in the laboratory coordinate system.

$$I(\theta,\phi,\psi_0:\psi,\psi') = S/\eta d^2$$

$$= A(\theta,\phi,\psi_0:\psi)+B(\theta,\phi,\psi_0:\psi)\cos2\psi'$$

$$C(\theta,\phi,\psi_0:\psi)\sin2\psi' \qquad (4)$$

where d is the Raman tensor component. The Raman intensity for an arbitrary oriented crystal is measured as a function of the angle ψ' for a fixed polarization direction of the incident light. Using θ, ϕ and ψ_0 as adjustable parameters, the crystallographic orientation is determined by fitting Eq.(4) to the observed intensity as a function of ψ'.

    For special planes such as (001), (101) and (111) surfaces, Eq.(4) is reduced to simple expressions as described in Table 1. For zincblende type crystals, the long wavelength optic modes split into TO and LO modes. The polarization angle dependence of the intensities of the two modes is also given in Table 1.

Table 1

Diamond type	Zincblende type
(100) surface	
$I = \sin^2(\psi+\psi'+2\psi_0)$	$I_{LO} = \sin^2(\psi+\psi'+2\psi_0)$
	$I_{TO} = 0$
(110) surface	
$I = \sin^2(\psi+\psi'+2\psi_0)$ $+\cos^2(\psi+\psi_0)\cos^2(\psi'+\psi_0)$	$I_{LO} = 0$ $I_{TO} = \sin^2(\psi+\psi'+2\psi_0)$ $+\cos^2(\psi+\psi_0)\cos^2(\psi'+\psi_0)$
(111) surface	
$I = (1/3)[\cos^2(\psi-\psi')+2]$	$I_{LO} = (1/3)\cos^2(\psi-\psi')$ $I_{TO} = 2/3$

When $\psi=0°$ and $\psi_0=90°$ or $\psi=90°$ and $\psi_0=0°$, the intensity variation of the (101) surface is identical to that of the (001) surface. Hence the measurement of the intensity variation for two different polarization directions of the incident light is necessary to distinguish between the (101) and (001) surfaces. Since the intensity variation for the (111) surface is independent of ψ_0, crystal axes within the (111) plane can not be determined by this method.

Single crystalline silicon wafers with (100), (110) and (111) surfaces were measured to examine the accuracy of this method. We rotated an analyzer at intervals of 3.6° and measured Raman intensity as a function of the rotation angle. The result is shown in Fig.3. The solid lines are the best fit curves from which the orientation parameters θ, ϕ, ψ_0 are determined. This fitting procedure showed that the accuracy of the determination by this method is within ±2.0°.

We have applied the above method to laser-recrystallized Si films on insulator (SOI). The specimens used have a structure as shown in Fig.4(a). Polycrystalline silicon films deposited on oxide coated Si crystals were melted by a scanning cw Ar$^+$ ion laser and then recrystallized. Since the polycrystalline silicon is directly attached to the crystal Si at an opening of the SiO_2 films, the underlying Si crystal seeds the crystal orientation of the overlayer. The oriented crystallite extends along the direction of the laser scanning. The method is called "lateral epitaxy" or "zone melting recrystallization".

The films recrystallized by the lateral epitaxy usually contain twins, stacking faults, large- or small-angle grain boundaries. The control of the temperature profile in the silicon films during the laser recrystallization has been tried by using antireflecting stripes[9], M-shaped laser beam[10], etc. These techniques provided single crystal films elongated about 1mm from the seeded region[10]. However, it was found that the crystallographic orientation of the recrystallized films varies continuously with position although defects and grain boundaries were not detected.

The crystallographic orientations of the laser recrystallized stripe were determined by the Raman microprobe polarization measurement. For this sample, the annealing laser (Ar ion laser) beam was scanned along the <100> direction of the substrate[10]. The variation of the <001> axis with position is illustrated in Fig.4(b).

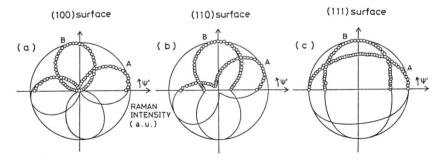

Fig.3 Raman intensity as a function of the rotation angle ψ' of the polarization analyzer for Si crystals of (100), (110) and (111) surfaces[8]. The curves A and B in each figure are results for $\psi=0$ and $\psi=90°$, respectively.

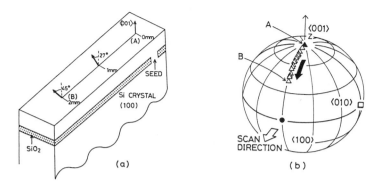

Fig.4 (a) Sketch of the silicon on insulator structure.
The single crystal film is grown by the lateral
epitaxy method and (b) the variation of the <001> axis
rotates about the <010> axis.

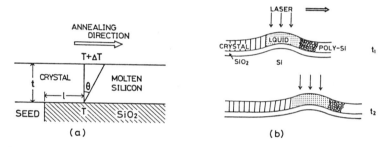

Fig.5 Mechanisms for the rotation of the crystal axis in
the laser-recrystallized Si film on insulator. (a)
The effect of vertical thermal gradient in the silicon
film and (b) The effect of the film bending during the
recrystallization.

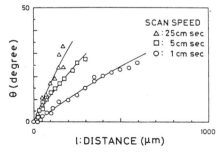

Fig.6 The angle between the <001> axis and the surface
normal is plotted as a function of the distance from
the seed for different scan speeds.

The surface orientation of the recrystallized layer at the opening is the same as that of the substrate. Going away from the seed, the <100> axis rotates forward about the <100> axis. The surface orientation passes through (n10), where n>1, and approaches the (110). When the surface is close to the (110) orientation, the rotation rate $d\theta/dl$ decreases and the growth of the single crystal films seems to stop and the grains with different orientations appear.

The mechanism for this rotation of the crystal axis has not been well understood. It is pointed out that the relation between liquid-solid interface and {111} planes plays an important role in the regrowth. The vertical temperature gradient existing in the films is also expected to cause the rotation of the orientation. The temperature difference between outer and inner surfaces of the film gives the elongation of the lattice at the outer surface as shown in Fig.5(a). If this non-uniform elongation is pinned by the surrounding silicon after the recrystallization, it would provide the forward inclination of the axis. However, the value of $d\theta/dl$, which is estimated from simulation of the thermal profile by use of a two-dimensional heat flow analysis, is one order of magnitude smaller than the experimental value. In order to get an agreement with the experimental result we must assume a large thermal expansion coefficient[10].

Another possible mechanism is bending of the surface layers arising from the local heating caused by laser irradiation (Fig.5(b)). When the film is recrystallized at the bending region resulting from the local expansion, the crystal axis would incline on cooling. This inclination of the axis is accumulated along the direction of the laser annealing. The rate of the inclination, $d\theta/dl$ depends on regrowth conditions such as scan speed, power of the annealing laser, Si film thickness, and substrate temperature.

Fig.6 shows the variation of the orientation with the distance for different scan speeds of the annealing laser. The faster the scan speed, the larger the rate of the axis rotation. The growth of the single crystalline films seems to stop at around (110) surface independently of the scan speed. It is to be noted that the angle rotation saturates as the surface orientation approaches the (110) orientation. The effect of the thermal gradient and bending can not explain this saturation. Thermodynamic effect or interaction of silicon and underlying SiO_2 films might be related to this saturation of the inclination and the growth interruption at around the (110) orientation.

FREE AND BOUND ELECTRONS IN GaP DIODES

In zincblende type crystals, LO phonons and free carriers are hybridized and form LO-phonon plasmon coupled modes. The upper branch of the coupled mode (L_+ mode) is located at the high frequency side of the pure LO phonon band and shifts towards higher frequency with increasing carrier concentration[11].

The characteristic feature of the coupled mode is that its band shape is asymmetric and depends strongly on the damping of the carrier. This fact has been used to evaluate the carrier concentration and mobility of doped semiconductors. It was found that the carrier concentration determined from the Raman measurements was in close agreement with that determined from Hall measurements in GaP[12] and SiC[13].

Nakashima et al. examined the Raman spectra of GaP light emitting diodes which were grown by liquid phase epitaxy[14]. The Raman spectra from various points in a cross section of a p^+-n-n^+ junction were measured with a Raman microscope in conjunction with an OMD.

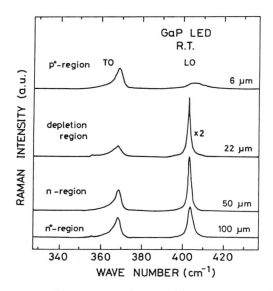

Fig.7   Raman spectra measured at typical points of the cross section of a GaP light emitting diode.

Figure 7 shows the Raman spectra taken at 300K for typical points in the diode cross section. The L_+ band varies in shape with the position. The peak frequency and intensity of the L_+ band at 300K are plotted as a function of the position in Fig.8. The peak intensity varies strikingly with position and has a maximum at the depletion region. The peak frequency of the L_+ band also changes with position. The frequency difference of the L_+ band between the p^+ and depletion regions is 2.5cm^{-1}.

The Raman scattering of the plasmon-LO phonon coupled mode has been analyzed by taking into account three mechanisms: charge density fluctuation (CDF) mechanism, deformation potential (DP) mechanism and electrooptic (EO) mechanism[11].

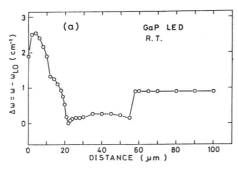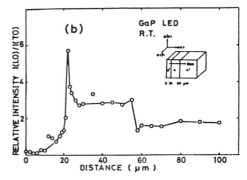

Fig.8 The frequency shift with respect to the LO phonon
frequency at the depletion layer (a) and peak
intensity (b) of the plasmon-LO phonon coupled mode
(L_+) mode as a function of the position[14].

Carrier damping in GaP is large, ie. $\omega_p \tau < 1$, where ω_p is the
plasma frequency and τ is the collision time of the free carrier. In
this case we may neglect the CDF mechanism in the calculation of the
band shape of the coupled mode because the CDF mechanism gives a very
broad band shape. Using a theoretical Raman cross section which
includes the phonon damping as well as plasmon damping, lineshape
fitting to the observed L_+ band was made. The carrier density and
mobility were determined from the best fit values of adjustable
parameters. The carrier concentration and mobility thus determined
are plotted as a function of the distance from the edge in Fig.9. The
carrier concentration in Fig.9 does not show a step-like
distribution, indicating interdiffusion of impurities around the
junctions and charge compensation induced by contamination[14].

We have observed the Raman spectra of this diode in temperatures
ranging from 80 to 300K. No separate peak was found at the low
frequency side of the LO band at low temperatures. Figure 10 shows
the peak frequency and intensity of the L_+ mode at 80K. The intensity
profile of the L_+ mode taken at 80K does not differ much from the
profile at room temperature except for the absolute value. However,
the frequency profiles at 80K and 300K are quite different, especially
in the p^+ region. At 300K the frequency of the L_+ mode in the p^+
region is higher than that of the depletion layer, but at 80K the
situation is reversed. The frequency of the L_+ mode at several points
was measured as a function of the temperature. As shown in Fig. 11,
the frequency increases with decreasing temperature in the n, n^+ and
depletion regions. The frequency of the L_+ mode in the high doped p^+
region varies slightly with temperature and lies below the frequency
of the other regions at 80K. The temperature dependence of the TO
phonon band is almost the same in every region. A similar result was
reported by Galtier and Martinez who observed that the frequency of
the LO band of sulphur doped GaP down-shifted with respect to the LO
phonon frequency in undoped GaP at 80K and it decreased with
increasing sulphur concentration[15].

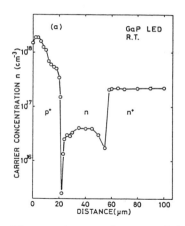
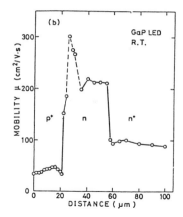

Fig.9 The profiles of (a) the free carrier concentration
and (b) mobility in the GaP diode deduced from the
Raman scattering analysis[14].

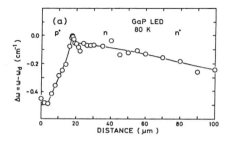
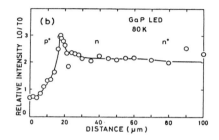

Fig.10 The frequency shift with respect to the value at the
depletion layer (a) and the peak intensity (b) of the
LO mode at 80 K are plotted as a function of the
position.

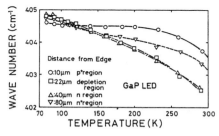

Fig.11 The temperature dependence of the frequency of the
L_+ mode at several points of the GaP diode.

The downshift of the LO phonon mode in the highly doped region is not considered to be due to the softening of lattice modes by doping of impurities, because the impurity concentration is too small to cause the softening and the frequencies of the TO phonons are the same in every region.

At 80K most of the electrons (holes) are bound to impurity centers and form neutral donors (acceptors) because the ionization energies are relatively large in GaP. The interaction of the LO phonon and the bound electron induces bound or localized LO modes in polar semiconductors, which have long been studied[16].

A dielectric sphere model was used by Barker to examine the behavior of the bound LO phonon[17]. At around impurity center (donors or acceptors) the polarization of the center perturbs the dielectric function locally, providing a frequency shift of longitudinal optic phonon.

If the contribution of the bound electrons to the dielectric function is restricted to the spheres with radius comparable to the effective Bohr radius, the dielectric function for the sphere is given by

$$\varepsilon_s(\omega) = \varepsilon_\infty[1 + \frac{\omega_L^2 - \omega_T^2}{\omega_T^2 - \omega^2 - i\gamma\omega}] + \varepsilon_\infty^{el} + \frac{S\omega_I^2}{\omega_I^2 - \omega^2 - i\omega\gamma_I}$$

$$= \varepsilon_L(\omega) + \varepsilon_B(\omega) \tag{5}$$

The first term arises from the phonon and the second and third terms represent the contribution from the bound electron. The dielectric function as a function of ω is shown in Fig.12, where we ignore the

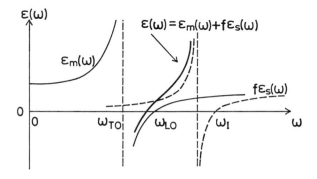

Fig.12 The average dielectric function for doped semiconductor at low temperature, in which electrons (holes) are bound to impurity centers.

damping constants γ and γ_I. When $\omega_I > \omega_L$, the dielectric function $\varepsilon_B(\omega)$ has a positive value at $\omega = \omega_L$. Hence $\varepsilon_S(\omega)$ becomes zero at frequencies smaller than ω_L. Because the shape of longitudinal optic modes is given by A $Im[-1/\varepsilon(\omega)]$, the longitudinal optic mode inside the sphere has a peak at frequency lower than ω_L.

For a bulk medium in which spheres with a different dielectric function are embedded, the effective dielectric function ε_{eff} is given by

$$\varepsilon_{eff} = \varepsilon_m + \frac{3\varepsilon_m(\varepsilon_S - \varepsilon_m)f}{\varepsilon_S + 2\varepsilon_m - (\varepsilon_S - \varepsilon_m)f} \qquad (6)$$

where f is the filling factor given by

$$f = \frac{4}{3} N_S r_S^3 \qquad (7)$$

Here N_S is the density of the spheres and r_S the sphere radius which is comparable to the effective Bohr radius of the bound electron. $Im(-1/\varepsilon_{eff})$ has peaks at the bulk LO phonon frequency ω_L and a frequency lower than ω_L, which gives the downshift of the LO phonon frequency.

In this model LO phonons outside the sphere are assumed to extend infinitely. However, this assumption would not be valid at impurity concentrations above $10^{18} cm^{-3}$, because the average interimpurity distance is less than about 100Å in this concentration range and the phonons may not propagate. This situation is similar to multilayer structures which consist of two alternating layers with different dielectric functions. The vibrational mode existing in the multilayer structure satisfies the following equation[18,19],

$$[\varepsilon_1^2(\omega) + \varepsilon_2^2(\omega)]\sinh(kd_1)\sinh(kd_2)$$

$$+ 2\varepsilon_1(\omega)\varepsilon_2(\omega)\cosh(kd_1)\cosh(kd_2)$$

$$= 2\varepsilon_1(\omega)\varepsilon_2(\omega)\cos(qD) \qquad (8)$$

where k is the wave vector component of the extended interface mode parallel to the layer and q is the component of the wave vector perpendicular to the layer. In the long wavelength limit, i.e. q=0, kd_1, $kd_2 \ll 1$, one gets two solutions for Eq.(8),

$$\varepsilon_1(\omega)d_1 + \varepsilon_2(\omega)d_2 = 0 \qquad (9a)$$

$$\varepsilon_1(\omega)d_2 + \varepsilon_2(\omega)d_1 = 0 \qquad (9b)$$

Equations (9a) and (9b) correspond to symmetric and antisymmetric interface phonon modes of the double heterostructure[20]. Equation (9a) is appropriate to our semiconductor medium doped with impurities. This equation implies that the frequency of the longitudinal phonon mode is given by the zero of average dielectric functions because

Eq.(9a) is rewritten as

$$\varepsilon_{av} = \frac{d_1}{D}\varepsilon_1(\omega) + \frac{d_2}{D}\varepsilon_2(\omega) = 0 \tag{10}$$

This result allows us to use a mean dielectric function to study the LO phonon modes in the heavily doped GaP. The volume average of the dielectric functions is defined as

$$\varepsilon^T(\omega) = (1-f)\varepsilon_m + f\varepsilon_S \tag{11}$$

where f is the filling factor given by Eq.(7). The zero of the equation gives the frequency of the longitudinal mode extending in the whole region. In Fig.13 the shift of the LO phonon frequency in the p^+ region relative to that of the depletion region at 80K is plotted as a function of impurity concentration. Here we assume that the carrier concentration at 300K is nearly the same as the impurity concentration. The open circles are the data of GaP:S by Galtier and Martinez[15].

Using Eq.(11) we calculate the LO phonon frequency as a function of the impurity concentration in which the values obtained by Barker are used for the parameters S and ε_∞^{el}. The calculated result is shown by the solid line in Fig.13. The calculated and experimental results show a qualitative agreement.

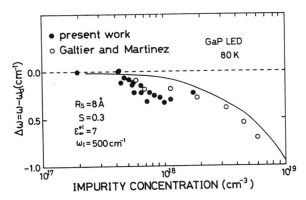

Fig.13 The frequency shift with respect to the pure LO phonon band is plotted against the impurity concentration. The data taken from Ref.[15] are shown by open circles.

RAMAN IMAGE MEASUREMENTS

The measurements of Raman images are useful not only for checking the inhomogeneity of the physical quantity but for process characterization in device fabrication. In this section we will describe several examples of one- and two-dimensional Raman images for semiconductor materials.

Atomic Composition of GaAlAs Diodes

The spatial resolution in the lateral direction attained in the Raman microprobe is comparable to the wavelength of laser light used. The resolution in depth is not good compared with that in the lateral direction for transparent materials. The depth resolution could be improved by placing a diaphram at the focal plane of the scattered light and rejecting signals from undesired regions. However, the resolution obtained thereby ($\sim$10μm) is not enough to examine inhomogeneous distribution of quantities in thin epitaxial films.

For opaque materials the penetration depth of probe light dominates the spatial resolution. The depth profiles of stress and ion-implantation damage were obtained by using laser lines with different wavelengths and deconvoluting the observed Raman spectra[21]. However, the Raman measurement with various laser lines is more or less tedious. Instead, Raman microprobe measurements of angle-lapped specimens are used to obtain depth profiles[22-25].

Using a scanning Raman microscope, Hattori et al. observed the Raman spectral profile of AlGaAs laser diodes having a double hetero structure[26]. The laser diodes were ball-lapped and then slightly stain-etched. Raman spectra of the beveled surface were measured by successively translating the sample so that the laser spot crosses the double hetero layers. A one-dimensional image of the Raman spectra taken at intervals of 0.4μm is shown in Fig.14. The spatial variation of the atomic composition is clearly seen in this figure. In the active layer the GaAs-like LO mode is intense and the AlAs-like LO mode is weak. The peak frequency of GaAs-like modes is shifted to the higher frequency side compared with that of other layers. The

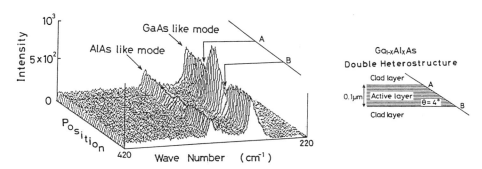

Fig.14 One-dimensional Raman spectral profile of a GaAlAs laser diode measured by Hattori et al[26]. The beveled surface is measured at intervals of 0.4μm. The spatial resolution in depth for this method is 0.1μm.

aluminum concentration in the clad and active layers was determined
from the data. These results indicate that the compositional analysis
is possible with a depth resolution better than 0.1μm in laser diodes.

Thermal Conversion of SiC Polytypes

It is well known that the SiC crystal has a number of polytypes
whose stacking sequences differ along the c-direction. Raman spectra
have been used to identify the polytype structure[27]. While at
elevated temperatures (∿2000°C), α-SiC (hexagonal or rhombohedral
structure) is stable, at lower temperatures β-type or 3C-type
(zincblende structure) is stable. Hence, thermal annealing of β-SiC
at high temperatures causes transformation into α-SiC. This polytype
conversion has been studied by the Raman microprobe measurement.
Figure 15 shows variation of the Raman spectrum of β-SiC annealed at
2080 and 2200°C, which was measured at intervals of 50μm. For short
thermal annealing time, most parts consist of β-SiC which shows only
TO band at 798cm^{-1}. For longer annealing time, Raman spectra
corresponding to the 6H-polytype are observed at a number of places.
The Raman intensity of the 6H component increases with increasing
annealing time although the intensity ratio of the 3C to 6H component
differs from place to place. The conversion into the 15R polytype at
some places is recognized from the spectra. In Fig.15 we compare the
spectra of β-SiC samples annealed at 2200°C and 2080°C. The spatial
variation of the 6H component is smaller for higher temperature
annealing. The width of the 6H-bands in the annealed specimens is
large compared with that of the pure 6H polytype. The result
mentioned above leads to the conclusion that the thermal polytype
conversion occurs randomly, and at the initial stage of the

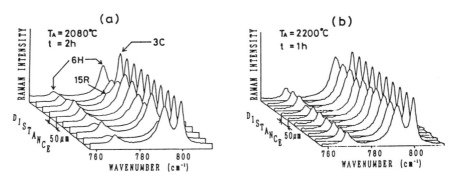

Fig.15   The Raman spectral distribution of β-SiC heated (a) at
         2080°C for 2 hour and (b) heated at 2200°C for 1 hour.

conversion, small domains with the 6H structure occur and grow up as
the annealing time is increased.

Recovery of Crystallinity of Ion Implanted Si by Flash
Lamp Annealing

Ion implantation is an important technology in making p-n
junctions in shallow surface regions. The implantation of high energy
ions produces highly disordered regions in surface layers of
semiconductors. At high dose of ions these individual disordered

regions begin to overlap and are eventually amorphized. The crystal quality of damaged or amorphized semiconductors can be recovered by annealing processes which follows the ion implantation: thermal, laser, flash lamp, and electron beam annealings. The Raman spectroscopy is a useful technique available to characterize damaged surface layers and subsequently annealed layers. As the damage is increased, the intensity of the crystal Raman bands is decreased and the amorphous bands grow up. This implies that the Raman intensity depends on the degree of the damage. Morhange et al. have proposed using a normalized intensity, $I_N = (I_c - I_i)/I_c$ to evaluate the disorder quantitatively, where I_c is the Raman intensity of the unimplanted crystalline region and I_i is that of the implanted region[28]. The normalized intensity is zero for defect free crystals and is unity for completely amorphized crystals. The degree of the damage is related to the dose of implanted ions. The precise determination of ion implantation damage or dose requires the precise measurement of Raman intensities. The Raman microprobe is convenient to obtain accurate normalized intensity because the measurement under the same condition is possible for different samples as mentioned before. Using Si crystals with alternatively aligned stripes of implanted and unimplanted stripes which are prepared by means of focused ion beams (FIB), Mizoguchi et al. measured the integral intensity of the $520cm^{-1}$ component of both regions by translating samples. The relation between the normalized intensity and the ion dose was determined for Au^{++}, Si^{++} and Be^{++} ions[29].

The variation of the damage with depth can be examined by measuring the normalized Raman intensity for different wavelengths of the exciting laser. Figure 16 shows the normalized intensity for three different laser wavelengths. For the Be^{++} ion, the normalized intensity decreases as the penetration depth of the light decreases at dose levels below $5 \times 10^{16} cm^2$. This result indicates that the crystalline layer remains near the surface region and that the damaged layer lies in deeper region around the projected range. For Au^{++} ions, on the other hand, the normalized intensity is almost the same for different wavelengths. From this result it is evident that the damaged or amorphous region is located at shallow region of the surface.

The kinetics of crystallinity recovery of ion implanted Si has been studied by Raman image measurements. Figure 17 shows Raman intensity profiles of a Si crystal which was amorphized by P^+ ion implantation and subsequently annealed by flash lamp irradiation for 10 sec at 600°C. In this map, Raman intensity is weak in most parts but there are regions which show strong Raman intensity (Region A). Raman spectra of the regions A and B are shown in this figure. The broad amorphous band peaked at $480cm^{-1}$ appears clearly at the region B, but it is weak at the region A. This result implies that the recovery of the crystallinity at the initial stage of this annealing process is not uniform spatially and that it occurs at local areas and extends as the annealing time is increased.

ACKNOWLEDGMENTS

The author would like to acknowledge many participants in our laboratory at Osaka University. He is indebted to D. Hattori for providing his unpublished data.

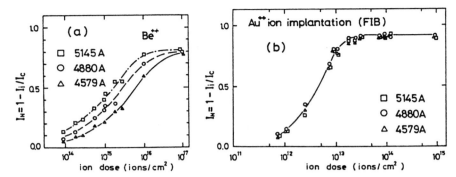

Fig.16 Normalized intensity of the silicon crystalline band as a function of ion dose for three different laser wavelengths. The implanted ion is (a) Be^{++} ion[29] and (b) Au^{++} ion.

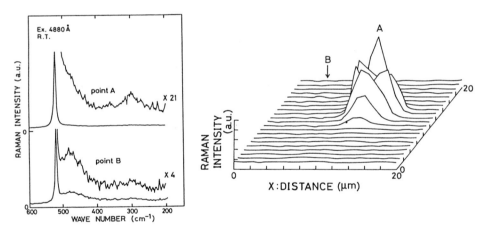

Fig.17 Two-dimensional Raman intensity image of ion implanted and subsequently annealed by a flash lamp for ten seconds. The Raman spectra at the points A and B are also shown.

REFERENCES

1. S. Nakashima and M. Hangyo, IEEE. J. Quantum Electron 25, 965 (1989).
2. J. C. Tsang, S. S. Iyer, P. Pukite and M. Copel, Phys. Rev. B39, 13545 (1989).
3. M. Delhaye and M. Dhamelincourt, J. Raman Spectroscopy 3, 33 (1975).
4. M. Bowden, D. J. Gardiner, G. Rice and D. L. Gerrard, J. Raman Spectroscopy 21, 37 (1990). 5. J. Laane and W. Kiefer, J. Chem. Phys. 72, 5305 (1980).
6. P. J. Treado and M. D. Morris, Applied Spectroscopy 43, 190 (1989).
7. J. B. Hopkins and L. A. Farrow, J. Appl. Phys. 59, 1103 (1986).
8. K. Mizoguchi and S. Nakashima, J. Appl. Phys. 65, 2583 (1989).
9. K. Sugawara, S. Kusunoki, Y. Inoue, T. Nishimura and Y. Akasaka, J. Appl. Phys. 62, 4178 (1987).
10. M. Maekawa and M. Koba, Proc. 5th Int. Workshop on Future Electron Devices, Niyagi-Zao, Japan (1988).
11. M. V. Klein, "Light Scattering in Solid" vol.1 Ed. by M. Cardona, Springer-Verlag, Berlin (1983) p.147.
12. G. Irmer, V. V. Toporov, B. H. Bairamov and J. Monecke, Phys. Stat. Sol. (b) 119, 595 (1983).
13. H. Yugami, S. Nakashima, A. Mitsuishi, A. Uemoto, M. Shigeta, K. Furukawa, A. Suzuki and S. Nakajima, J. Appl. Phys. 61, 354 (1987).
14. S. Nakashima, H. Yugami, A. Fujii, M. Hangyo and H. Yamanaka, J. Appl. Phys. 64, 3067 (1988).
15. P. Galtier and G. Martinez, Phys. Rev. B38, 10542 (1988).
16. See J. Monecke, W. Cordts, G. Irmer, B. H. Bairamov and V. V. Toporov. The references related to this subject are cited in this paper.
17. A. S. Barker, Jr., Phys. Rev. B7, 2507 (1973).
18. E. P. Pokatilov and S. I. Beril, Phys. Stat. Sol. (b) 119, K75 (1982).
19. P. E. Camley and D. L. Mills, Phys. Rev. B29, 1695 (1984).
20. L. Wendler and R. Pechstedt, Phys. Stat. Sol. (b) 141, 129 (1987).
21. H. Shen and F. H. Pollak, Appl. Phys. Lett. 45, 692 (1984).
22. G. Abstreiter, E. Bauser, A. Fischer and K. Ploog, Appl. Phys. 16, 345 (1978).
23. Y. Inoue, S. Nakashima, A. Mitsuishi, T. Nishimura and Y. Akasaka, Jpn. J. Appl. Phys. 25, 798 (1986).
24. Y. Huang, P. Y. Yu, M.-N. Charasse, Y. Lo and S. Wang, Appl. Phys. Lett. 51, 192 (1987).
25. K. Kakimoto and T. Katoda, Appl. Phys. Lett. 40, 826 (1982).
26. R. Hattori, K. Yamashita, Y. Ohta. S. Takamiya and S. Mitsui, presented at the Spring Meeting of Japan Society of Applied Physics (1986).
27. S. Nakashima and K. Tahara, Phys. Rev. B40, 6339 (1989).
28. J. F. Morhange, R. Beserman and M. Balkanski, Phys. Stat. Sol. (a) 23, 383 (1974).
29. K. Mizoguchi, S. Nakashima, A. Fujii, A. Mitsuishi, H. Morimoto, H. Onodera and T. Kato, Jpn. J. Appl. Phys. 26, 903 (1987).

SURFACE MODES IN MAGNETIC SEMICONDUCTOR FILMS AND MULTILAYERS

M. G. Cottam and Sudha Gopalan

Department of Physics
The University of Western Ontario
London, Ont., Canada N6A 3K7

1. INTRODUCTION

In this paper some calculations are described for surface and bulk magnetic excitations in magnetic semiconductors. We consider semi-infinite structures, thin films, multilayers, and superlattices. Results have been deduced for the excitation frequencies and, in some cases, for the magnetic Green's functions. Applications to light scattering are discussed. Some of the magnetic semiconductors of interest include europium chalcogenides (such as EuO and EuTe), spinels (such as $CdCr_2S_4$ and $CdCr_2Se_4$), other semiconductor materials with large single-ion anisotropy (such as $CrBr_3$), and diluted magnetic semiconductors (DMS) in their ordered phase (such as $Cd_{1-x}Mn_xTe$ with Mn concentration $x > 0.17$). Other magnetic semiconductors were listed, for example, in the book by Nagaev.[1] The calculations are presented here only in outline; details will be given elsewhere.

Part of this work is concerned with magnetostatic-type modes in ordered ferromagnets and antiferromagnets. In this regime the wave vectors of the excitations are taken to be sufficiently small that exchange effects can be neglected compared with the magnetic dipole-dipole terms, and this situation is often realized in inelastic light-scattering experiments (e.g., see Grünberg[2] and Cottam and Lockwood[3]). We include discussion of single-film and double-layer magnetic semiconductor systems and of semi-infinite superlattices of the structure ABABABAB... where one of the components (A or B) is a magnetic semiconductor. Examples of different geometries, with the magnetic ordering direction parallel or perpendicular to the surface, are considered. Some applications require the inclusion of a single-ion anisotropy that is non-uniaxial with respect to the direction of net magnetic ordering.

Another part of this work is devoted to a microscopic calculation of the surface excitations and spin correlations in magnetic semiconductors, which can be described in this approach by the s-f (or s-d) interaction model.[4] Here the localized spins are described by a Heisenberg Hamiltonian and the conduction electrons by a hopping Hamiltonian, and the coupling between these localized and itinerant spins is provided by a contact-type s-f (or s-d) interaction. A formalism is established to obtain the magnetic Green's functions for surface and bulk magnetic excitations, and to calculate their dispersion relations, in a semi-infinite ferromagnetic semiconductor.

Light Scattering in Semiconductor Structures and Superlattices   311
Edited by D.J. Lockwood and J.F. Young, Plenum Press, New York, 1991

2. SINGLE FILMS AND DOUBLE-LAYER MAGNETIC STRUCTURES

In recent years a considerable number of Brillouin scattering measurements have been reported for surface and bulk magnetostatic-type excitations in single thin films and in double systems where two coupled magnetic films are separated by a nonmagnetic spacer layer (e.g., see reviews by Cottam and Lockwood[3] and Grünberg[2,5]). Most of these studies have been for metallic systems, but results for Eu chalcogenide magnetic semiconductors have also been reported. For the excitation wave vectors typically encountered in inelastic light scattering ($k \sim 10^7$ m^{-1}) the magnetostatic approximation of neglecting short-range exchange interactions compared with the longer range magnetic dipole-dipole interactions gives a good description for many materials.[2,3,5]

Here we describe a calculation for magnetostatic modes in a ferro-magnetic semiconductor film where there is a single-ion anisotropy in the direction perpendicular to the film surface. This would be the case for a CrBr$_3$ film with the c-axis perpendicular to the film surface, as in the Brillouin scattering measurements of Sandercock[6] for bulk modes. For applications to surface modes in this material we consider the Voigt geometry, where the film normal is along the x-axis, the static magnetic field H_o is along the z-axis, and the propagation takes place along the y-axis (with in-plane wave vector k_y). As usual the starting point is the magnetostatic form of Maxwell's equations:

$$\nabla \cdot \mathbf{b} = 0, \qquad \nabla \times \mathbf{h} = 0 \qquad (2.1)$$

where $\mathbf{b}$ and $\mathbf{h}$ are the fluctuating parts of the total $\mathbf{B}$ and $\mathbf{H}$ fields and are related through $\mathbf{b} = \mathbf{h} + 4\pi\mathbf{m}$. Also we have the constitutive relation that the fluctuating magnetization $\mathbf{m} = \chi\mathbf{h}$ where the susceptibility tensor χ takes the form

$$\chi = \begin{pmatrix} \chi_1 & i\chi_2 & 0 \\ -i\chi_2 & \chi_3 & 0 \\ 0 & 0 & 0 \end{pmatrix} \qquad (2.2)$$

with $\chi_1 \neq \chi_3$ due to the single-ion anisotropy along the x direction. Expressions for χ_1, χ_2, and χ_3 can be derived by a straightforward gen-eralization of the well-known[3] result in the absence of the anisotropy. For example, if the static applied field and the anisotropy are such that $H_o > H_a - 4\pi M_o > 0$ (where H_a is an effective anisotropy field and M_o is the static magnetization) the static demagnetization field forces M_o to lie along the z direction, parallel to H_o. The susceptibility components at frequency ω are then of the form[7]

$$4\pi\chi_1 = \omega_o\omega_m / [\omega_o(\omega_o - \omega_a) - \omega^2] \qquad (2.3)$$

$$4\pi\chi_2 = \omega \omega_m / [\omega_o(\omega_o - \omega_a) - \omega^2] \qquad (2.4)$$

$$4\pi\chi_3 = (\omega_o - \omega_a)\omega_m / [\omega_o(\omega_o - \omega_a) - \omega^2] \qquad (2.5)$$

where we have defined the characteristic frequencies $\omega_o = \gamma H_o$, $\omega_m = 4\pi\gamma M_o$, and $\omega_a = \gamma(H_a - 4\pi M_o)$ with γ denoting the gyromagnetic ratio. For smaller H_o (corresponding to $\omega_o < \omega_a$) the static magnetization is in the xz-plane, and the results are easily generalized.

Following from Eq. (2.1) the h field can be derived from a scalar potential defined by $\mathbf{h} = \nabla\phi$, and utilizing the translational invariance property in the yz-plane we can write $\phi = \phi(x) \exp(ik_y y - i\omega t)$ for a Fourier component at frequency ω. It follows straightforwardly from the above results and from the torque equation for the magnetization that

$$\phi(x) \;=\; \begin{cases} A \exp\,(-k_y x) & (x > 0) \\ B \exp\,(ik_x x) \;+\; C \exp\,(-ik_x x) & (0 > x > -L) \qquad (2.6) \\ D \exp\,(k_y x) & (x < -L) \end{cases}$$

where the film surfaces are chosen to be at $x = 0$ and $x = -L$. The in-plane wave vector k_y has been taken as positive, and so we have decaying solutions outside the film. The quantity k_x inside the slab is complex in general and satisfies

$$k_x^2 \;=\; -k_y^2 \left(\frac{\omega^2 - \Omega_b^2}{\omega^2 - \Omega_B^2} \right) \qquad\qquad (2.7)$$

with frequencies Ω_b and Ω_B defined by

$$\Omega_b \;=\; [r\omega_o(\omega_o + \omega_m)], \qquad \Omega_B \;=\; [\omega_o(r\omega_o + \omega_m)] \qquad (2.8)$$

where $r = 1 - (\omega_a/\omega_o)$ in the present application to CrBr$_3$. Also, since $0 < r < 1$, we see that the wave vector k_x is real (corresponding to bulk modes) if $\Omega_b < \omega < \Omega_B$ and that any surface magnetostatic modes must occur outside that range when k_x is imaginary ($k_x = i\beta$ with β real). The bulk-mode frequencies are

$$\omega(k_x, k_y) \;=\; [(\Omega_B^2 k_x^2 + \Omega_b^2 k_y^2)/(k_x^2 + k_y^2)] \qquad (2.9)$$

This is a monotonically increasing function of k_x (for a fixed k_y). We note that the range of bulk frequencies shrinks to the usual value, which is independent of wave vector in the Voigt geometry, as $r \to 1$ (the case where the z direction is a uniaxis).

The application of the boundary conditions at $x = 0$ and $x = -L$ enables the surface-mode frequencies to be deduced. The appropriate boundary conditions here are that ϕ and b_x must be continuous across each interface. The full form of the surface mode dispersion relation is rather complicated and will be described elsewhere. However, a satisfactory approximation for most values of the parameters is that ω for a surface mode must satisfy

$$\omega^4 \;+\; \omega^2\,[2r\omega_o^2 + \omega_o\omega_m + \omega_m^2\{1 - \exp\,(-2\beta L)\}]$$

$$+\; \omega_o^2\,[r^2\omega_o^2 + r\omega_o\omega_m + \omega_m^2\{1 - \exp\,(-2\beta L)\}] \;=\; 0 \qquad (2.10)$$

where $\omega^2 > \Omega^2$ and the frequency-dependent β (>0) is found by replacing k_x with $i\beta$ in Eq. (2.7). Note that, in the limit of $r \to 1$, we have $\beta = k_y$, and the above equation factorizes to give $\omega = \pm\omega_s(k_y)$ as the only physical solution, where

$$\omega_s(k_y) \;=\; [(\omega_o + \omega_m)^2 - \omega_m^2 \exp\,(-2kL)] \qquad (2.11)$$

This is just the well-known dispersion relation for surface magnetostatic modes in the Voigt geometry for the case of a uniaxial material (see the reviews by Wolfram and DeWames[2] and Grünberg.[8] For the anisotropic type of material considered here, the solutions of Eq. (2.10) can be studied numerically. An example is shown in Fig. 1, taking approximate parameter values appropriate to CrBr$_3$: $H_o = 5$ kG, $4\pi M_o = 3.4$ kG, and $H_a = 6.0$ kG, implying that $r = 0.48$.

The above calculation of the mode frequencies can readily be extended to obtain the magnetic Green's functions, and hence the light-scattering intensities, for a magnetic film with non-uniaxial anisotropy effects. The method is analogous to previous theories for the uniaxial case[9,10] but with the generalization to $\chi_1 \neq \chi_3$ in the susceptibility tensor. The previous calculations of the magnetic Green's functions and the light-scattering intensities have also been extended, in the case of uniaxial anisotropy, to double-layer magnetic systems of the type investigated by Grünberg[2,5] where the two magnetic films are separated by a thin spacer layer of a non-magnetic material. Of interest here are the predictions for the Stokes/anti-Stokes intensity ratio (which is influenced by the nonreciprocal propagation of the surface modes) and an enhancement of the integrated intensities for particular choices of the spacer layer material (depending on its thickness and optical parameters). Details will be given elsewhere.

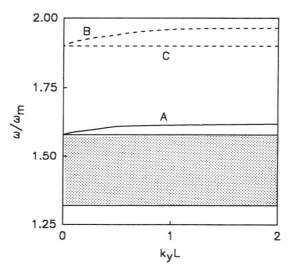

Fig. 1. The frequencies of magnetostatic-type modes plotted against $k_y L$ for the case of a ferromagnetic film with a single-ion anisotropy that is non-uniaxial with respect to the direction of the static magnet-ization. Parameters appropriate to $CrBr_3$ have been used (see the text). The bulk-mode region is shown shaded, and the line corre-sponding to the surface mode is labeled A. For comparison, the surface and bulk modes in the case of a uniaxial material (with the same values of H_0 and M_0 but with r = 1) are shown by the dashed lines B and C, respectively.

3. MAGNETIC SUPERLATTICES

The theory of magnetostatic excitations in superlattices of magnetic semiconductors has recently been discussed by Villeret et al.[11] Their work was mainly carried out in the context of light scattering from super-lattices involving DMS materials. References to the experimental work in this field are also given in the above paper.

Villeret et al.[11] gave a detailed derivation of the different types of excitations that may occur in infinitely extended superlattices. By contrast we consider here some examples relating to surface excitations in semi-infinite superlattices, generalizing some calculations due to Camley and Cottam.[12] For this we consider a semi-infinite magnetic superlattice in which the static applied field and the magnetic ordering direction are perpendicular to the surfaces of the layers. For simplicity we apply the theory only to uniaxial materials. In the case of a ferromagnet the susceptibility tensor corresponds to putting $\omega_a = 0$ in Eqs. (2.2)-(2.5) and replacing ω_0 by $\omega_1 = \gamma(H_0 + H_A - 4\pi M)$, which relates to the internal field for this geometry[1] (H_A denotes the uniaxial anisotropy field. For an anti-ferromagnet, on taking the special case of $H_0 = 0$, one has[13]

$$4\pi\chi_1 = 4\pi\chi_3 = \frac{2\omega_A \omega_m}{\omega_A(2\omega_E + \omega_A) - \omega^2} \, , \qquad 4\pi\chi_2 = 0 \qquad (3.1)$$

where ω_A and ω_E denote the effective anisotropy and exchange frequencies for the antiferromagnet, respectively. The denominator of χ_1 and χ_3 vanishes when $\pm\omega$ is equal to the antiferromagnetic resonance frequency. When $H_0 \neq 0$, χ_2 is nonzero because its contributions from the two sub-lattices of the antiferromagnet no longer cancel.

As in Camley and Cottam[12] we consider a semi-infinite superlattice composed of alternating layers of a magnetic material and a nonmagnetic spacer. The thickness of the magnetic layers is denoted by d_1 (except for the surface layer, which may have a different value L), and the thickness of the spacer layers is d_2. Hence the unit cell thickness of the super-lattice is $(d_1 + d_2)$, except at the surface. The calculation of the bulk and surface modes of the superlattice proceeds in a similar fashion to that described in the previous section, with the main difference being that the direction of magnetic ordering (the z direction) is now perpendicular to the surface and interfaces. It is still convenient to solve the problem in terms of a scalar potential ϕ defined as before, and the boundary conditions (which must be applied at the surface and at each interface) take the form that ϕ and $\partial\phi/\partial z$ must be continuous. We now quote the appropriate form of the results for the mode frequencies, using expressions deduced from Camley and Cottam.[12] The bulk superlattice modes are given by

$$\cos[Q(d_1 + d_2)] = \cosh[\alpha_0(\omega)d_1]\cosh[kd_2]$$
$$+ [\alpha_0(\omega)/k + k/\alpha_0(\omega)]\sinh[\alpha_0(\omega)d_1]\sinh[kd_2] \qquad (3.2)$$

where Q is a real wave-vector component describing the modulation of the envelope function of ϕ from layer to layer, k is the wave vector describing propagation parallel to the surface, and

$$\alpha_0(\omega) = k(1 + 4\pi\chi_1) \qquad (3.3)$$

For the bulk-like modes of the superlattice the coefficient α_0 is imaginary ($4\pi\chi_1 < -1$), and Eqs. (3.2) and (3.3) can be solved to obtain ω for a given k and Q. The surface modes of the superlattice correspond to real values of α_0, and we may solve for these using the following equation, which is a consistency condition for attenuation of the envelope function of the potential ϕ:

$$\exp(-2kd_2)\sin(\alpha_0 L)\,[(k^2 - \alpha_0^2)\sin(\alpha_0\Delta) + 2k\alpha_0\cos(\alpha_0\Delta)]$$
$$= \sin(\alpha\Delta)\,[(k^2 - \alpha_0^2)\sin(\alpha_0 L) + 2k\alpha_0\cos(\alpha_0 L)] \qquad (3.4)$$

315

where Δ denotes L-d_1. The surface mode frequencies are given in the present case by

$$\omega = \{ \omega_i + [\omega_i \omega_m k^2/(k^2 + \alpha_o^2)] \} \tag{3.5}$$

for a ferromagnet and by

$$\omega = \{ \omega_A (2\omega_E + \omega_A) + [2\omega_A \omega_m k^2/(k^2 + \alpha_o^2)] \} \tag{3.6}$$

for an antiferromagnet with $H_o = 0$.

The general features of the predicted bulk and surface modes for the semi-infinite superlattice in this geometry have been described by Camley and Cottam,[12] and there is a similar behavior found in the present applications. When the in-plane wave vector k is zero, there is a single band of bulk modes (both for the ferromagnetic case and for the $H_o = 0$ antiferromagnetic case). When k is nonzero the bulk modes break up into several branches, and for appropriate values of the surface-layer thickness L, surface modes can occur with frequencies above that of the bulk bands and within the gap regions between bulk bands. As an example of a ferromagnetic semiconductor we consider EuS with parameters M = 1.0 kG, H = 13.5 kG, $d_2 = d_1$, and L = 1.5 d_1. For a value of the in-plane wave vector giving $kd_2 = 1$ we find that the principal bulk-mode bands correspond to dimensionless ω/ω_o values in the ranges 0.093 to 0.100 and 0.144 to 0.200, while the discrete surface modes correspond to values 0.089, 0.133, and 0.211. For an example of an antiferromagnetic semiconductor we consider the case of $Cd_{0.3}Mn_{0.7}Te$ with parameters $H_E = 200$ kG, $H_A = 30$ kG, and M = 30 kG chosen as in reference 11. Also taking $d_2 = 2d_1$, L = $3d_1$, and $kd_2 = 1$, we find that the principal bulk bands correspond to ranges in ω/ω_R of 1.0003 to 1.0004 and 1.0018 to 1.0025 (where ω_R denotes the antiferromagnetic resonance frequency), and there are surface modes corresponding to the values 1.0002, 1.0009, and 1.0041. The modes are very close together in frequency in the latter example (compared with the ferromagnetic case) because M_o is very small compared with H_E.

4. THE s-f INTERACTION MODEL

In this section we present calculations for surface spin waves in a degenerate ferromagnetic semiconductor described by the s-f (or s-d) interaction model. The model and its application to bulk systems have been reviewed, for example, by Nagaev,[1] Vonsovskii,[4] and Krisement.[14] The full Hamiltonian $\mathcal{H}$ of the system can be expressed as the sum of three terms: a Hamiltonian $\mathcal{H}_E$ representing the conduction (s) electrons in the system, a Heisenberg exchange Hamiltonian $\mathcal{H}_M$ for the localized spins (of f or d electrons), and a s-f (or s-d) interaction term $\mathcal{H}_I$, with

$$\mathcal{H}_E = \sum_{i,j,\sigma} t_{ij} a_{i\sigma}^+ a_{j\sigma} \tag{4.1}$$

$$\mathcal{H}_M = - \sum_{i,j} J_{ij} S_i \cdot S_j - g\mu_B H_o \sum_i S_i^z \tag{4.2}$$

$$\mathcal{H}_I = - \sum_i I_i (S_i^+ a_{i-}^+ a_{i+} + S_i^- a_{i+}^+ a_{i-} - S_i^z a_{i-}^+ a_{i-} + S_i^z a_{i+}^+ a_{i+}) \tag{4.3}$$

where the $a_{i\sigma}^+$ and $a_{i\sigma}$ operators are creation and destruction operators for the electron at a particular site i and with a spin index σ ($\sigma = \pm1$ corresponding to up and down projections of the electron spins), and t_{ij} denotes a hopping term. Also S_i and S_j are spin operators for the localized spins at sites i and j, J_{ij} is the isotropic Heisenberg exchange interaction, and H_o is a static applied magnetic field in the z direction. Finally, I_i is a contact interaction energy at site i. We are neglecting here the direct Coulomb interaction between the itinerant electrons (as would be included in the Hubbard Hamiltonian) compared with the interaction via coupling to the localized spins.

There have been numerous calculations applying the above model to bulk (infinitely extended) ferromagnetic semiconductors. In addition to the usual "acoustic" spin-wave branch (e.g., as described by Woolsey and White[15]), it was shown by Babcenco and Cottam[16] that there is a higher-frequency (or "optical") spin-wave branch and the Stoner-like continuum of magnetic excitations. Estimates were made for the frequencies and relative intensities for light scattering from the two spin-wave branches. Here we outline the results of some calculations for the influence of a surface on the spin-wave excitation spectrum and the consequences regarding light scattering. We consider such a ferromagnetic semiconductor occupying the half-space $z \geq 0$ and with a free surface at $z = 0$. We further assume that the localized magnetic ions form a simple cubic structure (with a as the lattice parameter) and with the crystallographic axes parallel to the x-, y- and z-axes.

First we examine the special case of a narrow-band ferromagnetic semiconductor where the width W of the conduction band is much smaller than IS, I being the contact interaction energy and S the spin of the localized electrons. This case includes some chromium spinel materials such as $CdCr_2S_4$ (see Nagaev[1]). The width W is directly proportional to parameter t defined in Eq. (4.1), and hence in the limit $W \ll IS$ the Hamiltonian $\mathcal{H}$ becomes negligible compared with $(\mathcal{H}_M + \mathcal{H}_E)$. We introduce the frequency Fourier transform $\langle\langle S_i^+ ; S_j^- \rangle\rangle_\omega$ of the two-time retarded commutator Green's function for the localized spin operators. From its standard equation of motion, evaluated in the random phase approximation at $T \ll T_c$, we obtain

$$\left\{\omega - g\mu_B H_o - S\sum_\ell J_{\ell i} - I_i\rho_i - [I_i^2\rho_i S/(\omega - I_iS)]\right\} \langle\langle S_i^+ ; S_j^- \rangle\rangle_\omega$$

$$+ S\sum_\ell J_{\ell i} \langle\langle S_\ell^+ ; S_j^- \rangle\rangle_\omega = (S/\pi)\delta_{ij} \qquad (4.4)$$

where $\rho_i = (\langle a_{i+}^+ a_{i+} \rangle - \langle a_{i-}^+ a_{i-} \rangle)/2$ is a polarization factor for the conduction electrons. We now use the full translational invariance in the xy-plane to define the Fourier transform

$$\langle\langle S_i^+ ; S_j^- \rangle\rangle_\omega = \frac{1}{N}\sum_{k_p} \exp[ik_p \cdot (r_i - r_j)] G_{m,n}(k_p, \omega) \qquad (4.5)$$

where $k_p = (k_x, k_y)$ denotes a two-dimensional wave vector parallel to the surface, and m and n are labels of the lattice planes (parallel to the surface) that include the sites i and j, respectively.

We simplify Eq. (4.4) by considering nearest-neighbor exchange interactions and taking J_{ij} to have the value J_s only if both i and j are in the surface layer ($n = 1$) and otherwise to have the bulk value J. In addition we assume the contact interaction energy I_i to take the perturbed value I_s at the surface, but otherwise it has the bulk value I. Equation (4.4) can then be written, after using Eq. (4.5), in terms of infinite-dimensional matrices as

$$(A_o + D) \, G(k_p, \omega) \quad = \quad (-1/\pi J) I_o \tag{4.6}$$

where G is the Green's function matrix and I_o is the unit matrix. Also A_o is a tridiagonal matrix

$$A_o \quad = \quad \begin{pmatrix} d & -1 & 0 & \cdots & & \\ -1 & d & -1 & 0 & \cdots & \\ 0 & -1 & d & -1 & 0 & \cdots \\ & & \cdots & & & \end{pmatrix} \tag{4.7}$$

where

$$d \quad = \quad 2[3 - 2\gamma(k_p)] + [I^2 \rho/J(\omega - IS)] + [(g\mu_B H_o - \omega + I\rho)/SJ] \tag{4.8}$$

$$\gamma(k_p) \quad = \quad [\cos(k_x a) + \cos(k_y a)]/2 \tag{4.9}$$

The factor ρ is the averaged polarization factor over each layer, and we assume this to be independent of the layer number. Finally, the only non-zero matrix element of D is $D_{1,1} = \delta$, where

$$\delta \quad = \quad -1 - 4\left(1 - \frac{J_s}{J}\right)[1 - \gamma(k_p)] - \frac{\rho I}{SJ}\left(1 - \frac{I_s}{I}\right)$$

$$- [I^2 \rho/J(\omega - IS)] + [I_s^2 \rho/J(\omega - I_s S)] \tag{4.10}$$

We can now write the formal solution of Eq. (4.6) as

$$G(k_p, \omega) \quad = \quad (-1/\pi J) \, (I_o + A_o^{-1} D) \, A_o^{-1} \tag{4.11}$$

The inverse of A_o is known analytically, and the Green's function may be explicitly evaluated using a method that is formally similar to that employed for the semi-infinite Heisenberg ferromagnet.[17] The spin-wave frequencies can be obtained directly from the poles of G, and for this purpose we introduce a complex quantity ζ, defined to satisfy

$$\zeta + \zeta^{-1} \quad = \quad d, \qquad (|\zeta| \leq 1) \tag{4.12}$$

By analogy with the results of Cottam[17] the bulk modes correspond to $|\zeta| = 1$. Writing $\zeta = \exp(ik_z a)$, where k_z is the wave-vector component perpendicular to the surface, and using Eq. (4.8) we find that the bulk spin-wave frequencies are the solutions of

$$\omega - g\mu_B H_o - 2SJ[3 - \cos(k_x a) - \cos(k_y a) - \cos(k_z a)] - I\rho$$

$$- [I^2 \rho S/(\omega - IS)] \quad = \quad 0 \tag{4.13}$$

We note that this is consistent, as expected, with the result for the bulk spin-wave frequencies in an infinite ferromagnetic semiconductor[16] when we take the narrow-band case of $W \ll IS$. There are two branches to the bulk spectrum, and if $I \gg J$ (which is typically the case) they are well separated in frequency.

The surface spin waves correspond to real values of ζ with $|\zeta| < 1$. From the pole of $G(k_p, \omega)$ we have $\zeta = -1/\delta$ (see reference 17) where δ is given here by Eq. (4.10). This eventually leads to the following expression for the surface spin-wave frequencies

$$\omega - g\mu_B H_o - 2SJ[3 - 2\gamma(\mathbf{k}_p)] - SJ(\delta + \delta^{-1}) - I\rho$$

$$- [I^2\rho S/(\omega - IS)] = 0 \qquad (4.14)$$

provided the existence condition $|\delta| > 1$ is satisfied. We notice from Eq. (4.10) that δ is ω dependent in the general case of $I_s \neq I$, and Eqs. (4.10) and (4.14) then have to be solved self-consistently in order to obtain the spin-wave solutions. Some numerical examples to illustrate the results for the bulk and surface spin waves are shown in Fig. 2, where ω/SJ is plotted against $k_x a$ (taking $k_y = 0$) for various combinations of the ratios J_s/J and I_s/I. The bulk spin waves appear here as a continuum, with the upper and lower edges corresponding to $k_z = \pi/a$ and 0, respectively.

Next we consider the opposite limiting case of wide-band ferromagnetic semiconductors (such as EuO or EuS) where $W \gg IS$. We assume the same model as described before for J_{ij} and I_i, and we further take the hopping term t_{ij} to have the value t_s if both i and j are in the surface layer and the bulk value t otherwise. The matrix Eq. (4.6) now becomes generalized to

$$(\mathbf{A}_o + \mathbf{D}) \; \mathbf{G}(\mathbf{k}_p, \omega) = (-1/\pi J)\mathbf{I}_o - (I/J)\mathbf{R} \; \mathbf{G}'(\mathbf{k}, \omega) \qquad (4.15)$$

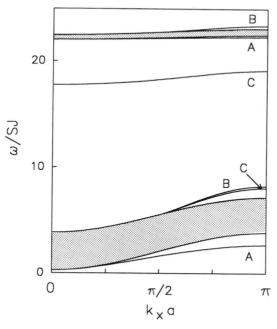

Fig. 2. The spin-wave frequencies plotted against $k_x a$ for a semi-infinite narrow-band semiconductor, taking parameters $I/J = 20$, $\rho/S = 0.1$, and an applied field strength of $g\mu_B H_o/SJ = 0.3$. The upper and lower bulk spin-wave regions are shown shaded, and the labeling of the surface spin-wave branches corresponds to
A, $J_s/J = 0.5$, $I_s/I = 1.0$;   B, $J_s/J = 2.0$, $I_s/I = 1.0$;
C, $J_s/J = 2.0$, $I_s/I = 0.8$.

where the quantities d and δ entering into the definitions of A_o and D are redefined as

$$d = 2[3 - 2\gamma(k_p)] + [(g\mu_B H_o - \omega + I\rho)/SJ] \qquad (4.16)$$

$$\delta = -1 - 4\left(1 - \frac{J_s}{J}\right)[1 - \gamma(k_p)] - \frac{\rho I}{SJ}\left(1 - \frac{I_s}{I}\right) \qquad (4.17)$$

Matrix R in Eq. (4.15) is a unit matrix except that the (1,1) element is replaced by I_s/I, and G' is a matrix whose elements are the wave-vector Fourier transform (defined as in Eq. (4.5)) of the Green's functions $<<a_i^+ ; S^->>_\omega$, where $a_i^+ = a^+ a$ is a spin operator describing the conduction electrons. The evaluation of G' was automatically included in our previous case of the narrow-band materials, but for the case of the hopping terms t and t_s nonzero we need to set up another equation of motion for G', and this may become complicated in general. However, there is a simplification in the wide-band limit of $W \gg IS$, and briefly the procedure is as follows. The interaction Hamiltonian $\mathcal{H}_I$ of Eq. (4.3) can be separated into a static (or non-spin-flip) part and a dynamic (or spin-flip) part. The static part can straightforwardly be incorporated into the Green's function equations, and we can use linear response theory for the dynamic part to write

$$a_i^+ = - \sum_j I_i <<a_i^+ ; a_j^->>_\omega S_j^+ \qquad (4.18)$$

Following now the work of Mills et al.[18] for a paramagnetic metal we can obtain the Green's function (response function) appearing in the above expression by defining its wave-vector Fourier transform $g_{m,n}(k_p,\omega)$ and using the formal expression

$$g_{m,n}(k_p,\omega) = -\frac{1}{N}\sum_{q_p}\int d\omega_1 d\omega_2 \left(\frac{f(\omega_1) - f(\omega_2)}{\omega_2 - \omega_1 - \omega + i\varepsilon}\right)$$

$$\times \rho_{m,n}(k_p+q_p,\omega_1;+) \rho_{n,m}(q_p,\omega_2;-) \qquad (4.19)$$

where ε is a positive infinitesimal, and the quantities of the form $\rho_{m,n}(k_p,\omega;\sigma)$ are one-electron spectral functions for the electrons with spin projection σ (= ±1). These spectral functions can now be evaluated for the electron system in a manner that is consistent with the wide-band limit of $W \gg IS$. We have included effects of $t_s \neq t$ by using matrix techniques similar to those described earlier in this section, thereby extending results due to Mills et al.[18] for a paramagnetic metal in the special case of $t_s = t$.

When the spectral functions have been evaluated they can be substituted into Eq. (4.19), and then by using Eqs. (4.15)-(4.18) we are able to deduce results for the bulk and surface excitations of the coupled system. When the above procedure is applied to an infinite ferromagnetic semiconductor, we recover the previous expressions[16,19] for the acoustic and optic bulk excitations, along with a Stoner-like continuum of electron-hole excitations. Details of these calculations for semi-infinite ferromagnetic semiconductors will be presented elsewhere.[20]

REFERENCES

1. E. L. Nagaev, "Physics of Magnetic Semiconductors," Mir Publishers, Moscow (1983).
2. P. Grünberg, Prog. in Surf. Sci. $\underline{18}$, 1 (1985).
3. M. G. Cottam and D. J. Lockwood, "Light Scattering in Magnetic Solids," Wiley, New York (1986).
4. S. V. Vonsovskii, "Magnetism," Vols. 1 and 2, Wiley, New York (1974).
5. P. Grünberg, Chap. 8, in: "Light Scattering in Solids V," M. Cardona and G. Güntherodt, eds., Springer-Verlag, Berlin (1989).
6. J. R. Sandercock, Solid State Commun. $\underline{15}$, 1715 (1974).
7. A. Caillé and C. Thibaudeau, Solid State Commun. $\underline{12}$, 939 (1986).
8. T. Wolfram and R. E. DeWames, Prog. in Surf. Sci. $\underline{2}$, 233 (1972).
9. M. G. Cottam, J. Phys. C $\underline{12}$, 1709 (1979).
10. M. G. Cottam, J. Phys. C $\underline{16}$, 1573 (1983).
11. M. Villeret, S. Rodriguez, and E. Kartheuser, Phys. Rev B $\underline{39}$, 2583 (1989).
12. R. E. Camley and M. G. Cottam, Phys. Rev. B $\underline{35}$, 189 (1987).
13. D. L. Mills, Chap. 3, in: "Surface Excitations," V. M. Agranovich and A. A. Maradudin, eds., North-Holland, Amsterdam (1984).
14. O. Krisement, J. Magn. Magn. Mat. $\underline{3}$, 7 (1976).
15. R. B. Woolsey and R. M. White, Phys. Rev. B $\underline{1}$, 4474 (1970).
16. A. Babcenco and M. G. Cottam, Solid State Commun. $\underline{22}$, 651 (1977).
17. M. G. Cottam, J. Phys. C $\underline{9}$, 2121 (1976).
18. D. L. Mills, M. T. Béal-Monod, and R. A. Weiner, Phys. Rev. B $\underline{5}$, 4637 (1972).
19. A. Babcenco and M. G. Cottam, J. Phys. C $\underline{14}$, 5347 (1981).
20. S. Gopalan and M. G. Cottam, Phys. Rev. B $\underline{42}$, in press (1990).

VIBRATIONAL, ELECTRONIC, AND MAGNETIC EXCITATIONS

IN II-VI QUANTUM WELL STRUCTURES

A. K. Ramdas and S. Rodriguez

Department of Physics
Purdue University
West Lafayette, Indiana 47907, U.S.A.

1. INTRODUCTION

The tetrahedrally coordinated II-VI compound semiconductors represent an important class of semiconductors in the physics of quantum-well structures and their opto-electronic applications. The majority of II-VI's are direct band gap semiconductors, with energy gaps spanning an impressive range of values (from zero for the Hg-based compounds to 3.8 eV for ZnS).[1] The plot of ionicity vs. bond-charge for the tetrahedrally coordinated semiconductors of the zinc blende or the wurtzite structure underscores the dramatic increase in ionicity of the II-VI's and their predisposition to assume a rocksalt structure.[2] Indeed, the first-order Raman spectrum of CdTe, which at ambient pressure exhibits the characteristic zone center LO-TO pair, disappears at a pressure of ~ 30 kbar; above that pressure, CdTe transforms to a rocksalt structure with the characteristic absence of the first-order Raman spectrum.[3] The increased ionicity translates to an increased Fröhlich interaction (polaron coupling constant). An examination of the lattice parameters and the band gaps characterizing the II-VI's and the III-V's reveals that a variety of lattice-matched, and hence strain-free, II-VI/II-VI as well as II-VI/III-V heterostructures can be visualized. They offer a wide range of total band off-sets. A judicious use of molecular beam epitaxy (MBE) also allows the fabrication of strained layer superlattices or quantum-well structures.[1] The successful incorporation of transition metal ions of the iron group (e.g., Mn^{2+}, Co^{2+}, Fe^{2+},...) into the II-VI's results in a diluted magnetic semiconductor (DMS) like $Cd_{1-x}Mn_xTe$. DMS's represent semiconducting alloys which, in addition to the 'tailoring' of physical properties typically feasible with alloying, exhibit dramatic magnetic phenomena.[4] MBE growth techniques have significantly increased the composition range of the II-VI alloys including DMS's; they have enabled the growth of epilayers and quantum well-structures not possible with in bulk growth. Even the 'magnetic' end members of DMS's like $Cd_{1-x}Mn_xTe$ and $Zn_{1-x}Mn_{1-x}Se$, i.e., MnTe and MnSe in the zinc blende phase[1] as well as new structures (cubic $Cd_{1-x}Mn_xSe$ rather than that with the wurtzite structure) have been grown with MBE.[5] They illustrate the remarkable expansion of new materials available for physical investigations.

The present paper focusses on the collective and localized excitations in Mn-based DMS superlattices investigated with Raman Spectroscopy.

Light Scattering in Semiconductor Structures and Superlattices
Edited by D.J. Lockwood and J.F. Young, Plenum Press, New York, 1991

323

2. EXPERIMENTAL

A standard computer-controlled double (or triple) grating spectrometer, equipped with holographic gratings; a thermoelectrically cooled GaAs photomultiplier; and a photon counting system followed by a computer-based data acquisition system constituted the Raman spectrometer. Discrete lines from a Ar^+, Kr^+ and a He-Ne laser as well as dye lasers with appropriate dyes provided the monochromatic lines for exciting the Raman spectrum. A variable-temperature optical magnet cryostat, incorporating superconducting coils providing a maximum field of 60 kG, allowed measurements down to temperatures as low as 1.8 K.[6]

3. VIBRATIONAL EXCITATIONS

In this section we focus on Raman scattering from vibrational excitations in DMS superlattices with frequencies comparable to those of optical phonons in the bulk DMS's as well as those of the acoustic branch accessible through special superlattice effects. In this context it is useful to recall some of the relevant features of the acoustic and optical phonons in the bulk and their modifications expected in a superlattice.

In order to identify some of the relevant features of lattice dynamics of heterostructures involving DMS's, consider, for example, $Cd_{1-x}Mn_xTe$. Due to the significant difference in the masses of Cd and Mn, $Cd_{1-x}Mn_xTe$ exhibits a "two-mode" behavior having two pairs of lines characteristic of zone center CdTe-like and MnTe-like LO-TO phonons.[7] The localized mode of Mn impurity in CdTe evolves into the LO and TO modes of the hypothetical zinc-blende MnTe while the LO and TO modes of CdTe merge into the gap mode of Cd in MnTe. The behavior of optical phonons in this alloy system has been interpreted by Peterson *et al.*[7] using a modified random element isodisplacement model. In contrast, optical phonons in $Zn_{1-x}Mn_xTe$ and $Zn_{1-x}Mn_xSe$[7,8] exhibit a mixed-mode behavior intermediate between two mode and one mode.

We now discuss the difference between the vibrational excitations for bulk crystals and those for the superlattices in the context of the matching of the optical phonon frequencies in the two constituent layers. As a relevant example, consider a superlattice consisting of alternating layers of CdTe and $Cd_{1-x}Mn_xTe$. The dispersion curves for LO phonons in bulk CdTe and $Cd_{1-x}Mn_xTe$ ($x = 0.25$) calculated using the linear-chain model with nearest-neighbor interaction only and assuming that Mn^{2+} ions randomly replace Cd^{2+} ions in the mixed crystal, $Cd_{1-x}Mn_xTe$, are shown in Fig. 1. This plot indicates that in $CdTe/Cd_{1-x}Mn_xTe$ superlattices, the MnTe-like phonon mode of the $Cd_{1-x}Mn_xTe$ layers cannot propagate into the CdTe layers because of the large attenuation of the vibration in that frequency region (region I). Similarly, in the frequency region between the CdTe zone-center LO phonon and the CdTe-like zone-center LO phonon of the $Cd_{1-x}Mn_xTe$ layer (region II), the CdTe LO mode cannot be sustained in the $Cd_{1-x}Mn_xTe$ layers. Therefore, these phonon modes can be considered to be confined to their respective layers, resulting in quantized optical phonons which are equivalent to vibrations in the bulk material whose wave vectors are given by[9] $q = \frac{m\pi}{d+\frac{a}{2}}$, where $d = d_1$ or d_2, the thickness of the CdTe (1) or the $Cd_{1-x}Mn_xTe$ (2) layer, and m is an integer. On the other hand, in the frequency region of the CdTe-like phonon modes in the $Cd_{1-x}Mn_xTe$ layers (region III), the vibrations of $Cd_{1-x}Mn_xTe$ layers and CdTe layers can propagate into both layers: the coupling between excitations originating in different layers may result in an average collective excitation as in the case of the acoustical phonons discussed in Sec. 3(a).

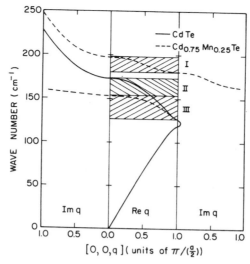

Fig. 1. Dispersion curves of LO phonons in bulk
CdTe and in $Cd_{0.75}Mn_{0.25}Te$, calculated using the linear chain model. Solid curve,
CdTe; dashed curves, $Cd_{0.75}Mn_{0.25}Te$. I and
II are frequency ranges of confined optical
phonons whereas III corresponds to propagating optical phonons.

We note that, as in the bulk crystal, in a superlattice one can obtain a selective
enhancement of the intensities of the Raman lines associated with optical phonons from
the well or the barrier layers by matching either the incident or the scattered photon
energy with electronic transitions;[10,11] in superlattices the relevant electronic transitions
can be associated with either the well or the barrier.

For finite crystals like thin films, or layers in superlattices and heterostructures, the
existence of surfaces or interfaces results in new vibrational excitations in addition to the
"bulk" vibrational excitations which we discussed earlier. These additional vibrational
modes are the surface modes in thin ionic slabs[12] or the "interface" vibrational (IF)
modes in superlattices. IF modes propagate along the interface planes and are highly
localized near them; the amplitude of an IF mode decays exponentially in the direction
perpendicular to the layer plane. One of the characteristics of IF modes is that their
Raman intensity is resonantly enhanced when the incident photon energy is close to the
electronic transitions of either the well or barrier layer since its vibrational amplitude
does not vanish in either layer. In addition, the frequencies of IF modes observed greatly
depend on the incident photon energy. These characteristics distinguish Raman lines
which are associated with IF modes.

Another consequence of the formation of the superlattice is the strain due to
the lattice mismatch between two constituent layers. The lattice parameter in bulk
DMS crystals changes almost linearly with the manganese concentration. For exam-

ple, the lattice parameter of bulk $Cd_{1-x}Mn_xTe$ is given by a = $(6.487 - 0.149x)$ Å,[13] whereas that of bulk (or epitaxial film) $Zn_{1-x}Mn_xSe$ shows a stronger dependence on x, viz. a = $5.666 + .268x$ Å.[14] Therefore, the lattice mismatch in $ZnSe/Zn_{1-x}Mn_xSe$ superlattices is significantly larger and the shift of the optical phonon frequencies due to such large strains must be considered. On the other hand, such strains in most of the $Cd_{1-x}Mn_xTe/Cd_{1-y}Mn_yTe$ superlattices are not significant and do not appear to be important in the interpretation of their Raman spectra.

a. Folded Acoustic Phonons

Since the acoustic dispersion curves of the two constituent materials of the superlattice overlap over a wide frequency range, acoustic phonons can propagate through both layers. For long-wavelength acoustic phonons one can apply a model in which the superlattice is considered as an elastic continuum composed of two alternating layers characterized by densities ρ_1 and ρ_2, and by bulk longitudinal acoustic velocities v_1 and v_2 along the superlattice axis.[9,15] The dispersion relation for the acoustic phonons propagating along the superlattice axis with wave vector q_z is given by[16]

$$\cos(q_z D) = \cos\left(\frac{\omega d_1}{v_1}\right)\cos\left(\frac{\omega d_2}{v_2}\right) - (1+\delta)\sin\left(\frac{\omega d_1}{v_1}\right)\sin\left(\frac{\omega d_2}{v_2}\right), \tag{1}$$

where d_1 and d_2 are the respective thicknesses of the well and barrier layers, $D = d_1 + d_2$, and $\delta = \frac{1}{2}\frac{(\rho_1 v_1 - \rho_2 v_2)^2}{\rho_1 v_1 \rho_2 v_2}$. From Eq. (1) one can deduce that the dispersion curve of the average bulk material is folded into the new Brillouin zone; in addition, small gaps open up at the zone center and the boundary when δ is different from zero. As a result of this zone folding, additional "zone center" modes which can interact with the electromagnetic radiation can be observed in Raman scattering. (For a short derivation of Eq. (1) see Suh et al.[17].)

Figure 2 shows the low-frequency Raman spectra for a [111] $Cd_{1-x}Mn_xTe/$ $Cd_{1-y}Mn_yTe$ superlattice at room temperature. All the spectra were recorded in the $z'(x'x')\bar{z}'$ scattering configuration: here x' and y' are in the plane of the layers with z' along the superlattice axis. The incident laser radiation has an energy $\hbar\omega_L$ intermediate between the band gap of the well and that of the barrier layers. Bulk crystals corresponding to the constituent materials of the superlattices show no Raman lines due to phonons in this spectral region. We attribute these lines to the longitudinal acoustic (LA) phonons folded into the new Brillouin zone, which arises from the additional periodicity of the superlattice.

Using the interpolated values of the densities and elastic moduli for various compositions of bulk $Cd_{1-x}Mn_xTe$, one can calculate the frequencies of LA phonons as a function of the wave vector $\vec{q}$. In the backscattering geometry, the wave vector of the scattered radiation is $(4\pi n/\lambda_L)$ where n is the refractive index of the sample. The solutions of Eq. (1) with the relevant parameters are 8.8 and 10.8 cm^{-1} for the first doublet and 18.6 and 20.5 cm^{-1} for the second doublet, in excellent agreement with the experimental values of 9.2, 11.0, 19.0, and 20.7 cm^{-1}, respectively. The strain associated with the lattice mismatch between the alternate layers is estimated to be less than ± 0.007 in both layers; with the Grüneisen constant estimated from the pressure derivative of the bulk elastic constants, the change in the frequencies of the LA phonons due to strain are expected to be insignificant in the context of the present measurements. Folded acoustic phonons have also been observed in $ZnSe/Zn_{1-x}Mn_xSe$ as well as in $ZnSe/Zn_{1-x}Cd_xSe$ superlattice structures.[17,18]

b. Optical Phonons

Figure 3 shows the Raman spectra of a [001] CdTe/Cd$_{0.75}$Mn$_{0.25}$Te superlattice (SSL-2) at T = 80 K in the frequency region of optical phonons with various wavelengths of laser excitation (λ_L). In Fig. 3(a), obtained with λ_L = 7525 Å, $i.e.$, with a photon energy ($\hbar\omega_L$) close to the lowest electronic transition of the CdTe well, the LO phonon of CdTe appears resonantly enhanced whereas those of the Cd$_{0.75}$Mn$_{0.25}$Te barrier are weak. As the incident photon energy is increased to a value lying between the energy gaps of the two layers, LO phonons from the CdTe well as well as those from the Cd$_{0.75}$Mn$_{0.25}$Te barrier are observed, as shown in Fig. 3(b). When the incident photon energy approaches the band gap of Cd$_{0.75}$Mn$_{0.25}$Te, only the phonons from the barrier layer are observed as shown in Fig. 3(c). These spectra illustrate the resonance enhancement of the intensities of phonons associated with the different layers as discussed earlier. By choosing the appropriate incident photon energy, we can thus obtain a selective resonance enhancement of the LO phonons in either superlattice layer.

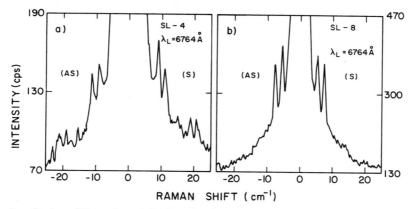

Fig. 2.   Stokes (S) and anti-Stokes (AS) components of the folded longitudinal acoustic phonons in Cd$_{1-x}$Mn$_x$Te/Cd$_{1-y}$Mn$_y$Te superlattices with $\hat{z}' \parallel [111]$; (a) $x = 0.11$, $y = 0.50$ (b) $x = 0$, $y = 0.24$. (Ref. 17.)

In Fig. 4, we show once more the Raman spectrum of SSL-2, recorded at 80 K. The peak at $\sim$ 200 cm^{-1} is the MnTe-like longitudinal optical (LO) phonon from the barrier layers. The peaks between 160 and 173 cm^{-1} are due to the CdTe-like and CdTe LO phonons of the barrier and well layers, respectively. The frequencies and shapes of these Raman lines are different from the LO phonons seen in the bulk crystals. The most prominent peak at 165 cm^{-1} is the CdTe-like LO phonon of Cd$_{0.75}$Mn$_{0.25}$Te, which propagates through both layers. Although this mode is expected to be folded, the Raman line does not resolve into multiple peaks since the frequencies of this mode do not change substantially with the variation of the wave vector. The peaks labeled $n = 2, 4$, and 6 are attributed to the LO phonons confined to the CdTe layers.

In the [001] superlattice, belonging to the point group D_{2d}, the phonons observed in Raman scattering have symmetry A_1 for $z(xx)\bar{z}$ scattering and B_2 for $z(xy)\bar{z}$ scattering.[17]

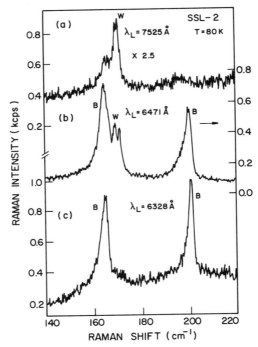

Fig. 3. Raman spectra from optical phonons in [001] CdTe/Cd$_{0.75}$Mn$_{0.25}$Te obtained at $T = 80$ K. Spectra were excited with three different laser wavelengths: (a) λ_L 7525 Å; (b) $\lambda_L = 6471$ Å, and (c) $\lambda_L = 6328$ Å. Phonons originating from the CdTe well layers and the Cd$_{0.75}$Mn$_{0.25}$Te barrier layers are labeled W and B, respectively. (Ref. 17.)

Here x, y, z are along the cubic axes, z being the superlattice axis. A_1 phonons, which dominate the Raman spectra of the well under resonance conditions, have wave vectors characterized by even m values, while B_2 phonons, which can be observed far from resonance, have wave vectors with odd m values. In Fig. 4, the confined LO phonons of CdTe layers appear in the (xx) polarization and hence these phonons should be assigned to the even m values, 2, 4, and 6. Their frequencies are given by $\omega(q_m)$, where $\omega(\vec{q})$ is the dispersion of the LO phonons in bulk CdTe and $q_m = \frac{m\pi}{d_1 + \frac{a}{2}}$. With this assignment, their measured frequencies are in excellent agreement with the dispersion curve calculated with the linear chain model for CdTe as shown in the inset in Fig. 4. Although not resolved into the additional peaks due to the absence of a significant variation of frequency with wave vector and the large layer thickness, the peak at the MnTe-like LO of Cd$_{1-x}$Mn$_x$Te layers has a frequency shifted to lower energy due to the confinement effect. Furthermore, this peak shows a marked low energy asymmetry which resolves into a weak additional peak labeled "IF" at the low energy side when the incident photon energy approaches the electronic subband transitions of the well layer.

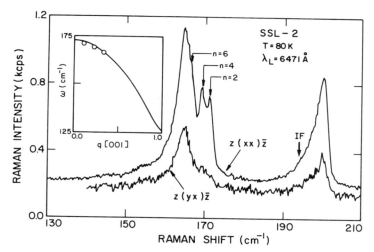

Fig. 4   Raman   spectra   from   optical   phonons   in   a
[001]CdTe/Cd$_{0.75}$Mn$_{0.25}$Te superlattice for different polarizations. Confined optical phonons are labeled with $n = 2$, 4, and
6.   Inset shows the observed frequencies of the confined LO
phonons plotted on the bulk CdTe dispersion curve calculated
with a linear chain model. (Ref. 17.)

In Fig. 5, we show the Raman spectrum of a [111] CdTe/Cd$_{0.76}$Mn$_{0.24}$Te superlattice, recorded at 80 K. The spectrum consists of only one Raman line in the frequency
region of the MnTe-like phonon (peak A) and one in the region of CdTe-like phonon (peak
B) with Raman shifts not uniquely identified with the LO or TO phonons of either the
well or the barrier layer; in contrast to the [001] superlattices, this is the case for all λ_L's.
Peaks A and B appear in $z'(x'x')\bar{z}'$ (parallel polarization) as well as in $z'(y'x')\bar{z}'$ (crossed
polarization), with a larger intensity in the former. Here x', y' and z' are along [01$\bar{1}$],
[2$\bar{1}\bar{1}$], and [111], respectively. A particularly noteworthy feature of these lines is their
marked frequency dependence on λ_L.

For [111] Cd$_{1-x}$Mn$_x$Te/Cd$_{1-y}$Mn$_y$Te superlattices, the point group symmetry is
reduced to C_{3v} from the higher symmetry T_d of the bulk crystal. LO phonons with A_1
symmetry can be observed in $z'(x'x')\bar{z}'$ whereas TO phonons with E symmetry are allowed
in both $z'(x'x')\bar{z}'$ and $z'(x'y')\bar{z}'$. Folded LO phonons referred to the new Brillouin zone
can show a small variation in its frequency as the energy of the incident laser radiation
is changed. However, the observed range of the frequency variation is too large for it to
be explained on this basis. Furthermore, the behavior of peak A is even more difficult to
understand in this manner because its frequency increases with the incident photon energy,
a behavior opposite to that expected for the folded LO phonons. Thus, there is a clear
difference between Raman scattering from LO phonons in [111] and [001] superlattices.

In the light of the above observations, it is useful to review other experimental
observations distinguishing these two types of superlattices. In Fig. 6, we show the photoluminescence peak at 5 K and magnetic fields of 0 and 60 kG in a [111] Cd$_{0.89}$Mn$_{0.11}$Te/
Cd$_{0.50}$Mn$_{0.50}$Te superlattice, SL-4. The position of the luminescence peak shifts with mag-

329

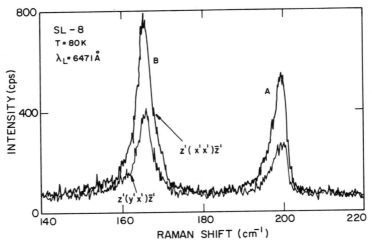

Fig. 5.   Raman spectra in the frequency region of optical phonons in a CdTe/Cd$_{0.76}$Mn$_{0.24}$Te superlattice recorded at $T = 80$ K, for different polarizations. A and B are the Raman lines in the MnTe-like and CdTe-like phonon regions, respectively. (Ref. 17.)

netic field towards lower energy with an effective g factor of ~ 100. This demonstrates that the s-d and p-d exchange-enhanced g factor characterizes the DMS superlattices just as it does the bulk DMS's.[17] This enhancement is both magnetic-field- and temperature-dependent. Furthermore, there is a marked difference between the magnetic-field dependence of the photoluminescence spectrum in [111] and [001] superlattices, as was first observed by Nurmikko and co-workers.[19] The luminescence spectrum of [111] superlattices exhibits one broad peak arising from the excitons localized at the heterointerfaces, while for [001] superlattices, a sharp peak due to intrinsic excitons as well as a weaker one arising from the localized excitons was observed. It appears that a significantly larger number of excitons are localized at the heterointerfaces in the [111] than in the [001] superlattice. A result of this localization is the observed shift of the exciton luminescence peak position as the magnetic field is changed from perpendicular to the superlattice axis to parallel with it. In Fig. 7, we show a similar result for SL-4. From the large spectral red shift of the luminescence peak seen in other [111] superlattices, the authors of Ref. 19 argued that the excitons are localized at the interfaces, conceivably due to the compositional fluctuation at the interface. In addition, a TEM evaluation of the superlattice structure indicates that the [111] samples have a surface dislocation density approximately an order of magnitude higher than in the [001] samples.[20] It also appears that the [001] interfaces are more abrupt than the [111] interfaces. We suggest that the quality of the interface results in the absence of confined or folded optical phonons in the [111] superlattices and the Raman lines A and B should be ascribed to vibrations associated with the interface. The imperfection of the interfaces suggests the possibility of the excitation of IF modes even in the backscattering experiments in [111] superlattices. Furthermore, the observation of the exciton localization at the interfaces suggests that the IF mode may be strongly enhanced, thereby dominating the Raman spectra in the [111] superlattices.

We can obtain a resonance enhancement of optical phonons or of IF modes by

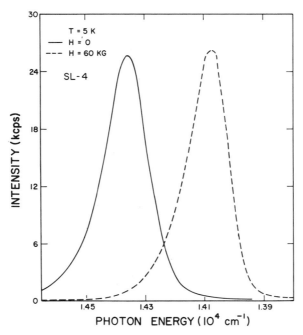

Fig. 6. Magnetic-field-induced shift in the position of the luminescence peak in $Cd_{0.89}Mn_{0.11}Te/Cd_{0.50}Mn_{0.50}Te$ superlattice, $\lambda_L = 5682$ Å. (Ref. 17).

matching the incident or the scattered photon energy with the electronic transitions in the superlattice. Exploiting the magnetic-field dependence of the photoluminescence peak with the large effective g factor as demonstrated in Fig. 6, we can achieve an "out resonance" either by magnetic-field or by temperature tuning. This is illustrated in Fig. 8 where we show a striking resonance effect observed for the Raman lines with a magnetic-field of 60 kG which shifts the onset of the photoluminescence peak to the region beyond 600 cm^{-1}. The clear observation of overtones and combinations of the two fundamental modes up to fourth order in these superlattices attests to the resonance enhancement. The striking "out resonance" seen for the IF phonons in these [111] superlattices suggests the coupling of these excitations with excitons localized at the interfaces via the Fröhlich electron-phonon interaction. The enhancement of the interface modes once more underscores the exciton localization at the interfaces.

4. MAGNETIC EXCITATIONS[21]

General Considerations

The theory of Raman scattering by magnetic excitations is similar to that by phonons except that, because of the axial nature of the magnetic field and of the magnetization $\vec{M}$, the selection rules for Raman scattering differ from those associated with symmetric polarizability tensors.

In a magnetic system, the electric susceptibility χ is a functional of the magnetization $\vec{M}$ as well as of other variables describing internal modes of motion. Thus, we

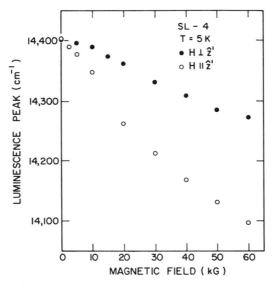

Fig. 7.   Magnetic-field-induced shift in the position of the luminescence peak in $Cd_{0.89}Mn_{0.11}Te/Cd_{0.50}Mn_{0.50}Te$ with the magnetic field directions parallel or perpendicular to the superlattice axis. (Ref. 17.)

write

$$\vec{P} \; = \; \chi(\vec{M}) \cdot \vec{E}_L, \tag{2}$$

where $\vec{E}_L$ is the electric field of the incident radiation and $\vec{P}$ the polarization vector. Several microscopic mechanisms for the dependence of χ on $\vec{M}$ can be envisioned of which exchange interactions with itinerant or localized electrons, having energies comparable to electrostatic interactions, are expected to be important. This suggests that the Raman features associated with magnetic excitations should exhibit strong resonance enhancement when $\hbar\omega_L$ is near the energy of an electronic transition, $e.g.$, the direct energy gap.

The modulation of $\vec{P}$ resulting from magnetic excitations is obtained from a Taylor series expansion of $\chi(\vec{M})$ where successive terms are independent of $\vec{M}$, linear in $\vec{M}$, second order in $\vec{M}, \ldots$. The restriction imposed by the Onsager reciprocity relations, the lack of absorption in the frequency region of interest and the cubic symmetry require that the contribution to $\vec{P}$, linear in $\vec{M}$, be of the form

$$\vec{P}^{(1)} \; = \; iG\vec{M} \times \vec{E}_L, \tag{3}$$

where G is a constant. The scattering cross-section for a Raman process is proportional to $|\hat{\epsilon}_S \cdot \vec{P}|^2$ where $\hat{\epsilon}_S$ is the direction of polarization of the scattered radiation. Thus, the scattering cross section involving a magnetic excitation in first order is of the form

$$\sigma \; = \; C|(\hat{\epsilon}_S \times \hat{\epsilon}_L) \cdot \vec{M}|^2, \tag{4}$$

where $\hat{\epsilon}_L$ is the polarization of the incident field and C is an appropriate function of ω_L and ω_S, the angular frequencies of the incident and scattered rays, respectively. Thus,

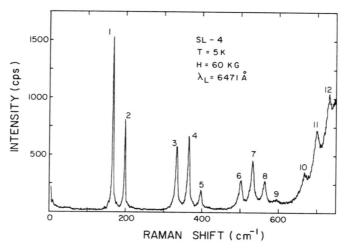

Fig. 8. Resonant Raman scattering from interface optical phonons, their overtones and combinations. The line labels denote the following assignments: (1) IF_1, (2) IF_2, (3) $2IF_1$, (4) $IF_1 + IF_2$, (5) $2IF_2$, (6) $3IF_1$, (7) $2IF_1 + IF_2$, (8) $IF_1 + 2IF_2$, (9) $3IF_2$, (10) $4IF_1$, (11) $3IF_1 + IF_2$, (12) $2IF_1 + 2IF_2$, where IF_1 and IF_2 are the Raman lines B and A in [111] superlattices. (Ref. 17.)

Raman scattering does not occur when the polarizations of the incident and scattered radiation are parallel. In the presence of a magnetic field $\vec{H}$, $\vec{M}$ evolves according to the Bloch equation which allows a determination of $\vec{M}$ as a function of the time.

We consider incident radiation propagating parallel to the z-axis selected along $\vec{H}$. For circularly polarized radiation $\hat{\sigma}_+$ and $\hat{\sigma}_-$, we find

$$\vec{P}^{(1)} = \pm \hat{z} G M_\perp E_o \exp[-i(\omega_L \pm \Omega)t]. \tag{5}$$

where Ω is the Larmor frequency ($geH/2mc$) and $M_\perp$ is the magnitude of the component of $\vec{M}$ normal to $\vec{H}$. This shows that in this geometry there is a Stokes line with polarization $(\hat{\sigma}_+, \hat{z})$ and an anti-Stokes line with $(\hat{\sigma}_-, \hat{z})$.

In a similar way, if the incident wave propagates at right angles to $\vec{H}$ but is polarized along $\vec{H}$,

$$\vec{P}^{(1)} = \frac{1}{2} G M_\perp E_o[(\hat{x} - i\hat{y}) \exp[-i(\omega_L - \Omega)t] - (\hat{x} + i\hat{y}) \exp[-i(\omega_L + \Omega)t]]. \tag{6}$$

Thus, Stokes and anti-Stokes lines occur in the geometries $(\hat{z}, \hat{\sigma}_-)$ and $(\hat{z}, \hat{\sigma}_+)$, respectively. The two cases described above are, of course, related to each another by time-reversal symmetry.

In the magnetically ordered phases originating in the antiferromagnetically coupled Mn^{2+} ions, the collective magnetic excitations give rise to larger Raman shifts than those associated with transitions between magnetic levels of individual atoms. These excitations are the spin waves (magnons). In superlattices, in addition to the bulk spin waves, other modes are present in rather close analogy to the situation prevailing for lattice vibrations.[22] Traveling and confined spin waves as well as interface spin waves occur. These modes have yet to be observed in DMS's.

a. Raman Electron Paramagnetic Resonance (Raman-EPR) of Mn^{2+}

We now consider Raman transitions between Zeeman sublevels of the individual Mn^{2+} ions in an external magnetic field, the sample being in its paramagnetic phase. In this phase the exchange interaction between Mn^{2+} ions is smaller than the thermal energy $k_B T$ and the ions can be considered as being independent of one another. The ground state of Mn^{2+} is $^6S_{5/2}$. In this section we discuss $Cd_{1-x}Mn_xTe$ as an illustrative example. The cubic crystalline field (site symmetry T_d) splits the sixfold degenerate ground state into a Γ_8 quadruplet at $+a$, and a Γ_7 doublet at $-2a$, where $3a$ is the crystal field splitting. This crystal field splitting is too small to be observed with the resolution of a standard Raman spectrometer and we treat the ground state of Mn^{2+} in $Cd_{1-x}Mn_xTe$ as an atomic $^6S_{5/2}$ level. The application of an external magnetic field results in the removal of the sixfold degeneracy of the ground state, the energy levels being $E(m_s) = g\mu_B H m_s$. Here m_s, the projection of $\vec{S}$ along $\vec{H}$, has the values $-5/2, -3/2, \ldots, +5/2$.

In the paramagnetic phase, Raman scattering associated with spin-flip transitions between adjacent sublevels of the Zeeman multiplet has been observed by Petrou et al.,[23] The results in $Cd_{1-x}Mn_xTe$ are shown in Fig. 9(a) for $x = 0.03$. As can be seen, a strong Stokes/anti-Stokes pair is observed with a Raman shift of $\omega_{PM} = 5.62 \pm 0.02$ cm^{-1} at 40 K and H = 60 kG. Taking $\vec{H}$ and the incident light parallel to $\hat{z}$, the Stokes line is observed in $(\hat{\sigma}_+, \hat{z})$, whereas the anti-Stokes line is seen in $(\hat{\sigma}_-, \hat{z})$. Within experimental error the frequency shift is linear in H. With the energy separation between adjacent sublevels of the Zeeman multiplet given by $\Delta E = g\mu_B H = \hbar\omega_{PM}$, it is found that $g = 2.01 \pm 0.02$. The Raman lines labeled "SF" in Fig. 9 will be discussed in Section 4(b).

Figure 10 shows the remarkable Raman-EPR spectrum observed in a $Cd_{0.9}Mn_{0.1}Se/ZnSe$ superlattice at 5 K, where $Cd_{0.9}Mn_{0.1}Se$ is the well.[24] The observed Raman lines labeled 1PM, ... and 5PM have their origin in the transitions with $\Delta m_s = 1, 2, 3, 4$, and 5 within the $J = \frac{5}{2}$ Zeeman multiplet of Mn^{2+}. Figure 11 shows the linear dependence of the Raman-EPR shift as a function of magnetic field where the solid lines correspond to $g_{Mn^{2+}} = 2$. In Fig. 10(b) we also observe peaks corresponding to 6PM and 7PM (see also Fig. 11). The spin-flip features higher than $\Delta m_S = 2$ in a DMS can be accounted for in terms of excitations within pairs of neighboring Mn^{2+} ions coupled antiferromagnetically and assuming an anisotropic exchange interaction between the ground-state multiplet of one and the excited state of the other. In fact, a pair of neighboring Mn^{2+} in their $^6S_{5/2}$ states in a magnetic field $\mathbf{H} \parallel \hat{z}$ can be described by the Hamiltonian

$$H = g\mu_B H S_z - J(S^2 - \frac{35}{2}), \tag{7}$$

where J is the $Mn^{2+} - Mn^{2+}$ exchange integral and $\vec{S}$ is the total spin of the pair. Note that in the virtual transitions involving an anisotropic exchange interaction, the total angular momentum need not be conserved.

In bulk Mn-based DMS's, at sufficiently high x and low temperatures, the spin-flip of isolated Mn^{2+} evolves into the high-frequency component of the antiferromagnetic magnon split into a doublet in the presence of a magnetic field. It appears that submicron heterostructures inhibit the formation of this long-range order.[17]

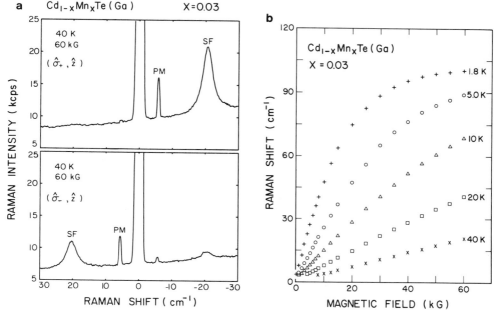

Fig. 9. (a) Raman spectra of $Cd_{1-x}Mn_xTe(Ga)$, $x = 0.03$, showing the $\Delta m_S = \pm 1$ transitions within the Zeeman multiplet of Mn^{2+} (PM) and the spin-flip of electrons bound to Ga donors (SF). kcps $= 10^3$ counts/sec. (b) Magnetic field and temperature dependence of the Raman shift associated with the spin-flip of electrons bound to donors in $Cd_{1-x}Mn_xTe(Ga)$, x = 1.8 K. (Ref. 25).

b. Spin-flip from Electrons Bound to Donors

Another Raman feature associated with magnetic excitations encountered in DMS's is the spin-flip of electrons bound to donors, enhanced by the sp-d exchange interaction. The spin splitting of the donor ground state in DMS's is determined by the macroscopic magnetization of the Mn^{2+} ions and the "intrinsic" Zeeman effect, $i.e.$,

$$\hbar\omega_{SF} = \frac{\alpha}{g_{Mn^{2+}}}M_O(H) + g^*\mu_B H = g_{eff}\mu_B H \qquad (8)$$

where α is the exchange integral characterizing the interaction between the spins of Mn^{2+} ions and those of the s-like Γ_6 electrons, μ_B the Bohr magneton, $M_O(H)$ the macroscopic magnetization, $g_{Mn^{2+}} = 2$ the g factor of Mn^{2+}, g^* the intrinsic g factor of the band electrons, and g_{eff} the effective g factor of the conduction band. Because of the strong s-d exchange interaction, the first term, characterized by the Brillouin function $B_{5/2}(g\mu_B H/k_B T)$, dominates the spin splitting.[25] The magnetic field and the temperature dependence of the feature labeled 'SF' in Fig. 9(a), displayed in Fig. 9(b), are consistent with Eq. (9).

Figure 12 shows the Raman spectrum of the $Cd_{0.9}Zn_{0.1}Se/Cd_{0.9}Mn_{0.1}Se$ superlattice, where a Raman line consistent with spin-flip is observed. The spectrum is obtained at 5 K in the backscattering configuration $z(xy)\bar{z}$, with a magnetic field of 50 kG along

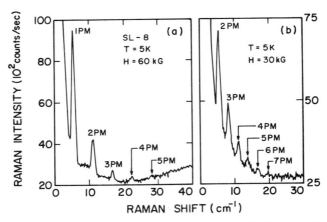

Fig. 10. Raman-EPR lines in a $Cd_{0.9}Mn_{0.1}Se/ZnSe$ superlattice at $T = 5$ K. The Raman lines in (a) correspond to 1PM, 2PM, 3PM, 4PM, and 5PM. They result from transitions within the Zeeman multiplet of Mn^{2+}, with $\Delta m_S = 1, 2, 3, 4,$ and 5, respectively. This spectrum is obtained in the crossed polarization $z(yx)\bar{z}$ with magnetic field $H = 60$ kG along x and incident wavelength of 6764 Å with 57 mW power. The 5PM line appearing superposed on the photoluminescence shows how closely the resonance condition is fulfilled. The spectrum in (b) shows the additional 6PM and 7PM lines observed at a lower magnetic field. In this spectrum the 1PM line is obscured by the parasitic laser light. (Ref. 24.)

x. As in a bulk DMS, the Raman shift of the donor spin-flip line exhibits a Brillouin-function-like behavior, as can be seen in Fig. 13. Since the spin-flip Raman mechanism also involves an interband electronic transition, the observed resonant enhancement for incident frequencies close to excitonic excitations is to be expected. From the slope of the linear portion of the spin-flip data shown in Fig. 13, we obtain $g_{eff} = 22$ at 5 K, which is comparable to that observed in bulk $Cd_{1-y}Mn_ySe$ with $y < 0.01$, but much lower than that for the actual value of y in this superlattice. The reason for the smaller Raman shift observed has to be sought in terms of the magnitude of the band offset which could result in electrons being in the quantum-well levels while the donors are in the barrier, the well or the interface, the Mn^{2+} ions being in the barrier.[24]

Finally we draw attention to the spin-flip Raman scattering from $Cd_{1-x}Mn_xTe:In$ epilayers and modulation-doped $Cd_{1-x}Mn_xTe:In/CdTe$ superlattices grown by photo-assisted molecular beam epitaxy reported by Suh et al.[26]

5. ACKNOWLEDGEMENTS

The authors acknowledge support from the National Science Foundation (DMR-86-16787) and (DMR-89-13706) and Defense Advanced Research Projects Agency (DARPA)–University Research Initiative (URI) Consortium administered by the U.S. Office of Naval Research (Contract No. N00014-86-K-0760).

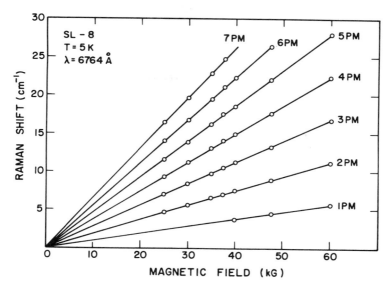

Fig. 11. Raman-EPR (PM) shift as a function of magnetic field in the $Cd_{0.9}Mn_{0.1}Se/ZnSe$ superlattice at $T = 5$ K. The solid lines correspond to the Raman-EPR shift given by $ng_{Mn^{2+}}\mu_B H$ with $g_{Mn^{2+}} = 2$ and $n = 1,2, \ldots,7$. (Ref. 24.)

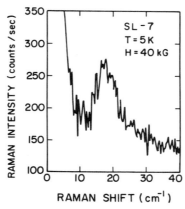

Fig. 12. Raman spectrum associated with the spin flip of electrons bound to donors in a $Cd_{0.9}Zn_{0.1}Se/Cd_{0.9}Mn_{0.1}Se$ superlattice. The spectrum is obtained at $T = 20$ K in the crossed polarization $z(yx)\bar{z}$ with $H = 40$ kG along x and incident wavelength of 6471 Å. (Ref. 24.)

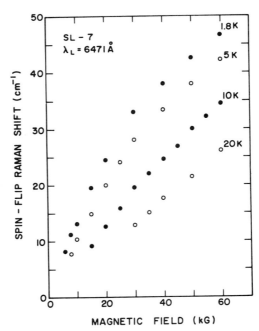

Fig. 13. Magnetic field and temperature dependence of the spin-flip Raman shift in the $Cd_{0.9}Zn_{0.1}Se/Cd_{0.9}Mn_{0.1}Se$ superlattice with external magnetic field in the plane of the (001) layers. The spectra were obtained in the cross polarization $z(xy)\bar{z}$ with incident laser wavelength $\lambda_L = 6471$ Å. (Ref. 24).

REFERENCES

1. R. L. Gunshor, L. A. Kolodziejski, A. V. Nurmikko, and N. Otsuka, *"Molecular Beam Epitaxy of II-VI Semiconductor Microstructures"* to appear in the special volume ed., T. P. Pearsall in the series *"Semiconductors and Semimetals"*, eds., R. K. Willardson and A. C. Beer (Academic Press, New York 1990).
2. See M. L. Cohen and J. R. Chelikowsky, *"Electronic Structure and Optical Properties of Semiconductors"* (Springer-Verlag, Berlin 1988), Fig. 8.66, p. 139.
3. A. K. Arora, D. U. Bartholomew, D. L. Peterson, and A. K. Ramdas, Phys. Rev. **35**, 7966 (1987).
4. J. K. Furdyna, J. Appl. Phys. **64**, R29 (1988).
5. N. Samarth, H. Luo, J. K. Furdyna, S. B. Qadri, Y. R. Lee, A. K. Ramdas, and N. Otsuka, Appl. Phys. Lett. **54**, 2680 (1989).
6. D. L. Peterson, Ph.D. thesis, Purdue University, 1984 (unpublished).
7. D. L. Peterson, A. Petrou, W. Giriat, A. K. Ramdas, and S. Rodriguez, Phys. Rev. B **33**, 1160 (1986). See also: S. Venugopalan, A. Petrou, R. R. Galazka, A. K. Ramdas, and S. Rodriguez, Phys. Rev. B **25**, 2681 (1982).

8. A. K. Arora, E.-K. Suh, U. Debska, and A. K. Ramdas, Phys. Rev. B **37**, 2927 (1988).

9. See, for example, C. Colvard, T. A. Gant, M. V. Klein, R. Merlin, R. Fischer, H. Morkoc, and A. C. Gossard, Phys. Rev. B **31**, 2080 (1985); A. K. Sood, J. Menendez, M. Cardona, and K. Ploog, Phys. Rev. Lett. **56**, 1753 (1986) and B. Jusserand and D. Paquet, *ibid.*, **56**, 1752 (1986).

10. A. S. Barker, Jr. and R. Loudon, Rev. Mod. Phys. **44**, 18 (1972).

11. D. L. Peterson, D. U. Bartholomew, A. K. Ramdas, and S. Rodriguez, Phys. Rev. B **31**, 7932 (1985).

12. R. Fuchs and K. L. Kliewer, Phys. Rev. **140**, A2076 (1965).

13. N. Bottka, J. Stankiewicz, and W. Giriat, J. Appl. Phys. **52**, 4189 (1981).

14. L. A. Kolodziejski, R. L. Gunshor, R. Venkatasubramanian, T. C. Bonsett, R. Frohne, S. Datta, N. Otsuka, R. B. Bylsma, W. M. Becker, and A. V. Nurmikko, J. Vac. Sci. Technol. B **4**, 583 (1986).

15. C. Colvard, R. Merlin, M. V. Klein, and A. C. Gossard, Phys. Rev. Lett. **45**, 298 (1980).

16. S. M. Rytov, Akust. Zh. **2**, 71 (1956). [Sov. Phys. Acoust. **2**, 68 (1956)].

17. E.-K. Suh, D. U. Bartholomew, A. K. Ramdas, S. Rodriguez, S. Venugopalan, L. A. Kolodziejski, and R. L. Gunshor, Phys. Rev. B **36**, 4316 (1987).

18. R. G. Alonso, Eunsoon Oh, A. K. Ramdas, N. Samarth, H. Luo, J. K. Furdyna, and L. R. Ram Mohan (unpublished).

19. X.-C. Zhang, S.-K. Chang, A. V. Nurmikko, L. A. Kolodziejski, R. L. Gunshor, and S. Datta, Phys. Rev. B **31**, 4056 (1985); S.-K. Chang, A. V. Nurmikko, L. A. Kolodziejski, and R. L. Gunshor, Phys. Rev. B **33**, 2589 (1986).

20. R. L. Gunshor, L. A. Kolodziejski, N. Otsuka, S. K. Chang, and A. V. Nurmikko, J. Vac. Sci. Technol. A **4**, 2117 (1986).

21. A. K. Ramdas and S. Rodriguez, in *"Diluted Magnetic Semiconductors"*, eds., J. K. Furdyna and J. Kossut, being vol. 25 of *"Semiconductors and Semimetals"*, eds., R. K. Willardson and A. C. Beer, Academic Press, New York (1988) pp. 345-412.

22. M. Villeret, S. Rodriguez, and E. Kartheuser, Phys. Rev. **39**, 2583 (1989).

23. A. Petrou, D. L. Peterson, S. Venugopalan, R. R. Galazka, A. K. Ramdas, and S. Rodriguez, Phys. Rev. B **27**, 3471 (1983).

24. R. G. Alonso, E.-K. Suh, A. K. Ramdas, N. Samarth, H. Luo, and J. K. Furdyna, Phys. Rev. B **40**, 3720 (1989).

25. D. L. Peterson, D. U. Bartholomew, U. Debska, A. K. Ramdas, and S. Rodriguez, Phys. Rev. B **32**, 323 (1985).

26. E.-K. Suh, D. U. Bartholomew, A. K. Ramdas, R. N. Bicknell, R. L. Harper, N. C. Giles, and J. F. Schetzina, Phys. Rev. B **36**, 9358 (1987).

ZINC BLENDE MnTe AS EFFICIENT CONFINEMENT LAYERS IN ZnTe AND CdTe

SINGLE-QUANTUM WELL STRUCTURES

A.V. Nurmikko

Division of Engineering and Department of Physics
Brown University, Providence, RI 02912, U.S.A.

ABSTRACT

The synthesis of wide-gap cubic zinc blende MnTe was recently realized and led to a range of optical studies on single-quantum wells of ZnTe/MnTe and CdTe/MnTe with strong electron-hole confinement. This leads to the possibility of studying exciton-longitudinal optical (LO) phonon coupling in designed structures.

INTRODUCTION

Today, a number of wide-gap II-VI heterostructures, prepared by advanced epitaxial methods, have been fabricated. Due to difficulties in versatile doping - which still persist - most of the characterization has been obtained from optical experiments, especially through interband spectroscopy. Among heterostructures that have been studied in some detail are CdTe/(Cd,Mn)Te,[1,2] ZnSe/(Zn,Mn)Se,[3] ZnSe/Zn(S,Se),[4] CdTe/ZnTe (or (Zn,Cd)Te/ZnTe),[5,6] ZnSe/(Zn,Fe)Se,[7] (Cd,Zn)Se/(Zn,Mn)Se,[8] and ZnSe/ZnTe.[9] Two aspects of these structures are nearly generic: (i) either the valence-band offset or the conduction-band offset is quite small (i.e., on the order of the exciton Rydberg), except for ZnSe/ZnTe, which appears to be a strongly type II superlattice system; and (ii) substantial lattice mismatch strains exist.

In this article we focus on two examples in the use of a very wide gap (>3 eV) barrier material in binary-quantum wells of CdTe/MnTe and ZnTe/MnTe where further insight into the electronic states of quasi-2-dimensional nature of lowest interband excitations has been obtained. Specifically, we have indications that a valence-band confinement sufficient to produce a quasi-2D exciton is possible. The particular physical aspect of these excitons we pursue here is the manifestation of their coupling to the optical phonons in different experimental circumstances.

While MnTe as bulk material crystallizes in the hexagonal layered NiAs structure, its incorporation in ZnTe and CdTe heterostructures has very recently permitted the synthesis of this antiferromagnetic semiconductor in zinc blende cubic form.[10] Optical measurements indicate that the s-p band gap occurs at 3.2 eV at low temperature, thus roughly following extrapolation from previously available data for (Cd,Mn)Te or (Zn, Mn)Te at low Mn composition. Such characteristically wide band gaps have also been recently reported for MnS in evaporated polycrystalline zinc blende films.[11,12]

Light Scattering in Semiconductor Structures and Superlattices
Edited by D.J. Lockwood and J.F. Young, Plenum Press, New York, 1991

341

The presence of anticipated lattice mismatch in a ZnTe/MnTe or CdTe/MnTe heterostructure (approximately 3% and 6%, respectively) implies immediately that suitably thin layered structures must be used in order to obtain pseudomorphic (coherently strained) heterostructures. The simplest of these is a "double barrier" kind, where a ZnTe or a CdTe single-quantum well is formed between thin MnTe barrier layers (<40 Å in the cases studied here). For such structures confinement of conduction and valence-band states is shown to take place below. The squarewave (SQW) section of the samples consisted typically of thin MnTe barrier layers (<50 Å) surrounding the CdTe or ZnTe quantum well; such thin barrier layers also allow observable tunneling effects in the quantum-well region. A 800-1000 Å thick ZnTe (or CdTe) overlayer and a 1 μm thick ZnTe (CdTe) buffer layer surrounded the SQW section.

HOT EXCITON LUMINESCENCE IN ZnTe/MnTe QUANTUM WELLS

It has been well established experimentally in a number of wide-gap II-VI semiconductors, notably ZnTe, that the recombination spectrum at photon energies at and above the band gap often consists of well-defined peaks that are separated by LO-phonon energies. These peaks follow the laser energy of excitation and are, in that sense, "Raman-like", even though sufficient scattering may have been present to cause the process of secondary emission to have little or no coherent relation to the incident electromagnetic wave. Many orders can be seen, with a strong resonance enhancement for the particular order that resonates with the band gap. The principal electron-phonon interaction arises from the Fröhlich interaction, which is substantially larger in II-VI compounds than in III-V compounds. Among several experimental and theoretical efforts in the 1970's whose aim was to delineate the details of such strongly coupled electronic and lattice excitations we note here especially the work by Klochikin et al.[13] in bulk ZnTe. The physical issues that were subject to "hot" debate were (i) the distinction between "hot luminescence" (HL) and resonance Raman scattering (RRS) and (ii) whether the energetically relaxing electron-hole pair maintained an exciton-like character throughout the entire secondary emission process. We show here how the availability of a wide-gap II-VI quantum well, together with advanced optical methods, gives us direct insight into this problem.

Our optical experiments were carried out on ZnTe/MnTe single-quantum-well structures with very thin MnTe barrier layers (22 Å). The heterostructural region was sandwiched between ZnTe buffer and cap layers.[10] Raman experiments in which the first-order scattering from LO phonons showed well-defined modes from the ZnTe quantum well and from the MnTe barrier layers demonstrated that coherent accommodation of the lattice mismatch strain is, to a good approximation, accomplished in such thin-layer structures[14] and that finite interdiffusion of the cations (Zn and Mn) no more than about a monolayer typically occurred. The key feature of such structures here was that due to the thin MnTe barrier layers, tunneling of electron-hole pairs out of the ZnTe quantum well strongly reduces the normal thermalized photoluminescence (PL) emission. Then, the entire secondary emission from the quantum well was dominated by hot luminescence. Figure 1 shows this emission ranging from the laser excitation (at $h\Omega = 2.602$ eV) to the region near the n = 1 heavy-hole (HH) excitonic band gap close to 2.5 eV (upper trace). Note how the fifth-order LO-phonon-related emission is strongly resonating at this gap. Its line width is considerably broadened in comparison with the lower orders; furthermore we observed a rapid loss of polarization memory with increasing order. The lower trace shows as a comparison spectrum taken for an epitaxial ZnTe thin film (same photon energy of excitation). Strong thermalized luminescence prevented the observation of phonon structure in the spectrum near the bulk exciton band gap (2.39 eV); however, well above the gap the LO-phonon features are clearly seen. In both cases, the lowest order LO-phonon features were distinctly polarized according to the selection rules for the Fröhlich interaction in our (100) oriented samples, suggesting a Raman-like coherent process.

The strongly resonating region of the n = 1 excitonic band gap in the quantum-well sample is the particular spectral region of interest here. The position of the gap was measured independently by reflectance and first-order RRS experiments. Figure 2 shows the reflectance spectra in the n = 1 resonance region in zero and finite magnetic field (for opposite circular polarization at $B_z = 4$ T and T = 2 K). Because finite interdiffusion

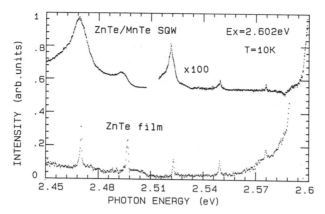

Fig. 1. Comparison of secondary emission spectra with multiple LO-phonon sidebands from a single ZnTe/MnTe quantum-well sample and a thin epitaxial film of ZnTe (T = 10 K). Note the large resonant enhancement near the n = 1 quantum-well exciton resonance.

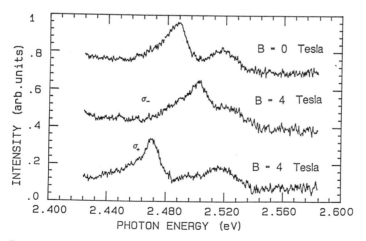

Fig. 2. Reflectance spectrum of the ZnTe/MnTe SQW near the n = 1 exciton resonance in zero and 4 T magnetic field (for two opposite circular polarizations).

diluts the antiferromagnetic semiconductor MnTe, one expects Zeeman effects similar to those in diluted magnetic semiconductors (DMS), such as $Zn_{1-x}Mn_xTe$. This allows us to identify the lowest two reflectance features to n = 1 HH (clearly spin split in Fig. 2) and light-hole (LH) transitions and provide input for band offset estimates. Note for comparison the bulk ZnTe exciton energy of 2.39 eV.

To investigate the question of exciton versus hot electron and hole character in the secondary emission, the spectra were measured in magnetic fields up to 23 T. Figure 3 shows the recombination spectra near the n = 1 resonance at fields B_z = 0, 10, and 23 T. Note how the Zeeman shifting of the excitonic gap from 0 to 10 T by the DMS effect causes the reduction in the fourth-order LO-phonon amplitude and the buildup of the sixth-order one. Beyond 10 T the DMS effect saturates (analogous to paramagnetic behavior), and the spectrum remains approximately constant for the highest fields available to us. The key point is that the main features of the spectrum are preserved, a direct indication of the exciton nature of the process of recombination. At B_z = 23 T the free electron cyclotron energy in ZnTe is approximately 25 meV, i.e., comparable to the LO-phonon energy and the exciton binding energy (the latter is not well known in our system but has been estimated for CdTe/MnTe SQW's on the basis of optical data and variational calculations to be in this range). If free electron and hole "energy relaxation" with LO-phonon steps were the dominant process, strong effects due to Landau quantization would have been expected to modify the spectra of Fig. 3.

The details of the HL spectrum near the n = 1 gap were found to be quite temperature dependent. Figure 4 shows the secondary emission at T = 10-100 K. Note how in the range 10-70 K, the fourth-order LO-phonon process is quenched and the fifth-order blue shifts slightly (in this range the gap shrinks by approximately 12 meV). At higher temperatures, the spectrum begins to lose the distinct phonon-related features and does so in a way that causes the energy separation between, e.g., the fifth and sixth-order significantly depart from the LO-phonon energy. By comparison, Fig. 5 shows the temperature dependence of the first- and second-order peaks (resonantly enhanced at the gap) where the LO-phonon energetic ordering stays intact.

Examination of the resonance enhancement for the HL and the n = 1 HH reflectance anomaly suggests that the enhancement actually occurs some 10 meV below the absorptive exciton resonance. This is further illustrated in Fig. 6, which shows the RRS spectra for the first-order LO- and second-order LO-phonon modes of ZnTe as well as the first-order LO-phonon mode of MnTe. The RRS spectra are dominated by the outgoing channels and are consistent with the assignment for the n = 1 resonance energy from the reflectance measurements. We expect the LH contribution in the Raman cross section to be about an order of magnitude smaller than that for HH's; consequently we assign the RRS spectrum to the n = 1 e-HH excitonic resonance. This is supported also by the HH-like Zeeman splitting in the RRS profile in the presence of a magnetic field (measured up to 4 T). A line width of 10-15 meV is inferred from the RRS profiles. The HL spectrum (also shown for reference) itself lies atop a broader background (~40 meV), which is probably the residue of thermalized luminescence at the n = 1 HH transition. We believe that the energetic difference between the "absorptive" and "emissive" excitonic band gaps originates from the disorder effects in our quantum well, i.e., a combination of geometrical disorder (well-thickness fluctuations in layer plane) and compositional disorder (interdiffusion). That is, it appears that localized exciton states in the density of states tail provide the most efficient means for the HL emission to be generated.

The strong enhancement in both HL and RRS (first-order LO and second-order LO) due to the n = 1 exciton resonance has the benefit of at least partially 2D enhancement of the exciton oscillator strength. The line-width broadening and the loss of "polarization memory" with increasing order of phonon sideband in the HL is interpreted by us as indicating that the coherence of a RRS process[8] is weakened in the SQW samples by disorder scattering (possibly including spin dephasing collisions by the Mn-ion d electrons) and perhaps also by partial tunneling effects, so that the process seen here in a quantum well is not identical to HL observations in bulk ZnTe.[13] One key argument from theoretical grounds separating the HL and RRS processes is that they should exhibit

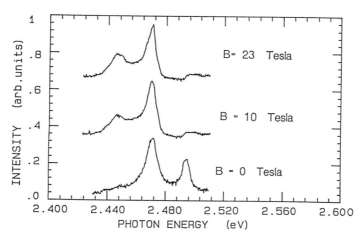

Fig. 3. The HL spectrum near the n = 1 resonance in magnetic fields of 0, 10, and 23 T (T = 2 K).

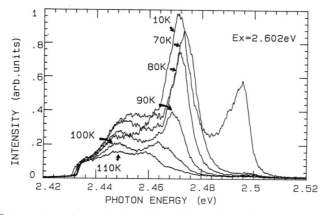

Fig. 4. Temperature dependence of the HL at the exciton resonance (fourth-, fifth-, and sixth-order LO-phonon sidebands) at T = 10-110 K. The low-energy side of the spectra are cut off by the spectrometer filter stage.

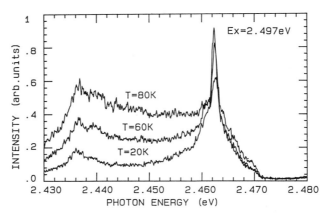

Fig. 5. Temperature dependence of the first- and second-order sidebands at the same resonance.

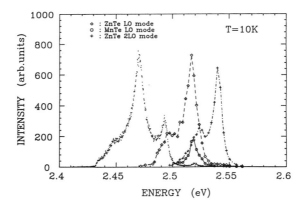

Fig. 6. RRS spectra at T = 10 K. The RRS spectra are shown for first-order LO- and second-order LO-phonon modes of ZnTe and first-order LO-phonon mode of MnTe (lines running through data points are a guide to the eye). For comparison, a portion of the HL spectrum is also included.

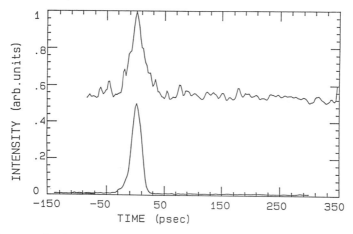

Fig. 7. Time-resolved emission from the fifth-order phonon sideband from the SQW sample at T = 10 K (upper trace). The system response is shown in the lower trace.

distinctly different time-dependent behavior,[15] a point of particular interest here for the 2D exciton systems in a quantum-well structure. We have performed initial time-resolved secondary emission experiments, an example of which is shown in Fig. 7. The upper trace is the transient intensity of the fifth-order LO peak of Fig. 1, while the lower trace indicates the time response of a monochromator/streak camera system to an ultrashort (~1 ps) laser excitation pulse. Clearly, the HL signal is not an instantaneous response to the laser pulse; rather, from this and other data a secondary emission time of approximately 10 ps is inferred. Hence, we have reached a direct experimental justification for the assumption that a HL - not a Raman process - is the mechanism for the secondary emission. An important experimental issue being presently pursued is concerned with the details of the LO-phonon intermediate energy relaxation of the "hot exciton", especially in terms of possible bottleneck processes within the localized states at the n = 1 resonance.

EXCITONS IN CdTe/MnTe SINGLE-QUANTUM WELLS

RRS techniques were employed to investigate the strain distribution in these single-quantum well structures. Linear extrapolation from the DMS alloy $CD_{1-x}Mn_xTe$ suggests a lattice mismatch exceeding 3% between CdTe and cubic MnTe (a = 6.48 and 6.28 Å, respectively). Figure 8 shows the LO-phonon spectrum, obtained in backscattering geometry under near-resonant conditions at the lowest n = 1 interband quantum-well transition (defined below) at T = 10 K from a SQW sample with L_w = 49 Å. The position of the MnTe LO-phonon mode in the SQW sample was seen to have shifted distinctly to lower frequencies when compared with an epilayer of MnTe, whereas the CdTe quantum-well LO-phonon mode is undisplaced from its bulk value. Complicating the quantitative evaluation is the fact that the elastic constants for zinc blende cubic MnTe are not known, so that accurate calculations about the magnitude of the lattice mismatch strain implied in the shift of the LO-phonon mode are not possible. Using the elastic constants of CdTe to the zeroth order, however, yields a strain from the Raman measurement of approximately 4.0%. Since the addition of Mn to CdTe is known to soften the lattice, this estimate presents an upper limit. These observations, including the

347

unshifted CdTe LO Raman peak, do suggest that the heterostructure is probably near a pseudomorphic limit, with the lattice mismatch strain accommodated in the thin MnTe barrier layers. Additionally, the spectral width of the MnTe LO-phonon mode in particular suggests that finite interdiffusion is present at the heterointerfaces. From the line width and for a simple Gaussian distribution of Mn concentration we estimate that a region of $Cd_{1-x}Mn_xTe$ with a range of composition of $x = 0.7-1.0$ occurs at each heterointerface within two monolayers. This estimate includes the use of the $n = 1$ exciton state in the resonant Raman process with a penetration of the exciton envelope function to the barrier calculated from the data discussed next.

From extrapolation of band-gap data of bulk or epitaxial (Cd,Mn)Te to the cubic MnTe limit, one expects a band-gap difference of approximately 1.6 eV between CdTe and MnTe. Strong confinement effects should then be possible, especially in a type I structure. This is demonstrated directly in Fig. 9, which shows PL spectra taken at $T = 10$ K for three SQW samples with approximate well thicknesses of 22, 15, and 10 Å. Note the remarkable degree of confinement, which shifts the quantum-well band gap from the infrared region characteristic of bulk CdTe across the visible spectrum into the blue!

Examination of the line widths of the individual peaks in Fig. 9 reveals that the typical values in our samples are on the order of 20 meV at low temperatures (narrowing for wider well samples). This is consistent with the role of finite interdiffusion (i.e., compositional disorder). We have made elementary calculations in which the data from RRS measurements (broadening of the MnTe LO-phonon mode) are used together with electron-hole confinement obtained from the band-offset determination. For a simple Gaussian distribution in the Mn concentration with a halfwidth of two monolayers, a satisfactory agreement can be reached with experimental data, assuming an exciton Bohr radius of 40 Å. The point we wish to pursue here is the behavior of the linewidth as a distinct function of temperature as shown in Fig. 10 for two CdTe/MnTe SQW samples with $L_w = 25$ and 49 Å from $T = 10$ K up to room temperature. Apart from an extrinsic contribution (concentration fluctuations), the temperature dependence can be used to estimate the exciton stability against dissociation by optical phonon scattering in a simple model. By using arguments of chemical balance for the exciton and free-carrier populations, one can directly compare the PL line width with that obtained from direct absorption measurements (not possible here for an optically thin SQW structure) if, in the temperature range of interest, the exciton density remains larger than the free-carrier density. In this case the line width contains a temperature-dependent contribution that is primarily due to exciton dissociation by LO-phonon absorption ($E_x < h\Omega_{LO}$). The free-carrier population in our structures is apparently subject to additional nonradiative recombination paths, this is, reflected by the unchanging PL line shape over the temperature range of our measurements. The dominance of exciton contribution to the observable optical signature in steady state can then be used in the spirit of a recent model where the LO-phonon-induced dissociation/formation rate yields the following temperature-dependent exciton line width in absorption: $\delta E(T) = \delta E_0 + \Gamma_{LO}/[\exp(E_{LO}/kT) - 1]$, where δE_0 is the inhomogeneous broadening (from concentration fluctuations at the heterointerfaces) and Γ_{LO} is proportional to exciton-LO-phonon coupling (Fröhlich interaction).[16] The fits to experimental data (solid lines) yield $\Gamma_{LO} = 42$ meV for the two samples in Fig. 10. This value is significantly larger than that measured for GaAs quantum wells (5-8 meV). Most of the difference is directly due to the larger polaron coupling in CdTe; while there appears to be no direct data on exciton-LO-phonon coupling, or on hole-LO-phonon coupling, an approximately five times larger coupling for electrons in bulk CdTe is expected. However, we also believe that additional localization of the quasi-2D excitons due to random disorder in the CdTe/MnTe SQW's provides further enhancement. Further work on this point is in progress.

SUMMARY

In summary, cubic zinc blende MnTe has been found useful as barrier material when employed in sufficiently thin layers within ZnTe- and CdTe-based quantum-well structures. Good confinement for electron and hole states manifests itself through pronounced quasi-

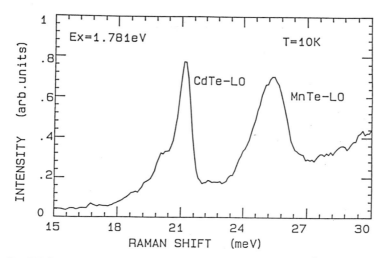

Fig. 8. RRS spectra of a CdTe/MnTe sample with $L_W = 49$ Å (excitation near resonance of the n = 1 excitons of the SQW).

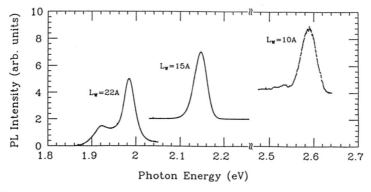

Fig. 9. PL spectra at T = 10 K of three CdTe/MnTe SQW's of different well widths (given in the figure). The bulk exciton band gap is 1.595 eV.

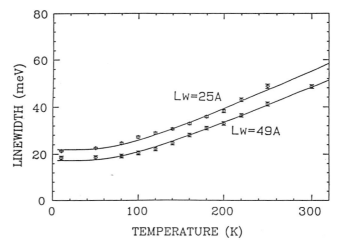

Fig. 10. Temperature dependence of the PL linewidth of the dominant peak for two CdTe/MnTe SQW samples. The solid line is a theoretical fit discussed in the text.

2D exciton effects, including strong coupling to the LO-phonon modes. A combination of experimental techniques, ranging from high magnetic field to time-resolved spectroscopy, has shown that a hot luminescence process involves Coulomb-correlated electron-hole pairs (hot excitons) in the ZnTe-based structures, which lose phase information upon scattering but surprisingly preserve the LO-phonon-imposed energetic structure even in the presence of finite disorder effects.

ACKNOWLEDGEMENTS

This research represents a close collaboration between two groups, that of Professor R. Gunshor at Purdue University, where all of the molecular beam epitaxy was carried out, and that of the author at Brown University, where most of the physical measurements described above were performed. I especially wish to acknowledge the following colleagues and students: S. Durbin, J. Han, Sungki O., D. Menke, and Professor M. Kobayashi (Purdue); and N. Pelekanos, Q. Fu, and J. Ding (Brown).

This research was supported by the Defense Advanced Research Projects Agency (DARPA), the National Science Foundation (NSF), and the Office of Naval Research (ONR).

REFERENCES

1. L.A. Kolodziejski, T. Sakamoto, R. Gunshor, and S. Datta, Appl. Phys. Lett. 44, 799 (1984).
2. R.N. Bicknell, R. Yanka, N. Giles-Taylor, D.K. Banks, E.L. Buckland, and J.F. Schetzina, Appl. Phys. Lett. 45, 92 (1985).

3. L.A. Kolodziejski, R.L. Gunshor, N. Otsuka, S. Datta, W. Becker, and A.V. Nurmikko, IEEE J. Quantum Electron QE-22, 1666 (1986).
4. K. Mohammed, D.J. Olego, P. Newbury, D.A. Cammack, R. Dalby, and H. Cornelissen, Appl. Phys. Lett. 50, 1820 (1987).
5. G. Monfroy, S. Sivananthan, X. Chu, J.-P. Faurie, R.D. Know, and J.L. Staudenmann, Appl. Phys. Lett. 49, 152 (1986).
6. N. Magnea, G. Lentz, H. Mariette, G. Feuillet, F. Dal'bo, and H. Tuffigo, J. Cryst. Growth 95, 584 (1989).
7. B.T. Jonker, G.A. Prinz, and J.J. Krebs, J. Vac. Sci. Tecnol. A6, 1946 (1988).
8. N. Samarth, H. Luo, and J. Furdyna, Proc. Conf. on Modulated Semiconductor Structures, Ann Arbor, Michigan (1989).
9. M. Kobayashi, M. Konagai, and K. Takahashi, J. Appl. Phys. 61, 1015 (1987).
10. S. Durbin, J. Han, Sungki O., M. Kobayashi, D. Menke, R.L. Gunshor, Q. Fu, N. Pelekanos, A.V. Nurmikko, D. Li, J. Gonsalves, and N. Otsuka, Appl. Phys. Lett. 55, 2087 (1989).
11. O. Goede, W. Heimbrodt, V. Weinhold, E. Schnurer, and H. Eberle, Phys. Status Solidi B: 143, 511 (1987).
12. O. Goede, W. Heimbrodt, V. Weinhold, E. Schnurer, and H. Eberle, Phys. Status Solidi B: 146, K65 (1988).
13. A. Klochikin, Y. Morozenko, and S. Permogorov, Sov. Phys. Solid State 20, 2057 (1978).
14. Q. Fu, N. Pelekanos, A.V. Nurmikko, S. Durbin, M. Kobayashi, R.L. Gunshor, Proc. Conf. on Modulated Semiconductor Structures, Ann Arbor, Michigan (1989).
15. Y.R. Shen, Phys. Rev. B9, 622 (1973).
16. D.S. Chemla and D.A.B. Miller, J. Opt. Soc. Am. B12, 1155 (1985).

RAMAN SCATTERING STUDY OF CdTe/CdMnTe SUPERLATTICES

L. Viña, F. Calle, J.M. Calleja, and F. Meseguer

Instituto de Ciencia de Materiales de Madrid-C.S.I.C.
and Departamento de Física Aplicada, C-IV
Universidad Autónoma, E-28049 Madrid, Spain

L.L. Chang, J. Yoshino, and M. Hong

IBM Thomas J. Watson Research Center
P.O. Box 218, Yorktown Heights, N.Y. 10598, U.S.A.

INTRODUCTION

Resonant Raman scattering (RRS) by LO phonons is an optical method which provides information on the relationship between lattice-dynamical and electronic properties of crystals. A resonant enhancement of the Raman-scattering efficiency is obtained whenever the energy of the incident photon (incoming channel) or the energy of the scattered photon (outgoing channel) approaches a singularity in the combined density of states of electronic interband transitions. Detailed information is obtained about the electron-phonon interactions, band structure parameters and optical deformation potentials.[1]

The substitution of the Cd element in the semiconductor CdTe by the magnetic transition-metal Mn leads to an interesting class of ternary compounds, known as diluted magnetic semiconductors. Mn^{2+} ions have large magnetic moments of 5.92 Bohr magnetons.[2] Therefore, they show a giant Zeeman splitting of the bands in external magnetic fields,[3] as a result of the exchange interactions between the localized $3d^5$ electrons of the Mn^{2+} ions and itinerant band electrons.[4] The energy gap of $Cd_{1-x}Mn_xTe$ increases with increasing Mn concentration, with a free-exciton energy dependence on Mn composition given by $E(x)=1.5976+1.564x$ (eV).[5] This can be used to build superlattices (SL's) and quantum wells by the successive deposition of CdTe and $Cd_{1-x}Mn_xTe$ layers. The lattice parameter of the alloy varies according to $a(x)=6.4807-0.1459x$ (Å),[2] therefore a strain occurs where the lattice mismatch is accommodated elastically in the different layers of the SL's.

The optical properties of these SL's have been studied extensively in the last few years.[6,7] The application of an external magnetic field also makes it possible to tune the electronic states, and therefore to investigate the effects of dimensionality on magnetic behavior.[8] The effects of external magnetic fields on the optical properties of SL's have been investigated by photoluminescence (PL) and PL excitation (PLE) up to

Light Scattering in Semiconductor Structures and Superlattices
Edited by D.J. Lockwood and J.F. Young, Plenum Press, New York, 1991

353

15T.[9] In $Cd_{1-x}Mn_xTe$, the Raman efficiencies for LO-phonon scattering near the fundamental gap, in the presence of magnetic fields, have been studied recently.[10] RRS studies of SL's have been performed by the Purdue[11,12] and other groups.[13-16] Here we present a RRS study of [111] $CdTe/Cd_{1-x}Mn_xTe$ SL's in the presence of an external magnetic field, and compare the Raman results with those obtained by PL and PLE spectroscopy.

EXPERIMENTAL DETAILS

The samples were grown by molecular-beam epitaxy on (100)-GaAs substrates. A 0.15μm (111)-oriented CdTe buffer layer was followed by a $CdTe/Cd_{1-x}Mn_xTe$ superlattice. A series of samples with x=0.21, a constant well width (d_1=86Å) and three different barrier thicknesses, d_2=86Å, 40Å and 20Å, with 25, 50 and 100 periods, respectively, were used to study the formation of SL's at zero magnetic field by decreasing the thickness between adjacent wells. The effects of the magnetic field were analyzed in a sample of 25 periods, with x=0.13 and wells and barriers of 100Å. The period and individual thicknesses were obtained from X-ray diffraction spectra and from small-angle X-ray interference.[17]

The Raman measurements were performed with a LD700 dye-laser pumped by a Kr$^+$-ion laser. The laser beam was focused onto the sample using a spherical lens and the power density was kept below 5 Wcm^{-2}. The scattered light was analyzed with a Jarrel-Ash 1-m double-grating monochromator and detected with standard photon-counting techniques. For the PL and PLE measurements the power was kept below 0.1 Wcm^{-2}. The incident light was circularly polarized by means of an achromatic λ/4 plate.[18] The emitted light was analyzed into its σ$^+$ and σ$^-$ components by means of a combination of a second achromatic λ/4 plate and a linear polarizer. The sample was immersed in a liquid-He bath-cryostat within a standard-coil super-conducting magnet, which enabled us to apply external magnetic fields, in the Faraday configuration, up to 13T. A bottom window on the cryostat allowed optical access to the specimens. All measurements were performed in a back-scattering configuration, with the directions of the incident and emitted light perpendicular to the (111) faces of the samples.

RESULTS AND DISCUSSION

1. Formation of Superlattices

We present, first, the study of the formation of SL's, at zero field, when the barriers thickness between adjacent wells is decreased. Figure 1 depicts unpolarized Raman spectra of a 86Å CdTe - 86Å $Cd_{0.79}Mn_{0.21}Te$ SL at different photon energies in the vicinity of the ground-state exciton. Previously, we have identified this exciton as the heavy-hole exciton (h_1).[15] However, as we will show later, this identification has to be viewed with caution, since a considerable admixture of 3/2 and 1/2 states seems to be present in the ground-state exciton. The character of this exciton depends strongly on the valence-band offset and the strain present in the samples. The problem of the band offset is still not clarified, although it is believed to be non-existent or very small.[19-21] The strain situation is difficult to deal with, since it depends strongly on the details of growth, such as temperature, the buffer layer and film thicknesses, and the sequence of the layers.[7] Above resonance (1.916 eV) four phonons are seen in the spectra. LO_1 and LO_2, at 166cm^{-1} and 198cm^{-1}, are assigned to the CdTe- and MnTe-like modes in the barrier,

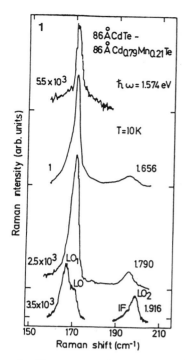

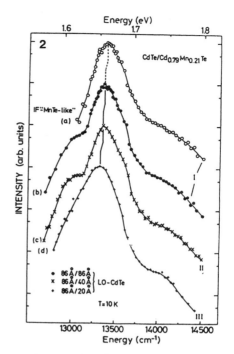

Fig. 1. Raman spectra of a 86Å CdTe / 86Å CdMnTe SL for different laser energies. The intensities of the spectra have been multiplied by the factors indicated on their left. LO_1 and LO_2 are the CdTe-like and MnTe-like phonons of the barriers. LO is the CdTe longitudinal optical phonon of the wells, and IF is the MnTe-like interface mode.

Fig. 2. Resonance Raman profiles for the MnTe-like IF mode and LO-CdTe phonon for a SL with 86Å barrier (curves a and b). Curves c and d depict the resonant behavior of the LO-CdTe phonon for samples with 40Å and 20Å barriers, respectively.

respectively, which presents a typical two-mode behavior. The other two peaks correspond to the LO phonon in the CdTe well ($171cm^{-1}$) and to a MnTe-like interface mode ($195cm^{-1}$).[11,15] Close to resonance, LO and IF are resonantly enhanced, LO_2, which is confined to the barriers, is no longer observed, while LO_1, which can propagate in the wells and barriers,[12] is seen as a shoulder on the low-energy side of LO. Below resonance (1.574 eV) only the LO phonon of CdTe is present in the spectra.

The two uppermost traces in Fig. 2 (a) and (b) show the resonance of the IF-mode and the CdTe LO-mode for the same sample as in Fig. 1. The incoming ($\sim13100cm^{-1}$) and the predominant outgoing channel ($13300cm^{-1}$) are clearly seen in curve (b). The similar behavior of both phonons confirms the interface character assigned to the IF-mode.[22] The small shift ($\sim25cm^{-1}$) between both resonances corresponds to the energy difference of their phonon frequencies.

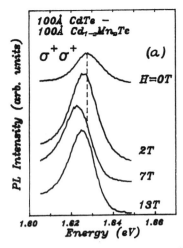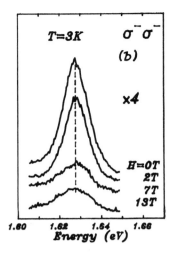

Fig. 3. PL spectra obtained in $\sigma^+\sigma^+$ configuration (left panel) and $\sigma^-\sigma^-$ (right panel) for different magnetic fields, at 3K for a sample with 100Å wells and barriers and a Mn concentration of 13%. The vertical scale for the right-hand spectra has been multiplied by a factor of 4.

The formation of minibands, as the barrier width is decreased, is suggested by a comparison between curves (b), (c), and (d), which correspond to three samples with the same well width (86Å) and Mn composition (x=0.21), but with different barrier thicknesses: 86Å (sample I), 40Å (II) and 20Å (III). A small shift towards lower energies and a concomitant broadening of the resonance is seen as the barrier width is decreased. The shifts, with respect to sample I, are $\Delta E=1(\pm1)$meV and $\Delta E=5(\pm2)$meV for samples II and III, respectively. The corresponding broadenings are $\Delta\Gamma=0(\pm1)$meV (sample II) and 9(±2)meV (sample III). An estimation, using a Kronig-Penney model with the parameters of Ref.14, gives a shift (broadening) of 2meV (4meV) and 7meV (12meV) for samples II and III, respectively, in agreement with the experimental results. This result from RRS experiments would be difficult to obtain with other optical techniques, such as photoluminescence, since the latter are less sensitive and can be hindered by extrinsic effects due to impurities and imperfections.

2. Photoluminescence in the Presence of Magnetic Fields

It is known that in $Cd_{1-x}Mn_xTe$ alloys the PL due to exciton recombination becomes circularly polarized when a magnetic field is applied in the Faraday configuration, and shows a saturation at fields as low as 0.5T. This behavior is explained as a result of the internal exchange field, due to the Mn^{2+} ions, which produces large Zeeman splittings of the bands.[5] A strongly circular polarized emission has been reported for CdTe/ $Cd_{1-x}Mn_xTe$ SL's, but only for the band-edge emission, not involving excitonic recombination,[9] and it has been attributed to the hole-Mn^{2+} exchange interaction.

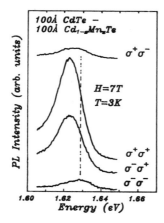

100Å CdTe –
100Å Cd$_{1-x}$Mn$_x$Te

PL Intensity (arb. units)

$\sigma^+\sigma^-$

H=7T
T=3K

$\sigma^+\sigma^+$

$\sigma^-\sigma^+$

$\sigma^-\sigma^-$

1.60   1.62   1.64   1.66
Energy (eV)

Fig. 4. PL spectra obtained at 7T and 3K for the four polarization configurations. The dashed line is a guide to indicate the shift of the σ^+ polarized emissions with respect to the σ^- ones.

We show, in Fig. 3, PL spectra for a SL with 100Å wells and barriers (25 periods) and a Mn concentration of 0.13, at different magnetic fields (from 0 to 13T). The spectra were excited with light of 1.76eV. The zero-field spectra correspond to the ground-state exciton of the SL. The left panel shows spectra obtained with σ^+ exciting light and analyzed in its σ^+ component. Similar spectra are depicted in the right panel for $\sigma^-\sigma^-$. The luminescence, which is unpolarized at zero field, becomes strongly polarized at small fields, indicating that the excitons interact, via the exchange interaction, with the Mn^{2+} ions in the Cd$_{1-x}$Mn$_x$Te layers. The σ^+ spectra show a red shift up to ~7T, and a small blue shift as the field is further increased. On the other hand, no appreciable shift is observed for the spectra obtained in the $\sigma^-\sigma^-$ configuration.

The strongest PL spectra are obtained with $\sigma^+\sigma^+$ polarizations. The strength of the emission is determined mainly by its polarization and it is nearly independent of the circular polarization of the exciting light. This is demonstrated in Fig. 4, which depicts PL spectra recorded in the four possible configurations at 7T. The $\sigma^+\sigma^+$ spectrum is approximately only twice stronger than the $\sigma^-\sigma^+$ one, but is ≈10 times greater than the σ^- polarized emissions. Likewise, appreciable shifts are only obtained for the right-handed polarized emission. The lower energy position of the right-handed circularly polarized emission should account partially for its greater strength. The pseudo-shift observed between the $\sigma^+\sigma^-$ and the $\sigma^-\sigma^-$ spectra is probably due to the presence of two peaks in the former spectrum. For the sake of simplicity, let us consider the emission process between electrons and holes in the $|J, m_J\rangle$ electron-hole representation, neglecting excitonic effects and interaction with the d-bands of the Mn^{2+}. The dominance of the right-handed polarized emission, together with further evidence obtained from PLE measurements in this sample (see below) and samples with different Cd$_{1-x}$Mn$_x$Te barrier widths,[23] indicates that the main emission takes place between $|1/2, \pm 1/2\rangle$ conduction- and nearly

357

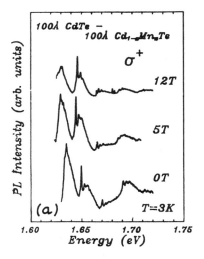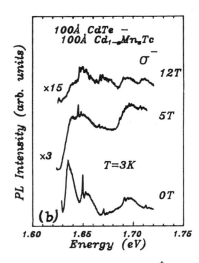

Fig. 5. (a) PLE spectra recorded at different fields with σ^+ exciting light at 3K. The sharp peaks are first- and higher-order LO-CdTe and IF-MnTe phonons.
(b) PLE spectra recorded at different fields with σ^- exciting light. The vertical scale of the spectra has been multiplied by the factors shown on their left.

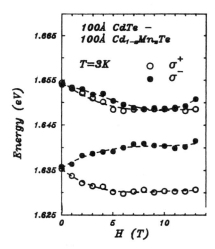

Fig. 6. Energy of the three lowest excitons of a CdTe/Cd$_{1-x}$Mn$_x$Te SL of 100Å wells and barriers (25 periods) and x=0.13 versus magnetic field. The open (full) circles are obtained from PLE spectra recorded with σ^+ (σ^-)-polarized exciting light.

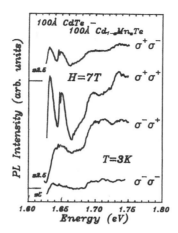

Fig. 7. PLE spectra at 7T and 3K for the four
polarization configurations, recorded
with the spectrometer set at 1.623
eV. The horizontal lines indicate the
offsets of the spectra, whose
intensities have been multiplied by
the factors shown on their left.

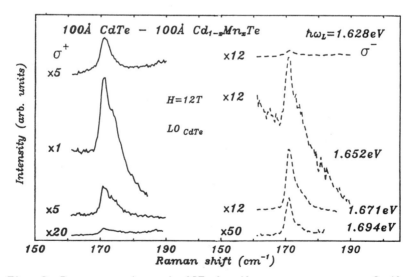

Fig. 8. Raman spectra at 12T in the energy range of the
the LO-CdTe phonon recorded at different laser
energies. Left (right) spectra were obtained with
$\sigma^+(\sigma^-)$ polarization. The intensities of the spectra
have been multiplied by the factors on their left.

degenerate $|3/2,-3/2\rangle$ and $|1/2,-1/2\rangle$ valence-band states. Since the valence-band offset is believed to be very small, the degeneracy of the heavy- and light-hole bands should be removed by the strain. However, our experimental results indicate that the strain in our [111] samples is negligible.

3. Excitation Spectroscopy in the Presence of Magnetic Fields

Figure 5 (a) shows PLE spectra recorded with σ^+ polarized light at three different magnetic fields. A red shift is clearly observed between the 0T and 5T spectra for the two lowest-lying excitons; at 12T the spectrum shifts again towards higher energies and new features appear in the spectral range. The additional structures due to the magnetic field will be discussed elsewhere.[23] The sharp peaks correspond to first and higher-order Raman scattering by LO and IF phonons. The Stokes shift of ~6meV between the emission and the lowest peak observed in the pseudo-absorption at 0T indicates that the excitons observed in emission are bound to some impurity[24] or defects localized near the interfaces.[12] The 1.655eV peak (0T) is probably related to excited states of the ground exciton[23] or to an exciton originating in the first confined electron band and the second confined hole band.[14] Corresponding spectra excited with left-handed, circularly polarized light are depicted in Fig. 5 (b). The pseudo-absorption obtained in this configuration is clearly smaller than that obtained with right-handed, circularly polarized light, as indicated by the enlargement factors shown to the left of the spectra. In this case a blue shift of the lowest-lying exciton and an increase in strength of the second exciton relative to the former one are observed.

The shift with magnetic field of the lowest transitions, observed in σ^+ and σ^- polarizations, are shown in Fig. 6. The lowest exciton splits symmetrically with magnetic fields up to ~7T, where a splitting of 10 meV is obtained. The shift of the exciton with H is about 1.6meV/T for weak fields. This corresponds to an effective g-factor of ~25, which indicates that the exciton interacts strongly with the Mn^{2+} ions. The presence of the Mn^{2+} ions in the $Cd_{1-x}Mn_xTe$ layers effectively amplifies the magnetic fields on the excitons,[25] due to the penetration of the excitonic wave-function into the paramagnetic barriers. No further change in the energy position is obtained above ~7T, indicating that the magnetization of $Cd_{1-x}Mn_xTe$ is saturated. The splitting of the second exciton is almost non-existent within experimental accuracy, and it behaves rather differently than the ground-state exciton, since a red shift is obtained that is independent of the polarization of the exciting light.

Finally, we present, in Fig. 7, excitation spectra at 7T, where the emitted light has also been analyzed into its circularly polarized components. Several facts are worth noting in this figure: the highest intensities are obtained, as we have pointed out before, for right-handed polarized emission, the largest one corresponding to the $\sigma^+\sigma^+$ configuration, and the smallest to $\sigma^-\sigma^-$. Let us now focus on the two lowest states of Fig.6, at 7T. The peak position depends only on the polarization of the exciting light. If we again consider a $|J,m_J\rangle$ electron-hole representation, a simplified scheme for the valence band can be obtained from the results of Fig. 7. We will use a picture with only two levels at different energies in the valence band (A, the lowest lying and B, the next one in energy). The fact that the 1.63eV peak (the lowest peak) is not observed in the $\sigma^-\sigma^-$ configuration indicates that the state $|3/2,3/2\rangle$ does not contribute appreciably to A. Similarly, the absence of the 1.638eV peak in $\sigma^+\sigma^+$ suggests the non-participation of $|3/2,-3/2\rangle$ in B. The relatively large intensity of the 1.638eV peak in $\sigma^-\sigma^-$ indicates

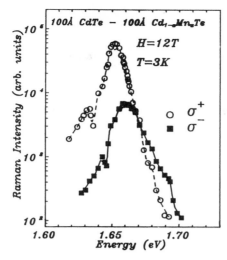

Fig. 9. Resonance Raman profiles at 12T for the LO-CdTe phonon obtained with σ^+ (open circles) and σ^- (full squares) polarized exciting light. The large (small) peak corresponds to the outgoing (incoming) resonance.

the participation of $|1/2,-1/2\rangle$ in A and/or $|1/2,1/2\rangle$ in B. On the other hand, the observation of the 1.63eV (1.638eV) peak in $\sigma^+\sigma^-$ ($\sigma^-\sigma^-$), but with low intensity, indicates some small admixture of $|J,+\rangle$ and $|J,-\rangle$ states. The situation is actually field-dependent,[23] but we propose that the A state is mainly an admixture of $|J,-\rangle$ and the B state an admixture of $|J,+\rangle$ states.

4. Resonant Raman Scattering

We will concentrate on the LO-CdTe phonon, since the behavior of the IF and the LO-CdTe phonons is quite similar. Raman spectra at 12T for σ^+ and σ^- incident polarized light and unanalyzed scattered light are presented in Fig.8 for different laser energies. The phonons are considerably larger for σ^+ polarization at the same laser energy. A large enhancement is observed when the scattered light approaches the excitons observed in PLE spectra at 1.631eV (1.641eV) for $\sigma^+(\sigma^-)$. The resonant behavior is depicted in Fig. 9 for both polarizations. The incoming and outgoing channels are clearly observed, with the outgoing mechanism being the most important, as in the zero-field case. The difference in the energy of the resonance maximum, ~10meV, agrees with that obtained in the excitation experiments. The resonance is ~10 larger for σ^+, the same factor which is also obtained in the excitation (compare the uppermost traces of Figs. 5 (a)and (b)).

Figure 10 compares the unpolarized resonance at 0T and that obtained with σ^+ polarization at 7T. Besides a narrowing of the resonance and a further enhancement when the magnetic field is applied, a red shift of ~6meV is obtained between 0T and 7T, again in very good agreement with the PLE results. The ratio of incoming to outgoing channels seems to be field-independent, in contrast with the results obtained in GaAs/GaAlAs SL's.[26]

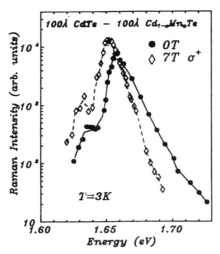

Fig. 10. Resonance Raman profiles for 0T
and 7T. The zero-field trace was
obtained from unpolarized Raman
spectra, while the 7T resonance
corresponds to σ^+ incident light.
Note the red shift of the
resonance when the magnetic field
is applied.

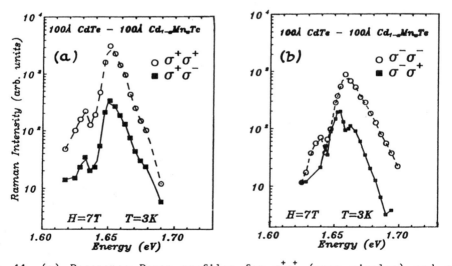

Fig. 11. (a) Resonance Raman profiles for $\sigma^+\sigma^+$ (open circles) and $\sigma^+\sigma^-$
(full squares) at 7T. The peak at 1.65 eV corresponds to the
outgoing resonance with the ground-state exciton, while the 1.63
eV peak is due to the incoming channel.
(b) Resonance Raman profiles for $\sigma^-\sigma^-$ (open circles) and $\sigma^-\sigma^+$
(full squares) at 7T. Note the dominance of the Fröhlich
interaction (parallel polarizations) over the deformation-
potential mechanism (orthogonal polarizations).

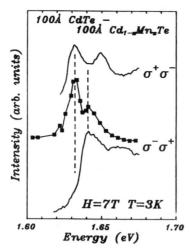

Fig. 12. Comparison between PLE spectra recorded in $\sigma^+\sigma^-$ and $\sigma^-\sigma^+$ with the resonance Raman profile for the latter configuration. The two parallel dashed-vertical lines are guides for the eye. The resonance profile has been red-shifted by a LO-CdTe phonon energy.

For a [111] surface and LO-phonon scattering, the Raman tensors for Fröhlich and deformation-potential mechanisms can be written as:

$$R_F = \begin{pmatrix} a & 0 & 0 \\ 0 & a & 0 \\ 0 & 0 & a \end{pmatrix} \qquad R_D = (1/3)^{1/2} \begin{pmatrix} 0 & d & d \\ d & 0 & d \\ d & d & 0 \end{pmatrix}$$

This implies that for circularly polarized incident and scattered light the Fröhlich mechanism should be observed for parallel polarizations, while for orthogonal polarizations the deformation-potential mechanism should be the most important. Figure 11 (a) shows the resonances at 7T for $\sigma^+\sigma^+$ (o) and $\sigma^+\sigma^-$ (∎) polarizations. Both resonances present a maximum at the same energy, and the one corresponding to the Fröhlich mechanism $(\sigma^+\sigma^+)$ is larger by a factor of ~10. However, the behavior is slightly different in the case of σ^- polarized incident light, as can be seen in Fig. 11 (b). A double peak (in the outgoing channel) is observed for $\sigma^-\sigma^+$, while only a broad peak is obtained for $\sigma^-\sigma^-$. The non-resolution of both structures makes this peak appear at an intermediate position between those obtained in orthogonal polarizations.

A comparison of the resonance Raman profile and excitation spectra obtained at 7T enables us to identify the origin of the double structure observed for $\sigma^-\sigma^+$ in RRS. This is shown in Fig. 12, where the resonant Raman profile has been rigidly shifted towards the red by a LO-CdTe phonon energy. The lower energy peak clearly corresponds to the ground-state exciton (at this field), while the high energy peak is associated with the exciton observed with σ^- polarized incident light. This double structure cannot be resolved if the scattered light is not analyzed, due to the dominance of the $\sigma^-\sigma^-$ configuration close to resonance.

SUMMARY

We have presented a resonant Raman study of [111] $CdTe/Cd_{1-x}Mn_xTe$ SL's, with special emphasis on the formation of minibands with decreasing barrier thickness and on the effects of external magnetic fields. Large effects of the Mn^{2+} ions of the barriers on the excitons have been measured, indicating a strong exchange interaction between the localized $3d^5$ electrons of the Mn^{2+} ions and the excitons. The shifts obtained in the resonance profiles and the enhancements, for different polarizations, agree with those measured by photoluminescence excitation spectroscopy. For the LO-CdTe phonon, the relative importance of the incoming and outgoing channels are shown to be field-independent. Using circularly polarized exciting light and analyzing the circular polarization of the scattered light, we have found the strongest resonances in parallel configurations, corresponding to the Fröhlich mechanism.

ACKNOWLEDGMENTS

We want to thank C. López for his help in some of the experiments. This work was sponsored in part by CICYT Grant No. MAT-88-0116-C02-02.

REFERENCES

1. M. Cardona, "Light Scattering in Solids, Topics in Applied Physics Vol.50", M. Cardona and G. Güntherodt, eds., Springer, New York (1982).
2. R.L. Harper, Jr., R.N. Bicknell, D.K. Blanks, N.C. Giles, J.F. Schetzina, Y.R. Lee and A.K. Ramdas, J. Appl. Phys. 65, 624 (1989).
3. D.L. Peterson, A. Petrou, M. Datta, A.K. Ramdas, and S. Rodriguez, Solid State Commun. 43, 667 (1982).
4. N.B. Brandt and V.V. Moshchalkov, Advances in Physics 33, 193 (1984).
5. D. Heiman, P. Becla, R. Kershaw, D. Ridgley, K. Dwight, A. Wold, and R. Galazka, Phys. Rev. B 34, 3961 (1986).
6. For a review, see, for example: A.V. Nurmikko, R.L. Gunshor, and L.A. Kolodziejski, IEEE J. Quant. Electr., QE22, 1785 (1986); L.A. Kolodziejski, R.L. Gunshor, N. Otsuka, S. Datta, W.M. Becker and A.V. Nurmikko, *ibid.* 1666 (1986).
7. L.L. Chang, Superlatt. and Microst. 6, 39 (1989).
8. D.D. Awschalom, G. Grinstein, J. Yoshino, H. Munekata, and L.L. Chang, Surf. Sci. 196, 649 (1988).
9. J. Warnock, A. Petrou, R.N. Bicknell, N.C. Giles-Taylor, D.K. Blanks, and J.F. Schetzina, Phys. Rev. B 32, 8116 (1985).
10. W. Limmer, H. Leiderer, and W. Gebhardt, "Growth and Optical Properties of Wide-Gap II-VI Low-Dimensional Semiconductors, NATO ASI Serie B, Vol 200", T.C. McGill, C.M. Sotomayor Torres and W. Gebhardt, eds., Plenum Press, New York (1989), p. 281.
11. S. Venogupalan, A. Petrou, R.R. Galazka, A.K. Ramdas, and S. Rodriguez, Phys. Rev. B 25, 2681 (1982).
12. E.-K. Suh, D.U. Bartholomew, A.K. Ramdas, S. Rodriguez, S. Venogupalan, L.A. Kolodziejski, and R.L. Gunshor, Phys. Rev. B 36, 4316 (1987).

13. S.-K. Chang, H. Nakata, A.V. Nurmikko, L.A. Kolodziejski, and R.L. Gunshor, Journal de Physique C5, 345 (1987).
14. L. Viña, L.L. Chang, and J. Yoshino, Journal de Physique C5, 317 (1987).
15. L. Viña, F. Calle, J.M. Calleja, C. Tejedor, M. Hong, and L.L. Chang, "Proceedings of the 19th International Conference on the Physics of Semiconductors", W. Zawadzki, eds., Institute of Physics, Polish Academy of Sciences, Warsaw (1988), p. 819.
16. L. Viña, L.L. Chang, M. Hong, J. Yoshino, F. Calle, J.M. Calleja, and C. Tejedor, "Growth and Optical Properties of Wide-Gap II-VI Low-Dimensional Semiconductors, NATO ASI Serie B, Vol 200", T.C. McGill, C.M. Sotomayor Torres and W. Gebhardt, eds., Plenum Press, New York (1989), p. 293.
17. L.L. Chang and E.E. Mendez, "Synthetic Modulated Structures, Chapter 4", L.L. Chang and B.C. Giessen , eds., Academic Press, Orlando, (1985).
18. Superachromatic λ/4 Plate. Bernhard Halle Nachf. GmbH. Huberstusstraße 10. D-1000 Berlin. F.R.G.
19. M. Pessa and O. Jylha, Appl. Phys. Lett. 45, 646 (1984).
20. S.-K. Chang, A.V. Nurmikko, J.-W. Wu, L.A. Kolodziejski, and R.L. Gunshor, Phys. Rev. B 37, 1191 (1988).
21. R.N. Bicknell-Tassius, N.C. Giles and J.F. Schetzina, "Growth and Optical Properties of Wide-Gap II-VI Low-Dimensional Semiconductors, NATO ASI Serie B, Vol 200", T.C. McGill, C.M. Sotomayor Torres and W. Gebhardt, eds., Plenum Press, New York (1989), p. 263.
22. A.K. Sood, J. Menendez, M. Cardona, and K. Ploog. Phys. Rev. Lett. 54, 2115 (1985).
23. L. Viña, F. Calle, C. Lopez, L.L. Chang and J. Yoshino, unpublished.
24. L. Viña, R.T. Collins, E.E. Mendez, and W.I. Wang, Phys. Rev. B 33, 5939 (1986).
25. S.M. Ryabchenko, O.V. Terletskii, I.B. Mizetskaya, and G.S. Olenik, Fiz. Tekh. Poluprovodn.15, 2314 (1981) [Sov. Phys.-Semic. 15, 1345 (1981)].
26. J.M. Calleja, F. Meseguer, F. Calle, C. Lopez, L. Viña, C. Tejedor, K. Ploog and F. Briones, this volume.

NONEQUILIBRIUM ELECTRONS AND PHONONS

IN GaAs AND RELATED MATERIALS

J. A. Kash

IBM Research Division
T. J. Watson Research Center
Yorktown Heights, NY 10598 USA

INTRODUCTION

Throughout most of this volume, light scattering is used to probe the excitations of a semiconductor or semiconductor microstructure which is in thermal equilibrium. When a sample is described as being in thermal equilibrium, then each excitation of the sample (carriers, phonons, etc.) can be described by an appropriate Fermi-Dirac or Bose-Einstein distribution function, where each distribution function is characterized by the same temperature. For these equilibrium studies, the excitation laser photons are presumed not to alter the distribution functions in a measurable way.

On the other hand, when a laser photon is absorbed by the sample (as is the case for visible photons since typical semiconductor band gaps are less than 2 eV) any excess photon energy above the band gap is given to the photogenerated carriers as kinetic energy. Such photoexcited carriers are commonly referred to as "hot carriers", although "nonequilibrium carriers" is a more precise description. In this article, we will trace the flow of this excess kinetic energy, as it goes from the nonequilibrium carriers to a general heating of the sample. It will be shown that most of relaxation of the carriers occurs in the first 10 psec after carrier generation, with interaction times in some cases shorter than 100 fsec. The eventual recombination of the electron-hole pairs, which occurs on the much longer nanosecond time scale, will not be discussed here.

Because of these short times, direct temporal investigation of these effects has not been possible until the development of picosecond and femtosecond laser pulses. For practical reasons (direct band gap in the near infrared and a light electron effective mass, plus the availability of high quality samples), most of the studies to date have focussed on GaAs and related materials. Because the phenomena to be studied involve both carriers and phonons, several related experimental probes are used in these studies. The most common optical techniques[1] used are absorption, reflection, Raman scattering, band gap luminescence, and hot luminescence. In this paper, Raman scattering will be examined as a probe of the nonequilibrium phonons generated as the hot carriers relax. To examine the distribution function of the relaxing carriers themselves, the related technique of hot luminescence will be explored. Other articles in this volume discuss the use of electronic Raman scattering to directly examine the nonequilibrium carriers. The majority of the experiments to be discussed will be on bulk samples. Although recent work[2] has begun to extend these techniques to microstructures, qualitative differences between bulk and microstructure samples have not generally been observed.

Light Scattering in Semiconductor Structures and Superlattices
Edited by D.J. Lockwood and J.F. Young, Plenum Press, New York, 1991

TIME RESOLVED RAMAN SCATTERING IN GaAs AND $Al_xGa_{1-x}As$

When a visible photon is absorbed in GaAs, the typical excess photon energy is 0.3 to 0.5 eV. This energy is divided between the electron and hole, with the lighter mass electron receiving most of the energy. In a polar semiconductor, for injected carrier densities less than about 2×10^{16} cm^{-3}, the dominant interaction for the hot electron is with small wavevector LO phonons through the polar (Fröhlich) coupling.[3] The range of LO phonon wavevectors excited in this process is illustrated in fig. 1(a). For an electron of kinetic energy $E_e = \hbar^2 k^2/2m^*$, $q_{min} < q < q_{max}$, where

$$q_{min} \simeq k \frac{\hbar \omega_{LO}}{2E_e} \quad , \quad q_{max} \simeq 2k \ , \qquad (1)$$

for $E_e >> \hbar \omega_{LO}$. This range of wavevector is determined only by the kinematics of the decay, i.e. by the available final states of the electron and LO phonon. Since the LO phonon dispersion is small, the energy of the emitted LO phonon will be taken as independent of its wavevector, so that the final states of the electron lie on the surface of constant energy $E_e - \hbar \omega_{LO}$, which is a sphere for a parabolic electron band. Because the polar interaction is stronger for small wavevector phonons, the LO phonon wavevector distribution for the scattering process is strongly peaked at q_{min}, as illustrated in fig. 1(b). By coincidence, the phonon wavevector for first-order Raman scattering in the backscattering geometry for visible photons is typically 6 to 8 $\times$ 10^5 cm^{-1}, quite near the peak of this nonequilibrium LO phonon distribution. Since the intensity of anti-Stokes Raman scattering is proportional to the mode occupation n of the phonons, it may be used to sample the nonequilibrium LO phonon population.

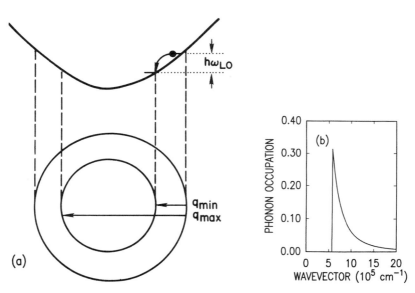

Fig. 1. (a) The kinematics of LO phonon emission in the Γ valley of GaAs. A spherical conduction band and no dispersion for the LO phonon are assumed. (b) The mode occupation of nonequilibrium LO phonons excited by 10^{16} cm^{-3} electrons with 0.35 eV kinetic energy in the Γ valley of GaAs as a function of the phonon wavevector. Each electron emits a single LO phonon.

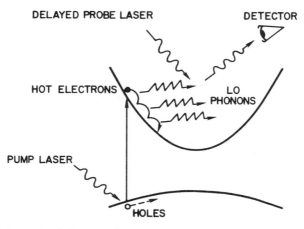

DELAYED PROBE LASER   DETECTOR

HOT ELECTRONS   LO PHONONS

PUMP LASER

HOLES

Fig. 2. Schematic of the experiment to measure nonequilibrium LO phonon
populations with time resolved Raman scattering.

The basic technique[4,5] of pump/probe Raman scattering to sample this nonequilibrium LO phonon distribution is illustrated in figure 2. A short laser pulse is split into orthogonally polarized pump and probe pulses which are then recombined on the sample with a suitable temporal delay. The pump pulse is used to photogenerate carriers. As these carriers relax, they generate nonequilibrium LO phonons. Anti-Stokes Raman scattering excited by the probe pulse then samples the nonequilibrium phonon population. A polarizer at the entrance slit of the monochromator is used to discriminate against Raman scattering excited by the pump pulse. A background spectrum, with Raman scattering from thermal phonons only, is taken with the probe preceding the pump by roughly 10 psec; both thermal and nonequilibrium phonons are seen when the probe follows the pump. Typical spectra are shown in fig. 3. The rise and decay of the nonequilibrium LO phonon population may be clearly seen.

Note that the width of the Raman peaks in this figure are much broader than the LO phonon peak ordinarily seen in cw Raman scattering. This width reflects the spectral width of the laser pulse[6]. In order to map out the phonon kinetics, the laser-pulse width must be of order or shorter than the LO phonon lifetime. Because of the uncertainty principle, the frequency width of the laser is therefore wider than the intrinsic homogeneous width of the LO phonon. The spectral width of such short laser pulses limits the ultimate resolution of this technique to times longer than the period of the phonon, about 120 fsec for LO phonons in GaAs.

We return now to the rise and decay of the nonequilibrium LO phonons. As shown in figure 4, the build-up of the LO phonon population occurs in about 2 psec, with a decay time of 3.5 psec at 300K. At 2K, the same rise time is observed, but the decay takes about 9 psec. The rise will be discussed first. In fig. 5 we show the theoretical polar scattering rate[3] in GaAs for an electron of energy E_e at 0K and at 300K. At 0K, only spontaneous phonon emission is allowed. At 300K, the LO phonon thermal mode occupation is $n_{th} = 0.31$ so that stimulated emission and absorption also occur. These two stimulated terms are nearly equal for $E_e/\hbar\omega_{LO} > 1$, so that the net rate of phonon generation (emission minus absorption) is virtually identical at 300K and at 0K. As a result, the rise of the nonequilbrium LO phonon population is independent of temperature. Further, note that for $E_e/\hbar\omega_{LO} > 3$ the net generation

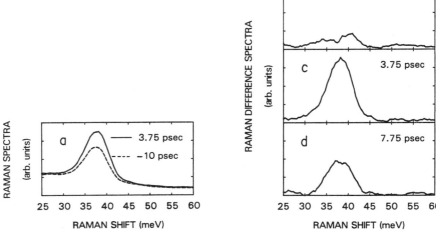

Fig. 3. (a) Anti-Stokes spectrum obtained with time resolved Raman scattering from GaAs at 300K. The dashed curve is the probe-excited
spectrum when the probe comes 10 psec before the pump. The solid
curve is obtained when the probe comes 3.75 psec after the pump.
The difference between these curves, which is the contribution due
to nonequilibrium phonons, is shown in (c) (to a different scale).
Difference spectra for other delays are shown in (b) and (d). For all
curves, the carrier density injected by the 0.5 psec laser pump pulse
at 2.09 eV is about 10^{16} cm^{-3}.

rate is essentially constant at about 200 fsec. Because the net generation rate is independent
of both sample temperature and electron energy, one can picture the electron relaxation as the
emission of a "cascade" of LO phonons*, where each step in the cascade should take about
200 fsec. With this model, interpretation of the observed rise time of the LO phonons is
straightforward. First consider the effect of reducing the laser photon energy from 2.09 to
1.91 eV. This corresponds to decreasing the initial electron kinetic energy by 0.15 eV, or four
LO phonons. Thus, there are four fewer steps in the cascade, so the observed decrease in the
rise time of 0.7 psec shows that the LO phonon emission time is about 175 fsec. A more detailed cascade model[7] accounts for the generation of electrons from both the heavy and light
hole valence bands, and recognizes that intervalley scattering will be much more important for
electrons generated from the heavy hole band. This model also includes the measured 3.5 psec
decay of the nonequilibrium LO phonon population at 300K. The results of the model, shown
as the dashed curve in fig. 4, indicates that the LO phonon emission time is 190 fsec, in good
agreement with fig. 5.

---

*This cascade model is valid only if carrier-carrier scattering can be ignored. As discussed
later, this assumption is valid at carrier densities below about 2×10^{16} cm^{-3}. However, even
if carrier-carrier scattering is significant, the net rate of energy loss to LO phonons is essentially unchanged and so the rise time of the nonequilbrium LO phonon population will be independent of carrier density until the coupling of the electron plasmon and the LO phonon
becomes significant, i.e. at densities exceeding 2×10^{17} cm^{-3} .

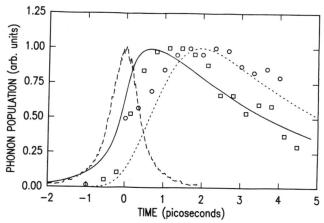

Fig. 4. The time dependence of the intensity of the pump-induced anti-Stokes
Raman scattering in GaAs at 300K under 2.09 eV (circles) and 1.91
eV (squares) excitation energies. The pump-injected carrier density
is $n = p = 10^{16}$ cm^{-3} for the 2.09 eV laser and
$n = p = 8 \times 10^{16}$ cm^{-3} for the 1.91 eV laser. The dashed curve
shows the autocorrelation of the 1.91 eV laser pulses. The solid curve
shows the predicted Raman intensity if the phonons were generated
instantly and decayed in 3.5 psec. The dotted curve shows the predicted response for the 2.09 eV laser excitation using the cascade
model with a 0.19 psec phonon emission time and a 3.5 psec phonon
decay time as explained in the text.

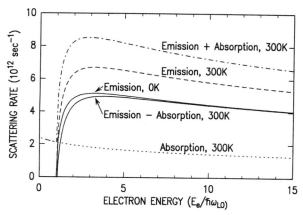

Fig. 5. Theoretical rates for the emission and absorption of LO phonons
through the Fröhlich coupling at 0K and 300K. The nonequilibrium
phonon mode occupations are assumed to be much less than one.

The decay of the nonequilibrium LO phonon population is 3.5 psec at 300K, and increases[8] to 9 psec at 2K. For injected carrier densities below 10^{17} cm^{-3}, the decay of the phonon population is due to the spontaneous decay of the nonequilibrium phonons. The temperature dependence has been successfully modeled[7] by assuming that these small wavevector phonons decay anharmonically into two acoustic phonons of opposite wavevector and half the LO phonon energy. For GaAs, it is found[4] that the inverse linewidth as measured with cw Raman scattering corresponds to the LO phonon lifetime, here measured directly with time resolved Raman scattering.

Studies comparable to the above have also been performed in Al$_x$Ga$_{1-x}$As for x < 0.3.[9,10] In Al$_x$Ga$_{1-x}$As, the presence of two LO phonon modes (GaAs-like and AlAs-like) divides the polar strength according to the spectroscopic ionicity of each mode[9]. For small x, most of the polar strength is in the GaAs-like mode, while for x approaching 1 most of the coupling is to the AlAs-like mode. This variation of the coupling strength with x has been experimentally observed by comparing the nonequilibrium mode occupation in each of the two modes as a function of x. In addition, the lifetime of each of the LO phonon modes has been measured.[9,10] The lifetime of both modes has been found to be the same, and in fact is the same as for pure GaAs[*]. Assuming that the decay mechanism is the same as in GaAs, it is somewhat surprising that the decay times are the same, inasmuch as the acoustic phonons generated as the LO phonons decay are not the same in each case.

NONEQUILBRIUM LO PHONON WAVEVECTOR DISTRIBUTIONS

As shown in fig. 1(b), the nonequilibrium LO phonon distribution is strongly peaked near q_{min} for the emission of a single LO phonon by a nonequilibrium electron. In fig. 6, the distribution of nonequilibrium LO phonons generated by the cascade of electrons starting at an initial energy of 0.5 eV is shown. This curve is the sum of a series of curves like fig. 1(b). As mentioned above, typical phonon wavevectors in Raman backscattering in GaAs are near the peak of this distribution. The backscattering geometry probes the largest possible wavevector of any scattering geometry, so unfortunately Raman scattering may not be used to experimentally measure the nonequilibrium phonon mode occupation at larger wavevector. One can, however, make an important test of the wavevector distribution by using the forward scattering geometry[11]. In exact forward scattering, q is very nearly zero. Allowing for a finite collection angle increases the range of q slightly, but it is still less than about 1.5×10^5 cm^{-1}. As can be seen in fig. 6, no nonequilibrium phonons are expected for this range of wavevectors. To compare the nonequilibrium LO phonon population in backscattering as opposed to forward scattering, the sample must be thin enough to be semitransparent to the laser light yet thick enough to be essentially bulk. For a layer of thickness L, we require $\pi/L < 5 \times 10^5$ cm^{-1}, or L > 2000Å.[**] To perform the comparison, a sample was epitaxially grown on a GaAs (100) substrate consisting of a 1 micron thick buffer layer of Al$_{0.70}$Ga$_{0.30}$As transparent to the laser and an active layer of 3300 Å of GaAs In addition, a 750 Å silicon nitride film was applied to the front surface of each sample as an antireflection (AR) coating. The AR coating reduced the reflection at normal incidence to less than one percent and limited the crosstalk between back and forward scattered light by reducing the intensity of the Raman scattered light reflected at the interior surface of the sample. After growth, a small area of the substrate was selectively etched away to reveal a semitransparent window of the epitaxial layers. Pump/probe Raman scattering was performed using 2.5 psec (autocorrelation FWHM) pulses at 2.10 eV. The average power in each pulse train was 2 mW, generating a peak carrier concentration of about 5×10^{16} cm^{-3} in a 10^{-4} cm^2 focal spot.

---

[*]Unlike pure GaAs, the measured cw Raman linewidth in Al$_x$Ga$_{1-x}$As is asymmetric and broader than the inverse lifetime. Thus the cw Raman line is not purely homogeneously broadened in Al$_x$Ga$_{1-x}$As. The next section has a further discussion of this point.

[**]Note that the absorption depth of the laser light must also be greater than 2000 Å in order to avoid uncertainty due to the finite absorption depth.

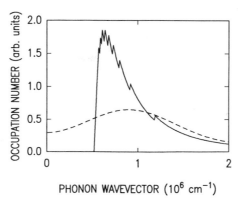

Fig. 6. Predicted wavevector distribution of nonequilibrium LO phonons created by a cascade of of electrons in the Γ valley of GaAs starting at an initial energy of 0.5 eV (solid curve). A parabolic band of mass 0.067 is assumed. Effects such as carrier-carrier scattering and intervalley scattering are ignored. Including such effects does not change the essential features of the curve. The dashed curve is a convolution of the solid curve with the function $\left[\sin(qL/2)/(qL/2)\right]^2$ for L = 500 Å, demonstrating the effects of wavevector uncertainty in a 500 Å layer.

Sample heating, monitored through the Stokes / anti-Stokes ratio, was unimportant for the samples, which were kept at room temperature. To allow a direct comparison of the back- and forward-scattered data, the backscattered light was collected by a lens and the forward-scattered light was collected with a mirror; the incident laser beams and sample were not disturbed. In all cases, anti-Stokes Raman spectra were taken with the probe preceding the pump by 17 psec (background spectrum with scattering from thermal phonons but no nonequilibrium phonons), and also with the probe following the pump by 2.5 psec, when the nonequilibrium phonon signal is maximum. The difference spectrum directly gives the absolute nonequilibrium phonon occupation n_{ne} when compared to the background signal from only the thermal phonons with known occupation $n_{thermal} = 0.31$. The comparison between forward and backscattering for this sample is shown in fig. 7. The nonequilibrium mode occupation seen in forward scattering is probably zero and is at most one tenth of the mode occupation observed in backscattering, verifying the prediction of the solid curve in fig. 6.

An interesting variation on this result[11] is seen when the GaAs layer thickness is much less than L = 2000Å. For such a thin layer, the phonon wavevector perpendicular to the layer plane becomes uncertain by an amount greater than q_{min}. (An alternative way of describing this wavevector uncertainty is to describe the phonon modes as slab modes instead of the bulk modes.) As the wavevector becomes uncertain, the measured distribution of hot phonon wavevectors is expected to spread out; the expected distribution for L = 500Å is shown in fig. 6 as the dashed curve. The comparison between forward and backscattering for a 500 Å sample is shown in fig. 8. Here we see essentially the same mode occupation in forward as in backscattering, reflecting the wavevector uncertainty in the growth direction for the thin layer.

Based upon cw Raman measurements of the lineshape of LO phonons in $Al_xGa_{1-x}As$, it has been suggested[12] that alloy disorder localizes optic phonons in $Al_xGa_{1-x}As$ with a localization length of typically 200 Å. If such localization occurred, then there should be nonequilibrium LO phonons observed in forward scattering in $Al_xGa_{1-x}As$ even for a thick layer. To test this suggestion, forward and backscattering were compared in a 3300 Å layer of $Al_{0.11}Ga_{0.89}As$. The results[11] were exactly as in the 3300 Å layer of GaAs, i.e. no nonequilibrium phonons

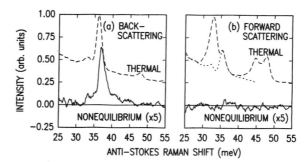

Fig. 7. Comparison of the anti-Stokes Raman scattering from thermal (dashed curves) and nonequilibrium (solid curves) LO phonons in back and forward scattering geometries for a 3300 Å thick GaAs layer. The spectrum for the thermal phonons includes Raman scattering from the 1 micron $Al_{0.7}Ga_{0.3}As$ support layer. The contribution from the GaAs layer LO phonon is shown as the dotted curve in forward scattering.

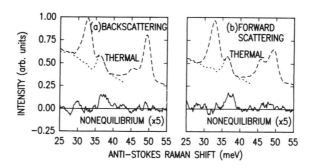

Fig. 8. Same as fig. 8, except the GaAs layer is 500 Å.

were observed in forward scattering. Therefore, the small wavevector LO phonons generated by relaxing electrons in $Al_xGa_{1-x}As$ are not localized, and the measured cw lineshape must be due to different reasons.

HOT (e,A^0) LUMINESCENCE AS A PROBE OF HOT ELECTRON DYNAMICS

So far we have described the use of Raman scattering to measure the dynamics of the nonequilibrium LO phonons, but have not directly studied the electrons which are generating the phonons. In principle, electronic Raman scattering can be used to probe the dynamics of the electrons directly[6]. As we have seen, however, the LO phonon emission time is 200 fsec in GaAs, so that extremely short laser pulses would be needed to monitor changes in the nonequilibrium electron distribution. The broad spectral width of such short laser pulses would make discrimination against the elastic light scattering very difficult. To avoid these difficulties, a different probe - hot luminescence - is used. When generating nonequilibrium carriers at low carrier densities the electrons will cascade down the conduction band as discussed above. When injecting the electrons with initial kinetic energy E_i with a cw laser, one expects the electron energy distribution to consist of a series of peaks at energies E_i, $E_i - \hbar\omega_{LO}$, $E_i - 2\hbar\omega_{LO}$, ..., plus a large pool of electrons near the conduction band minimum. If a small number of acceptors is introduced into the sample by doping, then at low temperatures (where the acceptors are neutral) some of the electrons can radiatively recombine at the

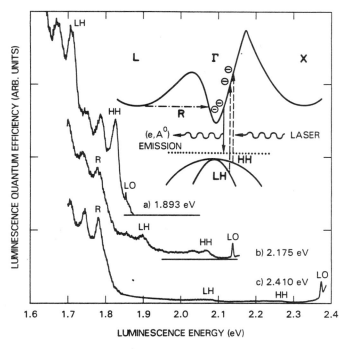

Fig. 9.   Hot (e,A^0) emission spectra at $T = 2K$ for GaAs:Mg, $p = 1.2 \times 10^{17}$ cm^{-3}. The spectra were excited at (a) 1.893 eV, (b) 2.175 eV, and (c) 2.410 eV with cw lasers. Injected carrier densities were below 10^{15} cm^{-3}, and the intensities of the spectra are normalized to constant laser photon flux. The inset shows the origins of the three sets of oscillations, as discussed in the text. The solid arrow indicates the radiative recombination of an electron at a neutral acceptor. The narrow peaks marked LO are Stokes Raman scattering from LO phonons.

neutral acceptors. Usually, one measures this (e,A⁰) luminescence from the electrons which are near the conduction band minimum. However, since the acceptors are localized in real space their wavefunctions spread out in k-space. As a result, the nonequilibrium electrons also have a small (but non-zero) probability of recombination. Typical spectra for this hot (e,A⁰) luminescence and a schematic illustration of the source of the emission are shown in fig. 9. As discussed by several authors[13,14], the spectra show three distinct sets of oscillations with periods close to the 37 meV zone center LO phonon energy. The set denoted HH involves electrons excited from the heavy hole valence band. A second set, denoted LH, involves electrons excited from the light hole band. The spectra from these transitions, which will both be called direct spectra, track the laser excitation energy. A third set, denoted R for re-entrant, is observed only for excitation energies greater than the threshold at which electrons are created that can scatter to the X and L valleys (i.e. curves b and c in fig. 1). These spectra, which are fixed in energy, result from carriers which reenter the Γ valley from the L minima. At excitation energies above 2.3 eV, warping of the valence and conduction bands broaden the hot electron distributions so much that distinct LO phonon oscillations become hard to observe in the direct spectra, although the re-entrant series remains sharp.

We have shown above that the LO phonon emission time is 200 fsec for these nonequilibrium electrons. Therefore, each peak in the hot (e,A⁰) luminescence corresponds to a specific time in the relaxation of the electrons. As a result, the luminescence can be used as an "internal clock" where the resolution is the LO phonon emission time. Other workers have used the lineshape[14] or magnetic field depolarization[13] of the luminescence to infer information about the electron dynamics. Here we will discuss the use of the intensity of the emission as a direct probe of the electron dynamics.

BREAKDOWN OF THE CASCADE MODEL

The spectra shown in fig. 9 were taken with cw lasers, with excited carrier densities below 10^{15} cm^{-3}. If one uses a picosecond laser pulse as an excitation source, an identical hot (e,A⁰) spectrum is seen for the same average power as the cw laser. This is not surprising, as the hot (e,A⁰) emission is measured with a cw detector in both cases and represents the time-averaged electron distribution. With a picosecond laser pulse, however, it is possible to increase the peak carrier concentration well above the levels that can be achieved with a cw laser without significant heating of the lattice. The effects of these high concentrations of carriers are shown in fig. 10, where it can be seen that the series of peaks at low excitations disappears into a broad background as the injected carrier density approaches 10^{17} cm^{-3}. The peaks disappear as carrier-carrier scattering begins to thermalize the electrons faster than they emit LO phonons, i.e. faster than 200 fsec. Because of the rising background, which is due to band-to-band recombination of hot electrons and holes, it is difficult to make a quantitative analysis of carrier-carrier scattering based on this data. Additional difficulties in analyzing this data result from the density variation over the laser focus spot and the increase in the carrier density during the 5 psec pulse.

A quantitative measurement of the carrier-carrier scattering rate has been made with a pump-probe variation of this experiment[15]. A schematic of the experiment is shown in figure 11. A 5 psec pump dye laser was operated close to the bandedge of GaAs, usually at 1.64 eV. The pump injected electrons thus have an initial energy of about 100 meV. A second laser was used as the probe laser, producing 5 psec pulses at 1.88 eV. This laser injected electrons at much higher energies, about 300 meV for the electrons excited from the heavy hole valence band. The cross correlation between the two lasers was under 10 psec. The lasers were combined on the sample, with the probe pulse temporally delayed 30 psec with respect to the pump. This delay allowed the pump injected plasma to thermalize and cool to about 100K. Cladding layers of $Al_{0.7}Ga_{0.3}As$ minimized interface recombination[7] and confined the pump injected plasma, insuring uniformity of the plasma density through the 1 micron depth of the sample during the 30 psec delay before the probe pulse. In addition, since the probe beam was focussed to to a spot less than half the size of the pump, the plasma which was sampled

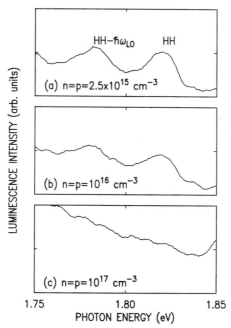

Fig. 10. Time-integrated hot (e,A⁰) luminescence excited by 5 psec, 1.88 eV
probe laser pulses for three injected carrier densities.   (a)
$n = p = 2.5 \times 10^{15}$ cm^{-3}.   (b) $n = p = 10^{16}$ cm^{-3}.   (c)
$n = p = 10^{17}$ cm^{-3}. The GaAs layer is 1 micron thick, Mg-doped
to 7×10^{16} cm^{-3}. A single laser was used to obtain this data. The
quoted density, which is an average over the beam profile, is only
approximate. The spectra are scaled so that the product of laser
power times integration time is constant.

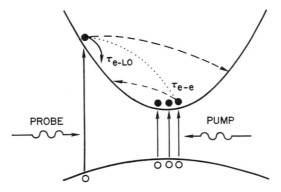

Fig. 11. Schematic of the pump/probe experiment to measure carrier-carrier
scattering rates. The interaction of the probe injected electron with
LO phonons and the electron plasma is indicated. This electron can
also interact with the holes.

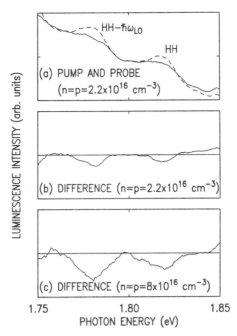

Fig. 12. (a) Hot luminescence spectra with the probe preceeding the pump by 30 psec (dashed curve), and with the probe following the pump by 30 psec (solid curve). The pump injected density is $n = p = 2.2 \times 10^{16}$ cm^{-3}. (b) The difference between the solid and dashed curves of (a). (c) Difference spectrum at a higher pump injected density, $n = p = 8 \times 10^{16}$ cm^{-3}. In both cases the probe injected density is below $n = p = 2 \times 10^{15}$ cm^{-3}. Fig. 10(a) is the (e, A^0) spectrum in the absence of the pump, to the same scale.

by the probe was laterally uniform as well. The plasma density was determined from the measured focal spot size, assuming a Gaussian beam profile. The maximum pump injected carrier density was 8×10^{16} cm^{-3}, while the carrier density injected by the probe laser was below 2×10^{15} cm^{-3}. As can be seen from fig. 10, this probe-injected density is low enough that carrier-carrier scattering can be ignored in the absence of the pump. Time resolved photoluminescence of the band edge emission showed that the pump-injected carriers recombined within 3 nsec of injection, so that no carriers remained from one pump pulse before the next pump pulse 13 nsec later.

When the probe pulse excites the sample 30 psec after the pump, the pump-generated plasma has thermalized and cooled to about 100 K. The density of the plasma in these experiments results in a plasma frequency less than 12.5 meV, and so is too low to cause significant screening of the polar coupling. Hence, the LO phonon emission process is unaltered. On the other hand, the plasma density is high enough to cause significant scattering of the probe-injected electrons. This scattering reduces the lifetime of the nonequilibrium probe-

---

*Because the weak, high energy tail of band-to-band luminescence excited by the pump laser produces a background, the decrease in the hot (e, A^0) emission is measured by subtracting the spectrum when the probe precedes the pump from the spectrum when the pump precedes the probe.

injected electrons in the states which contribute to the cascade luminescence. Since the intensity of the highest energy peak of the hot (e,A^0) luminescence is proportional to the time the electron remains at its initial energy, the amplitude of this peak is reduced when the plasma is present*. The same argument applies to the intensities of the other peaks in the hot (e,A^0) spectrum. The data (fig. 12) clearly show this reduction in the intensity of the hot (e,A^0) peaks. Writing the polar emission rate as Γ_{e-LO} and the electron-plasma scattering rate as Γ_{e-p}, straightforward analysis of this data indicates that the electron-plasma scattering rate scales linearly with pump-injected density n, with

$$\Gamma_{e-p}/\Gamma_{e-LO} = n/n_o \, , \quad n_o \simeq 8 \times 10^{16} \text{ cm}^{-3} \, . \tag{2}$$

This expression is valid for densities below 10^{17} cm^{-3} , but will be slightly modified by screening at higher densities. It should be noted that the carrier-carrier scattering rate derived from fig. 12 is the rate that a single nonequilibrium electron loses energy to the pump-injected electron-hole plasma. This may not be the same as the scattering rate for a high density of nearly monoenergetic electrons and monoenergetic holes, such as would exist for the first instant after optical carrier injection with sub-200 fsec pulses, Comparison of figs. 10 and 12 suggests that the two rates are not very different, however.

INTERVALLEY SCATTERING RATES FROM HOT (e,A^0) EMISSION

The intensity of the hot (e,A^0) luminescence may also be used to obtain intervalley scattering rates in GaAs as a function of electron kinetic energy[16]. Considerable controversy exists as to these rates,[13,14,16,17] which are important to accurately model hot electron devices fabricated from GaAs. Many of the optical measurements which have been made have been performed at carrier densities well above 10^{17} cm^{-3} where, as shown in the previous section, rapid carrier-carrier scattering will make interpretation of the data difficult. In addition, many of the experiments require complex modeling of the data to extract the intervalley scattering rates.

The measurements discussed here[16] are made at densities below 10^{16} cm^{-3}, and the intervalley scattering rates are derived in a very simple fashion. The hot (e,A^0) emission is measured for a series of laser photon energies and the quantum efficiency of the highest energy peak is measured in the limit that carrier-carrier scattering can be ignored. This data is shown in fig. 13. The efficiency of this peak depends only on the matrix element for recombination and the electron lifetime at its initial optically generated energy. The matrix element can be calculated from the ground state hydrogenic wavefunction of the neutral acceptor[18]. In the absence of carrier-carrier scattering, and ignoring intervalley scattering, the electron lifetime will be due only to polar scattering. Fig. 5 shows that for the range of electron energies here (0.06 to 0.6 eV), the polar scattering rate is essentially constant. The predicted intensity of the highest energy (e,A^0) peak for constant electron lifetime is shown in fig. 13 as the dashed curve. For laser photon energies below 1.88 eV (electron energies below 0.32 eV), this curve fits the data well. At higher laser photon energies, the electron lifetime is evidently shorter than the polar emission time, indicating the presence of additional scattering. As indicated in the figure, 1.88 eV laser photon energy is the threshold for electrons generated from the heavy hole valence band to scatter to the L valley. A second threshold is observed at a laser photon energy of 2.07 eV, corresponding to scattering to the X valley. We therefore ascribe the shorter electron lifetimes observed for laser photon energies above 1.88 eV to the presence of intervalley scattering. The intervalley scattering times can be directly obtained from this data in terms of the polar scattering rate. Taking the polar scattering time as 190 fsec (fig. 5) representative scattering times are $\tau_{\Gamma L} = 570$ fsec for 0.48 eV electrons and $\tau_{\Gamma X} = 190$ fsec for 0.58 eV electrons. These intervalley scattering rates have been independently verified[16] by analysis of the re-entrant peaks of the hot (e,A^0) emission, labeled R in fig. 9.

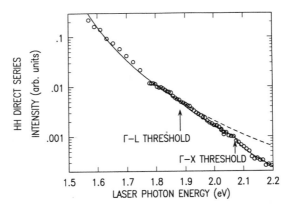

Fig. 13. Quantum efficiency of the highest energy peak of the hot (e, A^0) luminescence vs. laser photon energy. The dashed curve is the predicted intensity in the absence of intervalley scattering, while the solid curve fits the data above 1.88 eV to a model including intervalley scattering. The curves are identical for laser energies below 1.88 eV, where no intervalley scattering is possible.

CONCLUSIONS

Because Raman scattering involves a virtual intermediate state, it can be used in pump/probe experiments to sample nonequilibrium phonon mode occupations on a subpicosecond scale. As shown here, such studies reveal many details about the relaxation of hot carriers in semiconductors.

In principle, electronic Raman scattering can be used to obtain details about nonequilibrium electron distributions. To date,[19] only hot thermal electronic distributions have been probed in this way. In order to increase signal strength to detectable levels, electronic Raman scattering is usually done for a nearly resonant real intermediate state. Such resonant Raman studies call for a re-examination of the instantaneous nature of Raman scattering.

As an alternative to electronic Raman scattering, I have analyzed the results of hot (e, A^0) luminescence to obtain results about the nonequilibrium electron distribution. Using the internal clock of LO phonon emission, this hot luminescence gives quantitative results on carrier-carrier and intervalley scattering.

ACKNOWLEDGMENTS

Special thanks are due Dr. J. C. Tsang who has collaborated with me throughout this work. Additional collaborators in various aspects were Professors J. M. Hvam., S. S. Jha, and R. G. Ulbrich. Finally, I thank Dr. T. F. Kuech for providing many of the samples used here.

REFERENCES

1. See, for example, "Proceedings of the Sixth International Conference on Hot Carriers in Semiconductors", D. K. Ferry and L. A. Akers, eds., Solid-State Electronics 32, number 12 (1989).

2. K. T. Tsen, R. P. Joshi, D. K. Ferry, and H. Morkoc, Phys. Rev. B39, 1446 (1989); M. C. Tatham, J. F. Ryan, and C. T. Foxon, Phys. Rev. Lett. 63, 1637 (1989).

3. E. M. Conwell and M. O. Vassel, IEEE Trans. Electron. Devices 13, 22 (1966).

4. D. von der Linde, J. Kuhl, and H. Klingenberg, Phys. Rev. Lett 44, 1505 (1980).

5. J. A. Kash, J. C. Tsang, and J. M. Hvam, Phys. Rev. Lett 54, 2151 (1985).

6. S. S. Jha, J. A. Kash, and J. C. Tsang, Phys. Rev. B34,5498 (1986).

7. J. A. Kash and J. C. Tsang, Nonequilibrium Phonons in Semiconductors, in: "Spectroscopy of Nonequilibrium Electrons and Phonons," C. V. Shank and B. P. Zakharchenya, eds., North-Holland, Amsterdam, in press.

8. J. A. Kash and J. C. Tsang, Solid-State Electronics 31, 419 (1988).

9. J. A. Kash, S. S. Jha, and J. C. Tsang, Phys. Rev. Lett 58, 1869 (1987).

10. K. T. Tsen and H. Morkoc, Phys. Rev. B37, 7137 (1988).

11. J. A. Kash, J. M. Hvam, and J. C. Tsang, Phys. Rev. B38, 5766 (1988).

12. B. Jusserand and J. Sapriel, Phys. Rev. B24, 7194 (1981); P. Parayanthal and Fred H. Pollak, Phys. Rev. Lett 52, 1822 (1984).

13. D. N. Mirlin, I. Ja. Karlik, L. P. Nikitin, I. I. Reshina, and V. F. Sapega, Solid State Commun. 37, 757 (1981); D. N. Mirlin, I. Ya. Karlik, and V. F. Sapega, Solid State Commun. 65, 171 (1988).

14. G. Fasol, W. Hackenberg, H. P. Hughes, K. Ploog, E. Bauser, and H. Kano, Phys. Rev. B41, 1461 (1990),

15. J. A. Kash, Phys. Rev. B40, 3455 (1989).

16. R. G. Ulbrich, J. A. Kash, and J. C. Tsang, Phys. Rev. Lett. 62,949 (1989).

17. C. L. Collins and P. Y. Yu, Phys. Rev. B30, 4501 (1984); M. C. Nuss, D. H. Auston, and F. Capasso, Phys. Rev. Lett. 58, 2355 (1987); Jagdeep Shah, Benoit Deveaud, T. C. Damen, W. T. Tsang, and P. Lugli, Phys. Rev. Lett. 59, 2222 (1987); C. L. Tang, F. W. Wise, and I. A. Walmsley, Solid-State Electronics 31, 439 (1988).

18. W. P. Dumke, Phys. Rev. 132, 1998 (1963).

19. Y. Huang and P. Y. Yu, Solid State Commun. 63, 109 (1987).

SUBPICOSECOND RAMAN STUDY OF HOT ELECTRONS AND HOT PHONONS

IN GaAs

Dai-sik Kim and Peter Y. Yu

Department of Physics, University of California, Berkeley
and Materials and Chemical Sciences Division
Lawrence Berkeley Laboratory, Berkeley, CA 94720

INTRODUCTION

There has been much interest in the study of fast relaxations of hot electrons and hot phonons in the GaAs and related family of semiconductors because of their importance to the operation of high-speed devices. Optical techniques based on picosecond (ps) and femtosecond (fs) lasers have played an important role in these studies.[1-4] In particular, optical techniques such as time-resolved photo-induced absorption[1,3] and luminescence[2] have been used to study the relaxation and cooling of the hot electrons. The disadvantages of these optical techniques are that they do not provide information on the phonons that are excited by the cooling electrons. For example, Leheny et al.[5] have found that the cooling rate of photoexcited hot electrons in bulk GaAs decreased when the electron density was increased to above 10^{17} cm^{-3}. Leheny et al.[5] suggested that this effect can be explained by the screening of the Fröhlich electron-longitudinal optical (LO) phonon interaction. Subsequent work has shown that the observed carrier density was too low for screening effects to be significant. Instead, an alternative explanation known as the "hot-phonon effect", has been proposed.[6-8] In this model the hot electrons excite a large nonequilibrium population of LO-phonons during relaxation. When the hot electrons and the hot LO phonons reach thermal equilibrium, the LO phonons are not effective in cooling the electrons. Raman scattering has the advantage of being able to determine both the electron and the phonon populations. Using picosecond laser pulses, Huang and Yu[9] have determined the temperature of photoexcited hot electrons and hot phonons in GaAs with Raman scattering. Their results indicate that the electrons and LO phonons are in thermal equilibrium within a time shorter than their pulse length of about 4.5 ps. Recently, we have used Raman scattering with subpicosecond laser pulses to measure the electron and phonon temperatures in GaAs as a function of electron density.[10] For densities above 10^{18} cm^{-3} we found that the phonon temperature actually overshot the electron temperature. Our results have been explained by including scattering of electrons from the zone-center valley to the higher conduction-band valleys at the L and X points of the Brillouin zone. In this paper we will describe in detail the use of Raman scattering with subpicosecond laser pulses for investigating both the Fröhlich interactions between electrons and LO phonons and the intervalley electron-phonon interactions.

Light Scattering in Semiconductor Structures and Superlattices
Edited by D.J. Lockwood and J.F. Young, Plenum Press, New York, 1991

EXPERIMENTAL DETAILS

The source of subpicosecond laser pulses used in our experiment is a colliding-pulse mode-locked (CPM) dye laser pumped by a continuous-wave (cw) Ar ion laser. The construction of the CPM laser has been described by Valdmanis et al.[11] Without any tuning element inside the cavity this laser produces Fourier- transform-limited pulses of typically less than 100 fs duration. Unfortunately the corresponding spectral width of the laser is too large for observing the LO phonon in GaAs. We used a one-plate birefringent filter inside the cavity to narrow down the laser line width. To resolve the transverse optical (TO) and LO phonons in GaAs the line width of the laser was narrowed to 25 cm^{-1} while the full width at half maximum (FWHM) of the laser pulse length was increased to 600 fs. When birefringent plates of different thickness are used the FWHM of the CPM laser can be changed discretely between 200 and 600 fs. The photon energy of the CPM laser is about 2.0 eV, and the energy per pulse is typically 0.3 nJ. With a repetition rate of 120 MHz the average power of the laser is over 20 mW. The dye laser is passed through a set of Brewster angle prisms to reduce the dye fluorescence before being focussed on the sample.

The sample studied was a 0.3 μm thick intrinsic GaAs layer grown by molecular-beam epitaxy on a [001] oriented GaAs substrate. The GaAs layer was sandwiched between two AlAs layers. The AlAs layers act as barriers to prevent the photoexcited electrons from diffusing away from the GaAs layer. The top AlAs layer was protected by an 80 Å GaAs cap layer. The sample was maintained either at room temperature or around liquid nitrogen temperature during the experiment. The density of the photoexcited electron was varied by changing the size of the laser focus on the sample from about 100 μm to about 10 μm. The photoexcited carrier density was determined by analyzing the time-integrated photoluminescence spectra. Figure 1 shows a typical time-integrated luminescence spectrum. Laser heating of the sample was determined by measuring the luminescence and

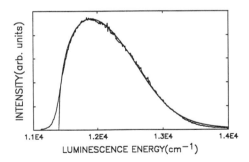

Fig. 1. The time-integrated photoluminesce spectrum of GaAs
        excited by laser pulses of 600 fs. The smooth curve is a
        theoretical curve for a degenerate electron gas with a
        quasi-Fermi energy of 155 meV or electron density of 5 x
        10^{18} cm^{-3} and electron temperature of 400 K.

Raman spectra with a cw dye laser of identical time-averaged power density and photon energy.[12] It was found that laser heating of the sample was minimal. The light-scattering experiment was performed in a backscattering geometry. Several combinations of polarization geometries with the incident and scattered radiations polarized along both the [1$\bar{1}$0] and [100] directions have been used. In all cases the observed scattering spectra are consistent with the spectra reported in the literature fromcw lasers. The results presented in this paper were obtained with both the incident and scattered light polarized along the [1$\bar{1}$0] directions. In this configuration scatterings by the LO phonons, the plasmon, and single particle excitations (SPE) of the photoexcited electron-hole plasmas are all allowed.

EXPERIMENTAL RESULTS

A typical scattered spectrum obtained on a sample at around 77 K is shown in Fig. 2. This scattered spectrum can be decomposed into three parts: (a) a broad background caused by luminescence extending all the way to the band gap of GaAs around 1.5 eV; (b) relatively sharp peaks caused by scattering from the coupled LO-plasmon modes on both sides of the laser line; and (c) a broader peak centered on the laser line caused by scattering from SPE. All three features change in their characteristic ways when the electron density is increased by decreasing the laser focus. At low electron density the photoluminescence is mainly centered around the band gap, and its intensity is negligible near the laser line. The temperature of the photoexcited electron can be determined from its line

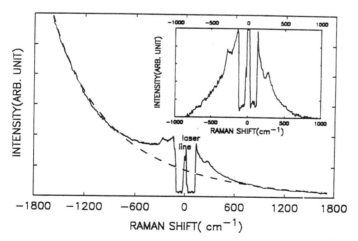

Fig. 2. Typical scattering spectrum in GaAs superimposed on the hot-luminescence background (broken line). The inset shows the Raman spectrum after subtraction of the hot-luminescence background. In both spectra note the change in scale in the vicinity of the laser line.

shape . At higher densities the electron gas becomes degenerate and hotter so the luminescence spectrum starts to extend to higher photon energies. Both the electron quasi-Fermi energy and the temperature can be deduced from the luminescence spectra as shown in Fig. 1. At low density the LO-phonon peaks dominate the Raman spectra. As the electron density increases, the phonon peak position shifts towards lower frequencies as shown in Fig. 3. The SPE spectrum is absent in the Raman spectra at low excitation intensities. It grows quadratically with laser intensity, since it is produced by scattering of light from the photoexcited electrons. At high intensities the SPE spectra dominate over phonon scattering.

The red shift of the LO-phonon frequency in Fig. 3 can be explained by the coupling between the plasmon and the LO phonon.[13] Normally this coupling produces two modes. The lower frequency mode approaches the TO-phonon frequency as the electron density is increased. The higher frequency mode starts with the LO frequency at low electron density and approaches the plasmon frequency at high densities. The photoexcited electrons, however, have a spatial-density profile determined by the laser- beam profile. As shown by Collins and Yu,[14] if the electrons have a Gaussian profile, the coupled modes will show up as two peaks in the Raman spectra: one at the LO-phonon frequency and the other at the TO-phonon frequency. The peak at the LO-phonon frequency results from scattering from unscreened LO phonons in the low-density ($<10^{17}$ cm^{-3}) region. The peak at the TO-phonon frequency results from the LO phonon screened by the free electrons in the high-density region ($>10^{18}$ cm^{-3}). As the light intensity is increased, the high-density region grows relative to the low- density region, so the relative intensity of the two peaks can be used to determine the peak electron density. Because of the larger laser width associated with our subpicosecond pulses, we cannot resolve the two peaks at the TO and LO frequencies. Instead we observe a continuous shift of the phonon peak as shown in Fig. 3. Thus the position of the phonon peak is another way to estimate the photoexcited electron density in the high density region. The densities obtained from both the

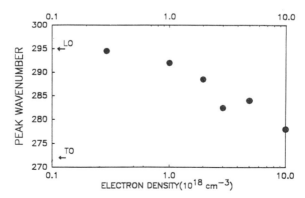

Fig. 3. Dependence of the coupled LO-phonon-plasmon mode on photoexcited electron density. The laser pulse FWHM is 600 fs.

luminescence and Raman spectra usually agree well with the densities
calculated from the number of photons absorbed by the sample. The
dependence of the coupled- mode frequency on density also allows us to
determine the phonon temperature in both the low- and high-density
regions.

We have deduced both the phonon occupation number N_q and electron
temperature T_e from the ratios of the anti-Stokes to Stokes Raman
intensities.[10] Care was taken to correct for the dispersion of the
scattering efficiencies by measuring the dispersion with a cw tunable dye
laser. For convenience, we define an "effective phonon temperature" T_q by
$N_q = [\exp (\hbar\omega/k_b T_q) - 1]^{-1}$ where ω is the LO-phonon frequency and k_b is the
Boltzmann's constant. In Fig. 4 we plot both T_e (open circles) and T_q
(closed circles) deduced from the Raman spectra of GaAs at 77 K as a
function of electron density excited by laser pulses of 600 fs FWHM. For
densities less than 5×10^{17} cm^{-3} the SPE spectra are too weak for T_e to be
determined reliably. Since the photoexcited hot electrons and hot phonons
and the Raman spectra from these excitations were obtained from the same
laser pulse, there is a question as to what these measured temperatures
mean with respect to the cooling curves of the hot electrons and phonons.
To answer this question, we divide the laser pulse into two halves. We
assume that the hot electrons and hot phonons are excited by the first
half of the laser pulse. These hot electrons and phonons then scatter the
second half of the laser pulse to produce the Raman signals. Therefore,
the measured temperatures represent the temperature of the hot electrons
and phonons after cooling for a duration of approximately less than half
of the laser pulse width. A more quantitative calculation based on a
sech^2(t) pulse profile showed that the measured temperature should be
equal to the electron temperature after cooling for a time equal to 0.4
times the FWHM of the laser pulse.[15]

It should be noted that Raman scattering measures a LO phonon with
a specific wave vector determined by the incident and scattered phonon

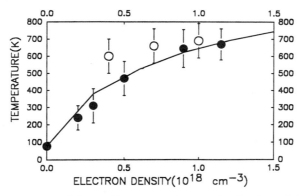

Fig. 4. Temperatures of hot electrons (open circles) and LO phonons
(closed circles) in GaAs at 77 K excited by laser pulses of 600
fs FWHM. The curve drawn through the experimental points has
been calculated from the multivalley model.

wave vectors. For the backscattering geometry and photon energy we use, the LO phonons have wave vectors $Q = 7.5 \times 10^5$ cm^{-1}. Fortunately the distribution of LO phonons generated by the relaxation of hot electrons in GaAs has a maximum very close to this wave vector.[16] Thus Raman scattering is a fairly sensitive probe of the nonequilibrium LO phonons generated by the relaxation of hot electrons in bulk GaAs.

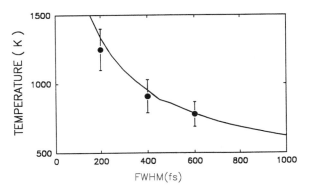

Fig. 5.   Cooling curve of photoexcited hot electrons in GaAs obtained by measuring the electron temperatures as a function of the FWHM of laser pulses. The solid circles are experimental points, while the curve drawn through the circles is calculated from the multivalley model discussed in the text.

The striking features of the curves shown in Fig. 4 are the strong increase in the phonon temperature for electron densities below 10^{18} cm^{-3} and the relatively low and constant value of the electron temperature around 700 K in the same range of densities. Since these hot electrons are excited with excess energy of over 400 meV, we cannot explain this low electron temperature as caused by cooling via emission of LO phonons only. With the short pulse width of 600 fs, the electrons had time to emit only one or two LO phonons and in the process lost less than 80 meV of energy. Thus our results suggest that the electrons are cooled very efficiently by another process, not associated with the LO phonons. To confirm the existence of this additional cooling mechanism we have determined the cooling curve of the hot electron at times shorter than 600 fs. By using different birefringent filters we have decreased the FWHM of the CPM laser pulses to 400 and 200 fs, respectively. The spectral width of these pulses were too broad for measuring the phonon peaks but they were adequate for observing the much broader SPE scattering. The electron temperatures T_e measured in GaAs for three different laser pulse widths are shown in Fig. 5. We notice that T_e increases appreciably as the laser pulse length is decreased. This reflects the fact that when the laser pulse is shortened, the electrons have less time to cool before they are probed by light scattering. For the shortest laser pulse of 200 fs FWHM the electrons will have time to emit at most one LO phonon so the low electron temperature must be attributed to another cooling mechanism.

MODEL CALCULATIONS

To model the cooling curve of the hot electrons and the increase in hot-phonon temperature with electron density, we have considered two possible models.

Model 1: Single Conduction-band Valley at Zone Center

In the first model we have neglected intervalley scattering between the conduction-band minimum at Γ and higher conduction-band minima at the X and L points of the Brillouin zone. The incident photons excite electron-hole pairs in the conduction and valence bands at zone center. The electrons and holes relax by emission of LO phonons via the Fröhlichinteraction. Within the subpicosecond time regime, other relaxation processes such as electron-TO-phonon interaction, electron-acoustic phonon interaction, and expansion of the electron-hole plasma are all negligble.[17] For electron densities below 10^{18} cm^{-3} it has been shown theoretically that electron-hole scattering is not as important as electron-LO scattering.[18] To simplify the calculation we further assume that the electrons thermalize to a Fermi-Dirac distribution instantaneously. This assumption is justified at electron densitites above 10^{18} cm^{-3}, based both theoretical considerations and experimental evidence.[19-21] Although this assumption may not be valid at lower densities, it does not affect the rate at which LO phonons are produced.

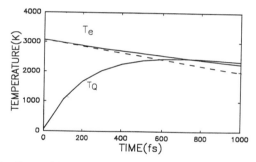

Fig. 6. Theoretical cooling curves (continuous lines) of hot electrons and phonons excited by laser pulses of 0 fs FWHM within the single conduction-band model where electrons interact with LO phonons only. The broken curve is the calculated cooling curve of electrons when there are no hot phonons.

The cooling of the electron and LO-phonon temperatures calculated within this single conduction-band valley model are shown continuous curves in Fig. 6. The electron density was assumed to be 10^{18} cm^{-3}. The electron-LO-phonon scattering time was fixed at 200 fs. The behavior of T_e and T_Q shown in Fig. 6 is easily understood. The electrons cool by emission of LO phonons. As the LO-phonon temperature increases, the cooling rate of the electrons slows down until at long enough time the two systems reach thermal equilibrium. The broken line in Fig. 6 shows the cooling curve of the electrons if the phonon temperature remains constant at 77 K. Thus Fig. 6 constains the essence of the "hot phonon effect". The predictions of Fig. 6, however, are not consistent with our experimental results. Both the electron and LO-phonon temperatures predicted by this model are far too high compared with the observed temperatures. We note that our T_e are comparable to those observed by Shah et al.[2] obtained with subpicosecond time-resolved photoluminescence. At higher electron densities we have found that T_Q overshot T_e, but the highest value of T_Q achieved was much lower than that predicted by Fig. 6.[10] Other energy-loss mechanisms, such as electron-hole scattering, can increase the cooling rate and lower T_e but will also predict a strong dependence of T_e on electron density. Experimentally we found that T_e is not very sensitive to electron density. At densities near 10^{19} cm^{-3}, T_e actually increases with densities while electron-hole scattering will predict a decrease in T_e.[10]

Model 2: Multiple Conduction-band Valleys

In this model we include the scattering of electrons from the Γ valley to the higher conduction-band minima at X and L. It is known from study of hot-electron transport in GaAs, such as the Gunn effect, that these scattering processes are much faster than intravalley scattering by LO-phonons.[17] For example, in GaAs the deformation potential $D_{\Gamma-X}$ for the Γ to X intervalley electron-phonon scattering is about 1×10^{19} eV/cm.[12] Because of the large density of states at the X valleys, electrons with high enough energy will transfer to the X valley in about 10-50 fs, depending on the electron energy. Thus the intervalley processes are more efficient than the Fröhlich interaction in cooling the hot electrons in the Γ valley because they remove the high-energy electrons in times shorter than 100 fs. Although eventually these electrons in the higher valleys will return to the Γ valley, the return processes take more than 1 ps because the Γ valley has smaller density of states.[2] Thus at times shorter than 1 ps intervalley scattering can account for the very fast cooling of the hot Γ valley electrons we observed.

In calculating the cooling curves of the hot electrons, we found that the L valleys also played a significant role. If only the X valleys are included in the model, the electron temperature will drop very fast within the first 100 fs. But once the electron temperature is below 1500 K, the cooling rate decreases rapidly because there are now very few electrons left in the high-energy tail of the distribution capable of transfering to the X valley. As a result the electron temperature in Fig. 4 will be about 1200 K, rather than 700 K as observed. The theoretical cooling curve (continuous line) in Fig. 5 is obtained by including both the X and L valleys in the model and by setting $D_{\Gamma-L}$ to 7×10^8 eV/cm. Once the intervalley-scattering cooling rates are known, the dependence of the LO-phonon temperature on electron density can be calculated. The theoretical curve (solid line) in Fig. 4 has been obtained with a electron-LO-phonon scattering time of 200 fs.

DISCUSSIONS

From the subpicosecond Raman experiments we conclude that intervalley scattering can play a dominant role in the cooling of the photoexcited hot electrons in GaAs within the first picosecond after excitation. The electron cooling rate depends strongly on time, density, and temperature. The so-called "hot-phonon effect" arising from electron-LO-phonon interaction is important only for times longer than 1 ps or for electrons with initial temperature lower than about 700 K. For higher electron temperatures the electron cooling rate within the first picosecond is dominated by intervalley scattering. The reason is because of the ultrafast rate of transfer of high-energy electrons from the Γ valley to the higher conduction-band minima in GaAs. The separation between the Γ and X valleys in GaAs is about 500 meV. With a scattering time of about 50 fs or less, the electron cooling rate due to Γ to X intervalley scattering is about 10 meV/fs and is about two orders of magnitude larger than electron-LO-phonon scattering. The Γ to L intervalley scattering plays a similar, albeit, smaller role. Both the deformation potential $D_{\Gamma-L}$ and density of states for the L valleys are smaller than the corresponding quantities for the X valleys. The Γ to L scattering time is about 100-200 fs. The separation between the Γ and L valley is only about 300 meV. As a result of these two factors, when T_e $\gg$ 1000 K the cooling rate due to Γ to L scattering is much smaller than that due to Γ to X scattering, although it is still larger than the electron-LO-phonon cooling rate. As the electrons cool to about 1000 K or lower the Γ to L scattering becomes more important, since there are very few electrons energetic enought to scatter to the X valleys. The observed T_e of about 700 K measured with laser pulse of 600 fs FWHM in Fig. 4 is determined mainly by the Γ to L scattering. At this electron temperature the Γ to L scattering rate is comparable to the electron-LO-phonon scattering rate. Thus electrons and LO-phonons can reach thermal equilibrium only after the electron temperature is lower than about 700 K.

CONCLUSIONS

In conclusion we have studied the cooling of hot electrons and heating of LO-phonons in GaAs excited by subpicosecond laser pulses as a function of photoexcited electron densities. For electron densities above 2×10^{17} cm^{-3} we found that the LO-phonon temperature rose rapidly with electron densities. On the other hand, the electron cooling rate was found to be too fast to be explained by the electron-LO interaction. Our results have been explained quantitatively by scattering of electrons from the zone-center conduction-band valley to the higher conduction valleys at the X and L points of the Brillouin zone. By fitting our results to model calculations we find that the electron-LO scattering time is about 200 fs and the Γ to L intervalley deformation potential is 7×10^8 eV/cm.

ACKNOWLEDGMENTS

We are grateful to Henry Lee and Professor Shyh Wang for providing us with the GaAs samples used in our experiment. This work was supported by the Director, Office of Energy Research, Office of Basic Energy Science, Materials Sciences Division of the U. S. Department of Energy under Contract No. DE-AC03-76SF00098.

REFERENCES

1. W.Z. Lin, L. G. Fujimoto, E. P. Ippen, and R. A. Logan, Appl. Phys. Lett. 50, 124 (1987).

2. J. Shah, B. Deveausd, T.C. Damen, W.T. Tsang, and P. Lugli, Phys. Rev. Lett. 59, 2222 (1987).
3. C.L. Tang, F.W. Wise, and I.A. Walmsley, Solid State Electron. 31, 439 (1988).
4. J.A. Kash, J.M. Hvam, and J.C. Tsang, Phys. Rev. Lett. 54, 2151 (1985).
5. R.F. Leheny, J. Shah, R.L. Fork, C.V. Shank, and A. Migus, Solid State Commun. 31, 809 (1979).
6. W. Potz and P. Kocevar, Phys. Rev. B 28, 7040 (1983).
7. W. Cai, M.C. Marchetti, and M. Lax, Phys. Rev. B 35, 1369 (1987).
8. W.W. Ruhle, K. Leo, and E. Bauser, Phys. Rev. B 40, 1756 (1989).
9. Y.H. Huang and P.Y. Yu, Solid State Commun. 63, 109 (1987).
10. D.S. Kim and P.Y. Yu, Phys. Rev. Lett. 6 , 946 (1990).
11. A. Valdmanis, R.L. Fork, and J.P. Gordon, Opt. Lett. 10, 131 (1985).
12. C.L. Collins and P.Y. Yu, Phys. Rev. B 30, 4501 (1984).
13. M.V. Klein in:"Light Scattering in Solids, Topics in:Applied Physics," Vol. 8, M. Cardona, ed., Springer-Verlag, New York (1975).
14. C.L. Collins and P.Y. Yu, Solid State Commun. 51, 123 (1984).
15. D.S. Kim and P. Y. Yu, Appl. Phys. Lett. 56, 2210 (1990).
16. C.L. Collins, Ph.D. thesis, University of California, Berkeley, (1983).
17. E.M. Conwell and E.O. Vassel, IEEE Trans. Electron Devices 13, 22 (1966).
18. M.A. Osman and D.K. Ferry, Phys. Rev. B 36, 6018 (1987).
19. W.H. Knox, C. Hirlimann, D.A. Miller, J. Shah, D.S. Chemla, and C.V Shank, Phys. Rev. Lett, 56, 1191 (1986).
20. D. Hulin, A. Migus, A. Antonetti, and J.L. Oudar, in:"Proceedings of the 18th International Conference on the Physics of Semiconductors", O. Engstrom, ed., World Scientific, Singapore (1987).
21. J.A. Kash and J.C. Tsang, in:"Proceedings of the 18th International Conference on the Physics of Semiconductors", O. Engstrom, ed., World Scientific, Singapore (1987).

TIME-RESOLVED RAMAN STUDIES OF THE TRANSPORT PROPERTIES

OF EXCITONS IN GaAs QUANTUM WELLS

K.T. Tsen and Otto F. Sankey

Department of Physics
Arizona State University
Tempe, AZ 85287

ABSTRACT

Time-resolved, space-imaged Raman spectroscopy has been employed to study the transport properties of excitons in GaAs quantum wells. For an injected exciton density of $n_{ex} \approx 1.5 \times 10^{11}$ cm^{-2}, the experimental results show that exciton transport can be well described by a simple diffusion model. Based on the temperature and well-width dependence of the deduced diffusion constant, we demonstrate that at low temperatures (i.e. T < 60K), the transport of excitons in GaAs quantum wells is primarily governed by two dominant mechanisms: acoustic phonon scattering and interface roughness scattering.

INTRODUCTION

Study of the transport properties of carriers in multiple quantum-well structures (MQWS) has attracted considerable interest. This is partly because the carrier-carrier or carrier-phonon interactions can be investigated in a physical system of lower dimensionality; and partly because it has important applications in high-speed electronic devices. Hopfel et al.[1] investigated two-dimensional minority-carrier mobilities in GaAs-Al$_x$Ga$_{1-x}$As MQWS by a time-of-flight (TOF) method. Hegarty and Sturge[2] observed the diffusion of excitons in a quantum well by using an optical transient grating technique. More recently, Hillmer et al.[3] have reported the transport of excitons in GaAs-Al$_x$Ga$_{1-x}$As quantum wells by using a TOF method with very high spatial resolution. Smith et al.[4] studied the diffusion of photo-excited carriers in GaAs quantum wells and observed anomalous lateral confinement at high densities. These studies have suggested that interface roughness scattering has an important effect on an exciton travelling parallel to the heterointerface at low temperatures; however, no specific information is available concerning the role acoustic phonons play on the transport properties of excitons in GaAs quantum wells.

In this paper, we have used time-resolved, space-imaged Raman spectroscopy to examine the transport properties of excitons in GaAs quantum wells at low temperatures (i.e., T< 60K) . This technique has been successfully used[5] to investigate the transport of electron-hole plasmas in

Light Scattering in Semiconductor Structures and Superlattices
Edited by D.J. Lockwood and J.F. Young, Plenum Press, New York, 1991

393

GaAs quantum wells. We demonstrate here that under our experimental conditions, <u>both</u> the interface roughness scattering and the acoustic phonon scattering <u>are</u> important, and both have to be taken into account to understand the transport of excitons in GaAs quantum wells.

At low temperatures, photo-excited electron-hole pairs quickly bind to form excitons. As is the case of the photo-excited electron-hole plasma, the density of excitons can be directly measured from the inelastic light scattering of the electronic intersubband excitations. The intersubband Raman signal is made up of two parts which can be separated by proper selection rules.[6] The first part consists of light scattering from spin-density fluctuations, also called single-particle excitations. The peak positions in the Raman spectra reflect the energy difference of the subbands involved in the transition. The second part comes from the scattering of light due to charge-density fluctuations, which are usually referred to as collective intersubband excitations. Its peak position will shift in proportion to the carrier concentration due to the depolarization field effect and coupling to LO phonons. In general, either of these two spectra can be used to determine the exciton density. However, because the spectra of single-particle excitations provide better resolution than those of collective intersubband excitations at low exciton densities, we have used the former for the determination of the exciton density profiles and the latter for the calculations of initial exciton density and as a consistency check.

EXPERIMENTAL TECHNIQUES AND RESULTS

The experiments were carried out with two independently tunable cavity-dumped dye lasers (DCM and Styryl 9) which are synchronously pumped by the second harmonic of a cw mode-locked YAG laser. The relative time jitter between the dye lasers is estimated to be about 20ps from the measurement of cross correlation. The pulse width of these ultrashort pulses is about 20ps and the repetition rate of the dye lasers is about 10MHz. In order to make use of the resonance enhancement, the dye laser which provides the probe pulse is chosen to operate at a photon energy very close to the optical gap separating the first conduction subband and the split-off valence band of GaAs. The spatial measurements are performed by moving the pump beam a distance Δx away from the probe beam, with the help of a calibrated periscope located at the entrance slit of the double monochromator. The pump and the probe beams are very carefully focused into a spot of full-width at half maximum $\simeq 5\mu m$ on the surface of the samples. The ratio of the intensity of the pump beam to that of the probe beam is suitably chosen to be $\simeq 50:1$, so that the perturbation created by the probe beam is minimized. The backscattered light is collected by a computer-controlled, gated (gate width $\simeq 2ns$) Raman system. A novel feature of the gated electronics in photon counting is that the background noise can be effectively reduced to a negligible level. A Janis Supervaritemp Dewar is employed to make measurements at various temperatures.

The undoped GaAs-Al$_x$Ga$_{1-x}$As MQWS studied in this work were grown by molecular-beam epitaxy on (100)-oriented undoped GaAs substrates. They consist of about 30 periods of 100Å-thick Al$_x$Ga$_{1-x}$As ($x \simeq 0.3$) and various thicknesses of GaAs quantum wells ranging from 106Å to 275Å (as determined from transmission electron microscope). In our experiments, the electron-hole pairs are generated by a near-infrared laser pulse whose penetration depth is typically > 1μm, and the absorption length of the probe pulse is about 5000Å. Therefore, the exciton density created in the quantum wells can be considered to be rather uniform.

We now demonstrate how the spectra of intersubband transitions of electrons can be used to calculate the exciton density. In Fig. 1, two Stokes-Raman spectra are plotted as a function of frequency shift for a sample with well-thickness ≈ 275Å at T=50K and time delay Δt= 200ps. We observed a peak at ≈ 145 cm^{-1} in the depolarized spectrum (solid curve) which is attributed to single-particle intersubband scattering involving transition of an electron from the first subband to the second subband. In the polarized spectrum (dashed curve), a peak at ≈ 154 cm^{-1} is observed, which comes from collective intersubband transitions of an electron between the first and the second subbands. In addition, there are three more peaks in the spectral range of higher frequency shift, which are the result of light scattering from GaAs-like LO phonons (≈284cm^{-1}, indicated by the arrow at left), GaAs LO phonons (≈296cm^{-1}, indicated by the arrow at right) and AlAs-like LO phonons (≈378cm^{-1}). We notice that relative exciton density is sufficient to determine the diffusion profiles in the present analysis. These can be obtained from the scattering intensity associated with the single-particle excitations since its intensity is directly proportional to the exciton concentration. An absolute exciton density can be obtained as a consistency check from the collective intersubband excitations. Here we follow the analysis of Pinczuck et al.[7] For the situation shown in Fig. 1, the photo-excited exciton density is found to be ≈ 6x10^{10} cm^{-2}. This value is very close to the exciton density (≈8 x10^{10} cm^{-2}) as estimated from the power density and the absorption coefficient of the pump pulse used in the measurements.

To be certain that we are studying excitons and not the electron-hole plasmas, we have carried out independent time-resolved photoluminescence measurements to determine how quickly the electron-hole pairs form excitons. Our experimental results indicate that the formation time of excitons from unbound electron-hole pairs is approximately 150ps. This result shows that what we are probing, in our time-resolved Raman scattering experiments of time delay > 200ps, is primarily the transport of excitons in GaAs quantum wells.

The exciton density profile in space and time, i.e., $n_{ex}(r,t)$, is compared with theoretical calculations based on a simple diffusion model. The diffusion equation has the following form,

$$\partial n_{ex}(r,t)/\partial t = D \nabla^2 n_{ex}(r,t) + G(r,t) - n_{ex}(r,t)/\tau .$$

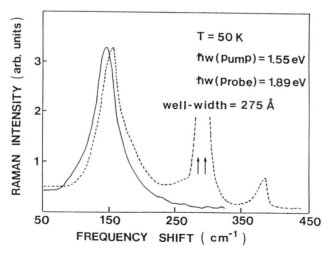

Fig. 1. Raman spectra of single-particle (solid curve) and collective (dashed curve) intersubband excitations for a MQWS sample with well-thickness ≈ 275Å taken at T≈ 50K; and the time delay between the pump and the probe is ≈ 200ps.

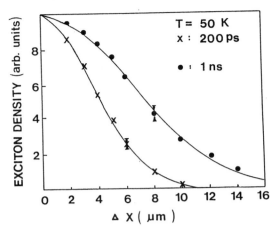

Fig. 2. Time-resolved spatial images of the exciton
distribution for the same sample as in Fig. 1, taken
at two different time delays and at T ≃ 50K. The exciton
distribution profiles have been normalized to their peak
intensities at Δx=0μm. Because they are essentially
symmetric with repect to Δx=0μm, we only display half
of the images for the sake of clarity.

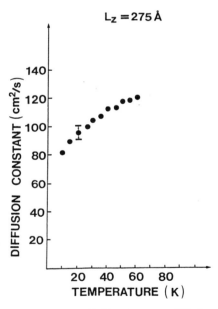

Fig. 3. Deduced diffusion coefficient of
excitons is plotted as a function of the
temperature for a sample with well-width
L_z ≃ 275Å.

The first term on the right-hand side of the equation corresponds to the ordinary diffusion, the second term represents the generation of excitons by laser excitations, and the last term considers the density loss as a result of recombination. The lifetime τ of an exciton has been determined to be $\approx$2ns from time-resolved photoluminescence measurements. The above diffusion equation is first solved numerically with the boundary condition $n_{ex}(\infty,t)=0$. The calculated $n_{ex}(r,t)$ is then used to obtain the observed exciton density by taking into account the spatial and the temporal profiles of the probe pulse.

Figure 2 shows the measured temporal and spatial profiles of excitons taken at T $\approx$ 50K and with initial density $n_{ex} \approx 1.5 \times 10^{11}$cm^{-2} at two different time delays Δt=200ps, and 1ns, respectively. The solid curves correspond to theoretical calculations with the parameter D=115 cm^2/s, which yields the best fit to the data. The procedure is repeated at various temperatures ranging from 10K to 60K and the deduced diffusion coefficient of the exciton with initial density $n_{ex} \approx 1.5 \times 10^{11}$ cm^{-2} is plotted as a function of temperature in Fig. 3.

The data in Fig. 3 cannot be explained in terms of acoustic phonon scattering, or interface roughness scattering, alone. In two dimensions, acoustic phonon deformation potential scattering predicts a diffusion constant which is independent of the temperature,[8] which clearly contradicts the data in Fig. 3. Defect scattering (most likely originating from the interface roughness) does have temperature dependence,[9] but by itself cannot satisfactorily account for the observed temperature dependence of the diffusion constant. We find that if we consider both the acoustic phonon scattering and the interface roughness scattering mechanisms on an equally important footing, we can interpret our experimental results very well.

The acoustic phonon scattering rate due to the deformation potential can be shown[8] (under our experimental conditions) to be given by
$$< 1/\tau >ac \approx A \, T/L_z,$$
where A is a constant independent of temperature (T) and well-thickness (L_z). Similarly, the interface roughness scattering rate is given by[9]
$$< 1/\tau >ir \approx B \, \sqrt{T}/L_z{}^2,$$
where B is a constant independent of temperature and well-width. Our model calculations of interface roughness scattering have assumed that (1) the interface roughness can be approximately simulated by localized square-well potentials, and (2) the wave-function of the exciton is almost constant over the size of the interface roughness.

Adding the two scattering rates to obtain the total scattering rate, $< 1/\tau >t = < 1/\tau >ac + < 1/\tau >ir$, and using the Einstein relation, we obtain D in terms of T and L_z:

$$D = \left(\frac{k_B \, L_z^2}{m^* \, B} \right) \left(\frac{1}{\frac{A}{B} L_z + T^{-1/2}} \right) \qquad (1).$$

where k_B is the Boltzmann constant, and m^* is the mass of the exciton.

We have attempted to fit the experimental data in Fig. 3 to Eq. 1 with two adjustable parameters A and B; in other words, the two-parameter fit is made to a set of temperature data where L_z is fixed at 275Å. The quality of the fit with parameter set A= 45.8 mK^{-1}s^{-1} , B=4.79x10^{-6} m^{2}K$^{-0.5}$s^{-1} is very good and is shown in Fig. 4.

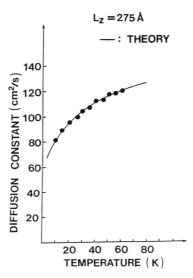

Fig. 4. Experimentally deduced diffusion constant of excitons as a function of temperature (solid circles) is fitted to a theory (solid curve) which takes into account both the acoustic phonon scattering and the interface roughness scattering mechanisms.

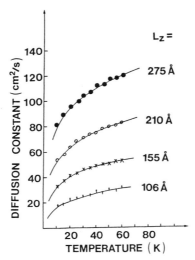

Fig. 5. Diffusion constant of excitons as a function of temperature for samples with four different well thicknesses $L_z \simeq 275\text{Å}$ (solid circles), 210Å (open circles), 155Å (x's) and 106Å (crosses), is fitted to a theory which considers the acoustic phonon scattering and the interface roughness scattering on an equally important footing.

Although Eq. (1) has only been demonstrated so far to be suitable to describe the transport of excitons for a sample with well-width $L_z = 275\text{Å}$, it <u>predicts</u> the behavior of samples with different well-widths. To test these predictions, we have carried out experimental measurements of the diffusion constant as a function of temperature for samples with well-thicknesses $L_z = 210\text{Å}$, 155Å and 106Å, and have compared them with Eq. (1). The results are shown in Fig. 5. The agreement of experiment with the theory is quite good. This strongly indicates that under our experimental conditions, both the acoustic phonon scattering and the interface roughness scattering are important and both have to be taken into account to properly understand the transport of excitons in GaAs quantum wells.

CONCLUSIONS

In conclusion, we have used time-resolved, space-imaged Raman spectroscopy to study the exciton transport in GaAs quantum wells. For an injected exciton density of $n_{ex} \simeq 1.5\times10^{11}$ cm^{-2}, our experimental results have demonstrated that the exciton transport can be very well described by a simple diffusion model. Based on the temperature and well-width dependence of the deduced diffusion constant, we have shown that at low temperatures (T< 60K), the transport of excitons in GaAs quantum wells is primarily governed by the acoustic phonon scattering and the interface roughness scattering.

ACKNOWLEDGEMENTS

The work by one of us (K.T.T) was supported by the National Science Foundation under Grant No. DMR-8718228.

REFERENCES

1. R.A. Hopfel, J. Shah, P.A. Wolff and A.C. Gossard, Phys. Rev. B37, 6941 (1988).
2. J. Hegarty and M.D. Sturge, J. Opt. Soc. Am. B2, 1143 (1985).
3. H. Hillmer, S. Hansmann, A. Forchel, M. Morohashi, E. Lopez, H.P. Meier and K. Ploog, Appl. Phys. Lett. 53, 1937 (1988).
4. L.M. Smith, D.R. Wake, J.P. Wolfe, D.Levi, M.V. Klein, J.Klem, T. Henderson and H. Morkoc, Phys. Rev. B38, 5788 (1988).
5. K.T. Tsen, O.F. Sankey, G. Halama, Shu-Chen Y. Tsen and H. Morkoc, Phys. Rev. B39, 6276 (1989).
6. A. Pinczuk, J. Shah, A.C. Gossard and W. Wiegmann, Phys. Rev. Lett. 46, 1341 (1981).
7. A. Pinczuk, J.M. Worlock, H.L. Stormer, R. Dingle, W. Wiegmann and A.C. Gossard, Solid State Commun. 36, 43 (1980).
8. N. Holonyak, Jr. and K. Hess, in:"Synthetic Modulated Structures", L.L. Chang and B.C. Giessen, eds, Academic press, Orlando (1985), p. 257.
9. K. Hess, in:"Advanced Theory of Semiconductor Devices", Prentice-Hall, Englewood Cliffs, New Jersey (1988).

NON-EQUILIBRIUM PHONON DYNAMICS IN Ge AND GeSi ALLOYS

Jeff F. Young, D.J. Lockwood, J.M. Baribeau and P.J. Kelly

Institute For Microstructural Sciences
National Research Council
Ottawa, K1A 0R6, Canada

A. Othonos and H.M. van Driel

Department of Physics and Ontario Laser Lightwave Research Center
University of Toronto
Toronto, M5S 1A7, Canada

INTRODUCTION

The ability of the Raman scattering technique to coherently probe specific excitation modes of a semiconductor, providing both spectral and thermodynamic information, makes it a powerful tool for time-resolving relaxation processes. Its coherent nature means that the temporal resolution obtained when a short, "probe" pulse is used to Raman scatter from a non-equilibrium system induced by a relatively strong "pump" pulse applied some time τ in the past, is limited only to the pulse duration. This is achieved without having to employ any ultrafast electronic, electro-optic, or non-linear mixing techniques which are required in time-resolved photoluminescence experiments. By relating the ratio of Stokes and anti-Stokes scattering strengths obtained from the probe pulses to the occupation number of the corresponding mode, it is possible to directly monitor the buildup and decay of non-equilibrium populations generated either directly or indirectly by the absorption of the pump pulse.

This tool can of course only be of use if the pump beam is capable of inducing a detectable perturbation of the equilibrium population of the particular mode being studied. As pointed out by Collins and Yu[1], it is rather fortuitous that in semiconductors such as GaAs and InP, the polar nature of the electron-LO phonon interaction results in a very efficient coupling of energy from the pump-excited plasma into LO phonon modes with relatively small wavevectors, near those probed by laser pulses in the visible part of the spectrum ($\sim 7 \times 10^5 \mathrm{cm}^{-1}$). This is a consequence of the dipole nature of the Fröhlich interaction, which manifests itself as an inverse wavevector, $1/q$ dependence in the electron-LO phonon matrix element, $M_F(q)$.

Numerous examples of how this fact has been exploited to study LO phonon dynamics in group III-V bulk and superlattice systems are discussed in other chapters of this volume, and elsewhere[1-10]. Suffice it to note here that the pump/probe,

Light Scattering in Semiconductor Structures and Superlattices
Edited by D.J. Lockwood and J.F. Young, Plenum Press, New York, 1991

time-resolved Raman scattering technique has yielded much valuable information concerning the electron-LO phonon coupling strength[3] and the lattice dynamics associated with optical phonon decay[2-4] in GaAs and GaAs based structures. In some cases, such as the phonon lifetime measurements, Raman scattering offers the only means of directly determining a physical parameter in the time domain. However, in direct bandgap materials such as GaAs, other information derived from Raman measurements, such as the electron-LO phonon coupling strength, can be and has been successfully compared with results obtained using other time-resolved optical techniques such as photoluminescence decay.

In indirect bandgap semiconductors such as Si and Ge, time-resolved photoluminescence is not a viable technique, and transient Raman scattering can therefore potentially play an even more important role than in the polar semiconductors. However, in these non-polar group IV materials energy is transferred to the lattice from a photo-excited plasma via deformation potential interactions alone, and to leading order, there is no wavevector dependence in the deformation potential carrier-optical phonon matrix elements. Hence the energy transferred from the plasma to the optical phonon modes of the lattice is spread over a larger region of the Brillouin zone than in polar materials, resulting in a reduced non-equilibrium population in any given mode.

In this chapter it will be shown that despite the absence of a Fröhlich mechanism in group IV materials, the same pump/probe Raman techniques used so successfully in group III-V materials can also be applied to provide valuable microscopic information concerning non-equilibrium electronic and phonon decay processes in both Ge and GeSi binary alloys. The first part of the chapter is devoted to results on intrinsic Ge[11,12] for which the temporal evolution of the non-equilibrium optical phonon population has been quantitatively modeled by solving a coupled system of Boltzmann equations for the optically excited electron, hole and phonon subsystems.

Having demonstrated the utility of time-resolved Raman scattering in the elemental group IV material Ge, it was of interest to extend the studies to the GeSi alloy system, which has already shown technological potential in the area of heterojunction bipolar Si/GeSi transistors[13]. In contrast to the findings of Kash et al.[4] in the ternary group III-V alloy system AlGaAs, where the LO phonon lifetime was independent of alloy composition, the optical phonon lifetime in the GeSi material system exhibits a strong dependence on composition. For Si compositions up to $\sim 10\%$ the phonon lifetime actually increases by nearly a factor of two, while at 25% it is close to half of the value in pure Ge. These interesting results are presented in the second part of this chapter, together with a number of the steady-state Raman properties of the same alloy samples, which are used to support tentative explanations for the anomalous compositional dependence of the phonon lifetimes.

NON-EQUILIBRIUM OPTICAL PHONON DYNAMICS IN Ge

In the first part of this section the experimental technique used to generate and detect non-equilibrium optical phonons in Ge is described together with the results obtained on both (100) and (110) oriented surfaces[11]. The latter half of this section outlines a coupled Boltzmann equation model[12] which quantitatively accounts for the observed optical phonon dynamics using material parameters found in the literature. This quantitative comparison of the experimental Raman results with a detailed microscopic model serves both as support for the somewhat unexpected experimental results given the deformation potential interaction in Ge, and also to demonstrate the

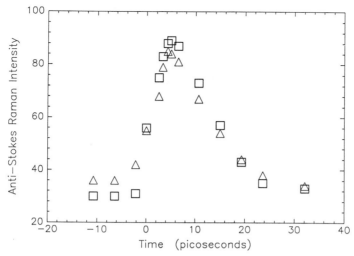

Fig. 1.   The peak anti-Stokes Raman intensity from the LO(□) and TO(△) phonons as a function of the pump/probe delay time at 77 K.

sensitivity of phonon dynamic measurements to different aspects of the hot-carrier relaxation problem. Two particular results from the comparison are that the majority of non-equilibrium optical phonons are actually generated via relaxation of hot holes as opposed to hot electrons, and that the non-equilibrium phonon population is very sensitive to the diffusion of the high temperature plasma away from the surface, before significant energy is transferred to the lattice.

In all of the pump/probe experiments used to obtain the results in this chapter, both the pump and the probe beams were derived from a single synchronously pumped, tunable dye laser using R590 dye. The individual pulse duration was 4 ps (full width half maximum), tunable from 560 to 590 nm, with a repetition rate of 76 MHz. Each pulse typically had an energy of 1.5 nJ. This single beam was first passed through a half-wave plate/polarizing beam splitter combination that resulted in two independent beams of orthogonal polarizations and continuously tunable relative intensities. One of these beams was passed through a variable delay line and then the two orthogonally polarized, delayed beams were superimposed using a second polarizing beam cube. This combined beam was then focussed to a spot size of $\sim 30\mu$m diameter, with the probe beam typically half to one third the intensity of the pump beam.

Chemically polished, single-crystal intrinsic Ge samples were mounted on the cold finger of a flow-through cryostat which had one warm window that preserved the polarization of the incident and scattered light. A backscattering geometry was used where for (100) oriented samples the pump beam and analyzer were polarized along the [010] direction, and the probe beam along the [001] direction. In this configuration only the probe beam resulted in a first order Raman signal, and it was associated with LO phonons in the crystal. For (110) oriented samples, TO phonons were detected by the probe pulse alone by polarizing it and the analyzer along the [001] direction. At a laser wavelength of 575 nm, phonon modes with wavevector $q = 1.2 \times 10^6 \mathrm{cm}^{-1}$,

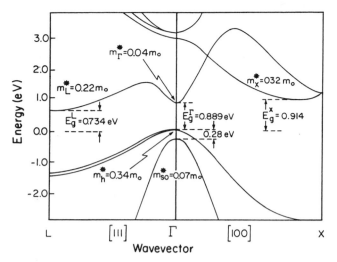

Fig. 2.   Schematic representation of the bandstructure of Ge at 77 K.

small compared to the Brillouin zone width, are being probed in the backscattering geometry. Experiments were carried out both at room temperature and at 77 K. A triple spectrograph (SPEX) was used to analyze the scattered light which was detected on a two dimensional Si charge coupled-device-detector cooled to −140°C.

Figure 1 shows the temporal evolution of the anti-Stokes signal (in arbitrary units proportional to the phonon occupation) observed for both LO and TO phonon modes with the samples at 77 K. The signals rise from background levels (delay independent) which are due to scattering of the probe beam from non-equilibrium phonons generated by the probe beam itself. The peak non-equilibrium populations induced by the pump beam occur well after the peak in the pump beam intensity (at the time origin), the non-equilibrium populations disappear after ∼ 25 ps, and the behaviour of the LO and TO phonons is essentially identical. Similar results are obtained at room temperature but the delay of the peak phonon population and the overall lifetime are both reduced by approximately a factor of two. These particular results were obtained for laser pulses at 575 nm, but there was no effect of excitation wavelength in the range 565 nm to 585 nm.

Even without a detailed analysis these results can be compared with those reported in GaAs based systems wherein the dominance of the Fröhlich interaction results in non-equilibrium LO phonons (polar), but no observable perturbation of the TO phonon populations (non-polar). From the results in Fig. 1, the deformation potential interaction in Ge transfers energy from the photo-excited plasma to small-wavevector LO and TO phonon modes with comparable efficiency. Furthermore, simply by measuring the decay time of the non-equilibrium population it is clear that the LO and TO phonon lifetimes are, within experimental error, the same in Ge, and are very similar to those measured in GaAs[3] using the same basic technique: 8 ± 1 ps at 77 K and 4 ± 1 ps at 300 K for Ge, and 9 ps at 2 K and 3.5 ps at 300 K for GaAs.

By measuring both anti-Stokes and Stokes intensities, and calibrating the detection system using a low power beam, it is possible to estimate the absolute non-equilibrium phonon populations reflected in Fig. 1 to within a factor of approximately two. Thus there are two quantitative features of this data which can be used to test the ability of microscopic models to accurately describe the non-equilibrium relaxation processes in Ge: the absolute non-equilibrium phonon populations induced at wavevectors of $\sim 1.2 \times 10^6 \mathrm{cm}^{-1}$, and the delay in the peak phonon population with respect to the peak of the pump pulse. Details of the model calculation will first be outlined, and comparisons with the data will follow.

The bandstructure of Ge is schematically illustrated in Fig. 2. For the wavelengths and sample temperatures used in the experiments described above, most of the incident radiation is absorbed via direct (no phonon participation) transitions near the Γ point at the center of the Brillouin zone. All of the light is absorbed within ~ 100 nm and ~ 30 nm at 77 K and 300 K respectively. This absorption process leaves the optically excited holes with ~ 0.4 meV of excess kinetic energy, with the exact amount depending on the bandgap, and hence the sample temperature. The optically excited electrons in the Γ valley scatter rapidly to regions of much higher density of states in the L and X valleys. The initial excess energy of these electrons, with respect to the indirect conduction band minima, is approximately 1 eV.

The pump laser pulse therefore acts as a source of energetic electrons and holes very close to the sample surface. This population of energetic carriers relaxes towards equilibrium through a variety of processes. Of primary importance here are diffusive transport processes, which carry both particles and energy away from the surface, and local carrier-lattice interactions which cool the carriers directly by transferring energy to the phonon modes of the crystal. All of these mechanisms, absorption, diffusion, and cooling can be treated microscopically by solving a coupled set of Boltzmann equations[14] for the temporal and spatial evolution of the electron and hole density, $N_e = N_h = N$, the electron and hole energy densities U_e and U_h, and the lattice energy density U_L,

$$\frac{\partial N}{\partial t} = -\nabla \cdot \vec{J} + \mathcal{G} \tag{1a}$$

$$\frac{\partial U_e}{\partial t} = -\nabla \cdot \vec{W}_e + S_e - L_e - L_{e \to h} \tag{1b}$$

$$\frac{\partial U_h}{\partial t} = -\nabla \cdot \vec{W}_h + S_h - L_h + L_{e \to h} \tag{1c}$$

$$\frac{\partial U_L}{\partial t} = -\nabla \cdot \vec{W}_{lat} + L_e + L_h \tag{1d}$$

where $\vec{J}$ is the carrier current away from the surface, $\mathcal{G}$ is the generation rate due to optical absorption, $\vec{W}_e$ and $\vec{W}_h$ are the thermal currents carried by the respective carriers, $\vec{W}_{lat}$ is the thermal current carried by the lattice, S_e and S_h are energy source terms associated with the excess kinetic energy given to the carriers by the photons, L_e and L_h are energy exchange rates between the carriers and the lattice, and $L_{e \to h}$ accounts for energy transferred between the electrons and holes due to direct Coulomb interactions.

In order to solve this set of equations it is necessary to parameterize the carrier distributions. In the present case the individual electron and hole populations are assumed to establish quasi-equilibrium Maxwellian distributions on timescales short compared to the laser pulse duration. These are reasonable assumptions since at the carrier densities achieved in these experiments, $\sim 10^{18}$ cm^{-3}, carrier-carrier scattering times should be sub-picosecond, and the effective temperatures of both distributions are in excess of ~ 2000 K during the time most of the optical phonons are generated. Under these conditions, detailed expressions for the transport terms in Eqns. 1 can be taken from Ref. 14. Of particular note here is the ambipolar diffusion coefficient, D, which relates the ambipolar particle current to gradients in the carrier density and temperatures, T, away from the surface,

$$J = -D \left[\nabla N + \frac{N}{T_e + T_h} \{ \nabla T_e + \nabla T_h \} \right] \qquad (2)$$

where D is given by

$$D = D_e^0 D_h^0 \left[\frac{T_e + T_h}{D_h^0 T_e + D_e^0 T_h} \right] \qquad (3)$$

with D_e^0 and D_h^0 being the individual, equilibrium diffusion coefficients at the ambient temperature of the lattice. The important point, which is illustrated explicitly below, is that because of the relatively small absorption depths and high carrier temperatures involved in these experiments, diffusion processes have a significant effect on the amount of energy that gets transferred to phonons within an absorption depth of the surface, and hence the magnitude of the non-equilibrium phonon population that can be probed by the Raman process.

The energy transfer terms associated with the lattice in Eqns. (1) were all treated using microscopic expressions for both intra- and inter-valley carrier-acoustic phonon and carrier-optic phonon processes. Explicit representations for these terms can be found in Ref. 15. Each expression relates the energy transfer rate to a complicated function of the carrier and lattice temperatures, with a proportionality constant given by the square of the corresponding deformation potential. Table 1 lists the deformation potentials associated with the various carrier-lattice energy transfer processes considered.

Energy transfer from the relatively hot electrons to the relatively cool holes via direct Coulomb scattering was also treated, using a static screening approximation. Details of the calculation can be found in Ref. 18. The intent of including these Coulomb processes was merely to obtain an estimate as to whether the two carrier systems are better thought of as being strongly or weakly coupled on the timescale of these experiments. Accurate calculations of carrier-carrier scattering processes are complex and time-consuming, so either the completely coupled, or completely uncoupled limits are often assumed for simplicity. It was found that based on this static screening approximation, the electrons and holes remain essentially weakly coupled on the few picosecond timescale associated with carrier-lattice relaxation, and that the Raman results are only weakly dependent on this coupling strength.

While numerically solving Eqns. (1) using the detailed expressions for transport and continuity outlined above, it is straight forward to obtain the population of individual optical phonon modes using

$$\frac{\partial n(q)}{\partial t} = \left(\frac{\partial n(q)}{\partial t} \right)_{gen} - \frac{n(q) - n(q)_{eq}}{\tau_{ph}}, \qquad (4)$$

where

$$\left(\frac{\partial n(q)}{\partial t} \right)_{gen} \qquad (5)$$

Table 1. Deformation potentials for germanium.[16,17]

Quantity	Symbol	Value	Units
Acoustic deformation at the Γ point	$\varepsilon_{ac}(\Gamma)$	5.0	eV
Acoustic deformation at the X point	$\varepsilon_{ac}(X)$	9.0	eV
Acoustic deformation at the L point	$\varepsilon_{ac}(L)$	11.0	eV
Acoustic deformation for the holes	$\varepsilon_{ac}(\text{hole})$	4.6	eV
Electron intravalley deformation	$\mathcal{D}_e$ $\mathcal{E}_{ph}$	4.0×10^8 37.0	eV/cm meV
Hole intraband deformation	$\mathcal{D}_h$ $\mathcal{E}_{ph}$	8.74×10^8 37.0	eV/cm meV
Intervalley deformation XX	$\mathcal{D}_{XX}^{(inter)}(LO)$ $\mathcal{E}_{ph}$	9.5×10^8 37.0	eV/cm meV
Intervalley deformation LL	$\mathcal{D}_{LL}^{(inter)}(LA,LO)$ $\mathcal{E}_{ph}$	3.0×10^8 27.6	eV/cm meV
Intervalley deformation LX	$\mathcal{D}_{LX}^{(inter)}(LA)$ $\mathcal{E}_{ph}$	4.1×10^8 27.6	eV/cm meV

represents the generation rate of optical phonons of wavevector q, due to intra-L valley (intra-X valley optical phonon interactions are symmetry forbidden to lowest order) and intra-valence band relaxation, while

$$\frac{n(q) - n(q)_{eq}}{\tau_{ph}} \qquad (6)$$

represents the decay of the non-equilibrium phonon population due to anharmonic lattice processes. The complete model can therefore be used to compare theoretical and measured non-equilibrium optical phonon dynamics, and is limited primarily by the simplifying assumptions[19] concerning the bandstructure, i.e., parabolic L and X valleys, parabolic and equivalent heavy and light hole bands, and non-existent Γ valley electrons and split-off valley holes.

It should be clear that the solution of the complete model provides much more information than that related just to the population of small wavevector optical phonons. The scope of this chapter does not permit discussion of independent, time-resolved reflectivity experiments,[12] the results of which have been used for additional quantitative tests of the model. In brief, these time-resolved reflectivity results provide a means of measuring the non-equilibrium plasma density at the sample surface as a function of pump/probe delay. Both the absolute value and the temporal behaviour of the experimentally inferred surface carrier density are in good agreement with the

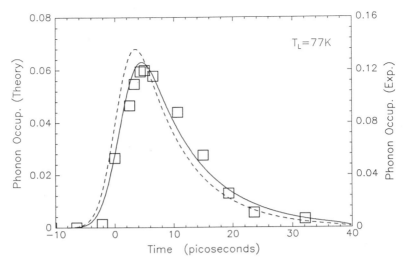

Fig. 3.   Temporal evolution of the actual non-equilibrium optical phonon population at 77 K ($\square$) together with the corresponding model results obtained with $\mathcal{D}_h = 8.7 \times 10^8$ eV/cm (solid line) and 12.5×10^8 eV/cm (dashed line).

above model, and in fact directly demonstrate the need to incorporate the generalized hot-carrier diffusion coefficient, Eqn. (3).

Table 2 contains those material parameters used in the solution of the above model which were not listed in Table 1. The results of the model calculation of the optical phonon population as a function of delay time between pump and probe beams are shown as solid curves in Figs. 3 and 4 at 77 K and 300 K respectively. The only variable parameter was the phonon lifetime, τ_{ph}, which turned out to be 8 ps and 4 ps at 77 K and 300 K respectively. Thus within the accuracy of the experimental data, and with material parameters taken from the literature at both temperatures, the coupled Boltzmann equation model yields good agreement for the delay in the peak non-equilibrium phonon population, and agreement within a factor of two for the absolute populations. The sensitivity of the model to the material parameters is illustrated by the dashed line in Fig. 3, where the results obtained with a hole deformation potential of 12.5×10^8 eV/cm[22] instead of 8.7×10^8 eV/cm as in Table 1, are shown. Note that the absolute values are not changed substantially, but the delay of the peak population is measurably different than that of the data. Since the accuracy of this delay is not limited by uncertainties in the absolute spot sizes or the spectrometer calibration factors, as is the accuracy of the absolute population scale, the excellent agreement of the model and experimental delay results strongly support the validity of the model. Even the factor of two discrepancy in the absolute population, consistent at the two temperatures, is within the uncertainty of the measured quantities and must be considered as supporting the model, given the necessary approximations incorporated in it.

The importance of incorporating diffusion, even on these picosecond timescales, is graphically illustrated in Fig. 5 where the surface carrier density is plotted as a

Table 2. Material parameters for germanium.[20,21]

Quantity	Symbol	Value	Units
Lattice constant	a	0.566	nm
Specific heat	C_v	0.32	J/gK
Density	ρ	5.32	g/cm^3
Auger coefficient	γ	2×10^{-31}	cm^6/sec
Electron diffusion	$D_e^0(77\,\mathrm{K})$	583	cm^2/sec
Hole diffusion	$D_h^0(77\,\mathrm{K})$	232	cm^2/sec
Electron diffusion	$D_e^0(300\,\mathrm{K})$	103	cm^2/sec
Hole diffusion	$D_h^0(300\,\mathrm{K})$	54	cm^2/sec
Absorption coefficient at $\lambda = 0.575\,\mu$m	$\alpha(77\,\mathrm{K})$	1.0×10^5	cm^{-1}
Absorption coefficient at $\lambda = 0.575\,\mu$m	$\alpha(300\,\mathrm{K})$	3.0×10^5	cm^{-1}
Electron effective mass at the L point	m_L^*	0.22	m_0
Electron effective mass near the X point	m_x^*	0.32	m_0
Hole effective mass	m_h^*	0.34	m_0
Mass of free electron	m_0	9.108×10^{-28}	g

function of time. The solid, dashed and dash-dotted curves correspond respectively to solutions of the model under the experimental conditions at 77 K including the full non-equilibrium ambipolar diffusion coefficient as in Eqn. (3), a constant ambient diffusion coefficient, and no diffusion at all. Ignoring diffusion altogether results in an order of magnitude overestimate of the peak surface carrier density, which translates directly into a similar overestimate of the non-equilibrium phonon population.

The solid curve in Fig. 5 also indicates that the peak carrier density achieved in the Raman experiments is of the order of 1×10^{18}cm^{-3}. This is approximately an order of magnitude higher than that at which similar experiments can be carried out in GaAs or InP, since in these polar materials the LO phonon is strongly renormalized[5,6] through coupling with the photo-excited plasma for densities above $\sim 1 \times 10^{17}$cm^{-3}. The relatively high carrier densities allowed in these Ge experiments partially explain how it is possible to observe non-equilibrium optical phonons in this non-polar material. However, if one assumed that the energy transferred from the plasma to the optical phonons was evenly distributed throughout the Brillouin zone due to wavevector independent matrix elements for the deformation potential interaction, higher carrier densities alone would not be sufficient to generate observable

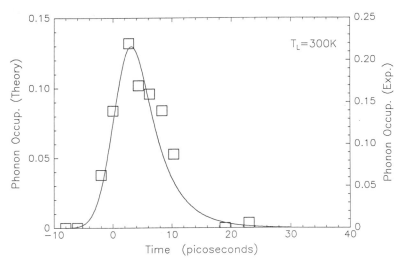

Fig. 4.   Temporal evolution of the actual non-equilibrium optical
           phonon population at 300 K (□) together with the corresonding
           model results obtained with $\mathcal{D}_h = 8.7 \times 10^8 \text{eV/cm}$ (solid line).

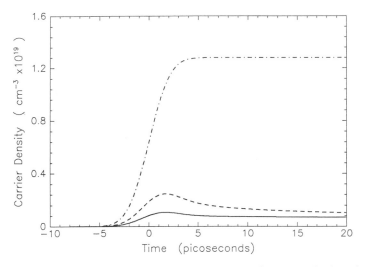

Fig. 5.   Temporal evolution of the surface carrier density calculated
           using a temperature dependent (solid line), constant
           (dashed line), and zero (dash-dotted line) diffusion coefficient at 77 K.

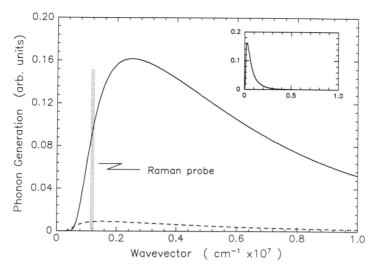

Fig. 6.   Optical phonon population generated by intra-band relaxation
          of heavy holes (solid line) and L valley electrons (dashed line) at
          77 K as a function of wavevector. The vertical line is at the
          wavevector probed in the Raman experiment. The inset shows
          the same plot on the scale of the full Brillouin zone.

non-equilibrium populations in any one mode. It is important to recognize that even
with no wavevector dependence in the matrix elements, the range of optical phonons
that couple to the hot carriers is still quite restricted due to the kinematic restraints
of energy and momentum conservation. This is illustrated in Fig. 6 where the model
result for the non-equilibrium optical phonon population generated at 77 K through
relaxation of the photoexcited plasma is plotted against phonon wavevector. The inset shows the same curve on the scale of the Brillouin zone. From these plots it is
clear that kinematic considerations alone result in a relatively strong focussing of the
energy from the plasma through relatively small wavevector optical phonon modes, despite the wavevector independence of the deformation potential interaction. Thus the
pump/probe Raman technique is capable of generating and detecting non-equilibrium
optical phonon populations in Ge because of both the ability to go to higher carrier
densities than in GaAs, and the kinematic restraints on the carrier-optical phonon
interaction itself.

OPTICAL PHONON DYNAMICS AND LINESHAPES IN GeSi ALLOYS

GeSi alloys, and GeSi/Si heterostructures in particular, are now being seriously
considered for use in the microelectronics field. In addition to their technological
potential, GeSi alloys grown on Ge substrates offer an interesting system in which
to study the effect of binary alloying on the phonon dynamics of an elemental semiconductor such as Ge. Phonon dynamic studies of LO modes in the AlGaAs ternary
alloy system by Kash et al.,[4] revealed that although the steady-state lineshape broadened substantially as Al was introduced, the phonon lifetime determined in the time
domain was unaffected.

Table 3. Alloy sample identification.

Sample	Si %	Thickness (μm)	$-\epsilon_\parallel$ ($\times 10^{-3}$)	$-\epsilon_\perp$ ($\times 10^{-3}$)
I	0			
II	4.5± 0.5	0.80 ± 0.05	0.6	2.9
III	10.0± 1.0	0.75 ± 0.05	3.2	4.4
IV*	10.0 ± 1.0	0.75 ± 0.05		
V	19.0± 1.5	0.35 ± 0.05	8.7	8.1
VI	25.0± 2.0	0.20 ± 0.05	6.3	11.2

* Sample III annealed for 30 seconds at 700°C.

Four, single-layer $Ge_{1-x}Si_x$ alloy samples were investigated. These layers were epitaxially grown on 50 mm (100) Ge wafers in a VG Semicon V80 molecular beam epitaxy system. The substrates were cleaned in situ by annealing for 10 min at 600°C prior to the growth. Films were deposited at 500 ± 50°C at a typical rate of 0.3 nm/s. All the epitaxial layers show good crystallinity and smooth surface morphology. The threading dislocation density in the epilayer as revealed by chemical etching was typically 10^6 cm^{-2}. Table 3 identifies the samples with their compositions, thicknesses and strain components perpendicular ($\epsilon_\perp$) and parallel ($\epsilon_\parallel$) to the surface (given relative to the Ge lattice constant). The compositions were obtained using X-ray, Auger, SIMS and Rutherford backscattering techniques. Strain values were obtained using double-crystal X-ray measurements in both non-dispersive (400) and dispersive (422) geometries. The residual strain component in the parallel direction measured on the alloy layers indicates that all are partially relaxed, except for sample V, which is almost fully relaxed. The steady-state (SS) Raman spectra were obtained in a quasi-backscattering geometry[23] using the 488 nm line of an Ar ion laser. The time-resolved data were taken using the identical setup described above for the Ge work.

Figure 7 shows the SS spectra from all as-grown samples in the vicinity of the Ge-Ge like optical mode near 300 cm^{-1}. The center frequency of the optical mode shifts monotonically down in frequency with increasing Si concentration, shifting by a maximum of 2% of the pure Ge frequency at x = 0.25. This downward shift in frequency is accompanied by a monotonic increase in linewidth and increase in line-asymmetry, always broadening more on the low energy side of the line. These frequency and width results are summarized and compared with published results[24-26] from bulk-grown alloys (unstressed) in Figs. 8 and 9.

Somewhat surprisingly, the frequency shifts for the strained epilayers are similar to those measured by others in (presumably unstrained) bulk alloys, except for those reported by Brya[24] at 300 K. The results of Brya[24] are closer to what one might expect for unstrained material and indeed, when sample III was annealed, its frequency shifted up, close to the value reported by Brya[24].

The asymmetric broadening of the line to lower energy is similar to that reported by others[27,4] in the ternary AlGaAs system. There it was attributed to a disorder-induced effect whereby a range of phonons with different wavevector were in effect being sampled by the Raman probe, and the asymmetry reflected the negative disper-

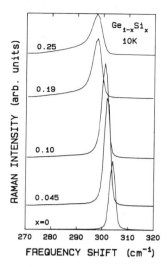

Fig. 7.   Steady state Raman spectra of samples I-III, V and VI
            at 10 K in the vicinity of the Ge-Ge like optical phonon mode.

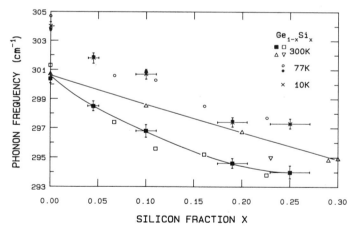

Fig. 8.   Ge-Ge optical phonon frequency versus x at different
            temperatures. The results from the present samples are
            shown at 10 K ($\times$), 77 K ($\bullet$) and 300 K ($\blacksquare$),
            and are compared with those from bulk alloys
            at 77 K ($\circ$)[25], and at 300 K ($\triangle$)[24], ($\square$)[25] and ($\triangledown$)[26].
            The solid lines are guides to the eye for different trends in the
            300 K data.

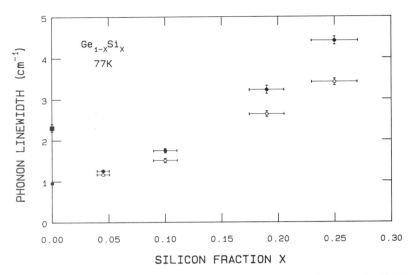

Fig. 9.   Ge-Ge optical phonon linewidth versus x for samples I-III,
V and VI at 77 K. The actual full width at half maximum,
$\delta\nu$, (•) is shown along with twice the high-energy-side
half width at half maximum (○). The 300 K result for
pure Ge is also shown (■).

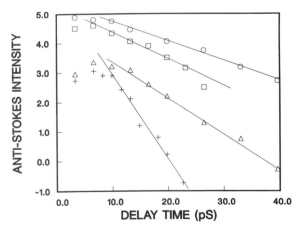

Fig. 10.   Logarithm of the anti-Stokes Raman intensity from the
Ge-Ge like optical phonon as a function of pump/probe
delay. Data is shown for samples II (□), III (○), IV (△),
and VI (+), all at 77 K.

414

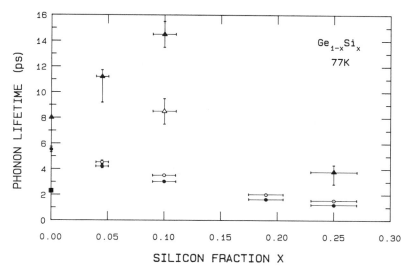

Fig. 11.   Summary of the optical phonon lifetimes at 77 K in samples
I-VI obtained in the time domain (▲ , unannealed),
(△ , annealed) and as defined in terms of the inverse linewidths
(see Fig. 9), $\tau_{ph} = (2\pi\delta\nu)^{-1}$, using for $\delta\nu$ the FWHM (•)
and twice the high-energy-side HWHM (o).

sion of the optical phonon modes. Si and Ge of course also exhibit negative dispersion
away from the Γ point of the Brillouin zone, and this could therefore also be the
explanation for the behaviour illustrated in Fig. 7.

Semi-logarithmic plots of the anti-Stokes Raman signal as a function of time delay
between pump and probe beams are shown in Fig. 10 for samples II-IV and VI. In
contrast to the AlGaAs system,[4] the lifetime of the Ge-Ge optical mode obtained in
the time domain is strongly dependent on the alloy composition, and on the effect of
annealing. These lifetime results are summarized together with the inverse linewidth
results (taken from Fig. 9) in Fig. 11. There is clearly no correlation between the two
sets of lifetimes. Those obtained in the time domain increase by close to a factor of
two upon the introduction of 10% Si, but it has decreased by close to a factor of two
from the Ge value at 25% Si. The time-resolved measurements on sample IV, (the
annealed version of sample III), yield a decreased lifetime, close to that of pure Ge.

The physical reason for these large variations in optical phonon lifetime has not
yet been identified. In the following discussion, additional SS Raman spectra of the
samples taken in a non-allowed geometry are presented as evidence in support of
plausible arguments as to possible causes for these interesting and unexpected results.

In ordered crystals the zone-center optical phonon lifetime is related to anharmonic terms in the lattice potential which couple optic and acoustic phonons.[28] To
lowest order, the value of the lifetime depends on the magnitude of an anharmonic
matrix element, and the joint two-phonon density of states (JTPDOS) into which the
optic phonon can decay conserving energy and wavevector. In a random alloy crystal,
the effects of disorder can in general change both the nature of the anharmonic matrix
elements, and even the concept of kinematic constraints. In the absence of a detailed

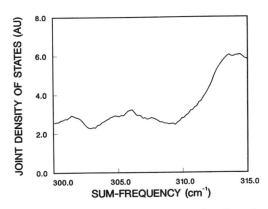

Fig. 12.   Joint two-phonon density of states for Ge calculated using the technique of Ref. 33 with the neutron scattering data in Ref. 32.

formalism for the lattice dynamics of concentrated GeSi alloys,[29] general arguments based on extensions of ordered crystal concepts and the qualitative effects of disorder will be used to suggest possible ways in which the introduction of Si might influence the lifetime of the phonons.

For low concentrations of Si it is perhaps possible to retain the basic concepts of phonon decay in ordered crystals and consider the influence of Si incorporation on the JTPDOS. Neglecting any influence of alloying on the matrix elements, the introduction of Si could influence the phonon lifetime by changing the JTPDOS at the energy of the Ge-Ge like optical phonon mode. There exists a very clear example[30] of how external perturbations such as pressure can change the JTPDOS seen by an optical phonon. In GaP the SS linewidth of the TO phonon mode can be changed by a factor of four upon the application of 87 kbar of hydrostatic pressure. This large change was attributed to relatively sharp features in the JTPDOS of unstrained GaP near the TO phonon energy, which can be moved with respect to the (also shifting) optical phonon energy by the application of pressure.

Although the quantitative details of calculated[28,31] JTPDOS in Ge differ depending on the theoretical treatment, there is general agreement that there exists a sharp feature in the JTPDOS at or just slightly above the optical phonon energy, whose magnitude increases by at least a factor of two towards higher energy, over a range of only ~ 5 cm^{-1}. The JTPDOS obtained using the neutron scattering data of Nilsson and Nelin[32] in a Brillouin zone integration scheme[33] (see Fig. 12) also exhibits a sharply rising feature in the JTPDOS $\sim 3\%$ above the optical phonon energy, which changes by a factor of ~ 2.5 over a range of only 3 cm^{-1}. This feature can be explicitly traced in the Brillouin zone integration case to a combination of acoustic and optic modes along the Q branch of the zone, as independently deduced by Menéndez and Cardona[31] from temperature dependent lineshape studies. Although none of these three techniques places the rapidly changing feature in the JTPDOS exactly at the optic phonon energy in Ge, there are obvious sources of uncertainty in all techniques, and it remains a possibility that in pure Ge the optic phonon energy is very close to

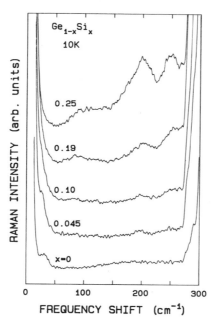

Fig. 13.   Non-allowed steady state Raman spectra from samples I-III,
V and VI at 10 K. The absolute intensities can be directly
compared for all spectra.

a rapidly varying structure in the JTPDOS. This being the case, a relative shift of
just ~ 3 cm^{-1} between the first-order optic phonon energy and the JTPDOS in the
GeSi epilayers could account for the observed increase in phonon lifetime by close to
a factor of two.

Review of the SS spectra shown in Fig. 7 shows that the first order Ge-Ge line
does in fact decrease monotonically by approximately 1% (~ 3 cm^{-1}) of its original
frequency upon the addition of 10% Si. In order to infer the influence of alloying
on the JTPDOS, non-allowed SS spectra were obtained from the samples in Table
3. These non-allowed spectra are shown in Figs. 13 and 14. The spectra in Fig.
13 were all taken with the beam incident at Brewster's angle, and the unanalyzed
scattered light was collected at 90° to the incident beam under identical conditions
so that the relative intensities can be compared. The spectra in Fig. 14 were taken in
a polarized backscattering geometry with the exposure time varied to obtain similar
signal to noise for the non-allowed features. The obvious features below 300 cm^{-1} in
the spectra from all samples except pure Ge are attributed to non-allowed first-order
scattering enabled due to the random nature of the alloy. (The structure in the pure
Ge spectrum, which is also evident as shoulders in the alloy spectra, is due to allowed
two-phonon scattering). This view is supported by comparison of these features with
those in the one-phonon density of states (OPDOS) curve obtained by Nilsson and
Nelin[32] from their neutron scattering data in Ge at 77 K, which is shown at the
bottom of Figs. 14 and 15. Although the non-allowed features in the alloy spectra
are broad and contain noise fluctuations, within experimental error all of the OPDOS
features appear at the same energy in the as-grown samples, while they shift slightly
to higher energy in the annealed sample. The JTPDOS at the optic phonon frequency
reflect combinations of OPDOS at lower energies, hence the invariance of the OPDOS
as evidenced in Fig. 14 suggests that the JTPDOS of the unannealed alloys might

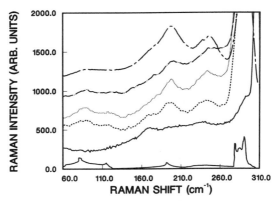

Fig. 14.   Non-allowed steady state Raman spectra from samples
I-IV and VI at 10 K. The absolute intensities cannot be directly
compared. The curve at the bottom shows the one phonon density
of states of Ge from Ref. 32.

remain essentially fixed, independent of Si concentration up to x = 0.25. If the sharp
feature in the JTPDOS of pure Ge remains essentially frozen while the first order
frequency decreases by ~ 3 cm^{-1} up to 10% Si, the effective JTPDOS into which
the optic mode can decay could be decreasing by a factor of two due to this relative
shift. Of course this argument ignores changes in the anharmonic matrix element,
and incorporates an indirect measure of the actual JTPDOS in the alloys, and thus
is only meant to suggest a possible explanation for the anomalous increase in phonon
lifetime. Much more detailed theory and experimental work would be required to
provide a complete explanation.

The above argument in terms of the effect of alloying on the JTPDOS at the
optic phonon frequency cannot explain the decrease in phonon lifetime at 25% Si.
The only feature in the SS spectra that can be correlated with this behaviour is the
disproportionate rise in the amplitude of the non-allowed features in the spectrum
as compared to the gradual increase in scattering strength up to 19% Si (see Fig.
13). If this increase in non-allowed Raman strength reflects an increasing effect of
the random potential fluctuations in the alloy on the Raman selection rules, there
might be a similar influence on the selection rules for optical phonon decay. Naively,
a breakdown of the selection rules for phonon decay would open up a larger density
of final states, decreasing the phonon lifetime. As with the above speculation, this
argument is meant only as a suggested implication of correlations in the SS and time-
resolved spectra of the samples. A full understanding of all the phonon dynamics
clearly requires the further development of models for lattice dynamics in random,
binary alloy systems.[34]

CONCLUSIONS

Picosecond time-resolved Raman scattering can be used to study the dynamical
properties of optical phonons generated by the relaxation of photo-excited plasmas in
both elemental and binary-alloy group IV semiconductors Ge and GeSi. By modelling

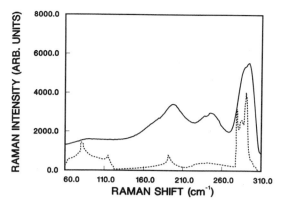

Fig. 15.　Non-allowed steady state Raman spectra from sample VI. The curve at the bottom is the one phonon density of states of Ge from Ref. 32.

the experimental conditions using coupled Boltzmann equations for the photo-excited electrons, holes and lattice, it is possible to quantitatively reproduce the observed buildup and decay of the LO and TO phonon populations using material parameters from the literature. In contrast to the AlGaAs ternary III-V alloy system, there is a strong influence of alloy composition on the Ge-Ge like optical phonon mode lifetime in the GeSi binary alloy system. This lifetime increases from its value in pure Ge by close to a factor of two at 10% Si, and is almost half of the pure Ge value at 25% Si. The reason for this influence of alloy composition on the phonon lifetime is not understood. However, there is evidence from steady state Raman spectra and phonon density of states calculations that the increase in lifetime could be associated with an alloy-induced shift of a sharp feature in the joint two phonon density of states with respect to the first order Ge-Ge optic frequency. The very low lifetime measured in the 25% Si sample is correlated with an abrupt increase in the strength of non-allowed first order Raman scattering below 300 cm^{-1}. This suggests that in the 25% sample the random potential fluctuations in the alloy might become very strong, leading to a complete breakdown of selection rules for the optical decay process and hence increasing the available density of states.

ACKNOWLEDGMENTS

The assistance of T.E. Jackman in the measurements of the alloy concentrations is gratefully acknowledged, as is the technical assistance of H.J. Labbe and B. Kettles.

REFERENCES

1.　C.L. Collins and P.Y. Yu, Phys. Rev. B 30, 4501 (1984).

2.　D. von der Linde, J. Kuhl and H. Klingenber, Phys. Rev. Lett. 44, 1505 (1980).

3.　J.A. Kash, J.C. Tsang, and J.M. Hvam, Phys. Rev. Lett. 54, 2151 (1985).

4. J.A. Kash, S.S. Jha, and J.C. Tsang, Phys. Rev. Lett. 58, 1869 (1987).

5. J.F. Young, and K. Wan, Phys. Rev. B 35, 2544 (1987).

6. J.F. Young, K. Wan, A.J. SpringThorpe, and P. Mandeville, Phys. Rev. B 36, 1316 (1987).

7. K.T. Tsen and H. Morkoc, Phys. Rev. B 34, 4412 (1986).

8. D.Y. Oberli, D.R. Wake, M.V. Klein, J. Klem, T. Henderson, and H. Morkoc, Phys. Rev. Lett. 59, 696 (1987).

9. D. Kim and P.Y. Yu, Phys. Rev. Lett. 64, 946 (1990).

10. M.C. Tatham, J.F. Ryan, and C.T. Foxon, Sol. State Elec. 32, 1497 (1989).

11. J.F. Young, K. Wan, and H.M. van Driel, Sol. State Elec. 31, 455 (1988).

12. A. Othonos, H.M. van Driel, J.F. Young, and P.J. Kelly, Phys. Rev. B 43, 6682 (1991).

13. S.S. Iyer, G.L. Patton, J.M.C. Stork, B.S. Meyerson, and D.L. Harame, IEEE Trans. Electron Devices 36, 2403 (1989).

14. H.M. van Driel, Phys. Rev. B 35, 8166 (1987).

15. E. Conwell, "High Field Transport in Semiconductors", Solid State Physics Suppl. 9, Academic, New York (1967).

16. L. Reggiani in: "Proceedings on the 15th International Conference on the Physics of Semiconductors", J. Phys. Soc. Jpn. Suppl. A49, 317 (1980).

17. R. Brunetti, C. Jacoboni, F. Nava, L. Reggiani, G. Bosman, and R.J. Zijlstra, J. Appl. Phys. 52, 6713 (1981).

18. J.F. Young, P.J. Kelly, and N.L. Henry, Phys. Rev. B 36, 4535 (1987).

19. A. Elci, M.O. Scully, A.L. Smirl, and J.C. Matter, Phys. Rev. B 16, 191 (1977).

20. M. Neuberger, "Group IV Semiconducting Materials", Handbook of electronic materials Vol. 5, Plenum, New York, (1971).

21. Landolt-Borstein, "Numerical Data and Functional Relationships in Science and Technology", Vol. 17, O. Madelung, M. Schulz, and H. Weiss eds., Springer-Verlag, Berlin, (1982).

22. This value was chosen for comparison because of the suggestion in Ref. 15 that the ratio of hole and electron squared deformation potentials in Ge might be ~ 10.

23. D.J. Lockwood, M.W.C. Dharma-wardana, D.C. Houghton, and J.M. Baribeau, Phys. Rev. B 35, 2243 (1987).

24. W.J. Brya, Sol. State Comm. 12, 253 (1973).

25. J.B. Renucci, M.A. Renucci, and M. Cardona, Sol. State Comm. 9, 1651 (1971).

26. M.I. Alonso and K. Winer, Phys. Rev. B 39, 10056 (1989).

27. P. Parayanthal and F.H. Pollak, Phys. Rev. Lett. 52, 1822 (1984).

28. R.A. Cowley, J. Phys. 26, 659 (1976).

29. D.W. Taylor in: "Optical Properties of Mixed Crystals", R.J. Elliott and I.P. Ipatova, eds., Elsevier, Amsterdam (1988), Chapter 2.

30. B.A. Weinstein, Sol. State Comm. 20, 999 (1976).

31. J. Menéndez and M. Cardona, Phys. Rev. B 29, 2051 (1984).

32. G. Nilsson and G. Nelin, Phys. Rev. B 3, 364 (1971).

33. A.H. MacDonald, S.H. Vosko, and P.T. Coleridge, J. Phys. C, 12, 2991 (1979).

34. D.N. Talwar and K.S. Suh, Phys. Rev. B 36, 6045 (1987).

TIME-RESOLVED RAMAN MEASUREMENTS OF ELECTRON-PHONON

INTERACTIONS IN QUANTUM WELLS

J.F. Ryan and M.C. Tatham

Clarendon Laboratory, University of Oxford
Parks Road, Oxford, UK

1. INTRODUCTION

The dynamics of non-equilibrium carriers in semiconductors is currently the subject of intense theoretical and experimental study. The issues are both of a fundamental nature, in particular, the determination of the mechanisms responsible for establishing equilibrium, and also of considerable practical significance, since the speed and efficiency of many semiconductor devices depend critically on the basic relaxation processes. The most important processes in this respect are carrier-carrier and carrier-lattice interactions whose timescales are in the femtosecond range. In polar semiconductors, for example, the interaction of carriers with optical phonons is primarily responsible for restoring equilibrium with the lattice. Much of our present understanding of these processes is derived from direct time-resolved optical measurements of non-equilibrium carriers using techniques such as optical absorption and photoluminescence. Femtosecond optical absorption measurements have shown that electrons and holes thermalise on different timescales. Whereas electron thermalisation itself is established on the timescale of ≤ 50fs, energy exchange between electrons and holes is relatively inefficient, and electron-hole thermalisation takes several picoseconds even at moderate carrier densities. Recent picosecond photoluminescence studies have also revealed many of the essential features of hot electron dynamics: energy relaxation is now known to depend on effects such as inter-valley scattering and non-equilibrium phonon generation.

The form of these interactions is changed dramatically in low dimensional heterostructures. This happens because the electronic states are quite different from those of bulk material, energy subbands being created by the spatial confinement, and also because the lattice dynamics acquires important new characteristics. In quantum well structures phonons can be dispersive, i.e. they can propagate through the entire structure, or they can be confined to one type of material. In addition, there are excitations in which the atomic displacements are localised close to the interfaces, so-called interface modes. Whereas the complexities of the lattice dynamics of these structures are immediately apparent in the Raman spectrum, their importance in carrier-phonon dynamics is only partially revealed by the conventional time-resolved optical techniques which probe only the behaviour of carriers. To get a more complete understanding of the non-equilibrium dynamics it is essential to probe directly the behaviour of both the electronic and lattice vibrational excitations of the structure.

In this paper we discuss time-resolved Raman scattering measurements of non-equilibrium carriers and phonons in GaAs quantum well structures which shed new light on the electron-phonon interaction. To begin with, in §2 we review

Light Scattering in Semiconductor Structures and Superlattices
Edited by D.J. Lockwood and J.F. Young, Plenum Press, New York, 1991

briefly some of the basic properties of the dielectric continuum model which gives a convenient description of phonons in quantum well structures, and which predicts inter- and intra-subband relaxation rates. In §3 we describe resonant anti-Stokes electronic Raman scattering studies of inter-subband relaxation in GaAs quantum wells which measure the transition rate. In §4 we describe Raman measurements of phonon emission in intra-subband relaxation in narrow quantum wells which reveal the importance of interactions with confined and interface phonons.

2.   THE FRÖHLICH INTERACTION IN TWO-DIMENSIONAL SYSTEMS

The interaction between electrons and polar optical phonons in bulk semiconductors is conveniently expressed in terms of the Fröhlich interaction, which is the effective electrostatic potential associated with the ionic displacements u :

$$\phi = \int \frac{e^*}{V\epsilon_0} \, \underline{u}\,(\underline{r}).d\underline{r} \tag{1}$$

where e^* is the effective charge, V is the sample volume and ϵ_0 is the vacuum permittivity. This interaction favours longwavelength modes, and for this reason the lattice dynamics can be treated in terms of a dielectric continuum model. The dielectric function is written in terms of the $k = 0$ transverse optical (TO) and longitudinal optical (LO) phonon frequencies:

$$\epsilon(\omega) = \epsilon(\infty) \; \frac{\omega^2 - \omega^2_{LO}}{\omega^2 - \omega^2_{TO}} \tag{2}$$

where $\epsilon(\infty)$ is the high-frequency dielectric constant. When the continuum model is applied to an isolated thin dielectric slab ($0 < z < d$) the optical phonons are given by the electrostatic boundary condition:

$$\epsilon(\omega)\nabla.\underline{E} = 0 \tag{3}$$

The solutions obtained from eq. (3) correspond to bulk-like slab modes and surface modes[1,2]. When $\epsilon(\omega) = 0$ it yields slab modes whose frequencies are the bulk LO phonon frequency, and

$$k_z = \frac{m\pi}{d}, \qquad m = 1,2,3..... \tag{4}$$

A suitable potential is:

$$\phi_S = \Phi_0 \sin(k_z z), \qquad 0 < z < d \tag{5}$$

i.e. ϕ_S has nodes at the surfaces. The z components of the corresponding E field and ionic displacements have antinodes at the interfaces. The longitudinal (i.e. in-plane) ionic displacements for modes with odd m are symmetric, and those with m even are antisymmetric with respect to reflection in the plane $z = d/2$.

There are also solutions of eq. (3) for $\epsilon(\omega) < 0$ (i.e. for $\omega_{TO} < \omega < \omega_{LO}$) which correspond to modes propagating parallel to the surfaces. These modes have a characteristic dispersion[3]:

$$\omega^2_+ = \omega^2_{TO} + \frac{\epsilon(\infty)\,(\omega^2_{LO} - \omega^2_{TO})}{\coth(k_\parallel d/2) + \epsilon(\infty)} \tag{6a}$$

$$\omega^2_- = \omega^2_{TO} + \frac{\epsilon(\infty)\,(\omega^2_{LO} - \omega^2_{TO})}{\tanh(k_\parallel d/2) + \epsilon(\infty)} \tag{6b}$$

where $k_\parallel$ is the in-plane wavevector. The subscripts $+$, $-$ refer to symmetric and antisymmetric modes under reflection in $z = d/2$. For $k_\parallel \to 0$, $\omega_+ \to \omega_{TO}$ and $\omega_- \to \omega_{LO}$; for $k_\parallel d \gg 1$ the modes are degenerate. The potentials produced by them are:

$$\phi_+ = \Phi_+ \exp{(i\underline{k}_\parallel.\underline{r})}\cosh(k_\parallel(z-d/2)) \tag{7a}$$

$$\phi_- = \Phi_- \exp{(i\underline{k}_\parallel.\underline{r})}\sinh(k_\parallel(z-d/2)) \tag{7b}$$

For $k_\parallel d \gg 1$ these modes are highly localised, one at each interface.

This continuum model can be applied in a similar fashion to multiple quantum well (MQW) structures and superlattices. The bulk-like optical modes are confined to the layers of either material and have the frequency of the corresponding LO and TO phonons. Interface modes (IF$_\pm$) now exist in the frequency ranges where $\epsilon(\omega)$ has different signs for the adjacent layers, and the frequency dispersion is given by[3]:

$$\cos(k_z d) = \frac{\eta^2 + 1}{2\eta}\sinh(k_\parallel d_1)\sinh(k_\parallel d_2) + \cosh(k_\parallel d_1)\cosh(k_\parallel d_2) \tag{8}$$

where $d = d_1 + d_2$, and $\eta = \epsilon_1(\omega)/\epsilon_2(\omega)$. Fig. 1 shows the dispersion of GaAs modes obtained from the model for a GaAs/AlAs MQW with $d_1 = d_2$. A similar result is obtained for the AlAs modes. The situation is more complicated for GaAs/AlGaAs structures because of the two-mode behaviour of the alloy. In this case there are two bands of interface modes in the GaAs-like phonon region[4].

The validity of this theoretical treatment has been investigated by extensive Raman measurements of phonons in multiple quantum well structures[5-9], and also by comparing the results with those obtained from microscopic models of the lattice dynamics[10-15]. This subject has been reviewed in a number of recent articles[16-18]. Measurements of confined optical phonons in GaAs/AlGaAs and GaAs/AlAs structures show that dispersion of the LO phonon causes the confined modes to have different frequencies so that eq. (4) maps out the bulk phonon dispersion curve[8]. Raman spectra show even m confined modes at resonance: odd m modes are observed in the depolarised spectrum away from resonance. Interface modes are observed in the polarised spectrum at resonance, but a breakdown in momentum selection rules must be invoked in order to explain the appearance of $k_\parallel \neq 0$ modes in backscattering[9].

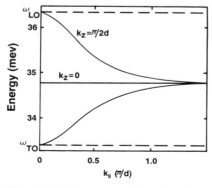

Fig. 1 Slab and interface mode dispersion for a GaAs/AlAs MQW obtained from the dielectric continuum model.

In view of the fundamentally different lattice dynamical properties of these quasi-two-dimensional structures compared to bulk it is important to consider the implications for electron-phonon interactions. From the outset we can see that the symmetry of the new modes imposes restrictions[19-23]: intra-subband electron relaxation requires even-parity phonons (i.e. potential), so that only odd m confined modes and ϕ_+ interface modes participate. Inter-subband relaxation between subbands of different parity requires even m confined modes and ϕ_- interface modes. Given that the wavevector of the confined modes has a minimum value determined by k_z, it is to be expected that the Fröhlich interaction involving such modes is weakened, given that long wavelengths are favoured. It is also evident that interface modes which are highly localised at the interfaces will couple only weakly to highly confined electrons whose wavefunctions have vanishingly small amplitudes at the interfaces.

The conventional continuum model approach to quantum well and superlattice structures follows essentially the same method as for the isolated slab: it adopts purely electrostatic boundary conditions, neglects retardation effects and treats the bulk modes as dispersionless. A difficulty with this approach is that the potential in eq. (5) is chosen to be zero at the interfaces (continuity of ϕ) which gives the unreasonable result of antinodes for u_z and E_z at the interfaces. Microscopic lattice dynamical calculations show, on the other hand, that displacement nodes occur at the interfaces; to be more precise, the interfacial As ions are displaced and the nodes occur between the As and Al or Ga ions, so that the effective layer thickness in a continuum model is slightly greater: $d_{eff} \rightarrow (d_1 + \delta a_0)$, where a_0 is a monolayer thickness. For perfect confinement $\delta = 1$ [24]; in general δ is found to depend on the details of the system and the model[13].

Furthermore, when the predictions of eq. (4) are compared to the results of microscopic lattice dynamical calculations the lowest-order (i.e. m = 1) confined mode is absent in the latter for $k_\parallel > 0$. Various calculations have shown that this mode is strongly dispersive for $k_\parallel$, similar to the surface mode (eq. (6))[13,25,26]. Huang and Zhu[26] have suggested that the discontinuities that occur in u_z and E_z at the interfaces is a fundamental limitation of the continuum model, containing as it does elements of modes with vanishingly small wavelengths - which are naturally permitted by the continuum model. In order to overcome this problem they proposed phenomenological potentials for the modes (referred to here as ϕ_{HZ}) which satisfy the boundary conditions that ϕ and $d\phi/dz$ (and so u_z) vanish at the interfaces. These potentials agree fairly well with those derived from microscopic lattice dynamical calculations for $k_\parallel \leq \pi/d$. An important result of their calculation is that interface phonons couple strongly to confined phonons when realistic LO phonon dispersion is included in the model. This may not be important for AlAs interface modes since the bulk LO mode is predicted to have very little dispersion[27]. However, the validity of the ϕ_{HZ} potentials remains to be tested experimentally.

Babiker[28] and Ridley[29] have used a radically different version of the continuum model that appears to overcome some of the difficulties raised above[30]. To begin with they include retardation effects and polariton modes are considered explicitly; they find that coupling to the surface modes (eq. (6)) is small, these modes being essentially transverse in character. Furthermore, they argue that the application of electromagnetic boundary conditions is fallacious, since both the electric displacement D and the magnetic field H are zero for LO phonons. The appropriate boundary condition should be that u_z is zero at the interfaces (for perfectly confined modes), and consequently, the electrostatic potential is of the form $\phi_c \sim \cos(k_z z)$ (c.f. eq. (5)). The consequencies of this result can be seen most readily by considering selection rules for electron-phonon scattering processes. Intra-subband relaxation now requires even m modes, and inter-subband relaxation occurs with odd m phonon emission: inter-subband scattering is faster (lower m), but intra-subband scattering is slower (higher m) in the Ridley-Babiker model. This modified continuum model also makes predictions about the importance of interface mode interactions. It is suggested that LO phonon dispersion, which is

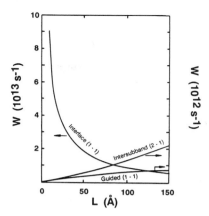

Fig. 2 Rates for electron inter- and intra-subband scattering in GaAs/AlGaAs quantum wells obtained from the dielectric continuum model [29].

different in both materials, provides the discontinuity that gives rise to interface phonons, and not a mismatch of the dielectric constant. This model reveals that coupling to ϕ_+ displacements is strong in the limit $\underline{k}_\parallel \to 0$; in fact, for narrow wells it dominates over confined mode relaxation processes. Fig. 2 shows the electron-phonon scattering rates obtained by this model for inter- and intra-subband transitions in a GaAs quantum well double heterostructure in the limit of infinite potential barriers[29]. For 150Å wells the $C_2 \to C_1$ scattering time is ~ 350fs, which is about half the $C_1 \to C_1$ relaxation time for confined mode emission. Interface mode processes are very rapid: at 50Å the $C_1 \to C_1$ time is ~ 50fs, and it diverges as $d_1 \to 0$. The rates shown in Fig. 2 are at $k_\parallel = 0$ for $C_2 \to C_1$, and at threshold for $C_1 \to C_1$; however the dependence on $k_\parallel$ is weak.

Rudin and Reinecke[31] have extended Ridley's calculation to obtain rates for the slab mode potential, ϕ_S, and also for ϕ_{HZ}. The $C_1 \to C_1$ rates for both potentials are found to be very similar, and about an order of magnitude greater than the rate for ϕ_C (see Fig. 2). ϕ_+ interface mode scattering rates were found to be similar to those shown in Fig. 2 for wide wells, with the dominant contribution arising from AlAs interface modes; however, for $d \lesssim 50$Å the rate does not show the divergence evident in Fig. 2. The $C_2 \to C_1$ rate for ϕ_S is very similar to that for ϕ_C shown in Fig. 2, but the ϕ_{HZ} rate is substantially smaller, giving, for example, a transition time of ~ 3ps for $d = 150$Å; when ϕ_- interface mode scattering is included this time is roughly halved.

3. INTER-SUBBAND RELAXATION IN QUANTUM WELLS

Inter-subband relaxation in GaAs quantum well structures has been studied using a variety of time-resolved optical techniques including inelastic light scattering[32], far infrared absorption[33] and inter-band absorption[34]. In these experiments electrons were excited into high-energy confined states, C_n, either by inter-band photoexcitation in undoped wells, or by excitation from the C_1 level in modulation-doped samples. Oberli et al[32] probed the Stokes electronic Raman spectrum of samples with well widths $d_1 = 96$Å (where $E_{21} > \hbar\omega_{LO}$) and 215Å (where $E_{21} < \hbar\omega_{LO}$) and measured the $C_2 \to C_1$ inter-subband relaxation time τ_{21}. In the former case the relaxation was faster than the instrumental resolution (~10ps), but in the latter case they measured $\tau_{21} \sim 325$ps. The fast relaxation in the narrower wells was attributed to LO phonon-assisted transitions; the slow relaxation in wide wells was interpreted as arising from acoustic phonon deformation potential scattering. Infrared absorption measurements[33] were made by tuning a picosecond laser to the E_{21} energy gap in a doped quantum well so as to excite electrons from the C_1 to the C_2 level; the relaxation back to C_1 was then detected by monitoring the absorption of a weak probe laser beam (at the same energy) as a function of time delay with respect to the pump. The samples had d_1 ~ 50Å with E_{21} ~ 150meV. Relaxation times for different samples were measured to lie in the range 10 - 15ps, consistent with the Raman data but more than an order of magnitude greater than theoretical estimates (see Fig. 2). More recent time-resolved inter-band optical absorption measurements of C_2-HH_2 excitons by Levenson et al[34] found that there was no sharp threshold for fast LO phonon emission as d_1 is varied: for $d_1 \lesssim 120$Å they measured $\tau_{21} \lesssim 12$ps, but even for wide wells with $d_1 = 240$Å the relaxation time was measured to be 20ps.

These results are somewhat contradictory, and they are in significant disagreement with theory. Although they attempted to measure electron-phonon interactions directly, the experiments share the same fundamental difficulty: the carrier density in each case, whether photoexcited or dopant, was relatively high, ~ 5×10^{11}cm^{-2}, and under such conditions carrier-carrier scattering is important. For example, Goodnick and Lugli[35] found in Monte Carlo simulations of the Raman experiment on wide quantum wells[32] that the non-equilibrium electron population thermalised by electron-electron scattering on a timescale that is significantly less than the electron-acoustic phonon emission time. Furthermore, intense photoexcitation gave rise to a non-equilibrium (NE) phonon population which caused the electron population in C_1 to remain high at long times at energies which overlap the C_2 subband. In other words, the simulation suggested that the Raman experiment measured energy relaxation of a thermalised electron distribution as opposed to inter-subband relaxation of unthermalised electrons. In the case of the infrared absorption experiment, there was the added complication that the C_2 level was not well confined, and it is anticipated[23] that the transition rate is reduced due to reduced C_2 - C_1 wavefunction overlap.

In this section we describe time-resolved anti-Stokes Raman measurements[36] of inter-subband relaxation by LO phonon emission. The experiment was designed to overcome many of the difficulties present in the experiments described above. In particular, the photoexcited carrier density was kept relatively low, ~10^{10}cm^{-2}, in order to minimise electron-electron inter-subband scattering. The well width was chosen so that $E_{21} \approx \hbar\omega_{LO}$; this has the advantage that the weak electronic Raman scattering intensity is enhanced by coupling to the phonons. Furthermore, under this condition the C_2 energy level is well below the confining barrier potential and so its wavefunction is highly confined to the well. The laser energy was tuned to an inter-band exciton transition so as to give resonant enhancement of the Raman signal, but it was kept below the threshold for electron scattering to the indirect conduction band minima at X and L.

3.1   Raman Scattering from Inter-Subband Excitations

The sample used in our measurements was a GaAs/Al$_{.35}$Ga$_{.65}$As MQW structure with $d_1 = 146$Å and $d_2 = 157$Å. The C_1 - C_2 energy separation, E_{21}, determined from photoluminescence excitation spectra is 52meV. This energy can also be measured directly in the Stokes Raman spectrum when photoexcited electrons are present in the wells[37]. Fig. 3 shows normalised time-integrated (the pulse width was 5ps) spectra obtained at 10K as a function of laser intensity. The scattering geometry was $z(x',x')\bar{z}$ (where $x' = (110)$ and $z = (001)$) which probes phonon and plasmon excitations. The spectrum shows intense peaks at the GaAs LO phonon frequency and at the AlAs-like LO phonon frequency of the barrier. The spectral resolution of 1.6meV is insufficient to allow the former to be resolved into distinct confined LO modes. There is in addition a well-resolved peak at 51meV which arises from C_1-C_2 transitions. This peak shows the same resonance enhancement as the GaAs LO phonon mode at C_n-HH_n inter-band exciton transitions. Fig. 3 was obtained at in-coming C_4-HH_4 resonance. Identification of the 51meV peak as inter-subband Raman scattering from photo-

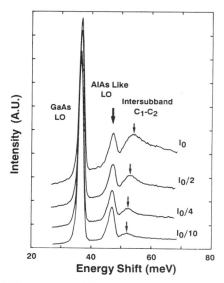

Fig. 3 Time-integrated Stokes Raman spectra measured as a function of laser intensity from a 146Å GaAs MQW structure at the C_4-HH_4 resonance. The broad peak at 51 meV arises from C_1-C_2 inter-subband transitions.

excited electrons is confirmed by the behaviour of its intensity as the laser intensity is increased: Fig. 3 shows that the scattering efficiency (the Raman intensity normalised to the laser intensity) increases with photoexcited electron density. In contrast to this behaviour the GaAs mode has approximately constant efficiency up to $n_S \sim 5 \times 10^{11} cm^{-2}$, but then decreases at higher density: the latter behaviour is is due to screening of the Fröhlich interaction mechanism responsible for the light scattering.

It is also clear from Fig. 3 that the energy of the E_{21} peak increases with increasing carrier density. This behaviour arises from the collective nature of the excitation. The coupled inter-subband plasmon-LO phonon modes have energies which depend on density in the following fashion[38]:

$$W_{\pm}^2 = \frac{\hbar^2}{2} (\omega_{12}^2 + \omega_{LO}^2 + \omega_p^{*2})$$

$$\pm \frac{\hbar^2}{2} [(\omega_{12}^2 + \omega_{LO}^2 + \omega_p^{*2})^2 - 4(\omega_{12}^2 \omega_{LO}^2 + \omega_{TO}^2 \omega_p^{*2})]^{1/2} \quad (9)$$

$\hbar\omega_{12}$ is the bare intersubband energy ($=E_{21}$), and ω_p^* is the inter-subband plasmon frequency given by:

$$\omega_p^{*2} = \frac{2n_S \omega_{12}}{\hbar} V_{12} \quad (10)$$

n_S is the difference in carrier populations of the two subbands: $n_S = n_1 - n_2$; V_{12} is the Coulomb interaction matrix element between electrons in the subbands. At low density:

$$W_+ \approx \hbar\omega_{12} \qquad : \qquad W_- \approx \hbar\omega_{LO} \tag{11a}$$

and at high density:

$$W_+ \approx \hbar(\omega_{12}^2 + \omega_{LO}^2 - \omega_{TO}^2)^{1/2}: \qquad W_- \approx \hbar\omega_{TO} \tag{11b}$$

Fig. 4 compares the measured energies with those obtained from eq. (9) using for n_S the estimated total photoexcited carrier density. The agreement is good at low densities, but there is some discrepancy at densities $\geq 5\times10^{11}\text{cm}^{-2}$. The origin of this discrepancy is not known, but it may result from the fact that for higher photoexcitation the population in C_2 remains high during the laser pulse ($\sim$ 5ps) making $(n_1 - n_2) < n_S$. Nevertheless, the conclusion is that inter-subband plasmon scattering can readily be detected in photoexcited quantum wells, and that measurement of its energy gives confidence in the estimate of carrier density.

Fig. 5 shows the time-integrated anti-Stokes spectrum obtained under the same conditions as described above. It shows a strong peak at the GaAs LO phonon frequency, indicating a non-equilibrium occupancy of the quantum well phonon modes due to energy relaxation of the photoexcited electrons. However, there is no evidence for AlAs-like modes in the anti-Stokes spectrum which shows that carriers excited within the wells do not couple strongly to barrier phonons. There is also strong scattering from the C_2-C_1 inter-subband excitations showing that there is substantial occupancy of the C_2 level induced by the laser pulse. The Raman scattering is again well polarised in the $z(x',x')\bar{z}$ configuration and is resonant with C_n-HH_n inter-band exciton transitions in the same manner as the GaAs LO mode. This result demonstrates that the anti-Stokes Raman spectrum can detect non-equilibrium electron populations in higher excited subbands. The time-resolved spectra presented below show that the time dependence of this signal can be used to measure the inter-subband relaxation time.

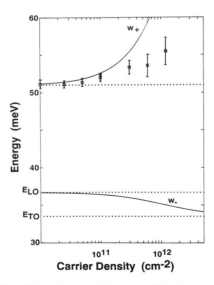

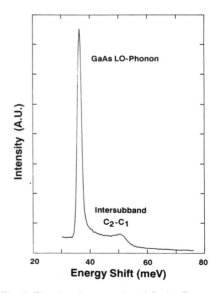

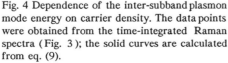

Fig. 4 Dependence of the inter-subband plasmon mode energy on carrier density. The data points were obtained from the time-integrated Raman spectra (Fig. 3); the solid curves are calculated from eq. (9).

Fig. 5 The time-integrated anti-Stokes Raman spectrum measured from the 146Å GaAs MQW The peaks at 36meV and 51meV are due to nonequilibrium GaAs LO phonons and C_2-C_1 inter-subband excitations respectively.

Before proceeding to discuss the time-resolved Raman data it is important to make some remarks about the Raman scattering mechanism itself, and the implications for measuring population relaxation. The electronic excitation measured here is a collective charge density mode as opposed to single-particle excitations. However, the inter-subband plasmon is a many-body state composed of single-particle states of c_1 and c_2 electrons, and so the decay of this mode reduces the c_2 population. At low electron density the collective mode is damped, i.e. plasmons and single-particle states are coupled, and both decay processes become equivalent. Consequently, the anti-Stokes Raman spectrum probes the c_2 population relaxation. However, it is important to realise that if there is a significant population of electrons in the final state then Raman scattering can be prevented by final state blocking. In general, the intensity is proportional to the initial and final state occupancies:

$$I_{nm}(t) \propto \sum f_n(\underline{k}_\parallel, t) \, (1 - f_m(\underline{k}'_\parallel, t)) \, R(\underline{k}_\parallel, \underline{k}'_\parallel) \tag{12}$$

where $f_n(\underline{k}_\parallel)$ denotes the occupancy of the state in subband c_n with in-plane wavevector $\underline{k}_\parallel$, and $R(\underline{k}_\parallel, \underline{k}'_\parallel)$ contains the wavevector dependence of the matrix elements for both inter- and intra-band steps in the Raman process. In the limit of low occupancies I_{21} depends approximately linearly on c_1 and c_2 populations for Stokes and anti-Stokes scattering respectively. (It was this dependence of intensity on occupancy that Oberli et al[32] used in the time-resolved Stokes measurements.) At higher occupancies the Stokes intensity saturates due to the high c_2 population, and the anti-Stokes intensity saturates due to large c_1 occupancy. Clearly, in the present experiment a low c_1 population is essential to prevent final state blocking.

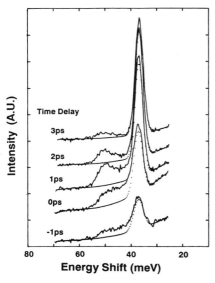

Fig. 6 Time-resolved anti-Stokes Raman spectra from the 146Å GaAs MQW structure. The dotted curves are fits to the phonon and luminescence components of the spectra.

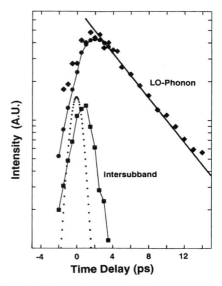

Fig. 7 Time-resolved anti-Stokes intensities of LO phonons and inter-subband excitations in the 146Å GaAs MQW structure at C_4-HH_4 resonance. The dotted curve is the laser pulse autocorrelation profile.

3.2   Time-Resolved Anti-Stokes Raman Scattering

The time-resolved anti-Stokes Raman spectrum obtained using 1ps pulses is shown in Fig. 6 for time delays ranging from -2ps to +3.5ps. For these measurements the laser was tuned to the C_4-HH_4 resonance so that all four electron subbands were directly photoexcited. The dominant feature of the spectrum is the GaAs LO mode which rises to a maximum at 2ps (i.e. within 1ps of the "end" of the pump pulse), and then decays relatively slowly at later times. The E_{21} line is substantially weaker, but it clearly shows much faster dynamics: I_{21} rises to a maximum after 1ps and then decays within 3ps. Measurements were also made at the C_3-HH_3 resonance: I_{21} was found to rise more rapidly, due to faster relaxation of electrons into C_2 from higher subbands, but the decay was similar to that measured at C_4-HH_4.

The time dependence of the integrated intensities of both peaks is shown in Fig. 7, together with the autocorrelation profile of the laser pulse which gives the instrumental time resolution. The E_{21} signal decays exponentially with a characteristic time of 1ps, which is therefore an upper limit for τ_{21}; as we will show below, when allowance is made for instrumental time resolution, and for the relaxation of electrons from higher subbands, the underlying value of τ_{21} is somewhat smaller. However, the relaxation time is clearly an order of magnitude faster than the previous measurements, and is closer to theoretical estimates (§2).

The theoretical analyses of inter-subband relaxation discussed in §2 show that there is only a weak dependence of the transition rate on in-plane wavevector, so that even although electrons are initially excited with considerable excess kinetic energy above the C_2 minimum the inter-subband dynamics probed in the Raman experiment is largely independent of intra-subband relaxation. Consequently, the population relaxation of electrons within the confined subbands can be expressed as a set of coupled rate equations:

$$\frac{dN_m(t)}{dt} = G_m(t) - \sum_{n<m} \frac{N_m(t)}{\tau_{mn}} + \sum_{n>m} \frac{N_n(t)}{\tau_{nm}} \tag{13}$$

$G_m(t)$ is the population generation rate for the C_m subband due to the pump laser pulse; τ_{mn} are the inter-subband relaxation times. For simplicity we assume that $G_m(t)$ is the same for all m-states excited, and that it follows the temporal profile of the laser pulse. This approximation is fairly reliable for the lower energy subbands, but it possibly overestimates the population generated in the subband in resonance.

Estimates of the relaxation times τ_{mn} can be obtained from the Fröhlich potentials given by the 2D electron-2D phonon models described in §2. Following Ridley[29] we obtain rates[36]:

$$\frac{1}{\tau^c_{21}} = \frac{1}{2} W_o \left\{ \frac{\hbar\omega_{LO}}{E_1} \right\}^{1/2} \left| \left\{ \frac{(64/15\pi)^2}{7 - \hbar\omega_{LO}/E_1} \right\} + \left\{ \frac{(128/105\pi)^2}{19 - \hbar\omega_{LO}/E_1} \right\} + ... \right| \tag{14a}$$

$$\frac{1}{\tau^s_{21}} = \frac{1}{2} W_o \left\{ \frac{\hbar\omega_{LO}}{E_1} \right\}^{1/2} \left| \left\{ \frac{1}{4 - \hbar\omega_{LO}/E_1} \right\} + \left\{ \frac{1}{12 - \hbar\omega_{LO}/E_1} \right\} \right| \tag{14b}$$

where τ^s and τ^c refer to the rates for the slab-mode and guided-mode potentials, i.e. $\sin(k_z z)$ and $\cos(k_z z)$ respectively. E_1 is the confinement energy of the C_1 level. W_o is the rate constant for the Fröhlich interaction:

$$W_o = (e^2/4\pi\hbar^2\epsilon_p)(2m^*\hbar\omega_{LO})^{1/2} \qquad (15)$$

where $\epsilon_p = \epsilon_o/\{\epsilon^{-1}(\infty) - \epsilon^{-1}(0)\}$. For GaAs $W_o \sim 7\times10^{12}s^{-1}$. We find that τ_{21} values differ by at most a factor of two between the models: they are 360fs and 630fs respectively for ϕ_C and ϕ_S. Transition rates between different subbands can be obtained in a similar manner; τ_n, the total population lifetime of the C_n level, increases with increasing n due to the increasing separation of the subbands which requires the emission of an LO phonon with large in-plane wavevector, whereas small wavevector transitions are favoured by the Fröhlich interaction.

The intensity of the anti-Stokes signal is given finally by the convolution of the probe laser intensity $I_{pr}(t)$ and the instantaneous C_2 population obtained from eq. (13):

$$I_{21}(t') \propto \int_{-\infty}^{+\infty} dt\, N_2(t)\, I_{pr}(t'-t) \qquad (16)$$

The dotted and full curves in Fig. 8 show the results of our calculations for both models. The lower curves (a) and (c) have three subbands excited, corresponding to C_3-HH_3 resonance, whereas the upper curves (b) and (d) have four subbands excited, corresponding to C_4-HH_4 resonance. Although there are clear differences between the calculations, there is overall very good agreement between

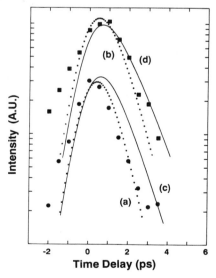

Fig. 8 Time-resolved anti-Stokes Raman intensities of C_2-C_1 transitions at the C_3-HH_3 (lower data set) and C_4-HH_4 (upper data set) resonances. Dotted curves (a) and (b) are calculated C_2 populations based on eq. (14a); the solid curves (c) and (d) are populations calculated from eq. (14b).

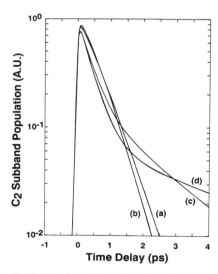

Fig. 9 Calculated time-dependent C_2 subband populations obtained from eqs. (13) and (16) for various values of τ_3: (a) 470fs (b) 360fs (=τ_{21}), (c) $5\tau_{21}$, (d) $10\tau_{21}$.

theory and experiment. When three subbands are excited the calculation based on ϕ_C is closer to the data, i.e. odd m guided phonon emission is favoured. However, when four subbands are excited both calculations compare favourably with the data. It should be recalled that at resonance even m guided phonons are detected in the Raman spectrum. This raises the question as to the origin of the non-equilibrium phonons observed in the anti-Stokes spectrum: are they confined modes which arise from intra-subband relaxation, or are they the phonons emitted in the inter-subband relaxation itself? In the latter case energy and wavevector conservation requires that the modes have considerable in-plane wavevector ~ $2 \times 10^6 cm^{-1}$. In order to observe such modes in backscattering there must be a mechanism that breaks wavevector conservation. The same consideration applies to the modes generated in intra-subband relaxation which have similar in-plane wavevectors. This issue will be discussed further in §4.2.

Finally, it is important to evaluate the extent to which the behaviour of the C_2 population depends on the population kinetics of the upper levels. Fig. 9 shows the calculated time dependence (neglecting instrumental broadening) of the C_2 population for various values of τ_3. Curves (c) and (d) show that when $\tau_3 \gg \tau_{21}$ the initial decay of C_2 is determined by τ_{21}, and that the slow decay characteristics of C_3 become apparent only at times > 2ps. The difference between curves (a) and (b) is relatively small and we would not expect to be able to resolve it experimentally. It is clear from this comparison that the observed inter-subband relaxation gives a reliable measure of τ_{21}.

4.   LO PHONON EMISSION IN INTRA-SUBBAND RELAXATION

The ratio of Stokes to anti-Stokes intensities, I^s/I^{as}, is proportional to $(1 + n(\omega)^{-1})$, where $n(\omega)$ is the Bose-Einstein factor, so that the Raman data can be used to determine $n(\omega)$ quantitatively. (This relationship between I^s/I^{as} and $n(\omega)$ applies strictly only to equilibrium conditions, although Jha et al[39] have shown that it applies to non-equilibrium excitations provided spectral broadening is taken into account.) Consideration of the kinematics of hot electron relaxation in bulk GaAs shows that when the electrons have an initial kinetic energy of 250meV, phonons are generated with wavevectors in the range $5 \times 10^5 cm^{-1}$ to ~ $10^7 cm^{-1}$; the peak of the NE phonon distribution occurs at a wavevector ~ $7 \times 10^5 cm^{-1}$, which is close to the wavevector probed in the backscattering Raman configuration (for a probe wavelength ~ 700nm). This method has been used to study NE phonons in bulk semiconductors, although in practice it is complicated by resonance effects and spatial non-uniformity of the photoexcitation. For quantum well structures, the small layer thickness and the periodicity along the growth direction modify the wavevector conservation conditions. For an isolated thin slab wavevector is uncertain to the extent $2\pi/d$, so that a large range of out-of-plane wavevectors is accessible in a Raman experiment. This has been verified by Kash et al[40] for GaAs layers with d ~ 500Å.

In exact backscattering geometry the Raman technique does not probe excitations with in-plane wavevectors; even at Brewster angle incidence, the maximum accessible $k_{\parallel}$ is ~ $6 \times 10^4 cm^{-1}$, so that it would appear that NE phonons produced in intra-subband relaxation, which have in-plane wavevectors $k_{\parallel} >$ $5 \times 10^5 cm^{-1}$, are inaccessible to probing by Raman spectroscopy. (This consideration does not apply to inter-subband relaxation, where $k_{\parallel} = 0$ phonons can be produced exactly at threshold when $E_{nm} = \hbar\omega_{LO}$.) Tsen et al[41] observed NE phonons in GaAs/AlGaAs quantum well structures when electrons were excited into continuum states above the confining barrier potential, but they were unable to identify the phonons either as barrier or well phonons. In a subsequent experiment[42], on a 50Å GaAs/AlGaAs MQW, confined carrier relaxation was observed when electrons were excited only within the lowest subband. In addition to observing peaks in the Stokes spectrum at 36.6meV and 35.2meV, which were assigned to GaAs confined phonons and GaAs-like barrier phonons respectively, an additional peak was observed at 35.5meV; this latter peak was also observed strongly in the anti-Stokes spectrum, and it was assigned to modes with $k_{\parallel} \neq 0$ emitted in the relaxation process.

In this section we describe a series of time-resolved Raman measurements of NE phonons in GaAs/AlGaAs quantum wells with different well thickness (and fixed barrier thickness 157Å). We observe NE confined phonons in relatively wide quantum wells, but the dominant feature in the anti-Stokes spectrum of narrow wells is interface phonons. We also describe experimental evidence which lends support to the suggested defect-induced Fröhlich mechanism which allows coupling to modes with in-plane wavevector.

4.1   Time-Resolved Raman Scattering from Confined and Interface Phonons

The time-resolved Raman spectra presented in §3 show that NE optical phonons can be observed in the anti-Stokes spectrum at remarkably low photoexcitation densities ($\sim 3 \times 10^{10} \text{cm}^{-2}$) due to the large enhancement in scattering efficiency that occurs at resonance. The data presented in Fig. 6 were obtained at the C_3-HH_3 resonance so that phonons were generated both in intra-subband and inter-subband relaxation. The spectrum shows only unresolved GaAs confined phonons; there is no obvious evidence of interface (IF) phonons in Fig. 6, but it should be recalled that the $k_\parallel = 0$ IF_+ modes are degenerate with the lowest-order confined TO and LO modes (see Fig. 1), and so would not be distinguished from them. It is only with substantially narrower wells that the individual confined modes can be resolved. Fig. 10 shows time-integrated Stokes and anti-Stokes spectra obtained from a GaAs/GaAlAs MQW structure with $d_1 = 56$Å using laser pulses tuned to the C_2-HH_2 resonance. The spectra show even m GaAs guided modes, which is expected for both the polarised and depolarised spectrum[43]. There are also weak features close to the GaAs TO mode (at 33.5meV) and at 35meV which corresponds to the GaAs-like LO mode of the AlGaAs barriers. The spectrum also shows a relatively strong line at 35.5meV which lies in the energy range expected for interface modes, but with $k_\parallel \neq 0$; tuning away from resonance its intensity decreases much more rapidly than for the

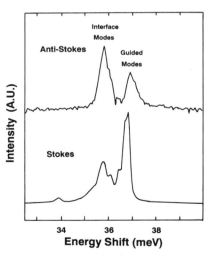

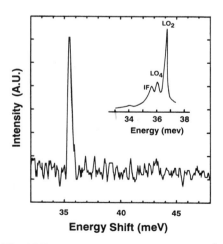

Fig. 10 Time-integrated Stokes and anti-Stokes Raman spectra measured from a 56Å GaAs MQW. The Stokes spectra show resolved even m guided GaAs modes and interface modes. The anti-Stokes spectrum shows non-equilibrium m = 2 confined and interface phonons.

Fig. 11 Raman spectra measured from a 25Å GaAs MQW. The inset shows the cw Stokes spectrum with well-resolved even m confined GaAs modes and interface modes; only non-equilibrium GaAs interface modes are present in the time-integrated anti-Stokes spectrum.

confined modes. For in-coming (IC) resonance, there is no detectable anti-Stokes spectrum. To observe this the laser was tuned to outgoing (OG) resonance; it is important to note that now only the C_1 electron subband is directly photoexcited, so that relaxation occurs entirely within the lowest subband. The anti-Stokes spectrum now reveals a remarkable new result, namely the presence of NE confined m = 2 and interface NE phonons.

In the case of very narrow quantum wells, where only one electron subband is confined, detection of a Raman signal is complicated by the very intense photoluminescence at the C_1-HH_1 gap. Although it is possible to achieve IC Stokes resonance at this energy, the laser energy at OG anti-Stokes resonance is lower than the bandgap and electrons are not photoexcited. However, rather special resonance conditions apply to wells with $d_1 \leq 30$Å where the LH_1 - HH_1 splitting at k = 0 is greater than $\hbar\omega_{LO}$. We have measured the anti-Stokes spectrum at C_1-LH_1 IC resonance from a GaAs/AlGaAs MQW with 25Å wells. The Stokes spectrum of this sample shows clearly resolved even m GaAs guided modes and interface modes (see Fig. 11 inset), and also AlAs-like barrier modes at 47meV (not shown). The time-integrated anti-Stokes spectrum shown in Fig. 11 reveals *only* NE GaAs interface phonons.

These measurements provide strong qualitative support for the theoretical models described in §2, which predict that interface phonon emission becomes the dominant intra-subband relaxation mechanism in narrow wells. The fact that the NE phonons are observed only at resonance is consistent with a Raman mechanism that breaks wavevector conservation, and so the spectrum probes an unknown range of $k_{||}$. However, since electrons are excited just above the threshold for phonon emission in the 25Å sample the only wavevector produced (neglecting thermalisation of the electrons) is 2.5×10^6cm^{-1}, so that $k_{||}d_1 \sim 0.6$. In this case we see from eq. (7) that the mode is not strongly localised at the interfaces but in fact penetrates into the barrier, as does the electron wavefunction.

An issue that remains unresolved by our measurements is the nature of the interaction of confined electrons with barrier phonons. For the wide quantum wells we found no evidence of NE barrier phonons, confined or interface, which suggests weak coupling. This is not unreasonable for AlAs-like confined modes which decay very rapidly close to the interface and so cannot couple strongly to GaAs-confined electrons. However, small-$k_{||}$ AlAs-like interface modes have a slowly decaying potential (eq. (7)) that should couple strongly to electrons confined in the GaAs layers, and therefore NE occupancy of these modes is expected. Our failure so far to detect these modes may be due to the fact that detection of interface modes in the anti-Stokes spectrum requires extreme resonance, which may require tuning to resonance at the AlGaAs bandgap.

Fig. 12 shows the time-resolved spectra obtained for the 56Å sample. In order to retain sufficient spectral resolution the pulse duration was set to 4ps. We find that the NE m = 2 and IF phonon populations build up very rapidly, on the timescale of the laser pulse, and then decay within 10ps. The risetimes of both modes are at most $\sim$ 1 - 2ps. The electrons initially have kinetic energy $\sim 5\hbar\omega_{LO}$, so that, assuming all five phonons are probed in the Raman spectrum, an upper limit for the mean intra-subband phonon emission time is $\sim$ 200 - 400fs (c.f. bulk GaAs where the time is 170fs [44]). Experimentally, we can expect the risetimes of the NE modes to be the same, and to be determined by the faster process; our estimate of the phonon emission time is considerably faster than the calculated confined mode relaxation time, $\sim$ 3ps (see Fig. 2), which suggests that IF mode interactions dominate. However, the mixing of confined and interface modes at finite $k_{||}$ [26] complicates the issue.

The overall decaytime of the anti-Stokes signal (measured with subpicosecond resolution, when both peaks are not spectrally resolved) is 4.5ps, which is significantly smaller than the bulk GaAs LO phonon lifetime of 8.8ps[45]. The latter is determined by decay into acoustic phonons near the Brillouin zone

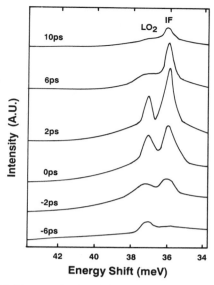

Fig. 12 Time-resolved anti-Stokes Raman spectra from the 56Å GaAs MQW. The laser pulse width was 4ps.

boundary. In the present case the shorter lifetime may indicate a larger anharmonic decay rate, due possibly to the two-dimensional phonon phase space. Alternatively, there may be dephasing caused by interface defects.

4.2   Resonance Profiles

The observation of "forbidden" Raman scattering in semiconductors at resonance has been studied extensively, and various mechanisms are known which can in principle break the constraint of wavevector conservation. It is now well established that the Fröhlich interaction in polar materials can give rise to "forbidden" scattering, which at resonance is more intense than the allowed deformation potential scattering. In two-dimensional systems there are two distinct resonance peaks for each electronic transition - in-coming (IC) and out-going (OG) - where either the laser energy or the scattered photon energy matches the transition energy. In the case that both photon energies correspond to real excited electronic states then double resonance can occur, provided energy and wavevector are simultaneously conserved. In the simplest (and most important) two-band case, where either the electron or the hole is scattered within the same band, this condition is difficult to meet. However, if the excited state is scattered by a defect or impurity then the process becomes allowed. In the case of Stokes scattering this effect gives rise to a stronger OG resonance: for bulk material it has been shown that ionised-impurity-induced forbidden scattering is important[46]. A similar mechanism has been invoked by Sood et al[9] to explain the observation of wavevector-forbidden IF mode Raman scattering from GaAs quantum wells. The calculation of Kauschke el al[47] showed that this mechanism can give a stronger OG resonance, but to obtain quantitative agreement with experiment required the introduction of screening of the impurity Coulomb potential. Asymmetries between OG and IC resonances in quantum wells were first reported by Zucker et al[48] who demonstrated the importance of confined excitons in the resonance process; they attributed the asymmetry between OG and IC strengths to intrinsic three-band resonance involving different subbands separated by an energy which is of the order of $\hbar\omega_{LO}$.

We have measured resonance profiles of the samples described above, including both Stokes and anti-Stokes spectra. We see strong IC and OG peaks of similar strengths at C_4-HH_4 in the Stokes spectrum of the 146Å sample. For the 56Å sample the profile is much more complicated. It shows a peak to the high-energy side of the C_2-HH_2 transition (IC resonance), but which also lies close to the parity-forbidden C_2-LH_1 transition (OG resonance); the latter transition has a weaker IC peak. An explanation of these results can be given in terms of resonant three-band processes involving the valence band: for this sample the LH_1 and HH_2 states are strongly mixed close to $\underline{k}_\parallel = 0$, and their separation is estimated to be ~30meV, which is close to $\hbar\omega_{LO}$, so that double resonance may occur.

The anti-Stokes resonance profiles for the same two samples are presented in Fig. 13. In (a) there is a stronger IC peak at C_4-HH_4 resonance. For the narrower wells, (b), there is a single, broad resonance associated with C_2-LH_1 IC

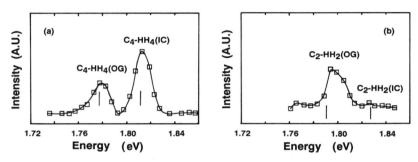

Fig. 13 Resonance profiles of the anti-Stokes Raman spectra of two GaAs MQW structures with well widths: (a) $d_1 = 146$Å, (b) $d_1 = 56$Å.

and C_2-HH_2 OG processes, which occurs about one phonon energy lower than the Stokes resonance. The peak is at least an order of magnitude stronger than either of the C_2-LH_1 OG or C_2-HH_2 IC peaks. The asymmetry of the profile for the 146Å sample is suggestive of the impurity-induced forbidden Fröhlich scattering mechanism. The single resonance observed for the 56Å sample is consistent with double resonance. Anti-Stokes resonance behaviour has not so far been discussed in the literature, although in principle the same considerations should apply as for Stokes. In the two-band case the requirement for real excited intermediate states means that the IC resonance will be stronger. However, since the anti-Stokes spectrum probes non-equilibrium phonon populations, there are additional considerations:

(i)   with increasing laser energy, each photoexcited carrier can emit more phonons;

(ii)   the NE phonon population is sensitive to the photoexcited carrier density, which changes as the optical absorption coefficient changes with laser energy;

(iii)   as the laser energy is tuned through a subband optical gap new inter- and intra-subband relaxation channels for phonon emission become possible.

For these reasons a quantitative interpretation of anti-Stokes resonance profiles is more difficult than in the Stokes case, and possibly even the simple test of IC/OG resonance symmetry is not sufficient. Nevertheless, it is evident from the resonance profiles that strong asymmetries exist which are most likely due to extrinsic mechanisms, and which would explain the observation of confined and interface modes with in-plane wavevector.

4.3   Non-Equilibrium Phonon Occupancies

It has been established by both experimental and theoretical studies that NE phonon effects are important in reducing hot carrier energy loss rates at high excitation. When the occupancies of phonon modes to which the carriers are coupled become high, i.e. the mode "temperature" approximately equals that of the carriers, then re-absorption of phonons occurs at a rate roughly equal to that of emission, so that in effect the energy loss rate is limited by the phonon decay rate. NE LO phonons in bulk GaAs have been studied extensively using time-resolved Raman spectroscopy by Kash and Tsang[45], who were able to determine mode occupancies by measuring Stokes-anti-Stokes intensity ratios. In this section we briefly describe measurements of GaAs quantum wells.

To begin with we present measurements of bulk GaAs. Fig. 14(a) shows the LO mode occupancy, $n(\omega)$, as a function of photoexcited carrier density. The data were obtained from measurements of time-integrated I^S/I^{AS} ratios using 1.3ps pulses. The laser energy was 1.946eV, which is far from resonance. The sample temperature was 38K, at which temperature the thermal equilibrium LO mode occupancy is 1.7×10^{-5}. The data show an approximately linear increase in occupancy with density up to $\sim 10^{17} \mathrm{cm}^{-3}$; at this density $n(\omega) \approx 1$. At higher densities $n(\omega)$ appears to decrease; this is consistent with plasmon-phonon interactions. It should be noted that the values of $n(\omega)$ given in Fig. 14 have not

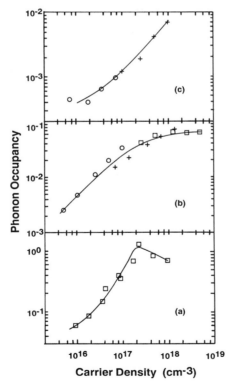

Fig. 14 Nonequilibrium phonon occupancies determined from Stokes/anti-Stokes intensity ratios: (a) bulk GaAs, (b) $d_1 = 146\text{Å}$ GaAs MQW, (c) $d_1 = 112\text{Å}$ GaAs MQW.

been corrected for spatial inhomogeneity of the photoexcitation nor for the effects of finite laser pulse duration (the electrons possibly have not fully relaxed by the end of the short laser pulse), both of which would tend to increase the measured value of $n(\omega)$.

Figs. 14 (b,c) show measurements of occupancy versus density for two quantum well samples: $d_1 = 146\text{Å}$ and 112Å. Here the laser energy was 1.82eV, so that only the wells were excited, and the sample temperature was 55K. The 3D equivalent density indicated in the figure is the 2D sheet density divided by the well width. The most important result is that $n(\omega)$ is more than an order of magnitude smaller than for bulk GaAs at the same density. Also, we find a marked *decrease* in Stokes efficiency with increasing density. For the 56Å quantum wells we measure the occupancy to be close to the thermal value, $\sim 5\times10^{-4}$, although the time-resolved data (Fig. 12) clearly show transient behaviour of NE phonons.

These observations are relevant to the discussion of forbidden $\underline{k}_\parallel$ scattering:

(i)   The phonon occupancy is low due to the fact that at resonance both Stokes and anti-Stokes scattering becomes allowed for a range of wavevectors $\delta\underline{k}_\parallel$; all of these modes contribute to the Stokes intensity, whereas only a small number (those emitted in the hot carrier relaxation) contribute to the anti-Stokes intensity. This argument is valid so long as $\delta\underline{k}_\parallel$ is greater than the range to which electrons are coupled.

(ii)   As the dominant Raman scattering process involves the Fröhlich mechanism it can be expected that, as shown by Kauschke et al[47], screening is important. In fact, Gammon et al[49] have demonstrated that the efficiency of forbidden IF mode scattering is reduced on increasing the laser intensity. The reduction in efficiency that we observe is consistent with this interpretation.

It appears from these measurements, therefore, that quantitative measurements of NE phonon occupancy in quantum wells are not possible for Raman measurements made at resonance.

5.   CONCLUDING REMARKS

The results presented here provide important new information about electron-phonon interactions in quantum well structures. Inter-subband relaxation has been observed and the transition rate has been measured directly: the optical phonon-assisted process occurs on the timescale of ~ 500fs in 146Å wells, which is in good agreement with the predictions of continuum models of the lattice dynamics. Intra-subband relaxation is found to be more rapid than the inter-subband process. In narrow quantum wells the experiments show that relaxation is dominated by interactions with GaAs interface modes. However, many issues are unresolved. There is little or no evidence of non-equilibrium AlAs interface phonons, although this interaction is predicted to be the dominant relaxation mechanism. The very nature of the Raman scattering process that probes the phonons at resonance prevents us from obtaining information about the wavevectors of the modes, and so there remains uncertainty in the intra-subband relaxation rate. Finally, although we have stressed the importance of the continuum models in making estimates of the scattering rates, it is clear that there are significant differences between the models that are difficult to resolve experimentally. Future work in this area will undoubtedly approach the problem from the viewpoint of microscopic lattice dynamical models.

ACKNOWLEDGEMENTS

This work has resulted from a close collaboration with C.T. Foxon, Philips Research Laboratories, Redhill. We also wish to thank A.C. Maciel and A.M. de Paula who participated in some of the experiments.

REFERENCES

1. R. Fuchs and K.L. Kliewer, Phys. Rev. 140, A2076 (1965)
2. W.E. Jones and R. Fuchs, Phys. Rev. B4, 358 (1971)
3. R.E. Camley and D.L. Mills, Phys. Rev. B29, 1695 (1984)
4. A.K. Arora, A.K. Ramdas, M.R. Melloch and N. Otsuka, Phys. Rev. B36, 1021 (1987)
5. J. Zucker, A. Pinczuk, D.S. Chemla, A. Gossard and W. Wiegmann, Phys. Rev. Lett. 53, 1280 (1984)
6. C. Colvard, R. Fisher, T.A. Grant, M.V. Klein, R. Merlin, H. Morkoç and A.C. Gossard, Superlatt. Microstr. 1, 81 (1985)
7. B. Jusserand, D. Paquet and A. Regreny, Superlatt. Microstr. 1, 61 (1985)
8. A.K. Sood, J. Menendez, M. Cardona and K. Ploog, Phys. Rev. Lett. 54, 2111 (1985)
9. A.K. Sood, J. Menendez, M. Cardona and K. Ploog, Phys. Rev. Lett. 54, 2115 (1985)
10. S. Yip and Y.C. Chang, Phys. Rev. B30, 7037 (1984)
11. E. Richter and D. Strauch, Solid State Commun. 64, 867 (1987)
12. S. Ren. H. Chu and Y.C. Chang, Phys. Rev. B37, 8899 (1988)
13. T. Tsuchiya, H. Akera and T. Ando, Phys. Rev. B39, 6025 (1989)
14. E. Molinari, A. Fasolino and K. Kunc, Superlatt. Microstruct. 2, 397 (1986)
15. E. Molinari, A. Fasolino and K. Kunc, Proc. 18th Int. Conf. Phys. Semicond., Ed. O. Engström (World Scientific, Singapore, 1987) p 663
16. M.V. Klein, IEEE J. Quantum Electron. QE-22, 1760 (1986)
17. B. Jusserand and M. Cardona in "Light Scattering in Solids V", Ed. M. Cardona and G. Güntherodt (Springer, Berlin, 1989)
18. J. Menendez, J. Lum. 44, 285 (1989)
19. R. Lassnig, Phys. Rev. B30, 7132 (1984)
20. F.A. Riddoch and B.K. Ridley, Physica 134B, 342 (1985)
21. N. Sawaki, J. Phys. C19, 4965 (1986)
22. L. Wendler and R. Pechstedt, Phys. Status. Solidi, B141, 129 (1987)
23. J.K. Jain and S. Das Sarma, Phys. Rev. Lett. 62, 2305 (1989)
24. B. Jusserand and D. Paquet, Phys. Rev. Lett. 56, 1756 (1986)
25. H. Chu, S.F. Ren and Y.C. Chang, Phys. Rev. B37, 10476 (1988)
26. Kun Huang and Bangfen Zhu, Phys. Rev. B38, 2183 (1988); B38, 13377 (1988)
27. S. Baroni, P. Giannozzi and E. Molinari, Proc. 19th Int. Conf. Phys. Semicond., Ed. W. Zawadski (Inst. of Physics, Polish Acad. Sci. Wroclaw, 1988) p795
28. M. Babiker, J. Phys. C19, 4965 (1986)
29. B.K. Ridley, Phys. Rev. B39, 5282 (1989)
30. B.K. Ridley and M. Babiker, to be published.
31. S. Rudin and T.L. Reinecke, Phys. Rev. B41, 7713 (1990)
32. D.Y. Oberli, D.R. Wake, M.V. Klein, J. Klem, T. Henderson and H. Morkoç, Phys. Rev. Lett. 59, 696 (1987)
33. A. Seilmeier, H.J. Hübner, G. Abstreiter, G. Weimann and W. Schlapp, Phys. Rev. Lett. 59, 1345 (1987)
34. J.A. Levenson, G. Dolique, J.L. Oudar and I.A. Abram, Solid State Electronics, 32, 1869 (1989)
35. S.M. Goodnick and P. Lugli, Superlatt. Microstruct. 5, 561 (1989)
36. M.C. Tatham, J.F. Ryan and C.T. Foxon, Phys. Rev. Lett. 63, 1637 (1989)
37. A. Pinczuk, J. Shah, A.C. Gossard and W.Wiegmann, Phys. Rev. Lett. 46, 1341 (1981)
38. E. Burstein, A. Pinczuk and D.L. Mills, Surf. Sci. 98, 451 (1980)
39. S.S. Jha, J.A. Kash and J.C. Tsang, Phys. Rev. B34, 5495 (1986)
40. J.A. Kash, J.V. Hvam, J.C. Tsang and T.F. Kuech, Phys. Rev. B38, 5776 (1988)
41. K.T. Tsen and H. Morkoç, Phys. Rev. B34, 4412 (1986); B38, 5615 (1988)
42. K.T. Tsen, R.P. Joshi, D.K. Ferry and H. Morkoç, Phys. Rev. B39, 1446 (1989)

43. T.A. Gant, M. Delaney, M.V. Klein, R. Houdre and H. Morkoç, Phys. Rev. B39, 1696 (1989)
44. J.A. Kash, J.C. Tsang and J.M. Hvam, Phys. Rev. Lett. 54, 2151 (1985)
45. J.A. Kash and J.C. Tsang, Solid State Electr. 31, 419 (1988)
46. J. Menendez and M. Cardona, Phys. Rev. B31, 3696 (1985)
47. W. Kauschke, A.K. Sood, M. Cardona and K. Ploog, Phys. Rev. B36, 1612 (1987)
48. J.E. Zucker, A. Pinczuk, D.S. Chemla, A.C. Gossard and W. Wiegmann, Phys. Rev. Lett. 51, 1293 (1983); Phys. Rev. B29, 7065 (1984)
49. D. Gammon, L. Shi, R. Merlin, G. Ambrazevicius, K. Ploog and H. Morkoç, Superlatt. and Microstruct. 4, 405 (1988)

RESONANT THREE-WAVE MIXING VIA SUBBAND LEVELS

IN QUANTUM WELLS: THEORETICAL CONSIDERATIONS

E. Burstein and M.Y. Jiang

Department of Physics and
Laboratory for Research on the Structure of Matter
University of Pennsylvania
Philadelphia, PA 19104

INTRODUCTION

There has been considerable interest in recent years in the nonlinear optical properties of semiconductor quantum-well structures. Attention has largely been focussed on four-wave mixing [1] and on optical Stark effects [2-4], phenomena that involve the third-order electric susceptibility, $\chi^{(3)}$, which can be quite large at frequencies in the vicinity of an exciton resonance. There has also been increasing interest in second-harmonic generation (SHG) and three-wave mixing (3WM) via subband levels in quantum wells, phenomena that involve the second-order electric susceptibility, $\chi^{(2)}$ [5-11]. Theoretical estimates of the $\chi^{(2)}$ for processes involving interband and intersubband transitions in asymmetric quantum wells and in asymmetric coupled quantum wells indicate that the magnitudes of the second harmonic and linear optic coefficients can be very much larger than in bulk semiconductors. A resonant SHG involving three intersubband transitions has recently been observed in an electric field-biased GaAs symmetric quantum well using a CO_2 laser beam as the fundamental [10]. The magnitude of the SHG was two orders of magnitude greater than that in bulk GaAs.

In this paper we will focus our attention on resonant 3WM in quantum wells, with and without carriers, via processes that involve two interband transitions and an intersubband (or intrasubband) transition [11] both for its intrinsic interest as a nonlinear optical phenomenon and for its potential as a spectroscopic probe of the electronic states and excitations of quantum wells. We will also discuss the similarity between three-wave mixing via intersubband transitions and inelastic light scattering by intersubband excitations [12,13] in asymmetric quantum wells with carriers.

3WM (sum and difference frequency generation) involves the coherent interaction of two input EM waves, w_1 and w_2, which set up a nonlinear polarization $P^{NL}(w_3)$ that is the source of the w_3 output EM wave, where $w_3 = w_1 \pm w_2$ [14]. Second-harmonic generation is a special case of sum frequency generation in which $w_1 = w_2$ and $w_3 = 2w_1$. The components of the wave vectors of the input and output EM waves parallel to the walls of the quantum well are conserved in traversing the interfaces, and only the parallel component of the wave vector for the overall nonlinear interaction is conserved, i.e., $k_{3//} = k_{1//} \pm k_{2//}$ inside the quantum well.

Light Scattering in Semiconductor Structures and Superlattices
Edited by D.J. Lockwood and J.F. Young, Plenum Press, New York, 1991

441

The envelope functions of the confined electrons and holes in symmetric quantum wells have even, or odd, parity (within the effective mass approximation) and, as a consequence, 3WM processes in symmetric quantum wells that involve intersubband (or intrasubband) transitions are dipole forbidden, even in semiconductor quantum wells that lack a center of inversion. 3WM is allowed in symmetric quantum wells with an externally applied electric field and in asymmetric quantum wells which have a space-charge electric field perpendicular to the growth direction and which are, therefore, conceptually equivalent to symmetric quantum wells with an externally applied electric field perpendicular to the walls (E_{0z}). 3WM in a quantum well can also be induced by an externally applied electric field parallel to the walls ($E_{0//}$) [11]. The $E_{0//}$-induced 3WM in either a symmetric quantum well, or an asymmetric quantum well, involves basically the same physics as electric field-induced 3WM via exciton levels in a bulk semiconductor [15].

The elementary 3WM processes consist of three steps that involve three virtual electronic transitions in which a photon of one of the input EM waves is annihilated, a photon of the other input EM wave is either annihilated or created, and a photon of the output EM wave is created [16]. We will be specifically concerned with 3WM in which the frequency of the w_1 input wave is resonant with an interband transition, and the frequency of the w_2 input wave is in the infrared, i.e., $w_2 \ll w_1$. Moreover, we will consider two situations: (a) electric field-induced 3WM in a quantum well that does not have any appreciable density of free carriers, i.e., E_F lies within the forbidden gap, and (b) electric field-induced 3WM in a quantum well that has an appreciable density of free carriers in, for purposes of discussion, the lowest conduction subband, i.e., E_F lies between the c_0 and c_1 conduction subbands. In situation (b) the electric field-induced, resonant 3WM via inter-subband transitions has features in common with the electric field-induced, inelastic light scattering by intersubband charge-density excitations.

3WM IN THE ABSENCE OF FREE CARRIERS

In a symmetric quantum well (i.e., $Al_xGa_{1-x}As/GaAs/Al_xGa_{1-x}As$) interband transitions are optically allowed only between valence and conduction subbands whose envelope functions have the same parity, and intersubband transitions in the valence and conduction subbands are only allowed between subbands that have different parity [17]. Optically allowed interband transitions that flip spin occur as a result of spin-orbit interactions. In the absence of any appreciable density of carriers, the optically allowed interband transitions between valence subbands and conduction subbands create 1s excitons. (When carriers are present in the c_0 subband, the creation of excitons by interband transitions between a valence subband and the c_0 conduction band is suppressed by Pauli exclusion and by screening of the Coulomb interaction). Because of differences in the periodic part of the wave functions of light and heavy holes, the polarization selection rules for interband transition leading to light and heavy hole excitons are different [18]. Thus, for GaAs quantum wells grown in the [100] direction, heavy hole excitons can be created only by photons that have either an E_x, or E_y, component, whereas light hole excitons can be created by photons that have an E_z component, as well as by photons that have an E_x, or E_y, component. Intersubband transitions that flip spin are not induced by an EM electric field, but may be induced by an EM magnetic field. Intersubband transitions that do not flip spin, (i.e., a non-spin-flip $c_0 \rightarrow c_1$ transition) can only be induced by photons having an E_z component, i.e., by p-polarized photons [19]. The transition from a $1s(v_0c_0)$ exciton to a $1s(v_0c_1)$ exciton, which involves the non-spin-flip transition of an electron from the c_0 subband to the c_1 subband, can also be induced by photons having an E_z component.

In the case of 3WM via intersubband transitions in quantum wells that do not have any free carriers, the first step involves an interband transition of an electron from a subband state of the valence band to an empty state in a subband state of the conduction band, creating an "interband" electron-hole pair. The second step involves the coherent intersubband transitions of the excited electron and the excited hole. The third step involves the interband recombination of the excited electron and hole returning the quantum well to the electronic ground state (Fig. 1).

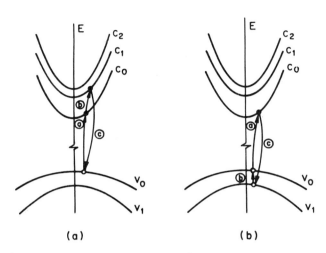

Fig. 1. The electronic transitions of sum frequency generation
        processes in which: (a) the excited electron undergoes
        an intersubband transition and (b) the excited hole under-
        goes an intersubband transition.

It follows, from the selection rules for the interband and intersubband transition in a symmetric quantum well, that 3WM via intersubband transitions are dipole forbidden. For example, the optically allowed interband transitions induced by one of the input EM waves generate excited electrons and excited holes whose envelope functions have the same parity. Since the allowed intersubband transitions of the electron (or hole) that are induced by a p-polarized, second input, EM wave lead to a change in the parity of the subband envelope function of the electron (or hole), the generation of an output EM wave by the radiative interband recombination of the excited electron and hole is forbidden, i.e., they have opposite parity. An electric field perpendicular to the walls of the quantum well (i.e., E_{0z}) polarizes the excited electron-hole pair (or exciton) along z and destroys the parity of the z part of the electron and hole subband envelope functions. It also causes a Stark shift in the subband energy levels [20-22]. Because of the confinement of the electrons and holes in the quantum well, the binding energy of the excitons in the quantum well is greater than that in the bulk and, as a consequence, the excitons are not dissociated even in moderately large E_{0z} fields. In the presence of the E_{0z} field the subband parity restrictions on interband transitions and on intersubband transitions are lifted. The normally forbidden interband transitions between valence and conduction subbands that have opposite parity in the absence of the E_{0z} field become allowed, and the intersubband transitions between subbands that have the same parity in the absence of the E_{0z} field are induced by photons that have an E_z component, i.e., by p-polarized photons.

In an asymmetric quantum well the interband transitions correspond to
Franz-Keldysh type transitions which create excited electrons and holes that
are spatially separated along z, i.e., along the direction of growth of the
quantum well [11]. This has the consequence that the intraband matrix element for the Fröhlich interaction of the z component of the electric field
of p-polarized photons with the spatially polarized electron-hole pairs
along z is non-zero [23,24]. In this situation, the second step may involve
either an intersubband transition, or an intrasubband transition, of the spatially separated excited electron and excited hole by a photon whose electric field has an E_z component. The third step involves a Franz-Keldysh
type recombination of the electron-hole pair (Fig. 2). 3WM processes in
which the second step involves an intersubband transition are termed three-
level processes. 3WM processes in which the second step involves an intrasubband transition are termed two-level processes. We note that the intersubband transitions, like the intrasubband transitions, only involve changes
in the envelope part of the electron and hole subband wave functions.

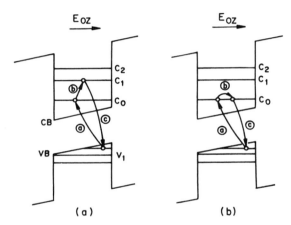

Fig. 2. The 3WM transitions in an asymmetric quantum well without carriers: (a) a three-level process which involves
an intersubband transition in the second step; (b) the
corresponding two-level process which involves an intrasubband transition in the second step.

The creation of virtual 1s excitons by the interband transitions give
rise to resonant peaks in the dependence of the w_3 output signal on the
frequency of the w_1 input EM wave. The exciton-enhanced contribution from
three-level 3WM processes to $X^{(2)}(E_z)$ involves a Franz-Keldysh optical interband matrix element for creating a virtual $1s(v_0c_0)$ exciton, an optical matrix element for the $1s(v_0c_0) \rightarrow 1s(v_0c_1)$ intersubband transition and a Franz-
Keldysh type matrix element for the E_z-induced recombination of the virtual
$1s(v_0c_1)$ exciton. Resonance peaks occur when either w_1 is resonant with the
$1s(v_0c_0)$ exciton, or w_3 is resonant with the $1s(v_0c_2)$ exciton (Fig. 3).
When w_1 is resonant with the $1s(h_0c_0)$ excitation energy and w_2 is resonant
with the $1s(v_0c_1) \rightarrow 1s(v_0c_2)$ excitation energy, the three-level 3WM process
is doubly resonant and therefore greatly enhanced. The corresponding contribution from two-level 3WM processes involves an E_z-induced matrix element for

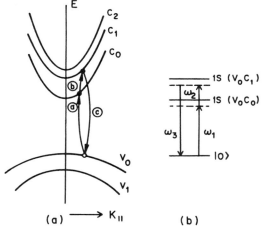

Fig. 3. (a) The 3WM interband and intersubband transitions in a
asymmetric quantum well without carriers; (b) the corresponding transitions that involve the creation of a
$1s(v_0c_0)$ exciton, its intersubband transition to, and
the subsequent recombination of, the $1s(v_0c_1)$ exciton.

the intrasubband excitation of the excited electron-hole pair in the intermediate state. Since $w_2 \ll w_1$, the two-level 3WM process is approximately doubly
resonant when $w_1 \approx w_3$ is resonant with the $1s(v_0c_0)$ exciton.

3WM IN THE PRESENCE OF FREE CARRIERS

The confined carriers in a quantum well (assumed for discussion to be
electrons) exhibit elementary single-particle and collective intersubband
excitations which can be probed by optical measurements, such as infrared
absorption [25,26] and inelastic light scattering [12,13]. The excitations
include single-particle spin-flip and non-spin-flip intersubband excitations
and collective intersubband charge-density and spin-density excitations.
Although forbidden by parity selection rules, resonant inelastic light scattering by single-particle excitations and by collective spin-density and
charge-density intersubband excitations is observed in symmetric quantum
wells. The inelastic scattering of light by single-particle intersubband
excitations that is observed in both polarized and depolarized spectra,
which have widths comparable with the Fermi energy, is attributed to two-
step processes. However, the scattering of light by the charge-density and
by the spin-density intersubband excitations that are observed, respectively, in polarized and unpolarized spectra, are attributed to three-step
processes [27]. Bajema et al. [28] have shown that the inelastic light scattering by the collective intersubband excitations of an "undoped" symmetric
well, in which carriers are photoexcited by the input EM laser beam, is
appreciably enhanced by an externally applied electric field perpendicular
to the wall of the well. The origin of the forbidden scattering in a symmetric quantum well has not been firmly established. It may simply be due
to a residual asymmetry in the quantum well that arises from the fact that
the epitaxial growth of GaAs on AlGaAs is not precisely identical to the
epitaxial growth of AlGaAs on GaAs.

Of particular relevance to our discussion of 3WM via intersubband excitations are the spectral data which Pinczuk et al. have obtained in their study of small-scattering wave vector inelastic light scattering by the intersubband excitations of high mobility electrons in modulation doped asymmetric quantum wells at low temperatures [29,30]. The spectra exhibit broad features corresponding to single-particle intersubband excitations, as well as very narrow peaks corresponding to collective charge-density and collective spin-density excitation modes (Fig. 4). Pinczuk et al. attribute the shift, to a lower energy from the single-particle transition energy, of the spin-density excitation peak to a sizeable exchange Coulomb interaction which had previously been assumed to be very small. They attribute the larger shift, to a higher energy from the single-particle transition energy, of the charge-density excitation mode to the both the exchange and direct Coulomb interaction. They suggest that the unexpected scattering by single-particle intersubband excitations, which is observed in both polarized and unpolarized small-scattering wave vector spectra, may be due to a disorder-induced breakdown of wave vector conservation.

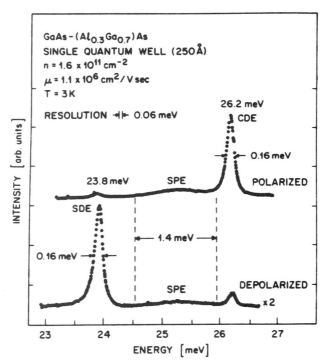

Fig. 4. Inelastic light-scattering spectra of the intersubband excitations of a modulation-doped, asymmetric quantum well showing the spin-density (SDE), charge-density (CDE) and single-particle excitation peaks. (From Pinczuk, et al.)

When carriers are present in an asymmetric quantum well, there are additional 3WM processes in which the intersubband (or intrasubband) transition occurs in either the first or the third step [27]. Two processes are of particular interest. The first is a sum frequency generation process which involves a $c_0 \rightarrow c_1$ (non-spin-flip) intersubband transition that creates an excited intersubband electron-hole pair as the first step, a $v_0 \rightarrow c_0$ interband transition of an electron from the valence band into the "excited hole" created in the c_0 subband as the second step, and an electric field-induced interband recombination of the c_1,v_0 electron-hole pair (Fig. 5a). The second is a difference frequency generation which involves a field-induced $v_0 \to c_1$ interband transition as the first step, a $c_0 \to v_0$ interband transition of an electron in the c_0 subband into the excited hole in the v_0 subband as the second step, and the recombination of the resulting c_1,c_0 intersubband electron-hole pair in the third step (Fig. 5b). We expect that difference frequency generation via the latter process, with w_1 and w_2 in the visible and $w_3 = w_1 - w_2$ resonant with the $c_0 \to c_1$ excitation in the infrared, should be a particularly effective one for generating infrared radiation.

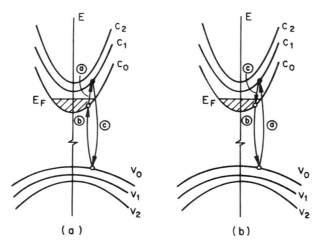

Fig. 5. The 3WM transitions in an asymmetric quantum well with carriers: (a) sum frequency generation in which the $c_0 \to c_1$ intersubband transition is the first step; (b) difference frequency generation in which the $c_1 \to c_0$ intersubband transition is the third step.

The 3WM processes in a modulation-doped, asymmetric quantum well that involve an intersubband transitions of the excited electron (or hole) in the second step, are similar to the processes that are involved in the inelastic scattering of light by intersubband charge-density excitations. In inelastic light scattering the interband transition in the first step is induced by an w_1 incident photon; the excitation of the virtual $1s(v_0c_1)$ exciton to the virtual $1s(v_0c_2)$ exciton in the second step is induced by the intersubband charge-density excitation; and the radiative recombination of the $1s(v_0c_1)$ exciton in the third step generates the w_s scattered photon. In resonant sum frequency generation the first step involves an w_1 photon-induced $v_0 \to c_1$ interband transition which generates a virtual $1s(v_0c_1)$ exciton; the second step involves an w_2 photon-induced transition of the virtual $1s(v_0c_1)$ exciton to a virtual $1s(v_0c_2)$ exciton; and the third step involves the radiative recombination of the $1s(v_0c_2)$ exciton creating an w_3 photon (Fig.6). (The presence of electrons in the c_1 subband suppresses the $1s(v_1c_0)$ exciton resonance, but does not suppress the exciton resonances associated with the "excited conduction subbands, providing the carrier density in the c_0 subband

447

is not too large.) Both the sum frequency generation and the inelastic
light scattering may be expected to exhibit an exciton-enhanced resonance
peak. either when the frequency of the the w_1 input wave (w_i incident wave)
is resonant with the $1s(v_0c_1)$ exciton, or when the w_3 output wave (w_s scattered wave) is resonant with the $1s(v_0c_2)$ exciton. The sum frequency generation will be doubly-resonant when the w_1 input wave is resonant with the
$1s(v_0c_1)$ exciton and the w_2 input wave is tuned to be resonant with the
energy separation between the $1s(v_1c_1)$ and $1s(v_0c_2)$ levels. This will not be
the case for inelastic light scattering, since the energy of the intersubband charge-density excitation will be somewhat larger than the energy separation of the exciton levels.

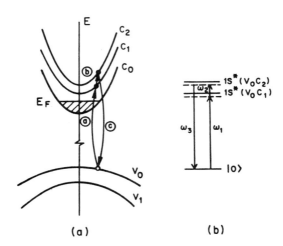

(a)   (b)

Fig. 6. (a) The interband and intersubband transitions that are
involved in 3WM and in inelastic light scattering by
intersubband charge-density excitations; (b) the corresponding processes that involve the creation of a
$1s(v_0c_1)$ exciton, its intersubband transiton to, and
the subsequent recombination of, the $1s(v_0c_2)$ exciton.

In inelastic light scattering, the intersubband transition is induced
by the charge-density intersubband excitation via the Coulomb and exchange
interactions. In 3WM in quantum wells that have no carriers, the intersubband transition is induced by the z component of the electric field of the
p-polarized infrared w_2 input EM wave. When the quantum well contains carriers, the p-polarized w_2 EM wave in the quantum well actually corresponds
to an intersubband charge-density excitation-polariton mode that involves
photons coupled to the intersubband charge-density excitations. Moreover,
when the frequency of the w_2 input EM wave is resonant with the intersubband
charge-density excitation mode, the charge-density excitation strength of the
polariton mode is close to unity. Under these circumstances, the "scattering"
of the virtual $1s(v_0c_1)$ exciton into the virtual $1s(v_0c_2)$ exciton in the
second step is actually induced by the intersubband charge-density excitation, and the matrix element for 3WM is essentially the same as that for the
corresponding inelastic light scattering. Accordingly, the w_3 output signal

may be expected to exhibit a resonance peak when the frequency of the w_2 input EM wave is tuned through the frequency of the intersubband charge-density excitation, even when the w_1 input wave is away from resonance with the $1s(v_0c_1)$ exciton.

We note that, within the quantum well, the incoming w_1 and w_i EM waves and the outgoing w_3 and w_s EM waves are, in fact, exciton-polaritons [31], and that the intersubband transition in the second step corresponds to the scattering of the exciton part of the incoming polariton into the exciton part of the outgoing polariton by the p-polarized w_2 input infrared EM wave in the sum frequency generation process, and by the intersubband charge-density excitation mode in the inelastic scattering process. The three steps involved in the two phenomena are reminiscent of exciton-enhanced resonance Raman scattering by LO phonons in bulk semiconductors in which the input EM wave corresponds to an exciton-polariton mode that is scattered by the Coulomb field of the LO phonons into another exciton-polariton mode which leaves the semiconductor as the scattered EM wave [32].

CONCLUDING REMARKS

Unlike inelastic light scattering, 3WM does not yield any fingerprint of the medium in which it occurs. The 3WM output signal is characterized by its magnitude and phase, and by the polarization of the input and output EM waves. When measurements are carried out at a single set of input EM wave frequencies, the output signal does not, per se, provide any information about the specific contributions from the different spatial regions (i.e., surface, interface, quantum well, substrate, etc.), nor about the specific 3WM processes that are involved. Such information can be obtained from data on the dependence of the output signal on the frequencies and polarizations of the input and output EM waves, and from data on the dependence of the output signal on the magnitude and sign of an applied E_{0z} (or $E_{0//}$) field. By measuring the electric field-induced resonant 3WM using a tunable source of the w_1 input EM wave of frequency close to that of the inter-band gap and a tunable source of the p-polarized w_2 input EM wave of frequency close to that of the intersubband excitation energies, it should be possible to obtain detailed information about the electronic struc-ture and excitations of quantum wells with and without free carriers.

ACKNOWLEDGEMENTS

We are pleased to express our appreciation to T.F. Heinz, R. Merlin, G. Pajer, A. Pinczuk and A. Yodh for valuable discussions.

This research was supported in part by the NSF/MRL Program at the University of Pennsylvania through Grant no. DMR-8216718 and by an IBM Grant.

REFERENCES

1. D.A.B. Miller, J.S. Weiner and D.S.Chemla, IEEE J. Quantum Electron. QE-22, 1816 (1986).
2. A. Mysyrowicz, D. Hulin, A. Antonette, A. Megus, W.T. Masselink and H. Morkoc, Phys. Rev. Lett. 56, 2748 (1986).
3. A. von Lehman, D.S. Chemla, J.E. Zucker and J.P. Heritage, Optics Lett. 11, 609 (1986).
4. D. Fröhlich, R. Wille, W. Schlapp and G. Weiman, Phys. Rev. Lett. 59, 1748 (1987).
5. M.K. Gurnick and T.A. DeTemple, IEEE J. of Quantum Electron. QE-19, 791 (1983).

6. E. Burstein, G. Pajer and A. Pinczuk, in "Proc. Int. Conf. on Physics of Semiconductors, Stockholm, 1986", O. Engstrom, ed., World Scientific Publ. 1987, p. 1609.
7. S.J.B. Yoo, M.M. Fejer, A. Harwit, R.L. Byer and J.S. Harris, Jr., J. Opt. Soc. Am. A 4, 27 (1987).
8. M.Y. Jiang, E. Burstein and A. Pinczuk, Bull. Opt. Soc. Am. 22, 40 (1987).
9. J. Khurgin, Appl. Phys. Lett. 51, 2100 (1987); Phys. Rev. B. 38, 4056 (1988).
10. M.M. Fejer, S.J.B. Yoo, R.L. Byer, A. Harwit and J.S. Harris, Jr., Phys. Rev. Lett. 62, 1941 (1989).
11. M.Y. Jiang and E. Burstein, in "Progress in Electronic Properties of Solids", R. Girlanda et al. eds., Kluwer Academic Publishers 1989, p. 395.
12. G. Abstreiter, A. Pinczuk and M. Cardona in "Light Scattering in Solids IV (Topics in Applied Physics)", M. Cardona and G. Guntherodt, eds., Springer, Berlin, Heidelberg, New York 54, 74 (1984).
13. G. Abstreiter, R. Merlin and A. Pinzuk, IEEE J. Quantum Electron. QE-22, 1771 (1986).
14. N. Bloembergen, "Nonlinear Optics", Benjamin, New York 1965; J. Opt. Soc. Am. 70, 1429 (1980).
15. G. Pajer, M.Y. Jiang and E. Burstein, Bull. Am. Phys. Soc. 33, 749 (1988).
16. P. M. Butcher and T.P. McLean, Proc. Phys. Soc. (London) 81, 219 (1963).
17. T. Ando and S. Mori, J. Phys. Soc. Japan 47, 1518 (1979).
18. R. Sooryakumar, A. Pinczuk, A.C. Gossard, D.S. Chemla and L.S. Sham, Phys. Rev. Lett. 58, 1150 (1987).
19. T. Ando, Solid State Communications 21, 133 (1967).
20. J.A. Brum and G. Bastard, Phys. Rev. B 31, 3893 (1985).
21. R.C. Miller, D.S. Chemla, T.C. Damen, A.C. Gossard, W. Weigmann, T.H. Wood and C.A. Burrus, Phys. Rev. B 32, 1043 (1985).
22. R.T. Collins, K.v. Klitzing and K. Ploog, Phys. Rev. 33, 4378 (1986).
23. A. Pinczuk and E. Burstein, Surf. Science 37, 153 (1973).
24. J.D. Gay, J.D. Dow, E. Burstein and A. Pinczuk, in "Proc. Int. Light Scattering in Solids", M. Balkansaki, ed., Flammarion, Paris 1971), p. 22.
25. L.C. West and S.J. Eglash, Appl. Phys. Lett. 46, 1156 (1985).
26. A. Harwit, J.S. Harris, Jr., Appl. Phys. Lett. 50, 685 (1987).
27. E. Burstein, A. Pinczuk and D.L. Mills, Surf. Science 98, 451 (1980).
28. K. Bajema, R. Merlin, F.Y. Juang, S.-C. Hong, J. Singh and P.K. Bhadhacharya, Phys. Rev. B 39, 5512 (1987).
29. G. Danan, A. Pinczuk, J.P. Valdhares, L.N. Pfeiffer, K. W. West and C.W. Tu, Phys. Rev. B 39, 5512 (1989).
30. A. Pinczuk, S. Schmitt-Rink, G. Danan, J.P. Valladares, L.N. Pfeiffer and K.W. West, Phys. Rev. Lett. 63, 1633 (1989).
31. D.L. Mills and E. Burstein, Rep. Prog. Phys. 37, 817 (1974).
32. E. Burstein, D.L. Mills, A. Pinczuk and S. Ushioda, Phys. Rev. Lett, 22, 348 (1969).

N-LAYER SUPERLATTICE PHONONS

L. Dobrzynski, A. Rodriguez,[*] J. Mendialdua,[*] D.J. Lockwood,[**] and
B. Djafari Rouhani

Laboratoire de Dynamique des Cristaux Moléculaires
CNRS, URA N[0] 801, UFR de Physique
Université de Lille 1
59655 Villeneuve d'Ascq Cedex, France

INTRODUCTION

N-layer superlattices are formed out of a periodic repetition of a unit cell containing N ($N > 2$) different slabs. Polytype superlattices made from three constituents (InAs - GaSb - AlSb) were proposed recently.[1] We review here the studies of phonons in N-layer superlattices.[2-9] A recent Raman determination[10] of folded acoustic phonons in a four-layer superlattice is then compared with the theoretical results. We discuss also the existence of surface phonons for three-layer superlattices and the possible extensions of these studies to more complex composite materials.

The vibrations in two-layer superlattices are beginning to be well known. As a detailed review[11] appeared recently we will not develop here the basic physical concepts underlying these studies. As a matter of fact, most of these concepts remain valid for $N > 2$ layer superlattices. However, the growth and the experimental and theoretical studies of these materials require more effort and provide a richer variety of properties. The interested reader will find a detailed description of the practical aspects of these studies in the original papers.[2-10] The present review paper intends only to summarize the main theoretical and experimental results on the acoustical vibrations of N-layer superlattices. We describe first the state of the art for the theory and then for the experiments. Finally, we turn to a discussion of surface vibrations in these novel materials.

THEORY OF BULK VIBRATIONS IN N-LAYER SUPERLATTICES

Elasticity Theory

The simplest approach to the acoustical vibrations of N-layer superlattices is to use the elasticity theory. The superlattice is formed out of an infinite repetition of the ensemble

---

[*] Departamento de Fisica, Facultad de Ciencias, Universidad de Los Andes, Merida, Venezuela.
[**] Division of Physics, National Research Council, Ottawa, Ontario, Canada K1A 0R6.

Light Scattering in Semiconductor Structures and Superlattices
Edited by D.J. Lockwood and J.F. Young, Plenum Press, New York, 1991

of N different slabs. Each of these slabs of width $2a_i$ is labeled by the index i ($1 \leq i \leq N$). All the interfaces are taken to be parallel to the (X_1, X_2) plane. In all the calculations known by the authors the slabs were formed out of isotropic media.

Transverse elastic waves. Consider first transverse elastic waves corresponding to displacements perpendicular to the sagittal plane, which contains $\hat{X}_3$ and the propagation vector $\vec{k}_{//}$ parallel to the interfaces. In such media, these waves decouple from the sagittal ones and implicit expressions[3-8] were obtained relating their frequency ω to the propagation vector $\vec{k} = (\vec{k}_{//}, k_3)$ for any value of N. Before giving one of these general relations, let us define first the bulk transverse sound velocity C_{ti}, such that

$$C_{ti}^2 = \frac{C_{44}(i)}{\rho_i}, \quad 1 \leq i \leq N \tag{1}$$

where $C_{44}(i)$ is the usual bulk elastic constant and ρ_i the density of medium i. We define then

$$\alpha_i = (k_{//}^2 - \frac{\omega^2}{C_{ti}^2})^{1/2} \tag{2}$$

$$a_u = 2 \sum_{i=1}^{N} a_i \tag{3}$$

$$C_1 = \cosh(2a_i\alpha_i) \tag{4}$$

$$S_i = \sinh(2a_i\alpha_i) \tag{5}$$

and

$$F_i = C_{44}(i)\alpha_i . \tag{6}$$

The general dispersion relation was found[3] to be

$$\cos k_3 a_u = \sum_{\{i_1...i_p\}\{i_{p+1}...i_N\}} C_{i_1} C_{i_2} ... C_{i_p} S_{i_{p+1}} S_{i_{p+2}} S_{i_N}$$

$$\times \frac{1}{2} \begin{vmatrix} F_{i_{p+1}} & F_{i_{p+3}} & & F_{i_{N-1}} \\ F_{i_{p+2}} & F_{i_{p+4}} & ... & F_{i_N} \end{vmatrix} + \begin{vmatrix} F_{i_{p+2}} & F_{i_{p+4}} & & F_{i_N} \\ F_{i_{p+1}} & F_{i_{p+3}} & ... & F_{i_{N-1}} \end{vmatrix} \tag{7}$$

where the right-hand-side summation provides 2^{N-1} different terms, adding one to each other. The numbers in each suite are in an increasing (or decreasing) order, and the suite of terms $\{i_{p+1}...i_N\}$ has to be even. The first term in the summation (corresponding to $p = N$) should be understood as $C_1 C_2 ... C_N$.

For example,

1) for N = 1, one has the trivial result of a homogeneous material

$$\cos k_3 a_u = C_1, \tag{8}$$

2) for N = 2, one obtains the well-known result[11]

$$\cos k_3 a_u = C_1 C_2 + \frac{1}{2} S_1 S_2 \left(\frac{F_1}{F_2} + \frac{F_2}{F_1} \right) \qquad (9)$$

of two-layer superlattices,

3) for $N = 3$,

$$\cos k_3 a_u = C_1 C_2 C_3 + \frac{1}{2} C_1 S_2 S_3 \left(\frac{F_2}{F_3} + \frac{F_3}{F_2} \right)$$

$$+ \frac{1}{2} C_2 S_1 S_3 \left(\frac{F_1}{F_3} + \frac{F_3}{F_1} \right)$$

$$+ \frac{1}{2} C_3 S_1 S_2 \left(\frac{F_1}{F_2} + \frac{F_2}{F_1} \right) \qquad (10)$$

4) for $N = 4$,

$$\cos k_3 a_u = C_1 C_2 C_3 C_4 + \frac{1}{2} C_1 C_2 S_3 S_4 \left(\frac{F_3}{F_4} + \frac{F_4}{F_3} \right)$$

$$+ \frac{1}{2} C_1 C_3 S_2 S_4 \left(\frac{F_2}{F_4} + \frac{F_4}{F_2} \right)$$

$$+ \frac{1}{2} C_1 C_4 S_2 S_3 \left(\frac{F_2}{F_3} + \frac{F_3}{F_2} \right)$$

$$+ \frac{1}{2} C_2 C_3 S_1 S_4 \left(\frac{F_1}{F_4} + \frac{F_4}{F_1} \right)$$

$$+ \frac{1}{2} C_2 C_4 S_1 S_3 \left(\frac{F_1}{F_3} + \frac{F_3}{F_1} \right)$$

$$+ \frac{1}{2} C_3 C_4 S_1 S_2 \left(\frac{F_1}{F_2} + \frac{F_2}{F_1} \right)$$

$$+ \frac{1}{2} S_1 S_2 S_3 S_4 \left(\frac{F_1 F_3}{F_2 F_4} + \frac{F_2 F_4}{F_1 F_3} \right) \qquad (11)$$

and so on according to Eq. (7).

These expressions were also given in another equivalent form,[8] together with the width of frequency gaps and the structure factors.

Longitudinal elastic waves. To our knowledge the study of sagittal elastic waves has still to be done. However, for $k_{//} = 0$, the sagittal waves decouple into a longitudinal and a transverse one. These transverse waves, polarized along x_3, have for $k_{//} = 0$ the same dispersion relation (Eqs. 7-11) as the transverse wave with polarization perpendicular to the sagittal plane presented above. The dispersion relations of the longitudinal wave corresponding to displacements along $k_{//}$ can also be obtained from Eqs. (7-11) just by replacing the elastic constants $C_{44}(i)$ by the $C_{11}(i)$ and then the transverse speeds of sound C_{ti} by the longitudinal ones C_{li} such that

$$C_{li}^2 = \frac{C_{11}(i)}{\rho(i)} \quad . \tag{12}$$

For these longitudinal waves (with $\vec{k}_{//} = 0$) and for the transverse ones (for any value of $\vec{k}_{//}$), closed form expressions for the response functions (Green functions) were given[7] for $N = 3$ and $N = 4$.

Lattice Dynamical Theory

Outside the limits of validity of the elasticity theory, proper lattice dynamical models have to be used. This becomes necessary for short wavelength phonons and even for long wavelength phonons when the number of atomic layers in the superlattice slabs is of the order of a few units. To our knowledge only one[5,6] lattice dynamical study has appeared for three- and four-layer superlattices. With the help of a simple cubic lattice with first nearest neighbour interactions, closed form results were obtained that can be qualitatively and even quantitatively useful for transverse polarized phonons and also for longitudinal ones with $\vec{k}_{//} = 0$. These expressions are functions of the masses M_i of the atoms, of the nearest neighbour force constants β_i, of the number L_i of atomic planes, and of the interface force constants. They give in the same manner as Eqs. (9) and (10) implicit expressions between the frequency ω and the propagation vector $\vec{k} = (\vec{k}_{//}, k_3)$.

As for the interpretation of the present experimental results[10] the expressions obtained within the framework of the elasticity theory seem to be sufficient. Thus we will not give here the lattice dynamical results[5,6] and instead turn directly to a discussion of the experimental state of the art.

EXPERIMENTAL STUDIES

Raman studies of three-layer[2,4] and four-layer[9] superlattices have appeared previously. The first of these studies was done with $GaAs/Al_{0.5}Ga_{0.5}As/AlAs$ superlattices. The zone folding effects on longitudinal acoustic (LA) phonons were interpreted with the help of Eq. (10) used for longitudinal phonons with $\vec{k}_{//} = 0$, as explained above. A similar Raman study was performed for three-layer a-Si:H/a-Ge:H superlattices.[4] The four-layer superlattice studied before[9] was made out of two different GaAs and two different AlAs layers.

We present in what follows a recent Raman study[10] of a four-layer superlattice. The unit cell of the sample,[10] grown with the molecular beam epitaxy (MBE) method, was made out of the following four layers: GaAs (thickness $2a_1$), $Ga_{0.9}Al_{0.1}As$ (thickness $2a_2$), GaAs (thickness $2a_3$) and $Ga_{0.9}Al_{0.1}As$ (thickness $2a_4$). The relevant parameters of the corresponding sample (numbered MBE180) are[10] $2a_1 = 48 \pm 3$ Å, $2a_2 = 19 \pm 3$ Å, $2a_3 = 24 \pm 3$ Å, and $2a_4 = 19 \pm 3$ Å. The total width of the four-layer unit cell is $a_u = 110 \pm 3$ Å.

The Raman scattering measurements were carried out using the quasi backscattering geometry previously described,[13] and with 4579 Å Ar laser light. The numerical values[10,12] necessary to interpret these measurements (the refractive indices n_i, the mass densities, the elastic constants $C_{11}(i)$, $C_{44}(i)$, and the longitudinal and transverse velocities C_{li} and C_{ti}) are given in Table 1.

Table 1. Numerical data for the four-layer superlattice.

Material	n_i	ρ_i (g/cm^3)	$C_{11}(i)$ (10^{11}dyn/cm^2)	$C_{44}(i)$ (10^{11}dyn/cm^2)	C_{li} (10^5cm/s)	C_{ti} (10^5cm/s)
GaAs	4.721	5.360	11.888	5.940	4.780	3.329
Ga$_{0.9}$Al$_{0.1}$As	4.579	5.200	12.000	5.980	4.803	3.390

Equation (11) was used to plot (see Fig. 1) the folded longitudinal branches for $\vec{k}_{//} = 0$ and for the above numerical values characterizing this four-layer superlattice.

We define

$$Q = \frac{k_3 a_u}{\pi} \tag{13}$$

and take into account the periodicity of the trigonometric functions

$$\cos k_3 a_u = \cos(\pi Q + 2\pi m) \quad \text{with} \quad m = 0, \pm1, \pm2, \pm3,\ldots \tag{14}$$

and the usual relation in this backscattering geometry[13] between the component k_3 of the scattering wave vector and the effective refractive index $n(\lambda)$

$$k_3 \simeq 4\pi \frac{n(\lambda)}{\lambda} \, , \tag{15}$$

where λ is the incident laser light wavelength.

The effective refractive index $n(\lambda)$ was calculated from the following relation:

$$n^2(\lambda) = \frac{2}{a_u} \sum_{i=1}^{4} a_i \, n_i^2(\lambda) \, , \tag{16}$$

which generalizes the one commonly used[14] for two-layer superlattices.

Table 2. Experimental and calculated phonon frequencies in the four-layer superlattice.

Peak Number	ω calc. (cm^{-1})	ω expt. (cm^{-1})
1	10.87	10.9 ± 0.5
2	17.23	16.8 ± 0.5
3	25.20	25.3 ± 0.5
4	31.57	31.6 ± 0.5
5	39.52	39.9 ± 0.5
6	45.88	45.8 ± 0.5

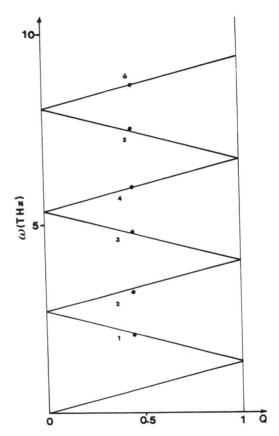

Fig. 1. Dispersion relation of the LA mode in the four-layer superlattice calculated using Eq. (11). The points correspond to the observed frequencies of these phonon modes. Q is proportional to the k_3 component of the wave vector (see Eq. (13)).

This provides finally

$$Q = 8 \, (a_u \sum_{i=1}^{4} a_i \, \frac{n_i^2(\lambda)}{\lambda^2})^{1/2} \; . \tag{17}$$

The corresponding numerical value for sample MBE180 was found to be $Q = 0.450$.

In Table 2 we compare the theoretical and experimental frequencies. These experimental values are also reported in Fig. 1, where one sees an excellent agreement with the folded LA branches obtained from the theoretical analysis described above.

SURFACE PHONONS IN N-LAYER SUPERLATTICES

In the gaps between the folded acoustical branches may appear localized vibrational modes and in particular modes corresponding to vibrations localized near a free surface of a N-layer superlattice. Such surface vibrations were shown to exist for two-layer superlattices (see, for example, Ref. 11) and also recently for three-layer superlattices,[5,6] for both the elastic and lattice dynamical models described above. Although closed-form implicit expressions for the dispersion curves were obtained in both models, we will give here only the simplest result, namely, the one obtained for transverse acoustic waves.[7]

For a given value of $\vec{k}_{//}$ the frequencies ω of the surface phonons can be calculated from a knowledge of the following quantities:

$$D(\omega^2) = F_1 S_1 C_2 C_3 + F_2 S_2 C_1 C_3 + F_3 S_3 C_1 C_2 + \frac{F_1 F_2}{F_3} S_1 S_2 S_3 \tag{18}$$

and

$$A(\omega^2) = C_1 C_2 C_3 + \frac{F_1}{F_2} S_1 S_2 C_3 + \frac{F_1}{F_3} S_1 S_3 C_2 + \frac{F_3}{F_2} S_2 S_3 C_1 \tag{19}$$

bearing in mind the definition of the S_i, C_i, and F_i given by Eqs. (4)-(6). The phonons localized near the surface of a semi-infinite superlattice terminated by the slab $i = 2$ followed by the slabs $i = 3$, $i = 1$, $i = 2$, $i = 3$ and so on were found[7] to be given by the implicit equation

$$D(\omega^2) = 0 \tag{20a}$$

Table 3. Numerical data for the three-layer superlattice.

i	Material	ρ_i (g/cm^3)	C_{ti} (10^5 cm/s)	$2a_i$ (10^{-5} cm)
1	Nb	8.57	1.83	1.0
2	Cu	8.92	2.905	0.5
3	Fe	7.8	5.13	0.33

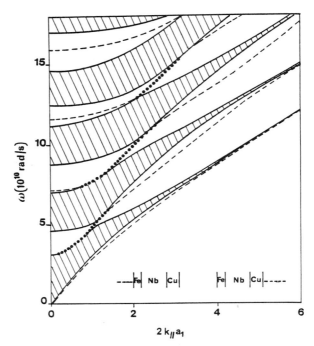

Fig. 2. Bulk and surface transverse elastic waves in a three-layer superlattice
formed out of Nb, Cu, and Fe layers. These curves represent the
frequencies ω as a function of $2k_{//}a_1$; $k_{//}$ is the propagation vector parallel
to the interfaces and $2a_1$ the width of the Nb layer. The shaded areas
represent the bulk phonons. The dotted lines appearing in the gaps
represent the surface phonons for the surface terminated by the Fe layer.
The dashed lines represent the surface phonons for the surface terminated
by the Cu layer. The parameters used in this calculation are given in
Table 3.

with

$$| A(\omega^2) | > 1. \tag{20b}$$

Figure 2 illustrates this result for a three-layer superlattice formed out of Nb, Cu, and Fe layers. The calculations were performed using the parameters listed in Table 3.

Closed form expressions for $D(\omega^2)$ and $A(\omega^2)$ were also obtained[7] for a four-layer superlattice.

Surface phonons may appear also in the gaps between optical confined modes, as has been shown for GaAs/AlAs superlattices.[15,16]

DISCUSSION

Studies of N-layer superlattices are still in their infancy. More experimental and theoretical results will certainly appear, bearing in mind that the introduction of more constituents provides additional degrees of freedom in controlling the physical properties of these superlattices. Let us mention also the beginnings of interest in the electronic properties of metallic[17] and semiconductor[18] N-layer superlattices.

It is clear that other types of lamellar composites will appear as interesting new materials in the near future. Let us mention first the Fibonacci superlattices, studies of which are reported in other papers of these Proceedings. Another interesting example worth mentioning is that of sandwich structures, namely one slab sandwiched between two other materials. Confined vibrational[19] and electronic[20] states may exist in such sandwiches. Such sandwich modes may have interesting practical properties in microdevices.

Finally, let us stress that the engineering of new materials and the construction of microdevices does not remain restricted to lamellar composites. Therefore one can expect the development of the study of more complex composite systems, for which a general theoretical method appeared recently.[21,22]

ACKNOWLEDGEMENT

We are grateful to R.L.S. Devine for the four-layer superlattice sample.

REFERENCES

1. L. Esaki, L.L. Chang, and E.E. Mendez, Jpn. J. Appl. Phys. 20, L529 (1981).
2. M. Nakayama, K. Kubota, S. Chika, H. Kato, and N. Sano, Solid State Commun. 58, 475 (1986).
3. B. Djafari Rouhani and L. Dobrzynski, Solid State Commun. 62, 609 (1987).
4. P.V. Santos and L. Ley, Phys. Rev. B 36, 3325 (1987).
5. T. Szwacka, A. Noguera, A. Rodriguez, J. Mendialdua, and L. Dobrzynski, Phys. Rev. B 37, 8451 (1988).
6. A. Rodriguez, A. Noguera, T. Szwacka, J. Mendialdua, and L. Dobrzynski, Phys. Rev. B 39, 12568 (1989).
7. J. Mendialdua, T. Szwacka, A. Rodriguez, and L. Dobrzynski, Phys. Rev. B 39, 10674 (1989).
8. S. Tamura and J.P. Wolfe, Phys. Rev. B 38, 5610 (1988).
9. B. Jusserand, D. Paquet, F. Mollot, F. Alexandre, and G. Le Roux, Phys. Rev.B 35, 2808 (1987).
10. D.J. Lockwood, R.L.S. Devine, B. Djafari Rouhani, A. Rodriguez, J. Mendialdua, and L. Dobrzynski, (to be published).
11. J. Sapriel and B. Djafari Rouhani, Surf. Sci. Rep. 10, 189 (1989).

12. A. Rodriguez, "Phonons dans les super-réseaux à N-couches", Thesis, Université de Lille 1, Villeneuve d'Ascq (1989).
13. D.J. Lockwood, M.W.C. Dharma-wardana, J.-M. Baribeau, and D.C. Houghton, Phys. Rev. B $\underline{35}$, 2243 (1987).
14. B. Djafari Rouhani and J. Sapriel, Phys. Rev. B $\underline{34}$, 7114 (1986).
15. B. Djafari Rouhani, J. Sapriel, and F. Bonnouvrier, Superl. Microstruct. $\underline{1}$, 29 (1985).
16. L. Sorba, E. Molinari, and A. Fasolino, Surf. Sci. $\underline{211/212}$, 354 (1989).
17. P. Masri and L. Dobrzynski, Surf. Sci. $\underline{198}$, 285 (1988).
18. M.D. Rahmani, P. Masri, and L. Dobrzynski, J. Phys. C $\underline{21}$, 4761 (1988).
19. A. Akjouj, B. Sylla, P. Zielinski, and L. Dobrzynski, J. Phys. C $\underline{20}$, 6137 (1987).
20. A. Akjouj, P. Zielinski, and L. Dobrzynski, J. Phys. C $\underline{20}$, 6201 (1987).
21. L. Dobrzynski, Surf. Sci. Rep. $\underline{6}$, 119 (1986).
22. L. Dobrzynski, Surf. Sci. Rep. $\underline{11}$, 139 (1990).

FAR-INFRARED AND RAMAN STUDIES OF SEMICONDUCTOR SUPERLATTICES

T. Dumelow, [a,b] A. A. Hamilton, [a] K. A. Maslin, [a]
T. J. Parker, [a] B. Samson, [b] S. R. P. Smith, [b]
D. R. Tilley, [b] R. B. Beall, [c] C. T .B. Foxon, [c]
J. J. Harris, [c] D. Hilton, [c] and K. J. Moore [c]

(a) Department of Physics, Royal Holloway and Bedford
    New College, Egham, Surrey TW20 0EX, England
(b) Department of Physics, University of Essex
    Colchester CO4 3SQ, England
(c) Philips Research Laboratories, Redhill, Surrey
    RH1 5HA, England

ABSTRACT

The experimental techniques of Raman spectroscopy and far-infrared
(FIR) Fourier-transform spectroscopy have been applied to a range of
semiconductor superlattice specimens. Both resonant and non-resonant
Raman scattering are available, and the Fourier-transform techniques
include normal-incidence dispersive Fourier-transform spectroscopy (DFTS),
attenuated total-reflection (ATR) spectroscopy, and oblique-incidence
power Fourier-transform spectroscopy (FTS).

We first present results on undoped long-period GaAs/Al$_x$Ga$_{1-x}$As
specimens that can be understood by means of a simple effective-medium
theory. A combination of Raman and FIR data on undoped short-period
GaAs/AlAs specimens yields detailed information on positions and dipole
strengths of confined optic phonons. Finally we report results on free-
carrier-related response, obtained on a GaAs superlattice δ-doped with Si
layers, and it is shown that the spectra are sensitive to the spatial
distribution of carriers within the specimen.

1. INTRODUCTION

This paper describes a program of work involving the use of a
combination of Raman spectroscopy and far infrared (FIR) Fourier-transform
spectroscopy (FTS) to study phonon- and plasmon-related properties of
semiconductor superlattices. So far, a fairly complete account can be
given of long-period specimens in the reststrahl frequency region,
considerable progress has been made with short-period specimens in the
reststrahl region, and some preliminary measurements and analysis have
been carried out on plasma properties. After an account of the
experimental techniques, we describe each part of the program in turn.

Light Scattering in Semiconductor Structures and Superlattices
Edited by D.J. Lockwood and J.F. Young, Plenum Press, New York, 1991

461

2. EXPERIMENTAL

The Raman measurements were performed using a Spex double monochromator and cooled photomultiplier in photon-counting mode. The exciting source was the 514.5 nm line of an argon ion laser (or a dye laser for resonant studies) with around 300 mW of power incident on the sample. The scattering geometry was near backscattering, with an external angle of incidence ~ 70° to the normal (internal angle ~ 10°). This probes phonons propagating along the z axis, normal to the (001) superlattice layers.

In pure GaAs, only the longitudinal optical phonon mode is observed in this geometry, with $z(xy)\bar{z}$ polarization. The superlattice has space group D_{2d}, in which z-propagating zone-centre phonons have Raman polarization tensors

$$T_{A_1} = \begin{pmatrix} a & 0 & 0 \\ 0 & a & 0 \\ 0 & 0 & b \end{pmatrix} ; \qquad T_{B_2} = \begin{pmatrix} 0 & d & 0 \\ d & 0 & 0 \\ 0 & 0 & 0 \end{pmatrix} . \tag{1}$$

The optic phonon modes have displacements that are confined primarily to one or other of the layer types (GaAs or AlAs) and are classified as LO_m, where the subscript m specifies how many displacement anti-nodes are fitted into the appropriate layer. The mode frequencies decrease with increasing m, and correspond to frequencies of phonons in the bulk material with wavevectors approximately $\pi m/a(N + 1)$, where N is the appropriate layer thickness in monolayer units a. For m odd, the modes have B_2 polarization, (similar to the bulk), whilst for m even they have A_1 polarization. The intensity of the latter modes is caused by the Fröhlich interaction, which is more intense under near-resonant excitation close to the GaAs well transitions. The interaction for the B_2 modes is the deformation potential, which does not show resonant behaviour in this region. The mode intensities for both types of mode decrease with increasing index m, as is expected from the behaviour of the overlap integral between the phonon eigenvectors and the essentially uniform field of the exciting radiation.

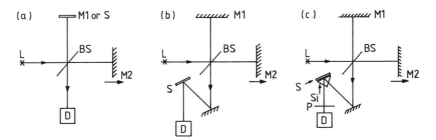

Fig. 1. Schematic drawings of the optical arrangement of the FIR interferometer for (a) DFTS, (b) FTS, and (c) ATR. L: mercury lamp source; BS: beam splitter; M1: fixed mirror; M2: moving mirror; S: sample; D: detector; Si: silicon prism; P: polarizer for ATR.

Far-infrared measurements were made with a modified NPL/Grubb Parsons cube Michelson interferometer,[1,2] with detection by means of a Golay cell with a diamond window or a liquid-He-cooled Ge bolometer. Three separate techniques are available: dispersive Fourier-transform spectroscopy (DFTS),[3] power reflection spectroscopy (FTS), and attentuated total reflection (ATR) spectroscopy.[4,5]

In reflection DFTS (Fig. 1a), the specimen is placed in the position of the fixed mirror of the interferometer so that it interacts with only one of the partial beams, and for the reference measurement the specimen is replaced by a fixed mirror. The measurements are made at normal incidence, and it can be shown[3,6] that both the amplitude r_0 and phase ϕ of the complex amplitude reflectivity $r = r_0 \exp(i\phi)$ can be determined from the ratio of the complex Fourier transforms of the two interferograms. The amplitude reflectivity is related to the power reflectivity R by $R = rr*$.

In FTS (Fig. 1b), generally applied in these experiments at oblique incidence, the specimen is placed in the output beam of the interferometer so that both partial beams from the arms of the interferometer interact with the specimen symmetrically. No phase information is obtained, and the measurement is directly of R.

For measurements on non-radiative surface polaritons and guided waves it is necessary to use a technique that advances the wave vector of the incident radiation so that coupling can occur. For this purpose we use ATR in the Otto configuration[7,8] with silicon prisms (refractive index 3.413), cut for various angles of incidence in the range 18° to 22°, which exceeds the critical angle (17.04°) for total internal reflection. Silicon has the advantage that it has no fundamental infrared-active lattice resonance, so that its refractive index is very nearly independent of frequency in the far infrared. However, there is sufficient absorption due to two-phonon processes to prohibit the use of long-path-length, multipass trapezoidal prisms of the type commonly used at higher frequencies, so we use simple triangular prisms. As shown in Fig. 1c, the prism is placed in the output beam of the interferometer; partly because of the high refractive index of silicon beam divergence inside the specimen is estimated to be less than 3°. Coupling to non-radiative modes requires adjustment of the vacuum gap between prism and specimen to an optimum value. This can be calculated and for the specimens, prisms and wavelengths used is generally of the order of 10 μm. It is set with the aid of spacers and a micrometer screw gauge. Polarization (s or p) is selected by means of a polarizer placed in front of the detector.

The wave-vector component of the incident light parallel to the specimen surface in ATR is

$$q = n_p(\omega/c) \sin \theta_I \qquad (2)$$

where n_p is the refractive index of the prism and θ_I is the angle of incidence. By elementary arguments about phase matching, it can be shown that this is the wave vector of any non-radiative mode to which coupling occurs. The scan over frequency that results from the FTS measurement is therefore seen to be a scan along the straight line in the ω-q plane given by (2).

3. PHONON RESONANCES IN LONG-PERIOD SPECIMENS

Here we consider a superlattice in which the component-layer thicknesses d_1 and d_2 are long compared with the interatomic spacing. In this case each layer may be described as having the dielectric function of the corresponding bulk material. The equations for electrodynamic propagation within this "bulk-slab" model have been known for a long time.[9,10] Since in the far infrared $\lambda \gg L$, where λ is the wavelength and $L = d_1 + d_2$ is the superlattice period, the propagation equations can be expanded to lowest order in L/λ. The result[11] is that the optical properties of the superlattice are those of a uniaxial medium with principal dielectric-tensor components ε_{xx} parallel to the interfaces and ε_{zz} normal to the interfaces given by

$$\varepsilon_{xx} = (\varepsilon_1 d_1 + \varepsilon_2 d_2)/L \tag{3}$$

$$\varepsilon_{zz}^{-1} = (\varepsilon_1^{-1} d_1 + \varepsilon_2^{-1} d_2)/L \tag{4}$$

where ε_1 and ε_2 are the dielectric functions of the two components. These results can also be derived from the fact that the field components E_x and D_z are continuous at the interfaces.[12]

Detailed discussions of the implications of (3) and (4), with particular reference to the GaAs/$Al_xGa_{1-x}As$ system, are given elsewhere.[13] The TO phonon in GaAs has a frequency about 269 cm^{-1}, while the corresponding phonon in $Al_xGa_{1-x}As$ is at 265 to 267 cm^{-1}. In consequence, (3) and (4) lead to striking frequency variation and anisotropy in ε in this reststrahl region[9].

Equations (3) and (4) have been applied to an analysis of DFTS measurements on a GaAs/$Al_xGa_{1-x}As$ specimen.[14] The specimen, grown on a 400 μm thick GaAs substrate, consists of 60 periods, each period comprising 5.5 nm of GaAs and 17 nm of $Al_{0.35}Ga_{0.65}As$, with 0.1 μm of $Al_{0.35}Ga_{0.65}As$ cladding on each side and a top layer of 20 nm of GaAs. The results, shown in Fig. 2, display a complicated resonance behaviour associated with the 267 and 269 cm^{-1} phonons and a simpler resonance behaviour associated with the AlAs-like $Al_{0.35}Ga_{0.65}As$ TO phonon at 361 cm^{-1}. The theoretical curves in Fig. 2 were calculated using expressions from standard multilayer optics, with the multi-quantum well

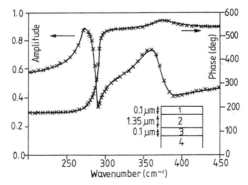

Fig. 2.   Amplitude and phase of the complex reflectivity of an MQW specimen. Crosses: experiment; curves: theory. The insert shows the specimen structure. Regions 1 and 3: $Al_{0.35}Ga_{0.65}As$; region 2: MQW structure; region 4: GaAs substrate. After Maslin et al.[14]

described by (3) and (4) and with phonon parameters adjusted to give the best least-squares fit to the data. Oblique-incidence (22°) p-polarized reflectivity measurements, with corresponding theoretical curves, have also been published for this specimen.[15]

Standard expressions for a uniaxial medium[16] can be applied, together with (3) and (4), to find the dispersion curves of surface polaritons on a semi-infinite specimen to which the effective-medium approximation applies. The result of such a calculation[17] is shown in Fig. 3. We distinguish there between virtual surface modes[18,19], which occur when ε_{xx} is negative and ε_{zz} is positive, and real surface modes, which occur when both components are negative. Real modes persist to $q \to \infty$ and therefore have an electrostatic counterpart, while virtual modes terminate at a finite value of q. Both real and virtual modes are p polarized.

Figure 4 shows the results of an ATR measurement in p polarization on the specimen used for Fig. 2. The dips S1, S2, S3, S4 agree well in frequency with the corresponding crossing points marked in Fig.3, and therefore are ascribed to coupling to these modes. Dip G1 is attributed to a guided-wave feature; this is supported by field-profile calculations.[17]

Applications of the effective-medium expressions to analysis of FIR power reflectivity measurements on semiconductor superlattices have been made by Perkowitz and collaborators[21-24], who consider the CdTe/HgTe system as well as GaAs/AlAs. In addition to the superlattice work reported here, we have also used a combination of DFTS and ATR for an analysis of CdTe layers on a GaAs substrate.[25]

4. PHONON RESONANCES IN SHORT-PERIOD SPECIMENS

The technique of molecular-beam epitaxy (MBE) is frequently used to produce specimens in which the component layers are only a few

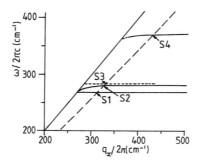

Fig. 3. Calculated dispersion curves for surface polaritons for a
GaAs/Al$_{0.35}$Ga$_{0.65}$As MQW-to-vacuum interface. Virtual modes are
indicated by broken lines and real modes by full lines. The
vacuum light line (full) and an ATR scan line (broken) for 20°
incidence in Si are also shown. The crossing points whose
frequencies are identified as those of the surface-mode dips S1 to
S4 in the ATR data of Figure 4 are identified. After El-Gohary et
al.[20]

monolayers thick. The bulk-slab model cannot apply to such specimens because of the close proximity of interfaces, and it is essential in this case to allow for the consequent modification of the phonon spectrum by the superlattice periodicity.[26]

Specifically what is required is to include the contributions of all the Q = 0 confined optic phonons to the dielectric function. Chu and Chang.[26] derive appropriate expressions for a superlattice such as GaAs/AlAs, in which the reststrahl bands of the two component media do not overlap, so that all the optic phonons are confined in one or other component. These expressions reduce to

$$\varepsilon_{xx} = \frac{1}{L} \sum_{\lambda=1,2} d\lambda \, \varepsilon_{\infty,\lambda} \left[1 - \sum_{m} \frac{r_{Tm}^{(\lambda)}}{\omega^2 - (\omega_{Tm}^{(\lambda)})^2} \right] \tag{5}$$

$$\frac{1}{\varepsilon_{zz}} = \frac{1}{L} \sum_{\lambda=1,2} \frac{d\lambda}{\varepsilon_{\infty,\lambda}} \left[1 + \sum_{m} \frac{r_{Lm}^{(\lambda)}}{\omega^2 - (\omega_{Lm}^{(\lambda)})^2} \right] \tag{6}$$

where λ indicates the component-layer type and m represents the confined-mode order; $\omega_{Tm}^{(\lambda)}$ and $\omega_{Lm}^{(\lambda)}$ are the frequencies of the confined TO and LO phonons respectively. Thus ε_{xx} has a series of poles at the TO frequencies and ε_{zz} has a series of zeroes at the LO frequencies, corresponding to oscillators with strengths $r_{Tm}^{(\lambda)}$ and $r_{Lm}^{(\lambda)}$.

Chu and Chang[27] relate the frequencies $\omega_{Tm}^{(\lambda)}$ and $\omega_{Tm}^{(\lambda)}$ and the strengths $r_{Tm}^{(\lambda)}$ and $r_{Lm}^{(\lambda)}$ to parameters of the 11-parameter rigid-ion model, on which their analysis is based. They also point out that (5) and (6) could be applied phenomenologically, with the frequencies and strengths determined from a fit to experiment. We have applied the latter approach to the analysis of a series of measurements on a number of short-

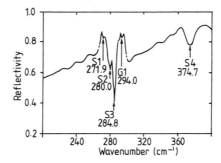

Fig. 4. Measured ATR spectrum for p-polarized radiation incident at 20° in a Si prism on specimen of Fig. 2. Near 280 cm^{-1} measurements were made with a liquid-He-cooled Ge bolometer; elsewhere they were made with a diamond-window Golay cell. After El-Gohary et al.[20]

466

period GaAs/AlAs superlattices. Most mode frequencies were taken from Raman spectra and first-order pole strengths from LO-TO splittings on the same spectra; other parameters were adjusted for best fits to the data.

Each sample had 1 μm of superlattice of the type $(GaAs)_n/(AlAs)_n$ (n is the number of monolayers per superlattice period) bounded by 0.1 μm GaAs cladding layers and grown by MBE on a GaAs substrate.

The Raman spectra were recorded on specimens with the top cladding layer removed, in a near backscattering geometry as described in section 2; the Stokes:anti-Stokes intensity ratio showed that the temperature of the scattering volume was 325 ± 25 K. Typical Raman spectra in the GaAs reststrahl frequency region of the $n = 6$ and $n = 4$ samples are shown in Fig. 5. The Raman tensors (1) imply that the modes LO_m are observable in $z(xx)\bar{z}$ polarization for even m and in $z(xy)\bar{z}$ polarization for odd m.[28] Although there are n confined optic phonon modes in the GaAs LO region, the fact that only three such modes have been identified is partly due to the decreasing intensity with increasing index m discussed in section 2 and partly due to the masking of the modes by a broad TO feature around 265 cm^{-1}. Although strictly forbidden in backscattering geometry, leakage of the TO mode may be caused by a number of features, including the finite aperture of the collecting lens, the departure from ideal backscattering geometry, and the presence of strain, impurities and sample imperfections. The spectra for the $n = 6$ sample show a feature at the LO_1 mode frequency in $z(xx)\bar{z}$ configuration, probably due to slight sample misalignment. In the AlAs LO region, we have only been able to identify clearly the first two confined modes ($m = 1$ and $m = 2$), due to the much decreased signal strength from AlAs as compared with GaAs.

The confined optic phonon frequencies determined by the Raman data are given in Table 1. Apart from the appearance of additional modes, the confinement has the effect that the frequency of a given mode decreases as n decreases, $\omega_1(n = 4) < \omega_1(n = 6)$ and so on. Confinement in a superlattice of shorter period means that the function describing the ionic displacements has a higher average curvature; this is equivalent to

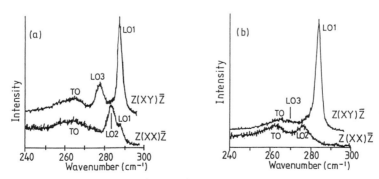

Fig. 5. Raman spectra of (a) a $(GaAs)6/(AlAs)6$ superlattice and (b) a $(GaAs)4/(AlAs)4$ superlattice in the GaAs reststrahl frequency region at room temperature in $z(xy)\bar{z}$ and $z(xx)\bar{z}$ polarizations.

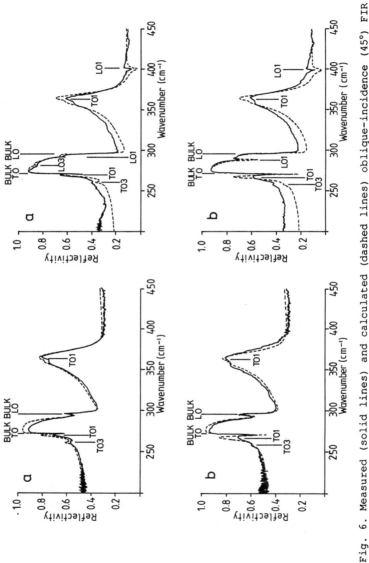

Fig. 6. Measured (solid lines) and calculated (dashed lines) oblique-incidence (45°) FIR reflectivitiy spectra at 77 K of $(GaAs)_n(AlAs)_n$ superlattices: (a) n = 6, and (b) n = 4. The drawings on the left are for s polarization; those on the right are for p polarization. All marked frequencies are those used in the calculations.

Table 1. Superlattice mode frequencies ($\omega_{Tm}^{(\lambda)}$ or $\omega_{Lm}^{(\lambda)}$) measured at 325 K by Raman spectroscopy on $(GaAs)_6(AlAs)_6$ and $(GaAs)_4(AlAs)_4$ samples, together with the values corrected to 77 K and the corresponding oscillator strengths ($r_{Tm}^{(\lambda)}$ or $r_{Lm}^{(\lambda)}$) that were used for the FIR reflectivity calculations. Asterisked values show estimates where no Raman data were available.

Phonon	$(GaAs)_6/(AlAs)_6$			$(GaAs)_4/(AlAs)_4$		
	325 K Mode Frequency (cm^{-1})	77 K Mode Frequency (cm^{-1})	77 K Oscillator strength (cm^{-2})	325 K Mode Frequency (cm^{-1})	77 K Mode Frequency (cm^{-1})	77 K Oscillator Strength (cm^{-2})
GaAs-like						
LO_1	288.0	291.3	12330	284.3	287.6	11800
LO_2	283.0			276.3		
LO_3	277.5	280.8	1600	270.5	273.8	400
TO_1	266.0	269.3	12330	263.0	266.3	
TO_3		260.8*	600		258.3*	200
AlAs-like						
LO_1	397.7	401.0	29470	396.2	399.5	28630
TO_1		362.4*	29470	357.0	360.3	28630

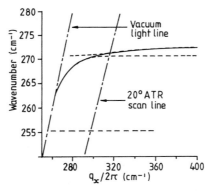

Fig. 7. Surface-polariton dispersion curves in the GaAs reststrahl region of a semi-infinite $(GaAs)_4/(AlAs)_4$ superlattice. The solid curve is calculated considering only first-order confinement, whereas the dashed curves take account of higher order terms.

a higher value of q in a comparable bulk mode, and therefore to a lower frequency ω, since ω decreases with increasing q for optic phonons.

Oblique-incidence (45°) IR reflectivity spectra for the two specimens of Fig. 5, measured at a nominal temperature of 77 K for both s and p polarizations, are shown in Fig. 6. In each case, we also show a theoretical curve derived from (5) and (6); the calculations were done with a standard 2 x 2 matrix formalism[20] and took account of cladding layers. In the theoretical curves we have used the phonon frequencies measured by Raman scattering (Table 1) corrected for the change in temperature (325 K to 77 K) simply by adding 3.3 cm^{-1}, a procedure justified by the known temperature dependence of the bulk modes.[29] First-order pole strengths were taken from the LO-TO splittings measured in Raman scattering, and higher order pole strengths were fitted. We have assumed a damping of 1.5 cm^{-1} on the GaAs-like modes, and 4 cm^{-1} on the AlAs-like modes. The overall good fit between theory and experiment gives strong support to the validity of (5) and (6). TO features correspond to the edges of reststrahl regions; LO features, to the dips in the spectrum in p polarization. In the GaAs reststrahl frequency region the spectra show strong features at the LO$_1$ and TO$_1$ frequencies, which are shifted downwards from the corresponding bulk values by a larger amount for the 4+4 sample than for the 6+6 sample; this results from the mode confinement. Features corresponding to the higher order confined modes are also seen.

The occurrence of a series of poles in (5) and (6) means that, as happens for bulk polaritons,[27] further branches appear in the surface-polariton dispersion curve.[30] The dispersion curves for a semi-infinite (GaAs)$_4$/(AlAs)$_4$ superlattice in the GaAs reststrahl region, calculated from (5) and (6) with use of the standard dispersion equation for a uniaxial medium,[16] are shown in Fig. 7. These are useful qualitative guides to the ATR spectra of the actual specimens, in which, however, the presence of cladding layers and the substrate leads to the appearance of additional features besides those associated with the crossing of the scan line and dispersion curves in Fig. 7. Room-temperature ATR spectra for n = 2 and n = 4 samples are shown in Fig. 8, which also shows theoretical curves, calculated with the phonon parameters used for Fig. 6, with the frequencies recorrected to room temperature. The cladding layers and substrate were included in the calculations. The prominent dip in the s-polarization data is a guided-wave feature, while the p-polarization data show a mixture of guided-wave and surface-polariton features. These room-temperature spectra are not, unfortunately, able to resolve the extra surface-polariton modes seen in Fig. 7, which arise from the higher order confined phonon modes.

5. PLASMA EFFECTS

The interaction between the far-infrared radiation and free carriers gives an additional contribution to the dielectric function; for example in the simplest case of a bulk material

$$\varepsilon(\omega) = \varepsilon_\infty (1 - \omega_p^2/\omega^2) \tag{7}$$

with plasma frequency ω_p given by

$$\omega_p^2 = ne^2/\varepsilon_0\varepsilon_\infty m^\star \tag{8}$$

where n is the density of carriers of effective mass $m^\star$ and ε_∞ is the high-frequency dielectric constant. In a superlattice with sufficiently thick layers, (7) might be used for the free-carrier response of the doped regions,

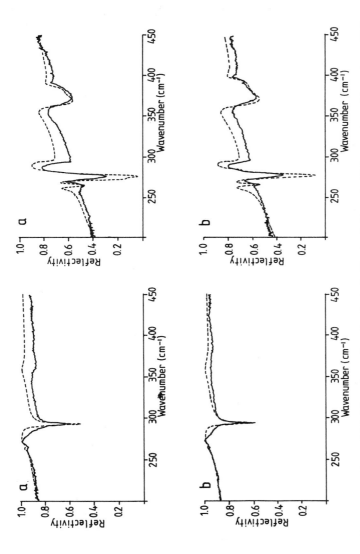

Fig. 8. Measured (solid lines) and calculated (dashed lines) ATR spectra Si prism, 20° angle of incidence) at room temperature of $(GaAs)_n(AlAs)_n$ superlattices: (a) n = 2, and (b) n = 4. The drawings on the left are for s polarization; those on the right are for p polarization.

and the effective-medium equations (3) and (4) might be applied for the dielectric tensor of the whole superlattice. Some discussion of the implications of this is given elsewhere.[13] However, this is only the simplest view, and in general much more sophisticated approaches to the calculation of dielectric properties are required. Clearly the experimental techniques we have applied to phonon resonances have considerable potential for investigating dielectric properties related to free carriers. We have made a start with a study of a Si δ-doped GaAs superlattice grown on a GaAs substrate. The sample contains 100 Si layers with a 2D density of the order of 10^{13} cm^{-2} Si atoms per layer spaced at 50nm intervals. The potential well in which the carriers move is not identical to the dopant profile: the potential due to the positive donor sites is screened by the carriers themselves, and the resulting potential well must be found by self-consistent solution of Schrodinger's equation and Poisson's equation. The results of this self-consistent calculation for the first five eigen-functions of a system of two δ planes are shown in Fig. 9.

At the expected carrier density in this specimen, the Fermi level is close to the top of the barriers, and we have used a simple model in which the carriers are divided between square well and barrier regions to give a qualitative analysis of the experimental data to be presented. Thus wells of width d_W with 3D carrier density n_W are separated by barriers of width d_B with 3D carrier density n_B. Such a model is not physically unreasonable, since diffusive broadening of the δ-doped planes into quasi-rectangular slabs has been observed.[31] The constraint $d_W + d_B = 50$ nm applies. The doping level requires $d_W n_W + d_B n_B \leq 10^{13}$ cm^{-2}, and in fact the theoretical curves appearing below were calculated with $d_W n_W + d_B n_B = 5.5 \times 10^{12}$ cm^{-2}. Equations (7) and (8) are used to find the plasma contributions to effective bulk-type dielectric functions for the well and barrier regions, the optic-phonon contributions being given by the familiar reststrahl-region expressions. The effective-medium equations (3) and (4) are applied to give a dielectric tensor for the superlattice, and theoretical reflectivity curves are calculated in the usual way.[20] For the sake of simplicity it is assumed that m* is constant and isotropic throughout the structure.

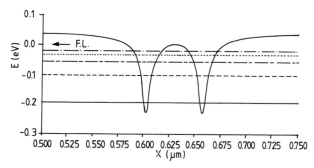

Fig. 9. Self-consistent potential and eigenvalues based on first five eigenfunctions for electrons generated by two sharp donor planes in GaAs. A total electron density of 10^{13} cm^{-2} is assumed, corresponding to the Fermi-surface position marked.

One of the most important parameters of this model is the division of carriers between the wells and the barriers, say n_B/n_W. We find that normal-incidence and s-polarization reflectivity, which probe only the component ε_{xx}, are insensitive to this parameter and effectively depend only on the averaged 3D carrier density $\bar{n} = (d_W n_W + d_B n_B)/(d_W + d_B)$. To illustrate this, Fig. 10 shows near-normal-incidence reflectivity in s polarization, together with a theoretical curve calculated assuming that the carriers are uniformly distributed with $\bar{n} = 1.11 \times 10^{18}$ cm^{-3}. This corresponds to a density per doping plane of 5.55×10^{12} cm^{-2}, which is about half the Si doping level, and also less than that indicated by Hall measurements. The fact that the carrier concentration is lower than the dopant concentration is consistent with secondary ion mass spectroscopy (SIMS) and capacitance-voltage (CV) profiling on samples of this kind.[31] The work[31] has shown that a fraction of the Si atoms are in an electrically inactive form. This is believed to be due to surface diffusion, which allows the aggregation of some of the deposited Si atoms into small clusters; these do not act as donors when incorporated into the GaAs crystal.

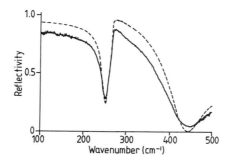

Fig. 10. Measured (solid line) and calculated (dashed line) room-temperature reflectivity in s polarization with angle of incidence 22° for the Si δ-doped GaAs specimen described in the text. The calculations are based on the assumption that the carriers are uniformly distributed with 3D density $\bar{n} = 1.11 \times 10^{18}$ cm^{-3}.

Although the fit to experiment in Fig. 10 is quite satisfactory, we find that similar theoretical curves are generated when n_B/n_W is taken to be different from unity, as long as the value of $\bar{n}$ is maintained. Similar effects have been observed in HgTe/CdTe superlattices by Perkowitz and co-workers.[32,33]

In contrast to normal-incidence and s-polarization oblique-incidence reflectivity, ATR depends upon both ε_{xx} and ε_{zz}. We find that for this reason, calculated ATR curves are sensitive to n_B/n_W. Figure 11 shows an ATR spectrum on the δ-doped specimen, together with the calculated curve that gives the best overall fit in the sense of reproducing the main features of the experimental curve. Given the simple nature of the model, closer agreement cannot be expected.

These far-infrared data on a doped superlattice demonstrate the value of the experimental techniques as probes of carrier distribution within the sample. It is clearly desirable to base analysis of the data on a more fundamental model, for example the application of the random-phase approximation or more accurate expressions, using the self-consistent wave functions and eigenvalues, to calculate the dielectric function.[34] Given the dielectric function, reflectivities can be calculated by means of more or less standard optical techniques.

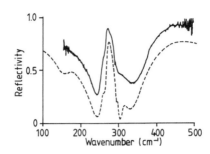

Fig. 11. Measured (solid line) and calculated (dashed line) room-temperature p-polarization ATR (Si prism, $\theta_I = 18°$) for the δ-doped specimen of Fig. 8. Parameters for the calculated curve are $n_W d_W = 4.6 \times 10^{12}$ cm^{-2}, $n_B d_B = 0.9 \times 10^{12}$ cm^{-2}, $d_W = 10$ nm, and $d_B = 40$ nm. The upper barrier layer is assumed depleted, $n_B = 0$.

6. CONCLUSIONS

A combination of Raman spectroscopy with a number of variants of far-infrared Fourier-transform spectroscopy yields detailed information about the dielectric function of superlattice specimens. The reststrahl-region properties of undoped long-period GaAs/Al$_x$Ga$_{1-x}$As specimens are satisfactorily described by the effective-medium approximation. The properties of undoped short-period GaAs/AlAs specimens are reasonably well understood at the level that the frequencies of the confined optic phonons as determined by Raman scattering are used within a phenomenological expression for the dielectric function to fit reflectivity data. This analysis could be carried further, since the pole positions and strengths that are used are related to the 11-parameter rigid ion model,[27] so that a microscopic analysis is possible. We have so far investigated the plasma response of only one specimen, a δ-doped GaAs superlattice. The high-quality data shown in Figs. 10 and 11 suggest that much useful information will become available once the techniques for analysis are fully established.

Our future program includes extensions to the phonon-related measurements, such as more systematic measurements at different temperatures. These should yield information on the temperature dependence of the resonance frequencies, strengths, and damping constants that are used as fitting parameters. Measurements so far have been largely restricted to the GaAs/AlAs system, and extension to many other systems should present no problems in principle.

The main part of the future program will be directed at free-carrier-related properties. We plan further measurements on both modulation-doped and δ-doped samples, and the installation of a 7 T superconducting solenoid opens up the possibility of studies of the magnetoplasma response. The research literature contains a very large number of theoretical studies of plasma modes of superlattices. Many of the surface modes that have been predicted lie in regions of the ω-q plane where the q value is beyond the ATR range (1), so we are currently developing the complementary technique of grating coupling to surface modes.

Far infrared ATR also opens up the possibility of investigating a far wider range of surface modes than those of semiconductor superlattices. For example, there is a long-standing prediction[35] that the surface magnetic polariton on a uniaxial antiferromagnet should be observable in ATR, and recently it has been shown[36] that the same holds for antiferromagnetic superlattices.

ACKNOWLEDGMENTS

The Science and Engineering Research Council is thanked for support under Grant GR/D/97841 and for studentships for A.A.H. and B.S.

REFERENCES

1. K. A. Maslin and T. J. Parker, J. SPIE 1240, 476 (1989).
2. K. A. Maslin and T. J. Parker, Int. J. Infrared and Millimeter Waves (in press).
3. T. J. Parker, Contemp. Phys. (in press).
4. T. Dumelow and T. J. Parker, SPIE J. 1240, 472 (1989).
5. T. Dumelow and T. J. Parker, Int. J. Infrared and Millimeter Waves (in press).
6. J. R. Birch and T. J. Parker, "Infrared and Millimeter Waves," Vol. 2, K. J. Button ed., Academic Press, New York (1979).
7. A. Otto, Z. Physik 216, 398 (1968).
8. A. Otto, Festkörperprobleme XIV, 1 (1974).
9. S. M. Rytov, Zh. Eksp. Teor. Fiz. 29, 605 (1955) [Sov. Phys. JETP 2, 466.]
10. A. Yariv and P. Yeh, J. Opt. Soc. Am. 67, 438 (1977).
11. N. Raj and D. R. Tilley, Solid State Commun. 55, 373 (1985).
12. V. M. Agranovich and V. E. Kravotov, Solid State Commun. 55, 85 (1985).
13. N. Raj and D.R. Tilley, The electrodynamics of superlattices, in: "The Dielectric Function of Condensed Systems," L. V. Keldysh, D. A., Kirzhnitz, and A. A. Maradudin, eds., Elsevier, Amsterdam (1989).
14. K. A. Maslin, T. J. Parker, N. Raj, D. R. Tilley, P. J. Dobson, D. Hilton and C. T. B. Foxon, Solid State Commun. 60, 461 (1986).
15. T. Dumelow, A. R. El-Gohary, A. A. Hamilton, K. A. Maslin, T. J. Parker, N. Raj, B. Samson, S. R. P. Smith, D. R. Tilley, P. J. Dobson, C. T. B. Foxon, D. Hilton, and K. J. Moore, Mater. Sci. Eng. B 5, 205 (1990).
16. M. G. Cottam and D. R. Tilley, "Introduction to Surface and Superlattice Excitations," (Cambridge University Press, Cambridge 1989).

475

17. N. Raj, R. E. Camley and D. R. Tilley, J. Phys. C 20, 5203 (1987).

18. V. N. Lyubimov and D. G. Sannikov, Fiz. Tverd. Tela 14, 675 (1972). [Sov. Phys. Solid State 14, 575.]

19. A. Hartstein, E. Burstein, J. J. Brian and R. F. Wallis, Solid State Commun. 12, 1083 (1973).

20. A. R. El-Gohary, T. J. Parker, N. Raj, D. R. Tilley, P. J. Dobson, D. Hilton, and C. T. B. Foxon, Semicond. Sci. Technol. 4, 388 (1989).

21. S. Perkowitz, D. Rajavel, I. K. Sou, J. Reno, J. P. Faurie, C. E. Jones, T. Casselman, K. A. Harris. J. W. Cook, Jr., and J. F. Schetzina, Appl. Phys. Lett. 49, 806 (1986).

22. S. Perkowitz, R. Sudharsanan, S. S. Yom, and T. J. Drummond, Solid State Commun. 62, 645 (1987).

23. R. Sudharsanan, S. Perkowitz, B. Lou, T. J. Drummond, and B. L. Doyle, Superlatt. Microstr. 4, 657 (1988).

24. B. Lou, R. Sudharshan, and S. Perkowitz, Phys. Rev. B 38, 2212 (1988).

25. T. Dumelow, A. R. El-Gohary, K. A. Maslin, D. R. Tilley, and S. N. Ershov, Mater. Sci. Eng. B 5, 217 (1990).

26. J. Sapriel and B. Djafari-Rouhani, Surf. Sci. Rep. 10, 189 (1989).

27. H. Chu and Y.-C. Chang, Phys. Rev. B 38, 12 369 (1988).

28. Z. P. Wang, D. S. Jiang, and K. Ploog, Solid State Commun. 64, 661 (1988).

29. B. Jusserand and J. Sapriel, Phys. Rev. B 24, 7194 (1981).

30. T. Dumelow, A. Hamilton, T. J. Parker, B. Samson, S. R. P. Smith, D. R. Tilley, D. Hilton, K. J. Moore, and C. T. B. Foxon, Proc. 3rd Int. Conf. on Phonon Physics, Heidelberg (1989).

31. R. B. Beall, J. B. Clegg, J. Castagné, J. J. Harris, R. Murray, and R. C. Newman, Semicond. Sci. Technol. 4, 1171 (1989).

32. S. Perkowitz, R. Sudharsanan, K. A. Harris, J. W. Cook, J. F. Schetzina, and J. N. Schulman, Phys. Rev. B 36, 9290 (1987).

33. S. Perkowitz, B. Lou, L. S. Kim, O. K. Wu, and J. N. Schulman, Phys. Rev. B 40, 5613 (1989).

34. S. R. Streight and D. L. Mills, Phys. Rev. B 40, 10 488 (1989).

35. R. E. Camley and D. L. Mills, Phys. Rev. B 26, 1280 (1982).

36. N. S. Almeida and D. R. Tilley, Solid State Commun. 73, 23 (1990).

OPTICAL PROPERTIES OF PERIODICALLY δ-DOPED GaAs

John M. Worlock,* A.C. Maciel,[†] M. Tatham,[†] J.F. Ryan,[†]
R.E. Nahory,* J.P. Harbison,* and L.T. Florez*

*Bellcore, Red Bank, NJ 07701-7020, U.S.A.
[†]Clarendon Laboratory, Oxford 0X1 3PU, U.K.

ABSTRACT

We have grown and studied the optical properties of a series of periodically δ-doped Si:GaAs structures. The sheet densities of Si range from 1×10^{11} to 5×10^{12} cm^{-2}, and the periods from 5 to 50 nm. Within this range of parameters, the electronic structure can exhibit a variety of behaviors, from planar-confined subbands through minibands to three dimensional freedom. Even in a given sample, we may hope to find such a variety of electrons.

Short-period and strongly doped samples give spectra, both in photoluminescence and in resonant electronic Raman scattering, which are similar to those of uniformly doped material. This implies not surprisingly, that the electronic overlap has triumphed over the Coulomb attraction of the donor layers. We are thus in the metallic portion of the metal-insulator (one-dimensional) phase diagram. At the other extreme, with the layers well separated, we expect insulating behavior in the direction of the axis perpendicular to the doping layers, and the electronic structure should be that of a collection of isolated δ-doping layers. Correspondingly, in the Raman spectra of longer-period structures we find new lines, which arise from intersubband or interminiband transitions. Small differences between spectra in different polarizations are taken to indicate collective effects such as resonant screening.

Future work in this area involves attempts to observe and interpret a variety of collective interactions arising from the variety of electron behaviors that are predicted to occur, and to observe Raman scattering from short-wavelength phonons made possible by the periodic electric fields.

Light Scattering in Semiconductor Structures and Superlattices
Edited by D.J. Lockwood and J.F. Young, Plenum Press, New York, 1991

477

NONLINEAR RESPONSE OF VIRTUAL EXCITATIONS IN SEMICONDUCTOR
SUPERLATTICES

Pawel Hawrylak

Institute for Microstructural Sciences
National Research Council of Canada
Ottawa, Ontario, Canada K1A 0R6

INTRODUCTION

There is currently a lot of interest in ultrafast modulation of optical properties of semiconductors by radiation in the transparency region.[1,2] This has been stimulated by the development of ultrashort pulses, and the subsequent discovery of the Dynamical Stark Effect in semiconductor quantum wells.[3] The Dynamical Stark Effect is a frequency shift of the absorption of a weak test beam in a sample illuminated by a strong pump beam with frequency below the exciton resonance. The changes in the absorption spectrum instantaneously follow the presence of the pump field. The pump field induces virtual transitions from valence to conduction states creating virtual electron-hole pairs. The time spent by electrons in the conduction band is inversely proportional to the degree of detuning of the pump frequency from the band gap. Electrons and holes are strongly correlated because of many-body effects. This results in vertex (excitonic) and self-energy (band-filling) effects. The effect of the short, strong pulse is to induce both changes in the occupation of valence and conduction states and macroscopic polarization. The presence of polarization is the main difference from the usual case of an electron-hole plasma in quasi-equilibrium created by strong and long resonant pulses. In such situations the pump field enters only via changes in the density of carriers. The current investigations[4-13] focused on the linear response of the semiconductor-pump-field system, measured via absorption of a weak test beam. Here we describe the nonlinear response of the semiconductor-pump-field system to the test field.

Some of the interesting effects associated with virtual carriers include the virtual photoconductivity[14] and ultrafast self-induced gaps in the transmittivity of the superlattice.[15]

The remaining parts of the paper are organized in the following way. In section 2 we present the description of the superlattice and the derivation of the optical Bloch equations including fermion exchange. Section 3 contains the transformation to excitonic basis. In section 4 we formulate a nonlinear response of a driven system to the test field, including boson and fermion exchange in a very simple way. The results are illustrated in section 5, which considers Wannier-Frenkel excitons in a coherently driven superlattice, and conclusions are contained in section 6.

THE OPTICAL BLOCH EQUATIONS

We consider a superlattice consisting of a periodic array of narrow quantum wells separated by thick barriers. The separation of the wells is denoted by "a". With a typical wavelength of radiation in the transparency region of ~3000 Å the pump field varies significantly over the sample size, providing that the periodicity "a" is of the order of hundreds of ångströms. At the same time the coupling of electronic states between different quantum wells can be neglected. Hence from the electronic point of view the superlattice is an array of isolated quantum wells. The superlattice is illuminated by a strong short pulse of the pump beam, with frequency ω_p and wave vector K_p, and a weak pulse of the test field, with frequency ω_t and wave vector K_t. The total field $E(z,t)$ can be written as $E(z,t) = \varepsilon(z,t) \exp(iK_pz - i\omega_pt) + \varepsilon^*(z,t) \exp(-iK_pz + i\omega_pt)$, and the slowly varying field $e(z,t)$ is a sum of a pump E_p and test E_t fields $e(z,t) = E_p(t) + E_t(t) \exp(i\Delta\omega t) \exp(-i\Delta Kz)$, with $\Delta\omega = \omega_p - \omega_t$ and $\Delta K = K_p - K_t$. Both the pump and test pulses have electric fields parallel to the plane of the quantum well and wave vectors parallel to the z axis, i.e., superlattice axis. The test pulse can be delayed with respect to the pump field by a delay time T_d. We consider the duration of pulses as shorter then most relaxation processes and the superlattice as coherently driven by the pump field. The description of collisions is not important in this situation, and we use the time-dependent Hartree-Fock approach to describe the coupled electron-field system. We restrict considerations to a single subband in conduction (c) and valence band (v), treated in the effective mass approximation. The coupling to the semiclassical electromagnetic field is treated in the dipole approximation. The Hartree energies can be neglected for narrow wells, but fermion exchange must be kept. We can think of the valence and conduction band states labeled by a wave vector q in the plane of the well and a quantum well index 1 as a two-level system coupled to the field. We shall use the quantum number k = (q,l) for the description of quantum-well states. The time evolution of the system is described in the time-dependent Hartree-Fock approximation via a two-band density matrix: intraband distribution functions $n_{c(v)}$ ($n_{c,k} = <c^+_kc_k>$) and off-diagonal elements describing polarization p_k, p_k^+ ($p_k = <v_k^+c_k>$) due to electron-hole pairs. The Hamiltonian of the system is now identical to that of an ensemble of noninteracting two-level systems labeled by wave vector k in the self-consistent time-dependent exchange field:

$$H = \sum_k \begin{pmatrix} c_k^+ & v_k^+ \end{pmatrix} \begin{bmatrix} e_{c,k} - \sum_{k'} V_{k,k'} \cdot n_{c,k'} & -(\mu E(t,z=la) + \sum_{k'} V_{k,k'} \cdot p_{k'}) \\ -(\mu E(t,z=la) + \sum_{k'} V_{k,k'} \cdot p_{k'})^+ & e_{v,k} - \sum_{k'} V_{k,k'} \cdot n_{v,k'} \end{bmatrix} \begin{pmatrix} c_k \\ v_k \end{pmatrix} \quad (1)$$

The exchange contributions are the terms proportional to the Coulomb interaction $V_{k,k'}$: they include the renormalization of energy levels due to band filling (diagonal terms) and the renormalization of the applied field by induced polarization (off diagonal terms). The off-diagonal terms include direct electron-hole interactions. The use of Eq. (1) and the equation of motion for an arbitrary operator A , $idA/dt = [A,H]$, yields Bloch equations for the density matrix.[7-9,12,13]

$$i\frac{\partial}{\partial t} n_{v,k} = [\Delta_k(t)p_k^+(t) - \Delta_k^+(t)p_k(t)] - i\Gamma n_{v,k} + i\Gamma$$

$$i\frac{\partial}{\partial t} n_{c,k} = -[\Delta_k(t)p_k^+(t) - \Delta_k^+(t)p_k(t)] - i\Gamma n_{c,k}$$

$$i\frac{\partial}{\partial t} p_k^+ = -(E_k + i\gamma_k)p_k(t)^+ + \Delta_k(t)^+(n_{v,k} - n_{c,k}) \quad (2)$$

$$i\frac{\partial}{\partial t} p_k = (E_k - i\gamma_k)p_k(t) - \Delta_k(t)(n_{v,k} - n_{c,k})$$

The total effective field Δ_k and the energy of the electron-hole pair E_k are modified by the exchange self-energy as

$$\Delta_k = \mu E(t,z=la) + \sum_{k'} V(k,k')p_{k'}$$

$$E_k = E_g^0 + e_{c,k} - e_{v,k} + \sum_{k'} V(k,k')(n_{v,k'} - n_{c,k'}) \quad (3)$$

The $e_{c,k} - e_{v,k} = e_k = k^2/2m$ is a sum of kinetic energies of an electron and a hole (m is a reduced mass), $V(k,k') = v(q-q') \exp(-/q - q'/ll-1'la)$ is a Fourier transform of the Coulomb interaction between particles with wavevectors q and q' in layers l and l'; $v(q) = 2\pi e^2/\varepsilon_0 q$ is the bare 2D coulomb interaction screened by a static dielectric constant ε_0. The bare gap is E_g^0, and $\mu = er_{vc}$ is the interband dipole matrix element . γ and Γ describe the transverse and longitudinal relaxation processes.[12] The scattering contributions are added by hand as they cannot be derived within the Hartree-Fock approximation. We have also assumed that the systems equilibrium state in the absence of applied field is a completely filled valence band ($n_{v,k} = 1$).

The optical Bloch equations Eq. (2) can be cast in a form familiar from the studies of two level atom . Using the rotating wave approximation (RWA) associated with the pump field we write polarization as $p_k = P_k \exp(iK_p la - i\omega_p t) + cc.$ Using the standard definition $u_k = P_k^+ + P_k$ of dispersive and $v_k = i(P_k^+ - P_k)$ as absorptive parts of polarization and $\delta n_k = n_{v,k} - n_{c,k}$ as the degree of inversion of state k , the optical Bloch equation for the three-dimensional pseudospin $S_k = (u_k, v_k, \delta n_k)$ can be written in a familiar form:

$$\frac{\partial}{\partial t}\vec{S}_k(t) = -[\vec{\Omega}_k^0(t) + \sum_{k'} V(k,k')\vec{S}_{k'}(t)] \times \vec{S}_k(t) - \hat{\gamma}\vec{S}_k(t) + \vec{L}$$

$$\frac{\partial}{\partial t}(\vec{S}_k(t))^2 = -2\vec{S}_k(t)\hat{\gamma}(t)\vec{S}_k(t) + 2\vec{S}_k(t)\vec{L}$$

(4a)

The optical Bloch equation describes a complicated motion of each pseudospin precessing around a self-consistent driving force, undergoing damping due to matrix γ, and moving toward an equilibrium state $L = (0,0,\Gamma)$ in the absence of the applied field. The total driving force Ω consists of an applied field Ω^0 and an internal field due to the fermion exchange, which is the sum over all pseudospins in different q states and different layers, weighted by the long-range Coulomb matrix elements V. The driving force $\Omega^0_k(t)$ can be written in terms of the total field $\varepsilon(t,z)$ and detuning as $\Omega^0_k(t) = [(\varepsilon^+_k + \varepsilon_k), i(\varepsilon^+_k - \varepsilon_k), (E_g^0 + e_k - \omega_p)]$. The applied field is the sum of a pump field $\Omega_p^0 = [2\mu E_p(t), 0, E_g^0 + e_k - \omega_p]$ and a time and space varying test field Ω_t^0. The field induces both absorptive and dispersive changes. The absorptive changes are related to the changes in inversion of population via stimulated absorption and emission. In the absence of dephasing and exchange each spin precesses on a unit sphere around a moving total selfconsistent driving force Ω, in a way similar to the standard Rabi oscillator. We note that eventhough the pump field does not posses the absorptive part, the effective field contains the absorptive part purely due to the exchange field.

The steady state solution S_k^* for times long after the switching of the pump field on the other hand depends strongly on the relaxation rates, which are treated very crudely here. The steady state solution S_k^* satisfies an integral equation:

$$[\vec{\Omega}_k^0 + \sum_{k'} V(k,k')\vec{S}_{k'}^*] \times \vec{S}_k^* - \hat{\gamma}\vec{S}_k^* = \vec{L}$$

(4b)

If we neglect Coulomb interactions, Eq. (4b) is identical to the Bloch equations for a two-level atom driven by a strong steady field, studied by Mollow.[16] We wish to distinguish the steady state solution from the stationary solution obtained from this equation by setting both γ and L to zero. The stationary solution has been investigated by Schmitt-Rink and Chemla.[7] It can be seen as the set of pseudospins S_k, which are parallel to the effective field. As the field is being switched on this solution follows the field adiabatically, i.e., there is no exchange of energy between the field and the system. Hence in the stationary solution the absorptive part of pseudospin $S_k(2) = v_k$ vanishes and the dispersive part $S_k(1) = u_k$ satisfies an integral equation:

$$(E_k - \omega_p)u_k - \delta n_k \sum_{k'} V(k,k')u_{k'} = 2\mu E_p \delta n_k$$

(4c)

subject to the constraint $\delta n_k^2 + u_k^2 = 1$. This equation is nothing else but a Bogoliubov transformation into a new set of quasiparticles dressed by pump photons. It is referred to as describing a driven Bose condensate. We will be interested here in short, intense pump-field pulses, lasting on a time scale shorter than the relaxation rates, hence the stationary solution will not be considered here. The proper treatment of the steady state solution requires the use of the non-equilibrium Green's function formalism due to Kieldysz.[15] The stationary solution might be acceptable at the peak of the pump field, but surely fails during the rise and fall of the pump pulse where energy is being exchanged between the field and the semiconductor.

BLOCH EQUATIONS IN THE EXCITONIC BASIS

Prior to the arrival of either pump or test pulse, the system is in its ground state, characterized by the initial pseudospin $S_k(0) = (0,0,1)$ for every state $k = (q,l)$. This means a large local self-consistent field present even in the absence of the driving field. It is therefore important to remove this effective field. This is done by writing Bloch equations for the change in the pseudospin system $\delta S_k(t) = S_k(t) - S_k(0)$. From Eq.(4) we have

$$\frac{\partial}{\partial t}\delta\vec{S}_k(t) + \sum_{k'}\vec{n}_3 H_{k,k'} \times \delta\vec{S}_{k'}(t) = \vec{n}_3 \times \vec{\Omega}_k^f - \sum_{k'}[\vec{\Omega}_k^f\delta_{k,k'} - V(k,k')\delta\vec{S}_{k'}(t)] \times \delta\vec{S}_k(t) \quad (5)$$

where the matrix $H_{k,k'}$ has the form

$$H_{k,k'} = (E_g^0 - \omega_p + e_k + \sum_{k''}V(k,k''))\delta_{k,k'} - V(k,k') \quad (6)$$

The driving force was separated into the time-independent component along the third axis and the field Ω_k^f, which contains the time-dependent amplitudes of the applied field, acting in the (1,2) plane. We see that $H_{k,k'}$ introduced long-range correlations that act as rotation of pseudospins in different k states. This is to be contrasted with Eq.(4) where total effective driving field was acting locally on each k state. We now introduce the eigenvectors f_k^i and eigenvalues ω_i of matrix $H_{k,k'}$. The eigenvalues ω_i are nothing else but the exciton energy levels measured from the pump frequency ω_p. We note that Eq. (6) explicitly shows the cancellation between vertex $(-V(k,k'))$ and self-energy $(\Sigma_{k''}V(k,k''))$ corrections. Using the set of exciton eigenvectors we write each pseudospin δS_k in a state k as a linear combination of psuedospins S_i describing the exciton state. The equation of motion for the exciton pseudospins S_i now takes a simple form:

$$\frac{\partial}{\partial t}\vec{S}_i(t) + \omega_i\vec{n}_3 \times \vec{S}_i(t) = \vec{n}_3 \times \vec{\Omega}_i^f(t) - \sum_m[\vec{\Omega}_{i,m}^f(t) - \sum_l V_{i,n,m}\vec{S}_n(t)] \times \vec{S}_m(t) - \hat{\gamma}\vec{S}_i(t) \quad (7)$$

The following definitions have been introduced in Eq.7:

$$\delta\vec{S}_k = \sum_i f_k^i\vec{S}_i , \quad V_{lmn} = \sum_{k,k'}(f_k^l)^* f_k^m V(k,k')f_k^n , \quad \vec{\Omega}_i^f = \sum_k (f_k^i)^*\vec{\Omega}_k^f , \quad \Omega_{m,n}^f = \sum_k (f_k^m)^*\vec{\Omega}_k^f f_k^n \quad (8)$$

Equation (7) describes a precession of the ith exciton state with frequency ω_i driven by the field Ω^f and the field of other exciton states. There is, however, no interaction of a state with itself. For a structureless driving field Ω^f, the fermion exchange field

$$\vec{\Omega}_{m,n}^{fexc} = \sum_l V_{m,l,n}\vec{S}_l(t)$$ introduces coupling between different exciton states. Equation (7) can be cast in the form of an integral equation, which is more suitable for iterative solutions:

$$\vec{S}_i(t) + \int_{-\infty}^{t} dt' \hat{G}_i(t,t')[\sum_m [\vec{\Omega}_{i,m}^f(t') - \vec{\Omega}_{i,m}^{fexc}(t')] \times \vec{S}_m(t') = \int_{-\infty}^{t} dt' \hat{G}_i(t,t') \hat{n}_3 \times \vec{\Omega}_i^f(t') \qquad (9)$$

The matrix $G(t > t')$ corresponds to a damped rotation by the angle $\omega_i(t - t')$ in the plane of polarization:

$$\hat{G}_i(t,0) = \begin{matrix} e^{-\varkappa}\cos\ (\omega_i t) & e^{-\varkappa}\sin\ (\omega_i t) & 0 \\ -e^{-\varkappa}\sin\ (\omega_i t) & e^{-\varkappa}\cos\ (\omega_i t) & 0 \\ 0 & 0 & 1 \end{matrix} \qquad (10)$$

In the absence of dephasing and for a time independent driving field Ω^f one can ask for a stationary state of the system. Setting the time derivative of S_i to zero in Eq.(7) we obtain a set of selfconsistent equations for the exciton pseudospins:

$$\sum_m [\omega_i \hat{n}_3 \delta_{i,m} + \vec{\Omega}_{i,m}^f - \sum_n V_{i,n,m} \vec{S}_n] \times \vec{S}_m = \hat{n}_3 \times \vec{\Omega}_i^f \qquad (11)$$

This equation is identical to Eq. (4c), derived in k space by many authors. It illustrates a strong coupling of all exciton states driven by a pump field.

RESPONSE TO THE TEST FIELD

We now wish to describe the time-dependent nonlinear response of the coupled pump-field-semiconductor system to a weak test field. The test field can be either a single beam or sum of incident and scattered fields as appropriate for light scattering. For definiteness we restrict ourselves to a single test field. We divide the field component of the driving force Ω into the time dependent pump field $\Omega^P(t) = (2\mu E_p, 0, 0)$ and a weak test field $\Omega^t(t,m) = \{E_t(t)[\exp (i\Delta\omega t - i\Delta Kma) + c.c.], E_t(t)[iexp (i\Delta\omega t - i\Delta Kma)+c.c.],0\}$ depending both on time and layer index m. In Eq. (9) we expand the pseudospin $(S^P + S^{(1)} + S^{(2)}+...)$ in powers of the amplitude of the test field. For the pump field pseudospin we find the equation driven by the pump field:

$$\vec{S}_i^P(t) + \int_{-\infty}^{t} dt' \hat{G}_i(t,t')[\sum_m [\vec{\Omega}^P(t')\delta_{i,m} - \vec{\Omega}_{i,m}^{fexc}(t')] \times \vec{S}_i^P(t') = \Psi_i(0) * \int_{-\infty}^{t} dt' \hat{G}_i(t,t') \hat{n}_3 \times \vec{\Omega}^P(t') \quad (12)$$

where the fermion exchange force $\vec{\Omega}_{m,n}^{fexc} = \sum_l V_{m,l,n} \vec{S}_l(t)$ is given by the pump-field exciton pseudospins, and $|\Psi(0)|^2 = |\Sigma_k(f_k^i)^*|^2$ is the probability of finding the electron and the hole in the same position in space. Only states with $\Psi(0)$ different from zero contribute. We now turn to nonlinear response.

In every order in the applied field there appears a term proportional to the changes in the fermion exchange field due to the test field i.e. $\delta\vec{\Omega}_{m,n}^{fexc} \times \vec{S}_i^P(t)$ where, up to third order,

$\delta\vec{\Omega}_{m,n}^{fexc} = \sum_l V_{m,l,n} \vec{S}_l^{(1)}(t)$. Interchanging the test field induced psueodospin $S_l^{(1)}$ with a pump-driven psuedospin S_i^P in the fermion exhange introduces a boson exchange term

$\vec{\Omega}_{m,n}^{bexc} = \sum_l V_{m,n,l} \vec{S}_l^P(t)$. The boson exchange field comes with the sign opposite to that of the fermion exchange as a consequence of the vector product between these two spins. The final equations for the induced changes in the exciton pseudospin $S_i^{(1)}, S_i^{(2)}, S_i^{(3)}$ are given below:

$$\sum_m \int_{-\infty}^t P_{n,m}(t,t')\vec{S}_m^{(1)}(t') = -\int_{-\infty}^t dt'\,\hat{G}_i(t,t')\sum_m [\vec{\Omega}_i^t(t') \times \hat{n}_3\delta_{i,m} + \vec{\Omega}_{i,m}^t \times \vec{S}_m^P(t')] \tag{13}$$

$$\sum_m \int_{-\infty}^t P_{n,m}(t,t')\vec{S}_m^{(2)}(t') = -\int_{-\infty}^t dt'\,\hat{G}_i(t,t')\sum_m [\vec{\Omega}_{i,m}^t(t') - \delta\Omega_{i,m}^{fexc}(t')] \times \vec{S}_m^{(1)}(t')] \tag{14}$$

$$\sum_m \int_{-\infty}^t P_{n,m}(t,t')\vec{S}_m^{(3)}(t') = -\int_0^t dt'\,\hat{G}_i(t,t')\sum_m [\vec{\Omega}_{i,m}^t(t') - \delta\Omega_{i,m}^{fexc}(t') + \delta\Omega_{i,m}^{bexc}(t')] \times \vec{S}_m^{(2)}(t')] \tag{15}$$

where the propagator for the pseudospin P and the effective frequency Ω are given by

$$\hat{P}_{m,n}(t,t') = \hat{1}\delta_{m,n}\delta(t-t') + \hat{G}_m(t,t')\vec{\Omega}_{m,n}^{eff}(t') \quad \text{and} \quad \vec{\Omega}_{i,m}^{eff}(t) = \vec{\Omega}^P(t')\delta_{i,m} - \vec{\Omega}_{i,m}^{fexc}(t') + \vec{\Omega}_{i,m}^{bexc}(t').$$

The left-hand sides of Eqs.(13-15) describe a free precession of the test pseudospin around a time-dependent effective frequency completely specified by the pump field. This frequency is different from the effective frequency seen by the pump driven system due to the additional boson exchange term. This precession takes place not with an exciton frequency but with two different Rabii frequencies. Due to the time dependence of the pump field there is a spectrum of Rabii frequencies. The pump field also introduces changes in the driving term on the right-hand side of Eqs. (13-15). The first order correction contains an effective test field, i.e., a test field screened by the pump field. The second-order correction is however driven almost entirely by the test field and the first order correction. The first-order fermion exchange appearing on the right-hand side of Eq. (14) is almost parallel to the first-order driving term and is expected to give a small contribution. Interestingly, the third-order correction is driven by a test field screened by both fermion and boson exchange fields. The boson exchange term appears due to the exchange of first- and second-order fields.

Equations (13-15) are the main result of this work. They allow for a systematic calculation of nonlinear response functions of a coherently driven semiconductor, including most important many-body effects. In the next section we apply our results to study a coherently driven superlattice.

THE RESPONSE OF A COHERENTLY DRIVEN SUPERLATTICE

We start with the exciton wave functions and eigenvalues. We expand the wave functions f_k^i in terms of exciton states ϕ_q^α of a single well and envelope functions g_l^α as

$f_{q,l}^i = \sum_\alpha g_l^\alpha \phi_q^\alpha$. Using Eq. (6) we obtain an equation for the envelope function g_l^α:

$$\omega_\alpha g_l^\alpha - \sum_{\beta,l'} V_{\alpha,\beta}(l,l')g_{l'}^\beta = \omega_i g_l^\alpha$$

$$V_{\alpha,\beta}(l,l') = \sum_{q,q'} (\phi_q^\alpha)^* V(q,q')\{\exp[-|q-q'||l-l'|a] - \delta_{l,l'}\}\phi_{q'}^\beta. \tag{16}$$

Despite the fact that electrons and holes do not tunnel between different quantum wells , the exciton as an excitation can propagate in the superlattice direction. Hence from the superlattice point of view we are dealing with Frenkel excitons.

In general, for a finite set of quantum wells, Eq. (16) has to be solved to determine exciton states in a superlattice. For a large number of wells we can use periodic boundary conditions. Then the exciton states of an isolated well form one-dimensional bands with exciton energies ω_i being characterized by a Bloch wavevector k_z and the original exciton index α. To make further analytical progress we restrict our attention to the lowest exciton band originating from the 1s exciton in each quantum well. The exciton wave function and energy is now simply given by

$$f^{1s,k_z}_{q,l} = \phi^{1s}_q \frac{\exp\,(ik_z la)}{\sqrt{N}}$$

$$\omega_{1s,k_z} = \omega_{1s} - \Delta(k_z) \tag{17}$$

$$\Delta(k_z) = \sum_{l'} V_{\alpha,\beta}(l-l')e^{-ik_z(l-l')} = \sum_{q,q'}\phi^{1s}_q V(q,q')\{F(|q-q'|,k_z)-1\}\phi^{1s}_{q'}$$

where N is the number of wells and $\Delta(k_z)$ is the Fourier transform of the inter-well exciton interaction. $F(q,k_z)$ is the structure factor of the Coulomb interaction $F(q,k_z)=\sinh(|q|)/[\cosh(|q|)-\cos(k_z)]$. Within the one band exciton approximation we can easily evaluate the coupling to the pump and test field and fermion and boson exchange terms(in the following all exciton states are only labeled by the wavevector $k_z=k$ and we omit the 1s index):

$$\vec{\Omega}^{fexc}_{k,k'}(t) = \upsilon(k-k')\vec{S}_{k-k'}(t)$$

$$\vec{\Omega}^{bexc}_{k,k'}(t) = \upsilon(k')\vec{S}_{k-k'}(t)$$

$$\vec{\Omega}^f_k = \Psi(0)\{\hat{n}_1 2\mu E_p(t)\delta_{k,0} + 2\mu E_t(t)[\delta_{k,-\Delta k}\hat{n}_+\exp\,(i\Delta\omega t) + \delta_{k,+\Delta k}\hat{n}_-\exp\,(-i\Delta\omega t)] \tag{18}$$

$$\vec{\Omega}^f_{k,k'} = \{\hat{n}_1 2\mu E_p(t)\delta_{k,k'} + 2\mu E_t(t)[\delta_{k,k'-\Delta k}\hat{n}_+\exp\,(i\Delta\omega t) + \delta_{k,k'+\Delta k}\hat{n}_-\exp\,(-i\Delta\omega t)]$$

$$\hat{n}_\mp = (\hat{n}_1 \mp i\hat{n}_2)/2$$

In Eq. (18) the exciton state S has been divided by $(N)^{1/2}$. The effective interaction $\upsilon(k)$ is the Fourier transform of the 3D Coulomb interaction averaged over the 1s state. It is given in Eq. (17) if we retain only the form factor F and discard -1. We see that the pump field couples only to the $k = 0$ exciton state S_0^p, i.e., it only excites excitons in phase in each quantum well. Starting from Eq. (12) we find that the superlattice exciton state S_0^p is completely uncoupled and satisfies a single Bloch equation:

$$\vec{S}^p_0(t) = \int_0^t dt'\, G_0(t,t')2\mu E_p(t')\left[n_2\psi_{1s}(0) + n_1 \times \vec{S}^p_0(t')\right] \tag{19}$$

This simplified picture is a consequence of treating only 1s exciton states in each quantum well. The coupling between exciton states generated by the pump field is due to the phase space filling by different exciton states in individual wells. This is included in the fermion exchange component of the driving field, which is exactly zero in Eq. (19). Hence the pump field builds up a macroscopic population (a condensate) of Frenkel excitons with zero momentum driven by a bare pump field. The test field probes this condensate. It is easy to see that the test field induces first-order changes in the exciton psuedospin $S_k^{(1)}$ only when the wave vector k matches the driving test-field wave vector Δk and the first-order excitation spectrum consists of two branches, i.e., $S_k^{(1)} = S_{\Delta k}^{(1)}\delta_{k,\Delta k} + S_{-\Delta k}^{(1)}\delta_{k,-\Delta k}$. Hence from Eq. (13) and Eq. (18) we obtain the equation for the first-order correction:

$$\vec{S}^{(1)}_{\pm\Delta k}(t) + \int_0^t dt'\,\hat{G}_0(t,t')\vec{\Omega}^{eff}_0(t') \times \vec{S}^{(1)}_{\pm\Delta k}(t') =$$

$$-\int_0^t dt'\,\hat{G}_{\pm\Delta k}(t,t')2\mu E_t(t')e^{\mp i\Delta\omega t'}\hat{n}_\mp \times [\hat{n}_2\Psi_{1s}(0) + \hat{n}_1 \times \vec{S}^p_0(t')] \tag{20}$$

The effective driving frequency is $\Omega_k^{eff}(t) = 2\mu E_p(t) - [\upsilon(0) - \upsilon(k)]S_0^P$. Interestingly, the effective frequency is modified by the difference between the fermion echange term $\upsilon(0)S_0^P$ and the boson exchange term $\upsilon(k)S_0^P$ despite the fact that the condensate itself is not modified by the fermion exchange. This is to be contrasted with a single quantum well situation. The second order spectrum consists of three branches: $S_k^{(2)} = S_{2\Delta k}^{(2)}\delta_{k,2\Delta k} + S_0^{(2)}\delta_{k,0} + S_{-2\Delta k}^{(2)}\delta_{k,-2\Delta k}$. The equation for the second order response in quadrature with the pump field $S_0^{(2)}$ and for the satellites at $k=+-2\Delta k$ is obtained from Eq. (14):

$$\vec{S}_0^{(2)}(t) + \int_{-\infty}^{t} dt' \hat{G}_0(t,t')\vec{\Omega}_0^{eff}(t') \times \vec{S}_0^{(2)}(t') =$$

$$-\int_{-\infty}^{t} dt' \hat{G}_0(t,t')2\mu E_t(t')\{e^{+i\Delta\omega t'}\hat{n}_+ \times \vec{S}_{+\Delta k}^{(1)}(t') + e^{-i\Delta\omega t'}\hat{n}_- \times \vec{S}_{-\Delta k}^{(1)}(t')\}$$

$$\vec{S}_{\pm 2\Delta k}^{(2)}(t) + \int_0^t dt' \hat{G}_{\pm 2\Delta k}(t,t')\vec{\Omega}_{\pm 2\Delta k}^{eff}(t') \times \vec{S}_{\pm 2\Delta k}^{(2)}(t') = \qquad (21)$$

$$-\int_0^t dt' \hat{G}_{\pm 2\Delta k}(t,t')2\mu E_t(t')e^{\mp i\Delta\omega t'}\hat{n}_{\mp} \times \vec{S}_{\pm\Delta k}^{(1)}(t')$$

We see that both fermion and boson exchange terms cancelled in the equation of motion for $S_0^{(2)}$ i.e., the effective driving frequency $\Omega_0^{eff}=\Omega P(t)$ is just the bare pump frequency. Also, no fermion exchange screening of the test field appears. However, for the satellites, both the effective frequency and effective field are affected by fermion and boson exchange. To understand how the pump field affects the nonlinear response, we consider a sech pulses studied in detail by McCall and Hahn.[19] For the pump field in the form $2\mu E_p T=2sech(t/T)$ an exact solution for the pump exciton exists for an arbitrary detuning ω_0 from resonance:

$$\frac{S^P_1(t/T)}{\Psi(0)} = \frac{2\omega_0 T}{1+(\omega_0 T)^2} \sec h(t/T)$$

$$\frac{S^P_2(t/T)}{\Psi(0)} = -\frac{2}{1+(\omega_0 T)^2} \sec h(t/T)\tanh(t/T) \qquad (22)$$

$$\frac{S^P_3(t/T)}{\Psi(0)} = -\frac{2}{1+(\omega_0 T)^2} \sec h^2(t/T)$$

From now on we shall understand S^P as measured in $\Psi(0)$. Exactly on resonance, there is no dispersive component $S_1 = 0$ and the exciton exchanges energy with the field via stimulated absorption and emission. Away from the resonance, the dispersive part is always non zero. The absorptive part vanishes at the center of the pump pulse. To obtain some inside into the dynamics of the linear response we approximate the pump pseudospin by its value at $t = 0$, neglect the frequency modulation, to arrive at the approximate expression of the linear response:

$$S_1^1(t) = -2\mu E_t(0)(1+S_3^P(0)) \, \exp \, (-i(\omega_{\Delta k} - i\gamma)t)f(t)$$

$$S_2^1(t) = i2\mu E_t(0)(1+S_3^P(0)) \, \exp \, (-i(\omega_{\Delta k} - i\gamma)t)f(t) \qquad (23)$$

$$S_3^1(t) = -i2\mu E_t(0)(S_2^P(0) + iS_1^P(0))[\exp \, (-i\Delta\omega t) - \exp(-i\Delta\omega T/2)]/\Delta\omega$$

$$f(t) = [\exp \, (-i(\Delta\omega - \omega_{\Delta k} + i\gamma)t) - \, \exp \, (i(\Delta\omega - \omega_{\Delta k} + i\gamma)T/2)]/(\Delta\omega - \omega_{\Delta k} + i\gamma)$$

For $t > ,T/2$ $f(t) = f(T/2)$ where T is the width of the pulse. Eq. (23) illustrates nicely the effect of the pump pulse on the linear response: the polarization components are modified by the changes in the exciton population, while the change in the population induced by the test field depends on the polarization due to the pump field. For the duration of the pulse the polarization oscillates with frequency changing from that of the driving field to that of the exciton. For times longer than the duration of the pulse polarization undergoes damped oscillations with the exciton frequency. The amplitude of oscillations depends strongly on the difference between the driving frequency and the natural frequency of the exciton. We note that there are no changes in the occupation of the exciton state in the absence of the pump field.

To assess the role of frequency modulation and the effect of fermion and boson exchange we evaluate the exciton dispersion and the fermion boson contribution to the effective frequency by retaining only nearest neighbor interaction in the Coulomb matrix elements $V(l,l')$ and evaluating all matrix elements to the lowest order in interlayer separation. The 2D exciton wave functions are used. The final result for the fermion boson exchange term is $[\upsilon(0) - \upsilon(k)]\Psi(0) = 4E_b[1/5 - 4a/a_0][1 - \cos(ka)]$, and for the exciton dispersion $\omega(k) = \omega_0 + 4E_b[1 - 4a/a_0][1 - \cos(ka)]$. The E_b and a_0 are the 2D exciton binding energy and Bohr radius respectively. Despite the fact that our approximation is rather poor and the layer separation "a" entering the amplitude of the exciton dispersion and fermion boson exchange term should be treated as effective separation, it does capture the essential physics and the relative magnitude of both terms.

We now illustrate the effect of the time dependence of the pump field and the many body effects on the response to the test field. Using our decomposition of the test field-induced pseudospin $S_k^{(1)} = S_{\Delta k}^{(1)}\delta_{k,\Delta k} + S_{-\Delta k}^{(1)}\delta_{k,-\Delta k}$ and writing $S_{\Delta k}^{(1)} = a\exp(-i\Delta\omega t)$, we can write the total changes (summed over all plane-wave states in each well) induced on layer l as:

$$S_l^{(1)} = |\Psi(0)|^2\{\vec{a}\,e^{-i\Delta\omega t}e^{i\Delta K l a} + \vec{a}^*e^{+i\Delta\omega t}e^{-i\Delta K l a}\} \tag{24}$$

Hence $\vec{a}$ is clearly a complex susceptibility in the rotating frame. Similar analysis for the second order response introduces three second-order susceptibilities corresponding to the driven exciton states with wave vectors $k = +\Delta K, -\Delta K, 0$. For time independent fields the susceptibilities are time independent and are a function of frequency only.

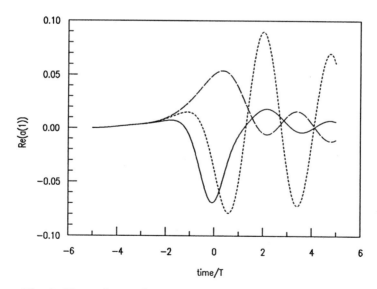

Fig. 1. The real part of the dispersive part of the linear, susceptibility Re(a(1)) as a function of time. Broken line, no pump field; dotted line, no frequency modulation; solid line, all effects. The pump field is on resonance, and other parameters are $\Delta Ka = \pi/2$, $\gamma T = 0.1$.

Figure 1 illustrates the time dependence of the real part of the dispersive component of the susceptibility for a weak sech test pulse of the same width as the pump pulse but with the amplitude reduced by the factor of 10. The pump field is exactly on resonance ($\omega_0 = 0$), and we took $\Delta Ka = \pi/2$. The broken curve shows the behaviour of the susceptibilty in the absence of the pump field, the dashed curve shows the effect of the pump field without frequency modulation, and the solid curve shows all effects. The effect of the pump field is to invert the

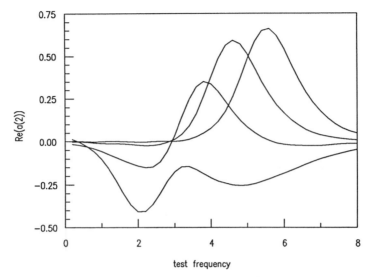

Fig. 2. The real part of the absorptive part of the linear response function Re(a(2)) as a function of test frequency (measured from pump frequency) for detunings of the pump field $\omega_0 T = 0,1,2,3$. All frequencies are measured by the width of the pump pulse T. The structure moves to the higher frequencies with increasing detuning.

population of the $\Delta Ka = 0$ state, and this has a dramatic effect on the probed exciton state with a different wave vector. This is because the test photons are dressed by the pump-field-driven condensate. The effect of frequency modulation is a shift in the peak of time response. This shift carries information about interaction with pump photons and fermion and boson exchange contributions.

The effect of the pump frequency on the absorptive part of the linear susceptibility is illustrated in Fig. 2. Here we show Re(a(2)) at time t = 0, which corresponds to the center of both the pump and test pulse for different test frequencies and different pump frequencies. For pump frequencies well below exciton resonance the broad peak in the absorptive part corresponds to absorption of test photons with a spectrum of Rabii frequencies. For the pump frequency on resonance test photons induce stimulated emission from the condensate, and gain occurs. In Fig. 3 we show the part of susceptibilty Re(a(3)) responsible for the changes in the inversion of population of a tested state at t = 0.These changes in the absence of the pump field occur in second order in the test-field intensity.We show both the first- and second-order contributions to Re(a(3)) for the pump frequency $\omega_0 T = 2$, $\Delta Ka = \pi/4$ as a function of test frequency measured from the pump frequency. The solid line corresponds to the satellite state with $2\Delta Ka = \pi/2$; the broken line corresponds to the second-order response with $\Delta Ka = 0$ i.e. it measures changes in the condensate directly; and the dotted line corresponds to the first-order response with $\Delta Ka = \pi/4$. The second-order response has been scaled by the square of the amplitude of the test field at t=0, and the first-order

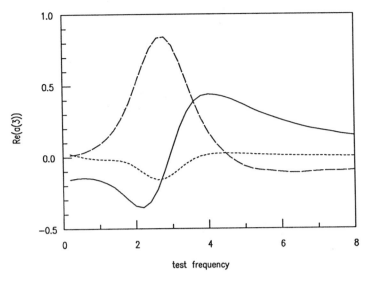

Fig.3. The real part of the inversion part of the linear and second order response Re(a(3)) at t = 0 as a function of test frequency (measured from the pump frequency) for the fixed detuning $\omega_0 T = 2$. Solid line, second-order response for $2\Delta Ka = \pi/2$; broken line, second-order response for $2\Delta Ka = 0$; dotted line, linear response for $\Delta Ka = \pi/2$.

response was scaled by the amplitude of the test field. The first-order response is nonzero due to the presence of the pump field. The second order response is much larger, as expected. The two peaks are shifted due the difference in exciton energies and the fermion boson exchange correction. Hence the measurement of this shift gives us direct information about the many body effects.

SUMMARY

The nonlinear optical response of a superlattice coherently driven by a strong, short, optical pulse is derived from the optical Bloch equations. The Bloch equations include both direct electron-hole Coulomb interactions and phase space filling effects due to the fermion exchange. When the nonlinear response is calculated, the changes in the fermion exchange field result in additional contributions, the boson exchange terms. The effect of many body effects is to renormalize the energy spectrum, effective interactions, and coupling to external field. They depend on the actual time dependence of the pump field. The general formalism is illustrated by considering a coherently driven superlattice of Frenkel excitons. We show that the many body contributions to the frequency renormalization can be extracted from the frequency spectrum of the second-order response. The Frenkel excitons should be realizable in semiconductor-insulator superlattices.

REFERENCES

1. S. W. Koch, N. Peyghambarian, and M. Linberg, J.Phys.C 21, 5229 (1988).
2. S. Schmitt-Rink, D. S. Chemla, and D. A. B. Miller, Adv. Phys. 38, 89 (1989).
3. A. Mysyrowicz, D. Hulin, A. Antonetti, A. Migus, W. T. Masselink, and H. Morkoc, Phys. Rev. Lett. 56, 2748 (1986).

4. A. Von Lehmen, D. S. Chemla, J. E. Zucker, and J. P. Heritage, Opt. Lett. 11, 609 (9186).
5. K. Tai, J. Hegarty and W. T. Tsang, Appl. Phys. Lett. 51, 152 (1987).
6. W. H. Knox, D. S. Chemla, D. A. B. Miller, J. B. Stark, and S. Schmitt-Rink, Phys. Rev. Lett. 62, 1189 (1989).
7. S. Schmitt-Rink, D. S. Chemla, and H. Haug, Phys. Rev. B37, 941 (1988).
8. C. Ell, J. F. Muller, K. El Sayed, and H. Haug, Phys. Rev. Lett. 62, 304 (1989).
9. R. Zimmermann and M. Hartmann, Phys. Stat. Sol. B150, 365 (1988).
10. M. Combescot and R. Combescot, Phys. Rev. Lett. 61, 117 (1988).
11. T. Hiroshima, Phys. Rev. B40, 3862 (1989).
12. M. Lindberg and S. W. Koch, Phys. Rev. B38, 3342 (1988).
13. I. Balslev, R. Zimmermann, and A. Stahl, Phys. Rev. B40, 4095 (1989).
14. E. Yablonovitch, J. P. Heritage, D. E. Aspnes, and Y. Yafet, Phys. Rev. Lett. 63, 976 (1989).
15. P. Hawrylak and M. Grabowski, Phys. Rev. B40, 8013 (1989).
16. B. R. Mollow, Phys. Rev. A5, 2217 (1972).
17. L. V. Keldysz, Sov. Phys. JETP 20, 1018 (1965).
18. H. Haug in: "Optical Nonlinearities and Instabilities in Semiconductors", H. Haug, ed., Academic Press, New York (1988).
19. S. L. McCall and E. L. Hahn, Phys. Rev. 183, 457 (1969).

SEQUENTIAL RESONANT TUNNELING IN SUPERLATTICES:

LIGHT SCATTERING BY INTERSUBBAND TRANSITIONS

S. H. Kwok, E. Liarokapis[*] and R. Merlin

The Harrison M. Randall Laboratory of Physics
The University of Michigan
Ann Arbor, MI 48109-1120, U. S. A.

K. Ploog

Max-Planck-Institut für Festkörperforschung
Heisenbergstraße 1
D-7000 Stuttgart 80, Federal Republic of Germany

INTRODUCTION

The phenomenon of sequential resonant tunneling (SRT) refers to a transport mode of superlattices where the usual resonant tunneling is followed by carrier decay through the emission of phonons.[1-9] When this form of transport is preferred, samples spontaneously break into electric-field domains characterized by the alignment of particular subbands in neighboring wells. This leads to considerable structure in the I-V (current-voltage) response associated with the motion of domain boundaries.[1-9] Following the pioneering work of Esaki and Chang on GaAs-AlAs superlattices,[1] there have been many studies of SRT involving transport[2-5] and, more recently, photoluminescence techniques.[6-9] The latter, in particular, allows one to directly probe the field in the domains and the domain pattern through measurements of Stark shifts.[8] Moreover, the carrier density can be easily varied by means of photoexcitation. This provides an additional experimental knob that is useful for understanding domain formation.[8]

In this work, we report on a Raman scattering investigation of intersubband excitations under SRT-conditions. The Raman spectra provide information on the resonant field, which complements that obtained from photoluminescence

---

[*] Permanent address: Department of Physics, National Technical University, Zografou Campus, GR-157 73 Athens, Greece.

Light Scattering in Semiconductor Structures and Superlattices
Edited by D.J. Lockwood and J.F. Young, Plenum Press, New York, 1991

studies,[8,9] as well as on the behavior of the charge density in the wells and, indirectly, on domain boundaries. Preliminary data suggest that resonant tunneling requires level-alignment for charge-density excitations.

EXPERIMENTAL DETAILS AND SAMPLE PARAMETERS

The superlattice, consisting of 100 periods of 131 Å GaAs and 79 Å $Al_x Ga_{1-x} As$ (x = 0.35), was grown by molecular beam epitaxy on a (100) n^+- GaAs substrate. Doped GaAs layers of thicknesses 0.5 μm ($n \sim 10^{18}$ cm^{-3}) and 0.8 μm ($p \sim 5 \times 10^{17}$ cm^{-3}) were deposited on the substrate and on top of the superlattice to form a p-i-n structure. To avoid depletion regions, the Al concentration was graded between the superlattice and the doped layers from x = 0.5 to x = 0 over 550 Å.[8] The diodes were processed into mesas of area $\sim$ 0.04 mm^2 with ohmic ring-shaped Cr/Au contacts on top. The superlattice was not intentionally doped (capacitance measurements give a residual doping of < 10^{15} cm^{-3}) and there is no evidence of SRT in I-V dark curves; carriers were generated by photoexcitation. The breakdown and the built-in voltage V_0 are approximately -20 V and 1.5 V. Here, the negative sign denotes reverse bias.

Raman measurements were performed at T = 4K using an Ar^+-laser-pumped DCM dye laser tuned to ω_L = 1.941 eV. This energy was chosen so as to resonate with the superlattice critical point derived from the $E_0 + \Delta_0$ gap of GaAs. At this resonance, electron-scattering (but not hole-scattering) is enhanced.[10] The laser beam was focused to a spot $\sim$ 30 μm in radius and powers were in the range 0.01-0.1 W. Spectra were recorded in the $z(x',x')\bar{z}$ and $z(x',y')\bar{z}$ configurations where z is normal to the layers and x', y' are along the [110] and [1$\bar{1}$0] directions. These geometries allow, respectively, scattering by charge-density and spin-density excitations.[10] For the purpose of our experiments, the intensity of the electronic scattering gives an approximate measure of the electron concentration in the wells.

Our sample is the same as the one previously studied by Grahn, Schneider and von Klitzing using photocurrent and photoluminescence techniques.[8] For the most part, the data discussed here are consistent with their results and analysis. In particular, their estimate of $\sim$ 56 meV (448 cm^{-1}) for the separation between the two lowest electron subbands (which they inferred from transport data) is in good agreement with our direct measurements giving $\sim$ 60-65 meV (480-520 cm^{-1}); see below.

Calculations using layer thicknesses from growth parameters indicate that our wells should exhibit three electron bound-states, i.e., three subbands, leading at least to three domains of resonant tunneling due to electrons (the data of Ref. 8 and our results show no evidence of hole-induced SRT). Let F be the magnitude of the electric field, d the superlattice period, N the number of wells and E_i the position of the bottom of the ith electron-subband c_i. At $F = 0$, effective-mass theory predicts $E_1 = 172$ cm^{-1}, $E_2 = 678$ cm^{-1} and $E_3 = 1473$ cm^{-1}. Grahn, Schneider and von Klitzing found single-domain regimes at applied voltages $V \cong -5V$ and $V \cong -12V$.[8] The former, referred to in the following as domain II, is associated with the alignment of the first and second electron-subband [$Fd = |V-V_0|/N = (E_2-E_1)/e$] while the latter bias corresponds to SRT involving the first and third states of the wells [domain III, with $|V-V_0|/N = (E_3-E_1)/e$]. The situation where both domains II and III are present, i.e., $12V > -V > 5V$, is represented in Fig. 1. For $5V > -V > -1.5V$, the

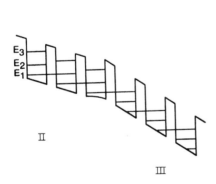

FIG. 1 – Schematic energy diagram showing coexistence of domain II and domain III.

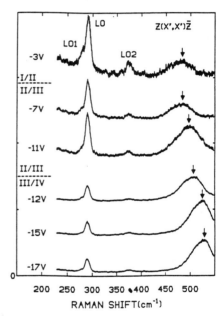

FIG. 2 – Charge-density-excitation spectra of the superlattice at various external voltages. Arrows denote the intersubband transition. The feature labelled LO is the confined longitudinal-optical phonon of GaAs. Other LO peaks are longitudinal-optical modes of the barrier layers. The horizontal dashed lines separate different domain regimes (e.g., II/III indicate coexistence of domains II and III). The power density is $P \cong 550$ W/cm^2.

superlattice breaks into domain II and the domain we labeled I resulting from carrier transport within the first miniband.[1,3,8] Finally, for reverse biases larger than ~ 12V, domain III is expected to coexist with a region where the field is not necessarily uniform and transport does *not* proceed through SRT.[8] This regime will be denoted as domain IV.

RESULTS AND DISCUSSION

Figure 2 shows $z(x',x')\bar{z}$ spectra of the superlattice as a function of applied voltage. Other than the longitudinal-optical (LO) phonon of the GaAs layers and those associated with the alloy (LO_1 and LO_2), the data reveal a feature in the range 480-520 cm^{-1} which we ascribe to the $c_1 \rightarrow c_2$ intersubband transition of photoexcited electrons. This assignment is consistent with the laser power and voltage-dependence of the scattering. Moreover, the position of the peak at low biases agrees well with a calculation based on growth-determined thicknesses predicting $(E_2 - E_1)$ = 506 cm^{-1} at $F = 0$ and, as mentioned above, with photocurrent measurements[8] giving $(E_2 - E_1)$ ~ 450 cm^{-1}. With increasing $|V|$, the intersubband scattering shifts to higher energies and its intensity increases considerably. The position of the maximum of the $c_1 \rightarrow c_2$ peak as a function of applied voltage is shown in Fig. 3 together with HH1-photoluminescence data (HH1 denotes the exciton associated with the ground electron and hole states of the wells). The latter, obtained at ω_L= 1.941 eV, are nearly identical to the results at ω_L= 1.916 eV reported in Ref. 8. The data of Fig. 3 are strong evidence of the existence of domains. Both the positions of HH1 and $c_1 \rightarrow c_2$ provide a measure of the magnitude of the field in the wells through the quantum-confined Stark effect[11] (notice that, unlike HH1, the energy of the intersubband transition *increases* with increasing F).[12] The observation of two HH1-peaks for 12V > -V > 5V reflects the coexistence of SRT-domains II and III. Values of F obtained from the measured HH1-energies are in reasonable agreement with those estimated from calculations of intersubband separations at $F = 0$ (the behavior of the HH1-intensities also supports this interpretation).[8] In the same range, the apparent presence of a $c_1 \rightarrow c_2$ singlet - as opposed to a doublet - does not contradict the domain picture for the width of the intersubband peak (~ 50 cm^{-1}) is larger than the expected doublet separation (~ 35 cm^{-1}).

In Fig. 4, we compare the V-dependence of the *normalized* intersubband scattering intensity with that of the photocurrent[8] (the fine structure in the latter is due to the motion of the domain boundary involving transfer of charges between neighboring wells).[1-9] The Raman-peak intensity is proportional to the density of electrons in the wells. Accordingly, the strong correlation between

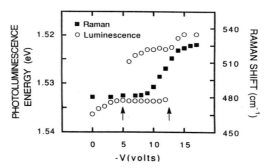

FIG. 3 - Maxima of the HH1 photoluminescence peaks (circles, left-hand scale) and the Raman peak (squares, right-hand scale) as a function of external voltage V. The data were obtained in the $z(x',x')\bar{z}$ configuration. Arrows indicate boundaries between different regimes; domains II and III coexist for $5V < -V < 12V$. $P \cong 550$ W/cm^2.

the two sets of results suggests that the overall transport behavior is determined primarily by the carrier concentration and not by the transit time. Moreover, the fact that the intensity increases with increasing V at the limiting voltages $V \sim -5V$ and $V \sim -12V$ indicates that the domain boundaries contain a deficiency of holes. The alternative, namely, a larger concentration of electrons, would have led to a sudden decrease in intensity as the boundary disappears at the cathode.

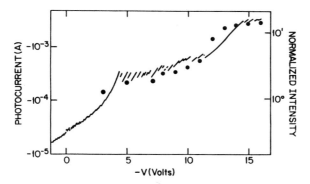

FIG. 4 - Intensity of the charge-density peak normalized to that of the GaAs LO-phonon (●, right-hand scale) and photocurrent (left-hand scale) versus external voltage. Experimental parameters are the same as in Fig. 3. The photocurrent data are from H. T. Grahn, H. Schneider and K. von Klitzing [*Phys. Rev. B* **41**, 2890 (1990)].

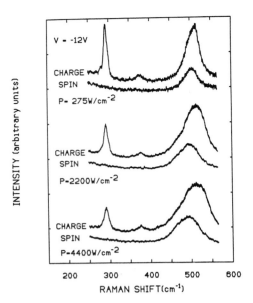

FIG. 5 - Raman spectra at $V = -12V$ as a function of power density P. The top and bottom traces correspond in each case to the charge-density and spin-density geometries. For the given value of the external voltage, there is a single domain (III) in the sample. The data show effects due to screening of the external field by the photoexcited carriers.

We conclude with a brief reference to carrier-induced screening of the external field. In the work of Helm *et al.* on n^+-n- n^+ (SRT) structures,[5,6] it was noticed that the field *inside* the wells was almost completely screened as shown by the fact that the subband spectrum hardly varied with the applied voltage. Although the differences in the behavior are not well understood at this time, it is apparent from the photoluminescence work[8] that screening effects are relatively minor for our sample, *i.e.*, the fields are such that $F \gg \sigma$ where σ is the areal carrier density. Albeit weak, it is nevertheless instructive to illustrate how these effects manifest themselves in the Raman spectra. In Fig. 5, we reproduce data as a function of power density at a voltage $V = -12V$ bringing the superlattice completely into domain III. With increasing excitation intensity, the spin-density scattering, probing essentially the *bare* intersubband energy,[10] exhibits a red shift and an increase in its width. The charge-density line also broadens, but it remains roughly at the same position. Recalling that larger fields lead to larger intersubband transition energies (and that the separation between the two peaks is a measure of the electron density),[10] it is clear that screening by photoexcited carrier is responsible for the observed behavior. That the charge-density excitation hardly moves with P suggests that its alignment is the one that matters for resonant tunneling. However, we do not know of any theoretical argument which could account for such a possibility.

ACKNOWLEDGMENTS

The authors have benefited from discussions with H. T. Grahn. This work was supported in part by the U. S. Army Research Office under Contract No. DAAL-03-89-K-0047 and the University Research Initiative Program under Contract No. AFOSR-90-0214. The work of E. L. was supported by the Fulbright Program.

REFERENCES

1. L. Esaki and L. L. Chang, Phys. Rev. Lett. $\underline{33}$, 495 (1974).
2. Y. Kawamura, K. Wakita and K. Oe, Jap. J. Appl. Phys. $\underline{26}$, L1603 (1987).
3. K. K. Choi, B. F. Levine, R. J. Malik, J. Walker and C. G. Bethea, Phys.
4. T. H. H. Vuong, D. C. Tsui and W. T. Tsang, Appl. Phys. Lett. $\underline{52}$, 981 (1988).
5. M. Helm, P. England, E. Colas, F. DeRosa and S. J. Allen, Jr., Phys. Rev. Lett. $\underline{63}$, 74 (1989).
6. M. Helm, J. E. Golub and E. Colas, Appl. Phys. Lett. $\underline{56}$, 1356 (1990).
7. T. Furuta, K. Hirakawa, J. Yoshino and H. Sakaki, Jap. J. Appl. Phys. $\underline{25}$, L151 (1986).
8. H. T. Grahn, H. Schneider and K. von Klitzing, Phys. Rev. B $\underline{41}$, 2890 (1990).
9. H. T. Grahn, H. Schneider, W. W. Rühle, K. von Klitzing and K.Ploog, Phys. Rev. Lett. $\underline{64}$, 2426 (1990).
10. For a review of electronic scattering in superlattices, see: A. Pinczuk and G. Abstreiter, in "Light Scattering in Solids V", ed. by M. Cardona and G. Güntherodt, Topics Appl. Phys. Vol. 66, Springer, Berlin (1989) p. 153.
11. D. A. B. Miller, D. S. Chemla, T. C. Damen, A. C. Gossard, W. Wiegmann, T. H. Wood and C. A. Burrus, Phys. Rev. Lett. $\underline{53}$, 2173 (1984).
12. K. Bajema, R. Merlin, F. -Y. Juang, S. -C. Hong, J. Singh and P. K. Bhattacharya, Phys. Rev. B $\underline{36}$, 1300 (1987).

ELEMENTARY EXCITATIONS IN LOW-DIMENSIONAL

SEMICONDUCTOR STRUCTURES

S. Das Sarma

Department of Physics
University of Maryland
College Park, Maryland 20742-4111, U.S.A.

ABSTRACT

Our current understanding of elementary electronic excitations in the low-dimensional semiconductor structures (e.g., quantum wells and heterojunctions, multiple quantum wells and superlattices, quantum wires and multiple quantum-wire lateral superlattices) is reviewed with an emphasis on the role of inelastic light scattering spectroscopy in elucidating the elementary excitation spectra. In particular, it is argued that linear response theories based on the random phase approximation give a good account of the existing experimental light-scattering results for the elementary excitation dispersion and spectral weights in $GaAs/Al_xGa_{1-x}As$-based, two-dimensional electron systems and multilayer superlattices. Theoretical predictions are made about plasmon dispersion in novel structures such as quasi-one-dimensional quantum wires and parabolic quantum wells.

1. INTRODUCTION

Elementary excitations in low-dimensional systems have been a subject[1] of interest in condensed matter physics for a long time. The first theoretical treatment[2] of the problem arose indirectly more than thirty years ago in the context of surface plasmons in metals. The existence of dynamically two-dimensional (2D) electron systems in semiconductor space charge layers (and in electrons bound on the liquid helium surface) gave further impetus[1] to the subject in the late 60's and the 70's, and, finally, in the mid-70's 2D plasmons were observed[3] in silicon inversion layers in far-infra-red spectroscopic experiments. While subsequent experimental work using far-infra-red spectroscopy has remained very useful[4-6] (particularly in silicon inversion layer systems where resonance Raman scattering cannot be used) in elucidating a number of aspects of low-dimensional plasmons, resonant inelastic light-scattering spectroscopy has increasingly become a very important experimental tool[7] for studying low-dimensional elementary excitations in free carrier systems occurring in polar (particularly III-V semiconductor-based) materials. The reason for the increasing use of light scattering in studying free carrier excitations in semiconductor microstructures and superlattices is almost obvious. In contrast to the far infra-red absorption measurements, light scattering is a direct spectroscopic tool because: (1) it is used to obtain both the energy and the momentum of the excitations; (2) the light scattering spectrum gives a direct measure of the response function of the electron gas, i.e., the dynamical structure factor; (3) one can, in principle, obtain the dispersion relation of a particular elementary excitation over a fairly wide range of wave vectors (without the problem of having lithographic gratings, etc. which are needed in coupling light to the elementary excitations in far-infra-red absorption measurements); (4) one can easily, just by changing the polarization configuration of the scattering experiment, obtain both single-particle ("spin density") and collective ("charge-density") excitations.

Resonant inelastic light-scattering spectroscopy has been extensively used[7,8] in the last dozen years or so to study electronic elementary excitations in GaAs/Al$_x$Ga$_{1-x}$As-based semiconductor structures and superlattices. In this paper, we will provide a brief review of our current theoretical understanding of the subject. A number of excellent reviews discussing various aspects of the subject matter exist. Here, we will only provide highlights, while continually referring to these reviews (and the original papers) for further details.[1,4-8] We will discuss in some detail new and future directions of research not covered in these earlier reviews, particularly in one-dimensional (1D) quantum-wire and other novel structures and emphasize, wherever appropriate, observable effects associated with many-body phenomena. Unless otherwise explicitly stated, all our discussions are for electrons in conduction subbands of GaAs/Al$_x$Ga$_{1-x}$As structures, even though most (if not all) of the theory should be of general validity for any free-carrier system with isotropic and parabolic effective-mass band dispersion. We use $\hbar = 1$ throughout this paper. We restrict our theoretical discussions to T = 0.

2. THEORY

2.1. Background

Elementary classical considerations show that the long-wavelength plasma frequency in the d(= 1,2,3)-dimension is given by:

$$\omega_p = (4\pi ne^2/\kappa m)^{1/2}, \quad d = 3$$
$$= (2\pi ne^2 q/\kappa m)^{1/2}, \quad d = 2 \qquad (1)$$
$$= (2ne^2/\kappa m)^{1/2}q|\ln(qa)|^{1/2}, \quad d = 1$$

where κ is the effective static background dielectric constant of the surrounding material, m is the electron effective mass, n is the electron density in the relevant dimension and q is the plasmon wave number. The length 'a' in d = 1 is a typical cut-off distance on the order of the lateral confinement size of the 1D structure; thus, the quantum-size correction shows up in the leading-order plasma dispersion in one dimension. Note that only in three dimensions is there a non-zero long-wavelength plasma frequency, because only in three dimensions is there a long-range restoring force for a charge-density fluctuation. The long-wavelength plasmon dispersions for d = 1,2 are linear (with logarithmic corrections) and square-root in the wave number, respectively.

In a general situation, where several subbands may be occupied by electrons, the generalized dielectric function approach to the elementary excitation spectrum has been a very successful theoretical tool[9] in obtaining the elementary excitation spectra. Within the random-phase-approximation (RPA) the electronic dielectric function of a confined semiconductor structure is given by a dielectric tensor ε:

$$\varepsilon = 1 - v\Pi , \tag{2}$$

where v and Π are the generalized Coulomb interaction and the generalized non-interacting irreducible polarizability of the system. In the subband representation the dielectric tensor has components ε_{ijlm} given by

$$\varepsilon_{ijlm} = \delta_{il}\delta_{jm} - v_{ijlm}\Pi_{lm} , \tag{3}$$

where i, j, l, m label quantum-level or subband indices associated with confinement. For a single 1D or 2D system, the indices i, j, l, m denote only the quantum subbands, whereas for multilayer or multiwire systems they also include the index labelling a specific layer or a specific wire.

The matrix elements of Coulomb interaction v are to be taken using the wave functions $\Psi_{i,j,l,m}(\underset{\sim}{r})$ in the usual manner:

$$v_{ijlm} = \int d\underset{\sim}{r} \int d\underset{\sim}{r}' \ \psi_i^*(\underset{\sim}{r}) \ \psi_j^*(\underset{\sim}{r}') \ v(\underset{\sim}{r},\underset{\sim}{r}') \ \psi_l(\underset{\sim}{r}) \ \psi_m(\underset{\sim}{r}') , \tag{4}$$

where

$$v(\underset{\sim}{r},\underset{\sim}{r}') = e^2/\kappa|\underset{\sim}{r}-\underset{\sim}{r}'| , \tag{5}$$

is the Coulomb interaction in the three dimensions with $\underset{\sim}{r}$, $\underset{\sim}{r}'$ denoting 3D position vectors. The wavefunctions $\psi_i(\underset{\sim}{r})$ are composite wave functions including free-particle motion in one (quantum-wire) or two (quantum-well or heterojunction) directions and confining bound-state wave functions in the restricted dimensions.

The non-interacting polarizability Π_{ij} is the so-called irreducible[10] bare bubble, given by

$$\Pi_{ij}(q,i\nu_m) = \left(\frac{g}{\beta}\right) \sum_{i\omega_1} \sum_{\underset{\sim}{p}} G_{ii}(\underset{\sim}{p},i\omega_1) \ G_{jj}(\underset{\sim}{p}+\underset{\sim}{q}, i\omega_1+i\nu_m) , \tag{6}$$

$$\equiv - g \int \frac{d^d p}{(2\pi)^d} \frac{n_j(\underset{\sim}{p}+\underset{\sim}{q})-n_i(\underset{\sim}{p})}{i\nu_m-[E_j(\underset{\sim}{p}+\underset{\sim}{q})-E_i(\underset{\sim}{p})]} , \tag{7}$$

where g is a degeneracy factor which is just 2 for electrons in GaAs structures, $\beta = (k_BT)^{-1}$ is the inverse temperature, ν_m and ω_1 are Matsubara boson and fermion frequencies, G_{ii} is the non-interacting Green's function for the i th quantum level (which is necessarily diagonal in the quantum index i), and E_i and n_i are, respectively, the one-electron energies and Fermi occupancies:

$$n_i(\underset{\sim}{p}) = \left[1+e^{\beta(E_i(\underset{\sim}{p})-\mu)}\right]^{-1} , \tag{8}$$

μ is the chemical potential of the system determined by the total density n and

$$n = g\sum_i \sum_{\underset{\sim}{p}} n_i(\underset{\sim}{p}) . \tag{9}$$

The elementary excitations of the system can be divided into collective modes or plasmons (also referred to as the charge-density excitations) and the single-particle or electron-hole excitations (the spin-density excitations). The collective excitation spectrum is given by the poles of the response function ε^{-1} which is the inverse dielectric tensor, i.e., they are given by the zeros of the determinant $|\varepsilon|$:

$$|\varepsilon| = 0 \ . \tag{10}$$

The single-particle excitations are given by the poles of the irreducible response function Π_{ij} or, equivalently, by the zeros of Π^{-1}:

$$\Pi^{-1} = 0 \ . \tag{11}$$

Equations (10) and (11) above, when solved for the frequency ω, give the elementary excitation spectra, $\omega(q)$, as a function of the relevant mode wave vector q, where q is 2D or 1D depending on the dimensionality of the system. The inelastic light-scattering spectrum for the charge-density excitations is given by the dynamical structure factor or the loss function, which is the imaginary part of the reducible response function of the system. For the single-particle spin-density excitations it is given by the imaginary part of the irreducible response function. The Raman intensity is thus given by the imaginary part of $\varepsilon^{-1}\Pi$ for the collective or the charge-density excitation and by $Im\Pi$ for the single-particle spin-density excitations. Note that the only difference between the two response functions is that the charge-density response is screened whereas the spin-density one is not.

Specific applications of the equations given above to inelastic light scattering by elementary excitations in specific systems will be discussed below. Here we note another salient feature of the dielectric formalism being discussed in this section. This is the distinction between intrasubband and intersubband elementary excitations in confined structures. In a confined system, elementary excitations associated with transitions between quantum subbands i, j, with i ≠ j (i.e., classical motion in the confinement direction) are referred to as intersubband excitations, whereas elementary excitations within individual subbands (i = j) corresponding to motion along the free dimensions are the intrasubband excitations. The dielectric formalism shows that this distinction is somewhat artificial for an arbitrary wave vector because there is a mode-coupling effect[9,11] between intra- and inter-subband excitations characterized by the off-diagonal terms in the dielectric matrix. At long wavelengths, this off-diagonal coupling is weak and intra- and inter-subband excitations can be thought of as distinct entities, but, in general, it is much more reasonable to think of the elementary excitations of the system as a whole without this distinction. If the mode-coupling effect is neglected (as is often done), then intra- and inter-subband excitations emerge as distinct excitations associated with parallel (to the free dimensions) and perpendicular (i.e., along the confinement) motion of the electrons.

Below we now consider a number of specific structures and situations in more detail.

2.2. A single 2D layer

A single 2D electron gas is an extensively studied[1], low-dimensional system occurring in electrons bound on the surface of liquid helium, in silicon MOSFET structures and in semiconductor heterojunctions and quantum wells. The discussion here deals with the high-mobility modulation-doped

GaAs/Al$_x$Ga$_{1-x}$As heterojunction and quantum-well system, but the general results apply to a wide variety of confined 2D electron systems.

In a single 2D layer with only one subband occupied by electrons, the collective excitations are given by Eq. (10), which can now be written as,

$$\left| \delta_{11}\delta_{jm} - v_{1j1m}\Pi_{1m} \right| = 0$$

with   (12)

$$\Pi_{1m} = 0 \quad \text{unless } l \text{ or } m = 1 \;,$$

where the subband index '1' denotes the lowest subband which is occupied by electrons. Neglecting the mode-coupling effect, the intrasubband plasmon mode is given by the well-known relation

$$1 - v_{1111}\Pi_{11} = 0 \;,$$ (13)

whereas the intersubband modes are many and correspond to transitions from the ground subband 1 to all possible excited subbands $j(= 2,3,4,\ldots)$. If we neglect coupling between these intersubband excitations, then the individual intersubband collective modes are given by[9,11,12]

$$1 - v_{1j1j}\chi_{1j} = 0 \;,$$ (14)

where $\chi_{1j} = \Pi_{1j} + \Pi_{j1}$. In these equations, v_{1j1m} is given by

$$v_{1j1m}(q) = \frac{2\pi e^2}{\kappa q} \int dz \int dz' \, \xi_i^*(z)\xi_j^*(z') \; e^{-q|z-z'|} \xi_1(z)\xi_m(z') \;,$$ (15)

Equations (13) – (15) have been extensively used in the literature to calculate the collective mode dispersion in 2D semiconductor systems such as silicon inversion layers, GaAs quantum wells and heterojunctions. The non-interacting 2D polarizability function (the "bare bubble" diagram) is easily calculated to be (at T = 0)

$$\Pi_{1j} = -\frac{m}{\pi} \left[\frac{k_{Fj}}{q} \{A_{+j} - (A_{+j}^2 - 1)^{1/2}\} \right.$$

$$\left. - \frac{k_{Fj}}{q} \{A_{-j} - (A_{-j}^2 - 1)^{1/2}\} \right] \;,$$ (16)

where

$$A_{\pm j} = \frac{\omega - E_j + E_1}{q v_{Fj}} \pm \frac{q}{2k_{Fj}} \;,$$ (17)

with k_{Fj} and v_{Fj} as, respectively, the Fermi wave vector and the Fermi velocity in the j th subband, and the square root of a complex quantity is always chosen to be the one with the positive imaginary part.

$$v_{Fj} = k_{Fj}/m \;,$$

$$k_{Fj} = \left(2m \, E_{Fj} \right)^{1/2} \;,$$ (18)

$$E_{Fj} = (E_F - E_j) \, \theta \, (E_F - E_j)$$

with

$$E_F = \frac{2\pi}{gm} \, N_S$$ (19)

where N_S is the 2D electron density per unit area. One can include effects of small amounts of impurity scattering by writing $\omega^2 \rightarrow \omega(\omega + i\gamma)$

503

where γ is the collisional broadening. For small γ, this substitution gives the same result as that of the full Mermin prescription[13] for the inclusion of collisional damping.

The leading-order collective mode dispersion given by Eqs. (13) and (14) are well-known:

$$\omega = \left(\frac{2\pi N_S e^2}{\kappa m} q \right)^{1/2} + 0 \; (q^{3/2}) \equiv \omega_p^{2D} \; , \tag{20}$$

and

$$\omega_{j1} = (E_{j1}^2 + \frac{2}{\hbar} E_{j1} N_S v_{j1 j1}(q=0))^{1/2} \left\{ E_{j1}^2 + \left(\frac{4\pi N_S e^2 E_{j1}}{\kappa} \right) S_j \right\}^{1/2} + 0(q) \; , \tag{21}$$

where

$$S_j = - \int dz \int dz' \; \xi_j(z) \xi_1(z) \; |z-z'| \; \xi_j(z') \xi_1(z') \; . \tag{22}$$

The wave functions $\xi_i(z)$ in Eq. (22) are the bound-state wave functions for z-confinement. The second term in Eq. (21) is the well-known[14] depolarization shift for the intersubband collective excitation whereas Eq. (20) gives the well-known intrasubband 2D plasmon dispersion. Note that in the leading order the 2D plasmon dispersion is independent of the confinement and is the same as the classical result. The full solution of Eq. (13) with the intrasubband polarizability Π_{11} gives the 2D plasmon dispersion, including the quantum size effect arising from the width of the layer. In Fig. 1 we display[11] the calculated 2D plasmon dispersion relation in a GaAs quantum well with the quantum width correction. It is clear that the finite width corrections are unimportant for 2D plasmon dispersion at long wavelengths.

The Raman charge-density excitation spectrum for light scattering from the collective excitations in a 2D layer are given by the imaginary part of the density-density response function

$$I_c(q, q; \omega) \propto \text{Im} \left[\int dz \int dz' e^{-iq_z(z-z')} \tilde{\Pi}(z, z') \right] \; , \tag{23}$$

$$\tilde{\Pi}(z, z') = \sum_{ijlm} \tilde{\Pi}_{ijlm}(q, \omega) \; \xi_i^*(z) \xi_j^*(z') \xi_l(z) \xi_m(z') \; , \tag{24}$$

$$\tilde{\Pi}_{ijlm} = \Pi_{ij} \delta_{il} \delta_{jm} + \sum_{k,n} \Pi_{ij} v_{ijkn} \tilde{\Pi}_{knlm} \; . \tag{25}$$

Note that Eq. (25) is just the equation for the reducible response function $\tilde{\Pi}$ in the subband representation, which in matrix notation becomes

$$\tilde{\Pi} = \Pi + \Pi v \tilde{\Pi}$$

or, equivalently, $\hspace{10cm}$ (26)

$$\tilde{\Pi} = (I - \Pi v)^{-1} \Pi = \varepsilon^{-1} \Pi \; .$$

We also note that the spin-density spectrum is given by the same expression as Eq. (23) except that the polarizability is the _irreducible_ polarizability:

$$I_S(q, q_z; \omega) \propto \text{Im} \left[\int dz \int dz' e^{-iq_z(z-z')} \Pi(z, z') \right] \; , \tag{27}$$

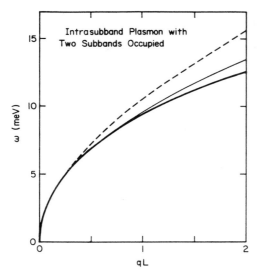

Fig. 1.   Intrasubband plasmon dispersion for a
           GaAs quantum well of 700Å width with
           2 subbands occupied (sample shown as
           an inset in Fig. 3).   Thick solid
           line: full RPA; thin line: RPA
           using   only   the   lowest   subband
           occupancy; dashed line: strict 2D
           approximation. [Ref. 11].

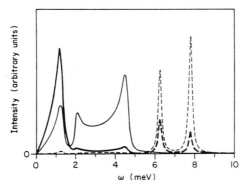

Fig. 2.   Spin-density   (solid   lines)   and
           charge-density (dashed lines) Raman
           spectra for $q_zL = 4.0$ (thin lines) and
           $q_zL = 1.0$ (thick lines) where L is the
           quantum-well width.   Sample parameters
           are the same as in Fig. 3 with q =
           $7 \times 10^4 cm^{-1}$ and $\gamma = 0.1$meV.   [Ref. 11].

where

$$\Pi(z,z') = \sum_{ijlm} \Pi_{ij}\delta_{11}\delta_{jm}\xi_i(z)\xi_j(z')\xi_1(z)\xi_m(z') \ . \tag{28}$$

From now on we use the fact that the confining bound-state wave functions are manifestly real and, therefore, $\xi \equiv \xi^*$.

In Fig. 2 we show our calculated[11] single-particle and collective Raman spectrum at fixed q and q_z as a function of the excitation frequency ω for a 2-subband model of a GaAs quantum well, assuming occupancy of both subbands and taking into account the mode-coupling effect between inter- and intra-subband excitations. The excitation spectrum itself is shown in Fig. 3, where the mode-coupling effect shows up prominently as an anti-crossing effect near the resonance region (i.e., where $\omega_{inter} \approx \omega_{intra}$). The results shown in Figs. 2 and 3 are based on a 2-subband model and direct calculations of the spectrum from Eq. (12), which is now a 4x4 determinant equation, and on Eqs. (23) and (27). (Note that the general dispersion relation[15] for the collective modes of the system defined by Eq. (12) is a $N^2 \times N^2$ determinantal equation where N is the number of subbands kept in the model.)

From the numerical results (Fig. 3) one concludes that the non-local mode-coupling effect is strong around the resonance region and not insignificant off-resonance. Writing out Eq. (12) for a 2-subband model and assuming, for simplicity, only subband '1' to be occupied, we get:

$$(1 - v_{1111}\Pi_{11})(1 - v_{1212}\chi_{12}) - v_{1112}^2\Pi_{11}\chi_{12} = 0 \ . \tag{29}$$

The last term in Eq. (28) is the mode-coupling term. Neglecting it, we get back the individual intra- and inter-subband modes discussed earlier. It turns out[9] that in the long-wavelength limit one can explicitly solve the mode-coupling equation, obtaining the following collective excitations:

$$\omega_\pm^2 = \tfrac{1}{2} \{\tilde{\omega}_{21}^2 + \omega_p^2 \pm \ [(\tilde{\omega}_{21}^2 - \omega_p^2)^2 + 4\omega_1^4]^{1/2}\} \ , \tag{30}$$

where

$$\omega_1^2 = \{2N_S^2 E_{21}[v_{1112}(q \to 0)]^2/m\} \ q \ , \tag{31}$$

and $\tilde{\omega}_{21}$ and ω_p are, respectively, the intersubband and intrasubband plasma modes given in Eqs. (20) and (21) with j = 2.

We conclude this section by pointing out that 2D intrasubband plasmons and the intersubband excitations of a single electron layer have been extensively studied in silicon inversion layers and, more recently, in GaAs heterojunctions using far-infra-red absorption spectroscopy. The theory outlined here has been quantitatively verified in these experiments, including the non-local mode-coupling effect.[16]

2.3. Multi-layer structures or superlattices

From the viewpoint of light-scattering spectroscopy, elementary excitations in multilayer or superlattice structures are more interesting than the single-layer case because of their higher spectral weights associated with the higher electron numbers involved.

The simplest possible model for a multilayer structure is an identical set of 2D layers stacked parallel to each other in the

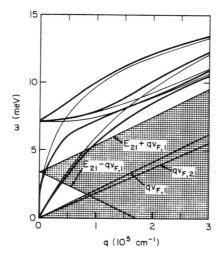

Fig. 3. Resonant coupling of inter- and
intra-subband plasmons in a GaAs
quantum well (sample shown as an
inset). Thick lines: coupled modes;
thin lines: ucoupled modes. Shaded
area shows the single-particle
spin-density excitation regime [Ref.
11].

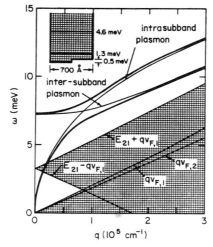

Fig. 4. Plasmon dispersion for two parallel
quantum wells of the type shown in
Fig. 3 and separated by a distance of
940Å. Shown are the two intersubband
and the intrasubband plasmon modes
with the shaded area being the
single-particle excitation regime.
Thick lines: coupled modes; thin
lines: uncoupled modes [Ref. 11].

z-direction with no electron hopping between the layers, i.e., the wave functions in the z-direction of individual layers do not overlap. For a finite number of layers, the formalism of the last chapter still applies, with the subband indices (i, j, l, m) also indicating the layer number index. If we keep n subbands in each 2D layer and there are N layers in the system, then one must solve an $(nN)^2$ x $(nN)^2$ determinantal equation in general. In Fig. 4 we show the calculated[11] elementary excitation spectra for a two-layer system. There are now four collective modes - two intrasubband plasmons and two intersubband plasmons in addition to the single-particle continua, It can be shown[15] that one of the intrasubband plasmon modes is acoustic, i.e., it becomes linear with q for q → 0 whereas the other mode goes as $q^{1/2}$. As one can see from Fig. 4, the mode-coupling effect between intra- and inter-subband plasmons can be substantial for a superlattice. As the number of layers increases, the regions between the two intrasubband and the two intersubband plasmon modes in Fig. 4 become plasmon bands and are filled up with plasmon spectral weight. This plasmon band formation in a superlattice is better understood on the basis of a simpler model which we discuss below.

Clearly, the generalized dielectric matrix approach emphasized so far in this article is not directly applicable in the limit of N → ∞, i.e., the infinite superlattice case, because the determinantal plasmon-dispersion relation (Eq. 12) becomes a meaningless infinite-dimensional equation. One can, however, exploit the periodic invariance in the z-direction to obtain the plasmon dispersion in such a situation. For a simple model of an infinite periodic superlattice with each layer an ideal (i.e., infinitely thin) 2D electron gas, the collective mode dispersion is given[17-19] by the zeros of the superlattice dielectric function:

$$\varepsilon(q, q_z; \omega) = 0 , \qquad (32)$$

where

$$\varepsilon(q, q_z; \omega) = 1 - v(q) \, f(q, q_z) \, \Pi(q, \omega) . \qquad (33)$$

In Eq. (33),

$$f = \frac{\sinh(qa)}{\cosh(qa) - \cos(q_z a)} , \qquad (34)$$

with 'a' as the superlattice period (i.e., the inter-layer distance) and $|q_z| (\leq \pi/a)$, the z-wave vector of the mode arising from the superlattice periodicity, is defined in the reduced zone scheme. (Note that in the strict 2D approximation employed for the layers here, there are no subbands and, therefore, no intersubband excitations.)

Collective modes defined by Eqs. (32) - (34) form a plasmon band for each value of q as q_z varies over the one-dimensional Brillouin zone. In the long-wavelength (q → 0) limit, we can use the following result for $q/k_F \ll 1$

$$\Pi(q \to 0, \omega) = \frac{N_s}{m} \frac{q^2}{\omega^2} + 0 \left(\frac{q^4}{\omega^4} \right) , \qquad (35)$$

to solve explicitly for the superlattice plasmon dispersion relation

$$\omega_p = \left(\frac{2\pi N_s e^2}{\kappa m} \frac{q \sinh qa}{\cosh qa - \cos q_z a} \right)^{1/2} . \qquad (36)$$

In the limit of qa → 0, Eq. (36) implies that the superlattice plasmon is an acoustic plasmon for all values of q except for the zone-edge mode at

$q_z = 0$ which has a finite gap at $q = 0$ (i.e., 3D plasmon-like behavior). The various long-wavelength limits of Eq. (36) are:

$$\omega_p = \left(\frac{2\pi N_s e^2 a}{\kappa m (1 - \cos q_z a)} \right)^{1/2} q \;, \qquad qa \ll 1, \; q_z a \neq 0$$

$$= \left(\frac{4\pi N_s e^2}{\kappa m a} \right)^{1/2} , \qquad qa \ll 1, \; q_z a = 0 \tag{37}$$

$$= \left(\frac{2\pi N_s e^2 q}{\kappa m} \right)^{1/2} , \qquad qa \gg 1 \text{ with any } q_z \; .$$

Equation (37) applies only in the long-wavelength limit for $q \ll k_F$.

Finite-layer-thickness and mode-coupling effects can be included in the superlattice plasmon dispersion by using the dielectric theory of II.2 and by incorporating the form-factor f (cf. Eq. (34)) which gives the phase modulation arising from the Coulomb interaction between different layers. The coupled intra-subband and inter-subband modes are given by a formula very similar to Eq. (28). One finds[9] that for $q \to 0$, the intersubband superlattice plasmon is only slightly modified by corrections of $O(q^2 a^2)$, whereas the intrasubband superlattice plasmon frequency is affected in the zeroth order. The intrasubband acoustic plasmon velocity decreases somewhat due to the mode-coupling effect and there is some experimental evidence for this phenomenon.[9]

Both intra- and inter-subband superlattice plasmons have been experimentally observed in GaAs/Al$_x$Ga$_{1-x}$As superlattices via light-scattering experiments.[7,20,21] The theoretical intersubband plasmon frequency (for a two-subband model) is

$$\tilde{\omega}_{21} = [E_{21}^2 + 4\pi N_s e^2 E_{21} L_{12}/\kappa]^{1/2} , \tag{38}$$

where

$$L_{12} = 2 \int dz \left[\int_0^z dz' \; \xi_1(z') \xi_2(z') \right]^2 , \tag{39}$$

and E_{21} is the energy separation between the subbands. Plasmon dispersions given by equations (36) - (38) are in excellent quantitative agreement with the measured excitation frequencies from Raman scattering spectroscopy and, in fact, the measured Raman intensities agree very well with the calculated Raman spectra.

The theoretical Raman spectrum for superlattice plasmons (charge-density excitations) is given by the density-density response function for the superlattice (ℓ, ℓ' are layer indices):

$$I(q, q_z; \omega) \propto \sum_{\ell, \ell'} \text{Im} \left[\frac{\Pi}{1 - \Pi v} \right]_{\ell \ell'} e^{iq_z(z_\ell - z_{\ell'})} , \tag{40}$$

where $\Pi(q, \omega)$ is the 2D polarizability (Eq. 16) and v is the 2D Coulomb interaction:

$$\Pi_{\ell \ell'} = \Pi \delta_{\ell \ell'}$$

$$v_{\ell \ell'} = \frac{2\pi e^2}{\kappa q} e^{-q|z_\ell - z_{\ell'}|} , \tag{41}$$

with $z_\ell = \ell a$ as the position of the ℓ the layer. Note that, as usual, the corresponding spin-density excitation spectrum is obtained by leaving out the denominator in Eq. (40).

For an infinite periodic superlattice, the expression for Raman intensity simplifies considerably due to the (periodic) translational invariance in the z-direction and one gets

$$I(q, q_z; \omega) \propto Im\ \Pi(q, \omega)/\varepsilon(q, q_z; \omega) \ . \tag{42}$$

The actual situation is more complicated[22] due to the decay of the photon inside the superlattice and also due to existence of surfaces in real superlattices. Results based on equations similar to Eq. (42) are in quantitative agreement[22] with the observed inelastic light-scattering spectra.[7]

Excellent reviews[7] of the subject of superlattice plasmons and its Raman spectra exist in the literature where much more detail can be found.

2.4.   One-dimensional quantum-wire and lateral multiwire superlattice structures

Submicron lithographic techniques have recently allowed fabrication of high quality, confined 1D electron systems where the electronic motion is quantized into subbands in two (y and z directions in this paper) directions and is free plane-wave-like along the third (x) direction. Elementary excitation spectra in such 1D quantum-wire structures is a subject of great current interest.   One can also construct multiwire lateral superlattices by putting a periodic array of these wires parallel to each other.   Usually, in these quantum-wire structures confinement along one direction (chosen to be the z direction here) is much tighter than in the other confining direction (i.e., the y direction).   Our model will be that of a delta-function confinement along the z-direction and rectangular (or parabolic) confinement along the y-direction.   Thus, the z-direction drops out of consideration and the problem becomes one in the x-y plane.

The 1D plasma dispersion in a quantum-wire structure is given, within the RPA, by

$$1 - v\Pi = 0 \ , \tag{43}$$

where v(q) is the Fourier transform of Coulomb interaction in 1D, $q \equiv q_x$ is the 1D wave-vector and Π is the 1D non-interacting polarizability function:

$$\Pi(q, \omega) = \frac{m}{\pi q}\ \ln\left[\frac{\omega^2 - \left(\frac{q^2}{2m} - qv_F\right)^2}{\omega^2 - \left(\frac{q^2}{2m} + qv_F\right)^2}\right] , \tag{44}$$

$v_F = k_F/m$ (with $k_F = \pi N_W/g$, where N_W is the 1D electron density) is the 1D Fermi velocity and the Coulomb interaction is

$$v(q) = \int_{-\infty}^{+\infty} dx\ e^{iqx}\left(\frac{e^2}{\kappa x}\right) \ . \tag{45}$$

It is clear that v(q) is logarithmically divergent in 1D and one introduces a typical length cut-off 'a' to obtain $v(q) \sim \ln(qa)$ for small q.   Solving Eq. (41) for small q then gives the 1D plasma dispersion[23]

$$\omega_p \sim q \left| \ln(qa) \right|^{1/2} . \tag{46}$$

Thus, 1D plasma dispersion is acoustic except for logarithmic corrections which show up at long wavelengths.

We can generalize the theory of elementary excitations in 1D systems by following II.2 and II.3 and introducing a generalized dielectric matrix $\varepsilon(q,\omega)$ defined within the RPA by

$$\varepsilon_{ijmn}(q,\omega) = \delta_{im}\delta_{jn} - v_{ijmn}(q) \, \Pi_{mn}(q,\omega) , \tag{47}$$

where $v_{ijmn}(q)$ is given by

$$v_{ijmn}(q) = \int dy \int dy' \, \phi_i(y) \, \phi_j(y') \, v(q,y-y') \, \phi_m(y) \, \phi_n(y') , \tag{48}$$

and

$$v(q,y-y') = \frac{2e^2}{\kappa} \int_0^\infty \frac{\cos qx}{[x^2+(y-y')^2]^{1/2}} \, dx , \tag{49}$$

$$= \frac{2e^2}{\kappa} K_0 \left(\left| q(y-y') \right| \right) \tag{50}$$

and $K_0(k)$ is the modified Bessel function of the second kind. The wave functions $\phi_i(y)$ are confining wave functions in the y-direction and we assume the z-confinement of electron charge density to be δ-function-like.

The collective modes are given, as usual, by the zeros of the determinant: $\left| \varepsilon_{ijmn}(q,\omega) \right| = 0$ which for a 2-subband model (assuming only the subband 1 to be occupied) becomes

$$(1 - v_{1111}\Pi_{11})(1 - v_{1212}\chi_{12}) - v_{1112}^2\Pi_{11}\chi_{12} = 0 , \tag{51}$$

where Π_{11} is the same as the 1D polarizability Π defined in Eq. (44), and the intersubband polarizability $\chi_{12} = \Pi_{12} + \Pi_{21}$ is given by

$$\chi_{12}(q,\omega) = \frac{m}{\pi q} \ln \left[\frac{\omega^2-(E_{21}-qv_F+q^2/2m)^2}{\omega^2-(E_{21}+qv_F+q^2/2m)^2} \right] , \tag{52}$$

where $E_{21} = E_2 - E_1$ is the energy separation between the 2 subbands.

We first ignore the mode-coupling term (the last term in Eq. 51) and consider the uncoupled intrasubband and intersubband excitations individually. In the long-wavelength limit the polarizabilities become

$$\Pi_{11}(q,\omega) = \frac{N_W}{m} \frac{q^2}{\omega^2} + O(q^4) , \tag{53}$$

$$\chi_{12}(q,\omega) = \frac{2E_{21}N_W}{\omega^2-E_{21}^2} + O(q) , \tag{54}$$

where N_W is the 1D electron density (per unit length). (Note that the long-wavelength forms of the polarizability functions are independent of the dimensionality of the system.)

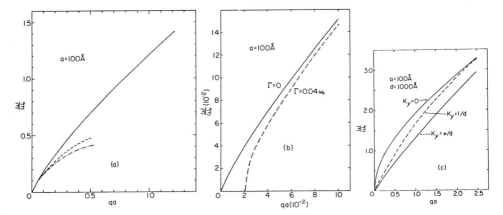

Fig. 5. Calculated plasmon dispersion in a 1D quantum wire as a function of qa (where a is the wire width). (a) Solid curve: Full RPA; dashed curve: long-wavelength Bessel function result; dash-dot curve: long-wavelength logarithmic result. (b) Plasmon dispersion with (dashed) and without (solid) impurity damping. (c) Plasmon dispersion in a lateral multi-wire superlattice for three values of $k_y d$ (where k_y is the wave vector in the superlattice direction and d is the period). [Ref. 23].

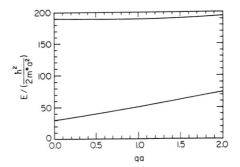

Fig. 6. Inter- (upper) and intra- (lower) subband plasmon dispersion in a single quantum wire for a two-subband model with a = 390nm and N_w = 1.66x10⁵cm⁻¹. [Ref. 24].

Putting (53) and (54) back in the plasmon dispersion relation (Eq. 51) and neglecting the mode-coupling effect we get the following[23,24] long-wavelength plasmon behavior in 1D:

$$\omega_p(q) = \left(\frac{2N_We^2}{\kappa m} \right) q |\ln(qa)|^{1/2} + 0 \, (q^2) \, , \tag{55}$$

and

$$\tilde{\omega}_{21}(q) = [E_{21}^2 + 2E_{21}N_Wv_{1212}(q \to 0)]^{1/2} \, , \tag{56}$$

where 'a' is the typical confinement width in the y-direction. The calculated full dispersions are shown in Figs. (5) and (6), respectively, for the intra- and inter-subband plasmons, assuming an infinite square-well-like confinement in the y-direction so that the confining wave functions $\phi_1(y)$ are trigonometric sine or cosine functions. In Figs. 5(a) and (b) we show the 1D plasmon dispersion with and without the collisional damping effect, whereas in Fig. 6 we show the intersubband single-particle and collective excitation modes in a two-subband model, neglecting the mode- coupling effect. Finally, in Fig. 7 we show results of a three-subband calculation which keeps the coupling between the intra- and inter-subband excitations by directly solving the determinantal equation $|\varepsilon_{ijmn}| = 0$ for a three-subband 1D model. Details can be found in the cited reference.[23,24]

Now, we consider a multiwire lateral superlattice formed by a periodic array of wires covering the x-y plane with the superlattice direction along y and the wire direction along x. At first we neglect tunnelling between the wires (but, of course, the electrons in different wires interact via the long-range Coulomb force). For a finite number of wires the collective excitation spectrum of a lateral superlattice can be found by solving the determinantal equation

$$|\varepsilon_{\ell\ell'}| = 0 \, , \tag{57}$$

where ℓ, ℓ' are now wire-number indices (we are assuming each quantum wire to have one subband only, neglecting intersubband modes for simplicity) and the dielectric matrix is given by

$$\varepsilon_{\ell\ell'}(q, \omega) = \delta_{\ell\ell'} - v_{\ell,\ell'}(q)\Pi(q, \omega) \, , \tag{58}$$

with

$$v_{\ell,\ell'}(q) = \frac{2e^2}{\kappa} K_0(|q(\ell-\ell')d|) \, , \tag{59}$$

where d is the period of the lateral superlattice, and $\Pi(q, \omega)$ is the 1D polarizability. The width of each wire can be taken into account by re-writing (59) as

$$v_{\ell-\ell'}(q) \equiv v_{\ell,\ell'}(q) = \int dy \int dy' \, |\phi(y-\ell d)|^2 |\phi(y'-\ell'd)|^2 v(q, y-y') \, , \tag{60}$$

where $\phi(y-\ell d)$ is the confining wave function for the ℓ th quantum wire located at $y = \ell d$ and the integral over y, y' in Eq. (60) runs over the width a of each wire.

For the infinite lateral superlattice, periodic invariance allows us to introduce the wave vector q_y (defined within the 1D Brillouin zone $0 <$

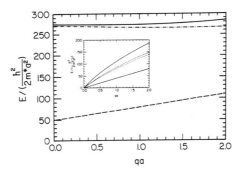

Fig. 7. Coupled inter- (main figure) and intra-
(inset) subband plasmon dispersion in a
three-subband model for a 1D quantum wire.
[Ref. 24].

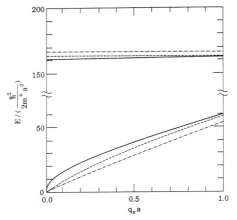

Fig. 8. Intrasubband (lower) and intersubband
(higher) plasmons of a quantum-wire
superlattice as a function of q_x for three
values of $q_y = 0$, $1/d$, π/d. The parameters
are a = 550nm, $N_w = 0.618 \times 10^5 cm^{-1}$ and d =
2a. [Ref. 25].

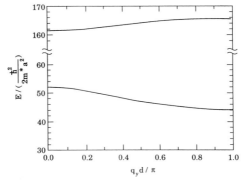

Fig. 9. Intra- (lower) and inter- (upper subband
plasmon dispersion for a fixed $q_x = 0.8/a$ as
a function of q_y for the quantum-wire
superlattice of Fig. 8. [Ref. 25].

$q_y < |\pi/d|$) and solve for the plasmon dispersion relation[23]

$$1 - \Pi(q,\omega) \sum_{\ell-\ell'} v_{\ell-\ell'}(q) \, e^{iq_y(\ell-\ell')d} = 0 .$$ (61)

The long-wavelength limit of Eq. (61) gives us,

$$\omega = \left(\frac{2N_W e^2}{\kappa m} \right)^{1/2} q \left[K_0(qa) + 2\sum_{\ell=1}^{\infty} K_0(\ell qd)\cos(\ell q_y d) \right]^{1/2} .$$ (62)

A numerical solution of Eq. (61) displaying the 1D plasmon band structure in a lateral multiwire superlattice is shown in Fig. 5(c).

Very recently,[25] the elementary excitation spectra of a lateral superlattice have been calculated including both intra- and inter-subband excitations and the mode-coupling effect between them. Results are shown in Figs. 8 and 9 where we depict, respectively, the plasmon band structure as a function of qa for a fixed q_y and as a function of q_y for a fixed qa.

The 1D plasmons have not yet been directly observed in Raman scattering spectroscopy. Their Raman spectra can be easily obtained by calculating $Im\Pi/\varepsilon$ or $Im\Pi$ for charge- and spin-density excitations respectively. Calculated results show that they should be observable by Raman scattering techniques.

2.5. Aperiodic systems

There has been a lot of recent interest[26-29] in the elementary excitation spectra of artificially structured aperiodic systems such as random and quasiperiodic multilayer structures. One can calculate the plasmon spectrum in aperiodic multilayer structures by following the method given in II.3 above and without assuming translational periodicity in the z-direction. (One can also consider random or quasiperiodic lateral superlattices by following the theory given in II.4.) There has been some recent experimental verification[30] of the theoretical results on plasmons in aperiodic structures. We show[28], in Figs. 10 and 11, two sets of calculated Raman scattering spectra for finite superlattices which are periodic and aperiodic, respectively. For details we refer to the original literature.[27-30]

2.6. Phonons and electron-phonon interaction

Longitudinal-optical (LO) phonons couple strongly to the collective plasmon modes in a polar semiconductor such as GaAs, giving rise to the (well-studied) plasmon-phonon coupled modes. Such plasmon-phonon coupling should be important for both intra- and inter-subband charge density excitations around the resonance regime $\omega_{LO} \approx \omega_p$, where ω_{LO} is the LO-phonon frequency and ω_p is the relevant plasma frequency. Away from resonance, the coupling effect is usually small.

The formal way to incorporate the LO-phonon coupling effect in the plasmon dispersion is to interpret the background lattice dielectric constant κ in the Coulomb interaction to be a dynamic quantity given by $\kappa(\omega)$:

$$\kappa(\omega) = \frac{\omega^2 - \omega_{LO}^2}{\omega_{LO}^2} \left(\frac{\kappa_0}{\kappa_0 - \kappa_\infty} \right) ,$$ (63)

where κ_0, κ_∞ are, respectively, the static and the high-frequency lattice

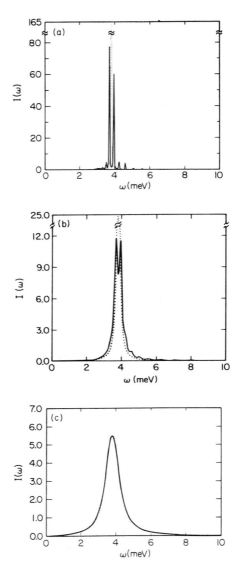

Fig. 10. Calculated Raman intensity for a finite (25-layer) multilayer superlattice with qa = 0.43 and $q_z a$ = 4.94 (the infinite system result is shown by the dotted line). (a) Γ = 0.02meV; (b) Γ = 0.2meV, and (c) Γ = 0.7meV. [Ref. 28].

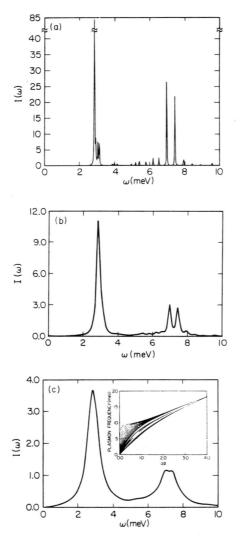

Fig. 11. The same as in Fig. 10 for a 34-layer Fibonacci superlattice (with the plasmon spectrum shown as an inset in part (c)). [Ref. 28].

dielectric constants (note κ has a zero at $\omega = \omega_{LO}$ corresponding to the LO-phonon mode). One can go through all the calculations of sections II.2 - II.5, including $\kappa(\omega)$ in the theory to obtain the Raman-active coupled plasmon-LO phonon modes in each situation.

The experimentally most relevant case is the coupling of LO-phonons to the intersubband collective excitations of a GaAs qauntum-well (or, multiquantum-well) structure. In the long-wavelength limit we can easily calculate the coupled intersubband plasmon-LO phonon frequencies by writing[31]

$$\Pi_{ij}(\omega, q \to 0) = \frac{2E_{ji}(N_i - N_j)}{\omega^2 + i\omega\Gamma - E_{ji}^2} , \tag{64}$$

where $E_{ji} = E_j - E_i$ is the subband energy separation, $N_{i,j}$ are the subband occupancies and Γ is a broadening. The Raman intensity is given by the corresponding reducible response function $\tilde{\Pi}_{ij}$ whose imaginary part is calculated to be

$$\text{Im}\tilde{\Pi}_{ij}(\omega) = \frac{2E_{ji}(N_i - N_j)\Gamma(\omega^2 - \omega_{LO}^2)}{[\{(\omega^2 - \omega_+^2)(\omega^2 - \omega_-^2)\}^2 + \omega^2\Gamma^2(\omega^2 - \omega_{LO}^2)^2]} , \tag{65}$$

where the coupled modes $\omega_\pm$ are given by

$$\omega_\pm^2 = \tfrac{1}{2}[(\omega_{LO}^2 + E_{ji}^2 + \omega_p^2) \pm \{(E_{ji}^2 - \omega_{LO}^2 + \omega_p^2)^2 + 4\omega_p^2(\omega_{LO}^2 - \omega_{TO}^2)\}^{1/2}] , \tag{66}$$

$\omega_{TO} = (\kappa_\infty/\kappa_0)^{1/2}\omega_{LO}$ is the TO-phonon frequency and ω_p is the depolarization shift given by

$$\omega_p^2 = 2E_{ji}(N_i - N_j)v_{ijij} . \tag{67}$$

The corresponding spin-density or the single-particle collective modes are given by $\text{Im}\Pi_{ij}(\omega)$:

$$\text{Im}\Pi_{ij}(\omega) = \frac{2E_{ji}(N_i - N_j)\Gamma}{(\omega^2 - E_{ji}^2)^2 + \omega^2\Gamma^2} . \tag{68}$$

Equations of the type (65) - (68) have been well-verified[7,32] by light-scattering experiments in GaAs quantum-well structures.

In addition to the above-mentioned macroscopic (and classical) electrodynamic coupling between LO-phonons and collective charge-density excitations, it is possible to have a non-local microscopic coupling[33] between LO-phonons and quasiparticles which should also be Raman-active, albeit with much smaller spectral weight. To see this microscopic many-body coupling effect, we consider the model of a 2D electron gas interacting with LO-phonons. The bare LO-phonon propagator is given by

$$D_{LO}(\omega) = \frac{2\omega_{LO}}{\omega^2 - \omega_{LO}^2} , \tag{69}$$

which has a pole at the bare LO-phonon frequency ω_{LO}. The bare LO-phonon propagator is renormalized by the many-body electron-phonon interaction and the renormalized LO-phonon propagator $D(\omega)$ is given by summing up the ring diagrams:

$$D(\omega) = \frac{2\omega_{LO}}{\omega^2 - \omega_{LO}^2 - 2\omega_{LO}M_q^2\Pi(q,\omega)} \; , \tag{70}$$

where M_q^2 is the Fröhlich electron-LO phonon interaction and $\Pi(q,\omega)$ is the 2D electron polarizability.

In Fig. 12 we show[33] the calculated Im $D(\omega)$ for two different fixed values of wave vectors by plotting the phonon spectral weight $A(q,\omega)$:

$$A(q,\omega) = -\frac{1}{\pi}\,\text{Im}\,D(q,\omega) \; . \tag{71}$$

Note that, for bare LO-phonons, the spectral weight is just a delta function of unit strength at $\omega = \omega_{LO}$ and $A(q,\omega)$ is a direct measure of the observed Raman-scattering spectra. From Fig. 12, we conclude that there is a small low-energy peak in the LO-phonon spectrum due to its coupling with the quasiparticles. Even though this peak is weak, typically 10^{-3} of the bare LO-phonon peak, it should, in principle, be observable in Raman-scattering experiments. The existence of this novel quasiparticle-coupled-LO-phonon mode is inferred[33] indirectly from low-temperature electron energy-loss measurements where these modes provide an additional loss channel for the hot electrons, thus enhancing the electron energy-loss rate compared with that of the bare LO-phonons alone.

2.7. Surface and edge plasmons

Since real superlattices are finite systems one expects surface[34] plasmons to be associated with their free surfaces. Such superlattice surface plasmons have been studied theoretically,[34-36] but their direct

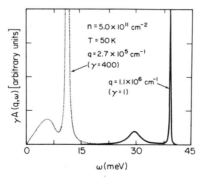

Fig. 12. Calculated LO-phonon spectral weight (which is proportional to the Raman intensity) for a 2D coupled electron-phonon system for two values of wave vectors (for $q = 2.7\times10^5\text{cm}^{-1}$, a δ-function peak at the bare LO-phonon frequency of 35.6meV is not shown). $N_s = 5\times10^{11}\text{cm}^{-2}$ (k $= 1.8\times10^6\text{cm}^{-1}$). [Ref. 33].

experimental observation via Raman-scattering spectroscopy still eludes us. One possibility is that surfaces of real superlattices may be emptied of electrons due to surface depletion layers which make it difficult for surface plasmons to exist.

Corresponding to surface plasmons in finite 3D systems, one could have edge plasmons in finite 2D systems. Such edge plasmons have been theoretically studied.[37] The Raman spectral intensity of such edge plasmons is rather small, but they have been observed[38] in experiments on 2D electron gases.

2.8. Magnetic field effects

Magnetic fields have profound effects on the elementary excitation spectra of low-dimensional systems. The subject has been extensively studied and is far too broad to be covered here.

2.9. Theoretical loose ends

The theoretical analyses of elementary excitations in low-dimensional semiconductor structures and superlattices have mostly been based on the random-phase approximation. In this review we have based our discussion exclusively on the RPA, where the dielectric function is expressed as $\varepsilon = 1 - v\Pi$ and Π is the irreducible non-interacting polarizability function. Because of the low electron effective mass (and rather high background dielectric constant) the effective r_s-parameter for GaAs systems is usually rather low (typically below unity for experimentally relevant electron densities), making RPA a fairly good[10] approximation in GaAs structures (RPA is exact for $r_s \ll 1$). Thus, it is not surprising that theoretical results based on RPA are in good quantitative agreement with light-scattering results on plasmons in GaAs structures. In silicon inversion layer structures effective r_s values are larger[1] and RPA is not such a good approximation. In the presence of a strong external magnetic field, kinetic energy effects are quenched and RPA becomes a poor approximation for a 2DEG (even in a GaAs structure). Attempts have been made[39] to go beyond RPA to calculate the elementary excitation spectra of a 2DEG under strong external magnetic fields. Recent light-scattering[40] experiments are in good agreement with these theoretical calculations; however, more work is clearly needed in this direction.

An important point, often unappreciated in theoretical discussions, is that RPA is a time-dependent Hartree calculation and the single-particle energies and wave functions entering the calculation should include the mean-field Hartree corrections. This is irrelevant for a translationally invariant system such as a pure 3D, 2D, or 1D electron gas because the Hartree self-energy correction, being a tadpole diagram, is identically zero in these systems. But, in a confined structure, the Hartree corrections arising from the self-consistent potential created by the electrons themselves are necessarily non-zero and the single-particle (subband) energies and wave functions are modified by the Hartree correction. These energies and wave functions should be used in the RPA plasmon calculation, particularly for the intersubband transitions. The Hartree correction typically reduces the subband separation whereas the depolarization shift tends to increase the resonance position above the subband separation. These two effects (i.e., Hartree and depolarization corrections), therefore, oppose each other and, in some experimentally relevant situations, may exactly cancel each other. In fact, this cancellation, where it exists, is valid to all orders in electron-electron interaction at long wavelengths. One famous example[41] is Kohn's theorem dealing with the cyclotron resonance in a pure 3D or 2D electron gas which, at long wavelengths, is necessarily at the bare cyclotron frequency due to an exact cancellation between the vertex and the self-energy corrections. More relevant in the current context of confined structures is the existence[42] of such an exact cancellation for any parabolically confined quantum wells which are of current experimental interest. In such wells, one must be careful about making conserving approximations so that the long-wavelength, intersubband charge density excitation resonance occurs at the bare frequency. Note that the single- particle spin-density excitations and the collective charge-density excitations in a parabolic well are still separated by a depolarization shift - only the long-wavelength charge-density resonance is at the bare resonance because there is an exact cancellation among the polarization, the vertex, and the self-energy diagrams. This issue is of considerable relevance in the optical properties of parabolic quantum wells[43] as well as the newly created zero-dimensional[44] quantum dot structures which are very well approximated by parabolic confinement.

We also note that collisional broadening due to impurity scattering has been treated in a phenomenological relaxation-time approximation in this paper. This seems to work well for GaAs-based high-mobility confined structures. The general theoretical issue of a microscopic treatment of plasmon broadening due to impurity scattering is a complicated one and has recently been treated in the literature.[45]

Very recently,[46] a formalism similar to the one discussed here has been used to calculate the collective excitation spectrum and the light scattering intensity of a layered superconductor structure. These results[46] show that inelastic light scattering could be a potentially important tool to study elementary excitations in layered superconductors.

CONCLUSION

We conclude by stating that we have made considerable progress in understanding the elementary excitation spectra in low-dimensional semiconductor structures. An RPA theory of intra- and inter-subband plasmons is in good quantitative agreement with the measured Raman-scattering spectra in GaAs quantum wells and superlattices. Future directions clearly lie in observing the elementary excitation spectra in quantum-wire and quantum-dot systems using light-scattering spectroscopy where, because of the reduced dimensionality of the system, one may have to go beyond RPA for a theoretical understanding. The situation in the presence of a strong external magnetic field, particularly in the fractional quantum Hall regime, also needs more work.

ACKNOWLEDGEMENTS

The author is grateful to the United States Army Research Office (US-ARO), the United States Office of Naval Research (US-ONR), and the United States Department of Defense for their continued support of his research.

REFERENCES

1. T. Ando, A.B. Fowler, and F. Stern, Rev. Mod. Phys. $\underline{54}$, 437 (1982), and references therein.
2. R.A. Ferrell, Phys. Rev. $\underline{111}$, 1214 (1958).
3. S.J. Allen, D.C. Tsui, and R.A. Logan, Phys. Rev. Lett. $\underline{38}$, 980 (1977).

4. T.N. Theis, Surf. Sci. 98, 515 (1980), and references therein.
5. J.P. Kotthaus, W. Hansen, H. Pohlmann, M. Wassermeir, and K. Ploog, Surf. Sci. 196, 600 (1988) and references therein.
6. D. Heitmann, Surf. Sci. 170, 332 (1986) and references therein.
7. A. Pinczuk and G. Abstreiter, p. 153 in "Light Scattering in Solids V," M. Cardona and G. Güntherodt, eds., Springer-Verlag, Berlin (1989) and references therein.
8. J.K. Jain and S. Das Sarma, Surf. Sci. 196, 466 (1988) and references therein.
9. S. Das Sarma, Phys. Rev. B29, 2334 (1984).
10. A.L. Fetter and J.D. Walecka, in "Quantum Theory of Many-Particle Systems," McGraw Hill, New York (1971).
11. J.K. Jain and S. Das Sarma, Phys. Rev. B36, 5949 (1987).
12. A.C. Tsellis and J.J. Quinn, Phys. Rev. B29, 3318 (1984).
13. N.D. Mermin, Phys. Rev. B1, 2362 (1970).
14. D.A. Dahl and L.J. Sham, Phys. Rev. B16, 651 (1977); S. Das Sarma, R.K. Kalia, J.J. Quinn, and M. Nakayama, Phys. Rev. B24, 7181 (1981).
15. S. Das Sarma and A. Madhukar, Phys. Rev. B23, 805 (1981).
16. S. Oelting, D. Heitmann, and J.P. Kotthaus, Phys. Rev. Lett. 56, 1846 (1986).
17. S. Das Sarma and J.J. Quinn, Phys. Rev. B25, 7603 (1982).
18. D. Grecu, Phys. Rev. B3, 2541 (1973).
19. A.L. Fetter, Ann. Phys. (NY) 88, 1 (1974).
20. D. Olego, A. Pinczuk, A.C. Gossard, and W. Wiegmann, Phys. Rev. B25, 7867 (1982); A. Pinczuk, M.G. Lamont, and A.C. Gossard, Phys. Rev. Lett. 56, 2092 (1985); G. Fasol, N. Mestres, H.P. Hughes, A. Fischer, and K. Ploog, Phys. Rev. Lett. 56, 2517 (1986).
21. R. Sooryakumar, A. Pinczuk, A.C. Gossard, and W. Wiegmann, Phys. Rev. B31, 2578 (1985).
22. J.K. Jain and P.B. Allen, Phys. Rev. Lett. 54, 947 (1985) and 54, 2437 (1985); Phys. Rev. B32, 997 (1985); P. Hawrylak, J.W. Wu, and J.J. Quinn, Phys. Rev. B32, 5169 (1985).
23. S. Das Sarma and W.Y. Lai, Phys. Rev. B32, 1401 (1985).
24. Q. Li and S. Das Sarma, Phys. Rev. B40, 5860 (1989).
25. Q. Li and S. Das Sarma, Phys. Rev. B (1990) to be published.
26. R. Merlin, K. Bajema, R. Clashe, F.Y. Juang, and P.K. Bhattacharya, Phys. Rev. Lett. 55, 1768 (1985).
27. S. Das Sarma, A. Kobayashi, and R.E. Prange, Phys. Rev. Lett. 56, 1280 (1986).
28. S. Das Sarma, A. Kobayashi, and R.E. Prange, Phys. Rev. B34, 5309 (1986).
29. P. Hawrylak and J.J. Quinn, Phys. Rev. Lett. 57, 380 (1986).
30. R. Merlin, p. 214 in ref. 7.
31. S. Das Sarma, Appl. Surf. Sci. 11/12, 535 (1982).
32. A. Pinczuk, J.M Worlock, H.L. Störmer, R. Dingle, W. Weigmann, and A.C. Gossard, Solid State Commun. 36, 43 (1980).
33. J.K. Jain, R. Jalabert, and S. Das Sarma, Phys. Rev. Lett. 60, 353 (1988); S. Das Sarma, J.K. Jain, and R. Jalabert, Phys. Rev. B (15 March, 1990).
34. G.F. Giuliani and J.J. Quinn, Phys. Rev. Lett. 51, 919 (1983).
35. J.K. Jain, Phys. Rev. B32, 5456 (1985).
36. J.K. Jain and S. Das Sarma, Phys. Rev. B35, 918 (1987).
37. A.L. Fetter, Phys. Rev. B32, 7676 (1985); J.W. Wu, P. Hawrylak, and J.J. Quinn, Phys. Rev. Lett. 55, 879 (1985); W.Y. Lai, A. Kobayashi, and S. Das Sarma, Phys. Rev. B34, 7380 (1986).
38. D.B. Mast, A.J. Dahm, and A.L. Fetter, Phys. Rev. Lett. 54, 1706 (1985).
39. C. Kallin and B.I. Halperin, Phys. Rev. B30, 5655 (1985) and B31, 3635 (1985).

40. A. Pinczuk, J.P. Valladares, D. Heiman, A.C. Gossard, J.H. English, C.W. Tu, L.N. Pfeiffer, and K. West, Phys. Rev. Lett. 61, 2701 (1988).

41. W. Kohn, Phys. Rev. 123, 1242 (1961).

42. L. Brey, N.F. Johnson, and B.I. Halperin, Phys. Rev. B40, 10647 (1989).

43. K. Karrai, X. Ying, H.D. Drew, and M. Shayegan, Phys. Rev. B40, 12020 (1989).

44. Ch. Sikorski and U. Merkt, Phys. Rev. Lett. 62, 2164 (1989).

45. D. Belitz and S. Das Sarma, Phys. Rev. B34, 8264 (1986).

46. H.A. Fertig and S. Das Sarma, Phys. Rev. Lett. 65, 1482 (1990).

ELECTRONIC PROPERTIES OF PARABOLIC QUANTUM WELLS

L. Brey, N.F. Johnson, Jed Dempsey, and B.I. Halperin

Lyman Laboratory of Physics
Harvard University
Cambridge, Massachusetts 02138

INTRODUCTION

The interacting three-dimensional electron gas in a uniform positive background has been widely studied[1] as a model for the electrons in metals and doped semiconductors. Some of the properties predicted by the model for low-density systems,[2-4] however, have not been observed experimentally. One reason may be that, in the n-doped semiconductors that have been used to achieve low densities, the electrons interact strongly with the charged impurities that are inevitably present.[5]

It has recently been proposed[2,6] that a high-mobility quasi-three-dimensional electron gas can be realized in a remotely-doped wide parabolic quantum well (WPQW). Poisson's equation implies that a parabolic potential $V(z) = Az^2$ is equivalent to the potential created by a uniform slab of positive charge of density $n_0 = A\epsilon/2\pi e^2$, where e is the electron charge and ϵ is the dielectric constant of the system. Electrons trapped in the well can be thought of as screening this fictitious positive charge density by forming a uniform layer of three-dimensional density n_0. Since the electrons are introduced into the WPQW by doping with donors located several hundred Angstroms from the center of the well, the electron mobility in these samples can be considerably higher than in bulk semiconductors. These WPQWs have recently been grown[7,8] by varying the aluminum concentration quadratically in $Ga_{1-x}Al_xAs$ heterostructures. Magnetotransport[7,8] experiments on these samples have verified the presence of a thick slab of electron gas. In addition, theoretical studies of the electronic properties of WPQWs have recently been reported.[9-11] In particular, calculations for wide parabolic quantum wells in the presence of strong in-plane magnetic fields support the idea that a spin-density-wave state may occur for an experimentally accessible range of parameters.[11]

Light Scattering in Semiconductor Structures and Superlattices
Edited by D.J. Lockwood and J.F. Young, Plenum Press, New York, 1991

525

Measurements[12,13] of far-infrared magnetotransmission in parabolic quantum wells reveal a coupling between the cyclotron resonance frequency and the frequency of the bare harmonic oscillator potential. The bare harmonic oscillator frequency, however, corresponds to an energy spacing that can be much larger than the energy separation between subbands calculated using the self-consistent Hartree approximation or more sophisticated techniques. Furthermore, the experiments suggest that the location of the peaks in the absorption spectrum is insensitive to the fractional filling of the well.

In this contribution we present a theoretical study of the optical properties of WPQWs. In the next Section, we discuss the infrared optical absorption of an ideal parabolic quantum well in the absence of a magnetic field and show that it depends only on the parabolic form of the bare potential and not on the electron-electron interaction or on the number of electrons in the well.[14] The second Section is dedicated to the study of the coupling that arises between the original excitations of the parabolic potential and the cyclotron motion of the electrons in the presence of a tilted magnetic field. In the third Section we report numerical calculations, done within the framework of the Local Density Approximation (LDA), of the optical absorption in non-ideal WPQWs with no magnetic field. We check that the LDA method reproduces the exact result in the case of a perfect parabolic potential and then use the method to study deviations from the ideal result when different imperfections are present. We finish the contribution with a short summary of our results.

INFRARED OPTICAL ABSORPTION IN ABSENCE OF MAGNETIC FIELD

Within the effective mass approximation and with the electron slab defining the xy plane, the electrons in the parabolic potential can be described by the following Hamiltonian:

$$H = H_{0,z} + H_{0,xy} + H_{e,e} \; , \tag{1}$$

where

$$H_{0,z} = \frac{1}{2m^*} \sum_{i=1}^{N} p_{i,z}^2 + \sum_{i=1}^{N} \frac{\omega_0^2 m^*}{2} z_i^2 \; , \tag{2}$$

$$H_{0,xy} = \frac{1}{2m^*} \sum_{i=1}^{N} \left(p_{i,x}^2 + p_{i,y}^2 \right) \; , \tag{3}$$

and where $\mathbf{p}_i$ and $\mathbf{r}_i$ are the momentum and position operators of the i-th particle, m^* is the electron effective mass in the host semiconductor, $\omega_0 = (2A/m^*)^{1/2}$ is the bare harmonic oscillator frequency of the parabolic well, and N is the number of electrons in the system. In equation (1), $H_{e,e}$ is the interaction between electrons which we assume depends only on the difference between the position vectors of the particles,

$$H_{ee} \equiv \sum_{i<j} u(\mathbf{r}_i - \mathbf{r}_j) \; . \tag{4}$$

526

In the previous Hamiltonian we have ignored spin, since it will play no role in our discussion.

We define the raising and lowering operators

$$\hat{c}_i^{\pm} \equiv \frac{1}{\sqrt{2\hbar\omega_0 m^*}} \left(m^* \omega_0 z_i \mp i p_{i,z} \right) \ , \tag{5a}$$

$$\hat{C}^{\pm} \equiv \sum_{i=1}^{N} \hat{c}_i^{\pm} \ . \tag{5b}$$

which obey the commutation relations

$$\left[\hat{c}_i^-, \hat{c}_j^+ \right] = \delta_{i,j} \ ,$$

$$\left[\hat{c}_i^+, \hat{c}_j^+ \right] = \left[\hat{c}_i^-, \hat{c}_j^- \right] = 0 \ ,$$

$$\left[\hat{C}^-, \hat{C}^+ \right] = 1 \ ,$$

$$\left[\hat{C}^-, \hat{C}^- \right] = \left[\hat{C}^+, \hat{C}^+ \right] = 0 \ . \tag{6}$$

We shall now show that $\hat{C}^+$ ($\hat{C}^-$) is an operator which creates (destroys) a collective excitation of the system with energy $\hbar\omega_0$. To do this, we calculate the commutator between $\hat{C}^{\pm}$ and the total Hamiltonian of the system, $H = H_{0,z} + H_{0,xy} + H_{e,e}$. Since $H_{0,xy}$ does not depend on the z coordinates of the particles, it follows that

$$\left[H_{0,xy}, \hat{C}^{\pm} \right] = 0 \ . \tag{7}$$

Writing the Hamiltonian $H_{0,z}$ as a function of the $\hat{c}_i^{\pm}$ operators,

$$H_{0,z} = \sum_{i=1}^{N} \left(\hat{c}_i^+ \hat{c}_i^- + \frac{1}{2} \right) \hbar\omega_0 \ , \tag{8}$$

and using the commutation relation (6), it is also straightforward to show that

$$\left[H_{0,z}, \hat{C}^{\pm} \right] = \pm\hbar\omega_0 \hat{C}^{\pm} \ . \tag{9}$$

Finally, we need the commutator of $\hat{C}^{\pm}$ with $H_{e,e}$:

$$\left[H_{e,e}, \hat{C}^{\pm} \right] = \mp \frac{i}{\sqrt{2\hbar\omega_0 m^*}} \sum_{i=1}^{N} [H_{e,e}, p_{i,z}] = \pm \frac{\hbar}{\sqrt{2\hbar\omega_0 m^*}} \sum_{i=1}^{N} \frac{\partial H_{e,e}}{\partial z_i} \ . \tag{10}$$

Whenever the interaction between particles i and j depends only on $(\mathbf{r}_i - \mathbf{r}_j)$, the interaction Hamiltonian $H_{e,e}$ does not change when the system is displaced as a whole, and equation (10) implies that

$$\left[H_{e,e}, \hat{C}^{\pm} \right] = 0 \ . \tag{11}$$

By adding (7), (9) and (11) we arrive at the desired commutator,

$$\left[H, \hat{C}^{\pm}\right] = \pm \hbar \omega_0 \hat{C}^{\pm} \ . \tag{12}$$

If Ψ_n is a eigenstate of H with eigenenergy E_n, equation (12) implies that

$$H\hat{C}^{\pm}\Psi_n = \left(\pm \hbar \omega_0 + E_n\right) \hat{C}^{\pm}\Psi_n \ . \tag{13}$$

Defining $\Psi_{n\pm 1} \equiv \hat{C}^{\pm}\Psi_n$, we see that $\Psi_{n\pm 1}$ is an exact eigenstate of our Hamiltonian with energy $E_{n\pm 1} \equiv E_n \pm \hbar \omega_o$.

When an electric field is applied in the z-direction, the only contributions to the dynamical polarizability of the system in the long-wavelength limit come from transitions between states connected by the operator

$$\sum_{i=1}^{N} z_i = \sqrt{\frac{\hbar}{2m^*\omega_0}} \left(\hat{C}^+ + \hat{C}^-\right) \ . \tag{14}$$

This operator will connect the state Ψ_n only with the states $\Psi_{n\pm 1}$ and thus, the dynamical polarizability of the parabolic well is different from zero only at the bare harmonic oscillator frequency ω_0, where it diverges. Because $H_{e,e}$ commutes with $\hat{C}^{\pm}$, this result is independent of the electron-electron interaction and of the number of electrons in the well. A sharp peak in the optical-absorption spectrum is therefore expected at ω_0. By noting that the motion of the center of mass of the system is governed by the Hamiltonian

$$H_{CM} = \hbar \omega_0 \left(\hat{C}^+ \hat{C}^- + \frac{1}{2}\right) \ , \tag{15}$$

one sees that the collective excitations created by $\hat{C}^{\pm}$ are excitations of the center-of-mass motion of the electron gas. This result is simply based in the separation of center of mass motion and the relative coordinates in parabolic confinement potentials.

It can be shown in general (see Appendix) that electrons confined in parabolic potentials of the form

$$V(x, y, z) = \frac{m^*}{2} \left(\omega_{0,x}^2 x^2 + \omega_{0,y}^2 y^2 + \omega_{0,z}^2 z^2\right) \tag{16}$$

will absorb long-wavelength light only at the frequencies $\omega_{0,x}$, $\omega_{0,y}$, and $\omega_{0,z}$.

By construction, the bare harmonic oscillator frequency ω_0 is equal to the plasma frequency of a three-dimensional electron gas of density n_0,[15]

$$\omega_0 \equiv \sqrt{\frac{2A}{m^*}} = \sqrt{\frac{4\pi e^2 n_0}{\epsilon m^*}} \ . \tag{17}$$

On the other hand, the appearance in optical absorption experiments of a peak at this frequency does not mean that the system has a three-dimensional character, but simply that the bare potential is parabolic. A three-dimensional system does not absorb light at the plasmon frequency, because the plasmon is a longitudinal excitation which cannot be created by the transverse electric field of the light. As it relates to optical behavior, the three-dimensionality of the electron slab in the parabolic well becomes relevant when the slab's thickness is comparable to the wavelength of light of frequency ω_0. In this case the dipole approximation is no longer adequate to describe the interaction of light with the system. Retardation effects[16] must be taken into account. The peak at ω_0 in the optical absorption disappears in the thick-slab limit, and the Drude-type behaviour of a bulk three-dimensional electron gas is restored.[15] In practice, the experimental parabolic wells are very narrow compared to the wavelength of the incident light and the dipole approximation is quite accurate.

INFRARED OPTICAL ABSORPTION IN TILTED MAGNETIC FIELDS

We now study the effect that a magnetic field $\mathbf{B} = (B \sin \theta, 0, B \cos \theta)$, applied at an angle θ with respect to the z-direction, has on the optical absorption of the parabolic well. Using the gauge $\mathbf{A} = (0, xB \cos \theta - zB \sin \theta, 0)$, the Hamiltonian of the system becomes

$$H = \frac{1}{2m^*} \sum_{i=1}^{N} (p_{i,x}^2 + (p_{i,y} + (eB/c)(x \cos \theta - z \sin \theta))^2 + p_{i,z}^2)$$

$$+ \sum_{i=1}^{N} \frac{\omega_0^2 m^*}{2} z_i^2 + H_{e,e} \ . \tag{18}$$

This is equivalent to the Hamiltonian for two coupled harmonic oscillators, and it can be simplified through a change of coordinates $x \to x'$ and $z \to z'$ corresponding to a rotation through an angle α about the y axis:[17,18]

$$H = \frac{1}{2m^*} \sum_{i=1}^{N} (p_{i,x'}^2 + p_{i,y}^2 + p_{i,z'}^2) + \frac{m^*}{2} \sum_{i=1}^{N} (\omega_1^2 x_i'^2 + \omega_2^2 z_i'^2) + H_{e,e}$$

$$+ \omega_c \sum_{i=1}^{N} (\sin(\alpha - \theta) z_i' p_{i,y} + \cos(\alpha - \theta) x_i' p_{i,y}) \ . \tag{19}$$

The angle of rotation α is obtained from

$$\tan (2\alpha) \equiv \frac{\omega_c^2 \sin 2\theta}{\omega_c^2 \cos 2\theta - \omega_0^2} \ . \tag{20}$$

The frequencies $\omega_{1,2}$ are

$$\omega_{1,2} \equiv \left[\frac{1}{2} (\omega_c^2 + \omega_0^2) \pm \frac{1}{2} \left(\omega_c^4 + \omega_0^4 - 2\omega_0^2 \omega_c^2 \cos 2\theta \right)^{1/2} \right]^{1/2} , \tag{21}$$

where $\omega_c = eB/m^*c$ is the cyclotron frequency. We define the raising and lowering operators

$$\hat{a}_i^{\pm} \equiv \frac{1}{\sqrt{2\hbar \omega_1 m^*}} \left(m^* \omega_1 x_i' \mp i p_{i,x'} + \frac{\omega_c}{\omega_1} \cos(\alpha - \theta) p_{i,y} \right) ,$$

$$\hat{b}_i^{\pm} \equiv \frac{1}{\sqrt{2\hbar\omega_2 m^*}} \left(m^* \omega_2 z_i' \mp i p_{i,z'} + \frac{\omega_c}{\omega_2} \sin(\alpha - \theta) p_{i,y} \right) ,$$

$$\hat{A}^{\pm} \equiv \sum_{i=1}^{N} \hat{a}_i^{\pm} ,$$

$$\hat{B}^{\pm} \equiv \sum_{i=1}^{N} \hat{b}_i^{\pm} . \tag{22}$$

These operators satisfy the same commutation relations as the operators $\hat{c}_i^{\pm}$ ($\hat{a}_i^{\pm}$ and $\hat{b}_i^{\pm}$) and $\hat{C}^{\pm}$ ($\hat{A}^{\pm}$ and $\hat{B}^{\pm}$). They also satisfy

$$\left[\hat{a}_i^{\pm}, \hat{b}_j^{\pm} \right] = 0 \quad ,$$

$$\left[\hat{A}^{\pm}, \hat{B}^{\pm} \right] = 0 . \tag{23}$$

With these operators, the Hamiltonian (18) can be written as

$$H =, \sum_{i=1}^{N} \left(\hat{a}_i^+ \hat{a}_i^- + \frac{1}{2} \right) \hbar\omega_1 + \sum_{i=1}^{N} \left(\hat{b}_i^+ \hat{b}_i^- + \frac{1}{2} \right) \hbar\omega_2 + H_{e,e} . \tag{24}$$

Using the commutation relations and the invariance of the $H_{e,e}$ with respect to translations of the center of mass of the electrons, it follows that

$$\left[H, \hat{A}^{\pm} \right] = \pm\hbar\omega_1 \hat{A}^{\pm} ,$$

$$\left[H, \hat{B}^{\pm} \right] = \pm\hbar\omega_2 \hat{B}^{\pm} . \tag{25}$$

If Ψ_{n_1,n_2} is an eigenstate of H with eigenvalue E_{n_1,n_2}, then

$$H \hat{A}^{\pm} \Psi_{n_1,n_2} = \left(\pm\hbar\omega_1 + E_{n_1,n_2} \right) \hat{A}^{\pm} \Psi_{n_1,n_2} ,$$

$$H \hat{B}^{\pm} \Psi_{n_1,n_2} = \left(\pm\hbar\omega_2 + E_{n_1,n_2} \right) \hat{B}^{\pm} \Psi_{n_1,n_2} . \tag{26}$$

Defining $\Psi_{n_1\pm1,n_2} \equiv \hat{A}^{\pm} \Psi_{n_1,n_2}$ and $\Psi_{n_1,n_2\pm1} \equiv \hat{B}^{\pm} \Psi_{n_1,n_2}$, we see that $\Psi_{n_1\pm1,n_2}$ and $\Psi_{n_1,n_2\pm1}$ are exact eigenstates with energies $E_{n_1\pm1,n_2} = E_{n_1,n_2} \pm\hbar\omega_1$ and $E_{n_1,n_2\pm1} = E_{n_1,n_2} \pm\hbar\omega_2$, respectively.

When the system is illuminated with long-wavelength light $\mathbf{E} = E_0 e^{-i\omega t} \mathbf{n}$ polarized in a direction $\mathbf{n} = \sin\beta \hat{\mathbf{x}} + \cos\beta \hat{\mathbf{z}}$, we must add to the Hamiltonian the term

$$H' = E_0 e^{-i\omega t} \sum_{i=1}^{N} \left(\sin(\beta - \alpha) x_i' + \cos(\beta - \alpha) z_i' \right)$$

$$= E_0 e^{-i\omega t} \left[\sqrt{\frac{\hbar}{2m^*\omega_1}} \sin(\beta - \alpha) \left(\hat{A}^+ + \hat{A}^- \right) \right.$$

$$+\sqrt{\frac{\hbar}{2m^*\omega_2}}\cos(\beta-\alpha)\left(\hat{B}^++\hat{B}^-\right)$$

$$-\frac{\omega_c}{m^*}\left(\frac{\cos(\alpha-\theta)\sin(\beta-\alpha)}{\omega_1^2}+\frac{\sin(\alpha-\theta)\cos(\beta-\alpha)}{\omega_2^2}\right)\hat{P}_y\right] \ , \ (27)$$

with

$$\hat{P}_y \equiv \sum_{i=1}^{N} p_{i,y} \ . \tag{28}$$

Because the parabolic potential does not depend on the y-coordinate of the particle positions, $\hat{P}_y$ commutes with the Hamiltonian (17). Thus the term proportional to $\hat{P}_y$ in equation (27) cannot connect eigenstates with different energies and only the first two terms of equation (27) will be relevant to the optical spectrum. The spectrum will have exactly two peaks, at frequencies ω_1 and ω_2, with intensities proportional to $\sin^2(\beta-\alpha)$ and $\cos^2(\beta-\alpha)$, respectively. Due once again to the parabolic form of the quantum well and to the translational invariance in the xy plane, this result is independent of the electron-electron interaction and of the filling of the well.

We obtain the limit of the interacting three-dimensional electron gas by taking $\omega_0=0$ and $\alpha=0$. In this limit, $\omega_1=\omega_c$, $\omega_2=0$ and we recover Kohn's theorem:[19] the cyclotron frequency ω_c in a bulk three-dimensional electron gas is unaffected by any electron-electron interaction that maintains translational invariance.

CALCULATED OPTICAL SPECTRUM IN IMPERFECT WPQWS

We now consider a WPQW with the geometry of Ref. 7, where the well has a finite width W, a depth Δ_1, and is bounded by an additional barrier of height Δ_2 (see figure 1). The electrons in the host semiconductor are described in the effective mass approximation. In the calculation, we take the static dielectric constant and the effective mass to be uniform across the well. The nonparabolicity of the conduction band and the band mixing effects induced by confinement are also omitted in the calculation. All these effects have only a small influence on the electronic properties of the system.

Working in the LDA[20,21] and assuming translational invariance in the xy plane, the eigenvalues and wavefunctions of the WPQW are given by

$$E_{n,k_x,k_y} = \varepsilon_n + \frac{\hbar^2 k_x^2}{2m^*} + \frac{\hbar^2 k_y^2}{2m^*} \ , \tag{29a}$$

$$\Psi_{n,k_x,k_y}(\mathbf{r}) = \frac{1}{\sqrt{S}} e^{i(k_x x + k_y y)} \varphi_n(z) \ , \tag{29b}$$

where we suppose spin degeneracy and omit the spin index. In the above expression, S is the sample area, $\mathbf{k}=(k_x,k_y)$ is the wavevector of the electron, m^* is the effective mass, and ε_n and $\varphi_n(z)$ are obtained from the one-dimensional Schrödinger equation

$$\left[-\frac{\hbar^2}{2m^*}\frac{d^2}{dz^2}+V(z)+V_H(z)+V_{XC}(z)\right]\varphi_n(z)=\varepsilon_n\varphi_n(z) \ . \tag{30}$$

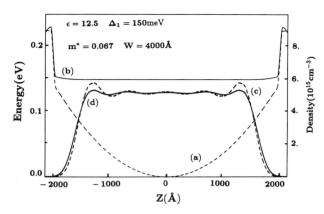

Fig. 1. Initial (a) and self-consistent potential in the Local Density Approximation (b) of a WPQW in the case $\eta = 0.8$. The charge-density profile obtained in the LDA, dashed line (c), is compared with that obtained in the Hartree approximation, continuous line (d). The parameters of the WPQW are given in the figure.

In this expression $V(z)$ is the bare potential of the WPQW, $V_H(z)$ is the Hartree potential arising from the electrostatic interaction of the electrons with the self-consistent electron density $n(z)$ and with the uniform distribution of positive charge necessary to maintain charge neutrality, and $V_{XC}(z)$ is the exchange correlation potential.[22]

For a given value of the two-dimensional density of electrons in the well, n_S, the self-consistent solution of equation (30) gives us the charge-density profile, the Fermi energy ε_F, the subband energies, and the total potential, which is defined as $V_T(z) \equiv V(z) + V_H(z) + V_{XC}(z)$.

The infrared optical spectrum is obtained by using the self-consistent field approximation[23] together with the LDA[24,25]. We neglect retardation effects[16] and we use the dipole approximation in describing the interaction between light and the electrons. In the Ando formalism,[24] the optical absorption per unit area is

$$P(\omega) = \frac{1}{2} Re \left[\tilde{\sigma}_{zz}(\omega) D^2 \right] , \tag{31}$$

where D is the external electric field directed in the z direction and $\tilde{\sigma}_{zz}$ is the modified two-dimensional dynamical conductivity

$$\tilde{\sigma}_{zz}(w) = \frac{e^2}{m^*} (-i\omega) n_s \sum_l \frac{\tilde{f}_l}{\tilde{E}_l^2 - (\hbar\omega)^2 - 2i\hbar\omega/\tau} , \tag{32}$$

where we have introduced a phenomenological relaxation time τ and

$$\tilde{f}_l = \left[\sum_{n,n'} \left(\frac{2m^*}{\hbar^2}(\varepsilon_n - \varepsilon_{n'}) \right)^{1/2} Z_{n'n} \left(\frac{N_n - N_{n'}}{n_s} \right)^{1/2} U_{n'n,l} \right]^2 , \qquad (33)$$

where $Z_{n'n}$ is the matrix element of the z-coordinate between the states $\varphi_{n'}$ and φ_n. In the above expression N_i is the number of electrons per unit area in the i-th subband. $\tilde{E}_l^2$ in equation (32) and $U_{n'n,l}$ in equation (33) are the eigenvalues and eigenvectors of the matrix

$$\Lambda_{nn',mm'} = E^2_{n'n}\delta_{n'm'}\delta_{nm} +$$

$$(N_n - N_{n'})^{1/2}(N_m - N_{m'})^{1/2}(E_{m'm}E_{n'n})^{1/2}(\alpha_{nn',mm'} - \beta_{nn',mm'}) . (34)$$

In the above expressions the indices n, n', m and m' refer to the eigenvalues and eigenfunctions of the one-dimensional Hamiltonian (30). The matrix elements α and β in equation (34) represent the depolarization and excitonic-like effects, respectively, and are given by the expressions

$$\alpha_{nn',mm'} = -2\frac{4\pi e^2}{\epsilon} \int_{-\infty}^{+\infty} dz \varphi_n(z)\varphi_{n'}^*(z) \int_{-\infty}^{z} dz' \int_{-\infty}^{z'} dz'' \varphi_m(z'')\varphi_{m'}^*(z'') , (35)$$

$$\beta_{nn',mm'} = -2 \int_{-\infty}^{+\infty} \varphi_n(z)\varphi_{n'}^*(z) \frac{\partial V_{XC}}{\partial n(z)} \varphi_m(z)\varphi_{m'}^*(z) dz . \qquad (36)$$

It is also easy to prove that the two-dimensional dynamical conductivity satisfies the sum-rule

$$\int \tilde{\sigma}_{zz}(\omega) d\omega = \frac{e^2\pi}{m^*} n_S . \qquad (37)$$

We are interested in the optical absorption properties of the model parabolic potential

$$V_1(z) = \frac{4\Delta_1 z^2}{W^2}\Theta(W/2 - |z|) + (\Delta_1 + \Delta_2)\Theta(|z| - W/2) , \qquad (38)$$

and in the effects that different perturbations, when added to this potential, have on those properties. In our calculations we use the following set of parameters, $\epsilon = 12.5$, $\Delta_1 = 150\text{meV}$, $\Delta_2 = 75\text{meV}$ and $m^* = 0.067$ times the free electron mass. The positively charged donor impurities are in two layers of equal charge density, 200Å thick, located just outside the well on either side. We have checked that our results are insensitive to the precise location of the positive charges for reasonable choices of the parameters.

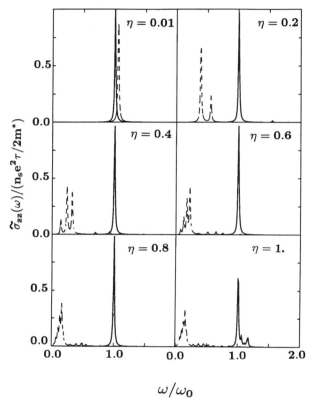

Fig. 2. Calculated real part of the dynamical conductivity $\tilde{\sigma}_{zz}(\omega)$ in a WPQW for different values of the fractional occupation η. The parameters of the WPQW are given in the text. The value of the phenomenological time τ is $0.01\omega_0$. The dashed lines correspond to the same function when depolarization and excitonic-like effects are neglected.

First we study the effects of the finite width, W, and the additional confining potential, Δ_2, on the properties of the system. We apply the method described above with the potential of equation (38), $V(z) = V_1(z)$, and we study the variation of the properties of the system with the fractional occupation, η, defined by $\eta = n_s/(n_0 W)$. We find, in agreement with reference (9), that the number of occupied subbands increases with the number of electrons in the well, that the separation in energies between occupied subbands decreases with n_s, and that the separation between the Fermi energy and the energy of the lowest occupied subband goes quickly to the three-dimensional value $\varepsilon_F = \hbar^2(3\pi^2 n_0)^{2/3}/2m^*$. We also find that the thickness of the electron slab increases linearly with the fractional occupation, η. Our calculations show that a very uniform slab of electrons is formed when only two subbands are occupied.

Size effects become important when the fractional occupation is near unity and the electrons feel the abrupt edge of the well. In figure 1 we plot the charge-density profile for $\eta = 0.8$. The charge is almost uniform in the occupied part of the well, with small Friedel-type oscillations at the edges of the slab similar to those which appear at metal-vaccuum interfaces.[26] Due to the attractive character of the exchange-correlation potential, its main effect on the charge-density profile is to increase the oscillations. In an overfilled WPQW ($\eta > 1$), the extra charge accumulates at the edges of the well and the distribution of electrons becomes more and more like two two-dimensional sheets.[27] The effect of V_{XC} on the energy separation between the occupied and the lowest empty subbands is very small, because in the spatial region where the wavefunctions for these subbands are different from zero the electronic charge distribution and the local-density potential are almost constant.

Figure 2 shows the infrared optical absorption for different values of the fractional occupation η. In accordance with the previous results, the system absorbs light only at the bare frequency $\omega_0 = (8\Delta_1/W^2 m^*)^{1/2}$ when the number of electrons in the well is small enough ($\eta \leq 0.8$) that the charge does not feel the edges of the well. In figure 2 we also plot, for different values of the fractional occupation, the response function obtained when the depolarization and the excitonic-like terms are omitted: $\alpha_{nn',mm'} = \beta_{nn',mm'} = 0$. By comparing the dashed spectra to the solid spectra, one may see clearly the importance of depolarization and excitonic effects in these wells. In particular, in a perfect parabolic potential, our calculations show that the depolarization and excitonic-like corrections to the optical spectrum cancel exactly the corrections arising from the Hartree potential and from the exchange-correlation potential, respectively.

The effects of the finite width of the parabolic well begin to appear in the infrared optical absorption spectrum when $\eta \geq 0.8$. For values of η smaller than one, the perturbation is still relatively small and has two main effects on the spectrum. First, the main peak is shifted to a slightly higher frequency because the extra confinement increases the frequency of oscillation of the center of mass of the electrons. Second, small satellites appear around the main peak because the electric field of the incident light can now couple to the internal motion of the electrons as well as to the motion of the center of mass.

Now we study the effect that a quartic term of the form

$$V_q(z) = \frac{16\Delta_q z^4}{W^4}\Theta(W/2 - |z|) \tag{39}$$

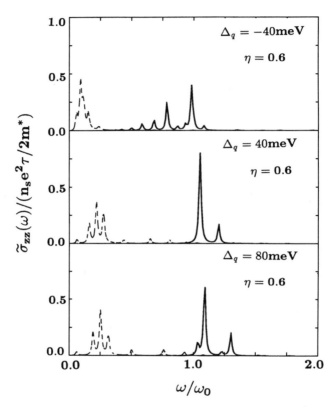

Fig. 3. Calculated real part of the dynamical conductivity $\tilde{\sigma}_{zz}(\omega)$ in a WPQW for different strengths of quartic term added to the potential. The fractional occupation η is 0.6. The parameters of the WPQW are given in the text. The value of the phenomenological time τ is $0.01\omega_0$. The dashed lines correspond to the same function when depolarization and excitonic-like effects are neglected.

has on the properties of the WPQW. We apply the method of calculation to the potential $V(z) = V_1(z) + V_q(z)$. To exclude effects due to the finite width of the well, we fix the fractional occupation at $\eta = 0.6$.

The effect of the perturbation depends on the sign of Δ_q. For positive values, the quartic term induces a convex parabolic correction to the nearly uniform charge distribution produced by the parabolic potential alone. For negative values, the quartic potential adds a concave parabolic component to the charge distribution which causes the charge to spread out in the well and, for even moderate values of Δ_q, destroys the uniform charge-density profile.[27] In figure 3 we show the real part of the dynamical conductivity for different values of Δ_q. For positive values, the effect of the perturbation is twofold: a small shift to higher frequencies of the main peak and the appearance of small satellites around it. For negative values, the perturbation produces a large change in the charge-density profile,[27] and the infrared spectrum shows several peaks of comparable intensity around the main peak, which is shifted down slightly from ω_0. In figure 3 we also plot the spectrum that is obtained when depolarization and excitonic-like effects are neglected ($\alpha_{nn',mm'} = \beta_{nn',mm'} = 0$).

Finally we apply the method to the potential $V(z) = V_1(z) + V_{fl}(z)$, where $V_{fl}(z)$ is given by

$$V_{fl}(z) = -\frac{4\Delta_1 z^2}{W^2}\Theta(W_{fl}/2 - |z|) - \frac{4\Delta_1}{W^2}\left(\frac{W_{fl}}{2}\right)^2 \Theta(|z| - W_{fl}/2) . \qquad (40)$$

$V_{fl}(z)$ cancels $V_1(z)$ in the center of the well, producing a flat segment of potential of width W_{fl} centered about $z = 0$. When added to the perfect parabolic potential, $V_{fl}(z)$ acts as a barrier in the center of the well which tends to separate the electron gas into two 2D systems. In figure 4 we plot the real part of the dynamical conductivity for different values of W_{fl}. We fix η at 0.6 as before to eliminate finite-width effects.

For small values of W_{fl}, such that the electron gas is not accumulated in two sheets at $\pm W_{fl}/2$, the infrared optical absorption is nearly unaffected. In this case, the effect on the spectrum is greater when the depolarization and excitonic-like terms are neglected. If W_{fl} is big enough to split the charge into two sheets, the optical absorption is strongly affected and different peaks, with comparable intensities, appear around the frequency ω_0.

SUMMARY

We have shown that an ideal parabolic quantum well absorbs light only at the bare harmonic oscillator frequency ω_0, independently of the electron-electron interaction and number of electrons in the system. In adition, the presence of a tilted magnetic field couples the cyclotron frequency with ω_0, and not with the frequencies corresponding to the self-consistent intersubband separation.

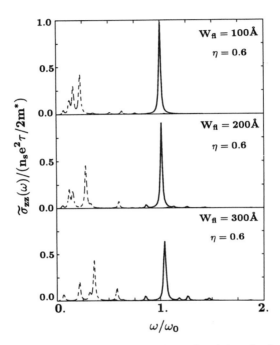

Fig. 4. Calculated real part of the dynamical conductivity $\tilde{\sigma}_{zz}(\omega)$ in a WPQW for different values of the width of the region of flat potential. The fractional occupation η is 0.6. The parameters of the WPQW are given in the text. The value of the phenomenological time τ is $0.01\omega_0$. The dashed lines correspond to the same function when depolarization and excitonic-like effects are neglected.

We have studied also, within the framework of the Local Density Aproximation, the changes that different perturbations induce on the infrared optical absorption spectrum of wide parabolic quantum wells. Three diferent types of effects have been considered: size effects, important when the well is overfilled; effects of a quartic term added to the confining potential; and effects of a region of flat potential in the center of the well. When the strength of any of these perturbations is small enough, the effect on the infrared absorption spectrum is the same, namely i) a shift of the main peak due to a shift inthe frequency of oscillation of the center of mass of the electrons, and ii) the appearance of new peaks around the main peak. The additional peaks had zero intensity in the ideal parabolic case, because they correspond to forbiden transitions. With the perturbations, the wavefunctions change and the transitions are not longer forbiden. The number of additional peak and their positions depend on the particular form of the perturbation.

ACKNOWLEDGEMENTS

We thank P.F.Hopkins, R.Westervelt, A.C.Gossard, H.Ehrenreich, and R.D.Meade for useful discussions. L. Brey wishes to acknowledge support from Spain's Ministerio de Educacion y Ciencia. This work was supported by the NSF through the the Harvard Materials Research Laboratory and grant DMR88-17291.

REFERENCES

1. See, for example, D.Pines, "The Many-Body Problem", Dunod-Willey, New York, (1959).
2. B.I.Halperin, Japan J. of Appl.Phys. **26**, Suppl. 26-3, 1913 (1987).
3. V.Celli and N.D.Mermin, Phys.Rev. **140**, A839 (1965).
4. Z.Tešanović and B.I.Halperin, Phys.Rev.B **36**, 4888 (1987); A.H.MacDonald and G.W.Bryant, Phys.Rev.Lett. **58**, 515 (1987).
5. See, for example, K.Ploog, Journal of Crystal Growth **81**, 304 (1987) and references therein.
6. A.C.Gossard, B.I.Halperin, and R.M.Westervelt (unpublished work).
7. M.Sundaram, A.C.Gossard, J.H.English and R.M.Westervelt, Superlattices and Microstructures **4** , 683 (1988); E.G.Gwinn, R.M.Westervelt, P.F.Hopkins, A.J.Rimberg, M.Sundaram and A.C.Gossard, Phys.Rev.B **39**, 6260 (1989); E.G.Gwinn, P.F.Hopkins, A.J.Rimberg, R.M.Westervelt, M.Sundaram and A.C.Gossard, in "*High Magnetic Fields in Semiconductors Physics II*", eds., G.Landwehr, Springer-Verlag, New York, (1989).
8. M.Shayegan, T.Sajoto, M.Santos, and C.Silvestre, Appl.Phys.Lett. **53**, 791 (1988); T.Sajoto, J.Jo, L.Engel, M.Santos and M.Shayegan, Phys.Rev.B **39**, 10464 (1989); M.Shayegan, T.Sajoto, J.Jo, M.Santos and H.D.Drew, *ibid.* **40**, 3476 (1989).
9. A.J.Rimberg and R.M.Westervelt, Phys.Rev.B **40** , 3970 (1989).
10. M.P.Stopa and S.Das Sarma, Phys.Rev.B **40**, 10048 (1989).
11. L.Brey and B.I.Halperin, Surface Science **229**, 142 (1990); Phys.Rev.B **40**, 11634 (1989).
12. K.Karrai, H.D.Drew, M.W.Lee, and M.Shayegan, Phys. Rev. B **39** , 1426 (1989).
13. K.Karrai, X.Ying, H.D.Drew, and M.Shayegan, Phys. Rev. B **40**, 12020 (1989).
14. L.Brey, N.F.Johnson and B.I.Halperin, Phys.Rev.B **40**, 10647 (1989).

15. See, e.g, N.N.Ashcroft and N.D.Mermin, *"Solid State Physics"* Holt, New York (1976) p.19.
16. D.A.Dahl and L.J.Sham, Phys. Rev. B **16**, 651 (1977).
17. J.C.Maan in *"Two-Dimensional Systems, Heterostructures and Superlattices"*, Vol.53 in "Solid State Sciences", eds., G.Bauer, F.Kuchar and H.Heinrich Springer-Verlag Berlin (1984).
18. R.Merlin, Solid State Commun. **64**, 99 (1987).
19. W.Kohn, Phys.Rev. **123**, 1242 (1961); see also S.K.Yip, Phys.Rev.B **40**, 3682 (1989).
20. F.Stern and S.Das Sarma, Phys.Rev, B **30**, 840 (1984).
21. T.Ando, A.B.Fowler and F.Stern, Rev.Mod.Phys. **54**, 437 (1982).
22. We use the exchange-correlation potential suggested by L.Hedin and B.I.Lundqvist, J.Phys.C **4** , 2064 (1971).
23. H.Ehrenreich and M.H.Cohen, Phys.Rev. **115**, 786 (1959); F.Stern, Phys.Rev.Lett **18**, 546 (1967); E.D.Siggia and P.C.Kwok, Phys.Rev.B **2**, 1024 (1970).
24. T.Ando, Z.Phys.B **26** , 263 (1977).
25. A.Tselis and J.J.Quinn, Surface Science **113**, 362 (1982); K.S.Yi and J.J.Quinn, Phys.Rev.B **27**, 1184 (1983); K.S.Yi and J.J.Quinn, *ibid* **27** , 2396 (1983).
26. W.Kohn and L.J.Sham, Phys.Rev. **137**, A1697 (1965); N.D.Lang, Solid State Physics, **28**, 225 (1973).
27. L.Brey, Jed Dempsey, N.F.Johnson and B.I.Halperin, Phys.Rev.B **42**, 1240 (1990).
28. Ch.Sikorski and U.Merkt, Surface Science **229**, 228 (1990).
29. A.V.Chaplik, in *"Proceedings of the Eight International Conference on the Electronic Properties of Two-Dimensional Systems"*, Grenoble, France, (1989).
30. H.L.Zhao, Y.Zhu and S.Feng, Phys.Rev.B **40** , 8107 (1989).

APPENDIX

In this appendix we study the infrared optical absorption of electrons in quantum dots with parabolic confining potentials. Consider the Hamiltonian

$$H = \frac{1}{2m^*} \sum_{i=1}^{N} \left(p_{i,z}^2 + p_{i,x}^2 + p_{i,y}^2 \right)$$

$$+\frac{m^*}{2} \sum_{i=1}^{N} \left(\omega_{0,x}^2 x_i^2 + \omega_{0,y}^2 y_i^2 + \omega_{0,z}^2 z_i^2 \right) + H_{e,e} \; , \tag{A1}$$

where $H_{e,e}$ is the interaction between electrons (equation (4)). This three-dimensional parabolic potential may be a reasonable approximation for describing electrons in some physically realized quantum dots[28,29] and (with one of the frequencies set to zero) quantum wires.[30]

We define the raising and lowering operators

$$\hat{c}_{i,\nu}^{\pm} \equiv \frac{1}{\sqrt{2\hbar\omega_{0,\nu} m^*}} \left(m^* \omega_{0,\nu} \nu_i \mp i p_{i,\nu} \right) \tag{A2}$$

540

and

$$\hat{C}_\nu^\pm \equiv \sum_{i=1}^{N} \hat{c}_{i,\nu}^\pm \quad , \tag{A3}$$

where ν can be x, y or z. These operators obey the commutation relations

$$\left[\hat{c}_{i,\nu}^-, \hat{c}_{j,\nu'}^+ \right] = \delta_{i,j} \delta_{\nu,\nu'} \quad ,$$

$$\left[\hat{c}_{i,\nu}^+, \hat{c}_{j,\nu'}^+ \right] = \left[\hat{c}_{i,\nu}^-, \hat{c}_{j,\nu'}^- \right] = 0 \quad ,$$

$$\left[\hat{C}_\nu^-, \hat{C}_{\nu'}^+ \right] = 1 \quad ,$$

$$\left[\hat{C}_\nu^-, \hat{C}_{\nu'}^- \right] = \left[\hat{C}_\nu^+, \hat{C}_{\nu'}^+ \right] = 0 \quad . \tag{A4}$$

It is possible to write the Hamiltonian (A1) as

$$H = \sum_{\substack{i=1 \\ \nu=x,y,z}}^{N} \hbar \omega_{0,\nu} \left(\hat{c}_{i,\nu}^+ \hat{c}_{i,\nu}^- + \frac{1}{2} \right) + H_{e,e} \quad . \tag{A5}$$

Using the commutation relations and the invariance of the electron-electron interaction with respect to translations of the center of mass of the system, it follows that

$$\left[H, \hat{C}_\nu^\pm \right] = \pm \hbar \omega_{0,\nu} \hat{C}_\nu^\pm \quad . \tag{A6}$$

If Ψ_{n_x,n_y,n_z} is an eigenstate of H with eigenenergy E_{n_x,n_y,n_z}, Eq. (A6) implies

$$H \hat{C}_\nu^\pm \Psi_{n_x,n_y,n_z} = \left(\pm \hbar \omega_{0,\nu} + E_{n_x,n_y,n_z} \right) \hat{C}_\nu^\pm \Psi_{n_x,n_y,n_z} \quad . \tag{A7}$$

Defining $\Psi_{n_x\pm1,n_y,n_z} = \hat{C}_x^\pm \Psi_{n_x,n_y,n_z}$, $\Psi_{n_x,n_y\pm1,n_z} = \hat{C}_y^\pm \Psi_{n_x,n_y,n_z}$, and $\Psi_{n_x,n_y,n_z\pm1} = \hat{C}_z^\pm \Psi_{n_x,n_y,n_z}$, we see that $\Psi_{n_x\pm1,n_y,n_z}$, $\Psi_{n_x,n_y\pm1,n_z}$, and $\Psi_{n_x,n_y,n_z\pm1}$ are exact eigenstates of the system with energies $E_{n_x\pm1,n_y,n_z}$, $E_{n_x,n_y\pm1,n_z}$, and $E_{n_x,n_y,n_z\pm1}$ respectively.

If the system is placed in an electric field that points in a direction n defined by angles α_x, α_y, and α_z with the x-, y-, and z-directions, respectively, the system will absorb light only at the frequencies $\omega_{0,x}$, $\omega_{0,y}$, and $\omega_{0,z}$, with intensities proportional to $\cos^2(\alpha_x)$, $\cos^2(\alpha_y)$, and $\cos^2(\alpha_z)$. This is a general result which is independent of the electron-electron interaction and of the number of electrons in the system.

ELECTRONIC RAMAN SCATTERING FROM MODULATION-DOPED

QUANTUM WELLS

D.Richards and G.Fasol*

Cavendish Laboratory, Madingley Road, Cambridge CB3 0HE, England

U.Ekenberg

Department of Physics, Uppsala University, S-75121 Uppsala, Sweden

K.Ploog

Max-Planck-Institut für Festkörperforschung
Heisenbergstrasse 1, D-7000 Stuttgart 80, West Germany

ABSTRACT

We show that electronic Raman scattering measurements of the plasmon dispersion, in combination with calculations of the RPA dielectric response and self-consistent electronic subband calculations, can be used to determine the subband structure and populations of modulation-doped GaAs/Al$_x$Ga$_{1-x}$As heterostructures. Thus we present a contactless optical alternative to measurements of Shubnikov-de Haas oscillations for the determination of carrier concentrations. We give an introduction to the computation of plasmons in quantum well systems and illustrate this for the case of a single modulation-doped quantum well, for which we predict the presence of a Fano-type resonance for plasmons in regions of Landau damping. From a Raman determination of the plasmon dispersion we obtain the subband carrier densities and wave functions of multi-quantum wells with multiple subband occupancy. The same technique can be also be applied for the characterisation of parallel conduction from doped AlGaAs layers in modulation-doped heterostructures. We report the Raman observation of the plasmon mode in a single heterojunction, with a δ-layer of acceptors in the GaAs buffer a well-defined distance from the interface. Under illumination above the band-gap of the AlGaAs barrier, a dynamic charge-transfer effect occurs in which the quasi-two-dimensional electron concentration of the heterojunction decreases. We use Raman measurements of the plasmon mode to directly determine the change in carrier concentration with excess illumination.

INTRODUCTION

One of the novel physical properties of quasi-two-dimensional electron gas systems of particular interest is the behaviour of the electronic collective-excitations or plasmons, which have been extensively studied both theoretically[1-8] and experimentally[9-11]. Plasmons in quantum well and quantum wire systems are important for electron scattering, such as the scattering of electrons crossing the heavily doped base region of hot electron devices[12]. In the present paper we show that the determination of the plasma dispersion of multilayered electron systems allows the quantitative determination of the electronic structure of such

Light Scattering in Semiconductor Structures and Superlattices
Edited by D.J. Lockwood and J.F. Young, Plenum Press, New York, 1991

543

systems. The methods we introduce here show that measurements of the plasmon dispersion of laterally structured and quantum wire systems[13,14] have much promise for a detailed determination of electronic structure.

The methods described in this paper are complementary to the more usual Hall and Shubnikov-de Haas determinations. The Raman techniques introduced here have certain advantages:

1. No contacts are necessary.
2. Electron subbands with similar carrier densities can be clearly distinguished, since dispersions of the coupled plasmon modes are split by the intra-well Coulomb interaction.
3. Micro-Raman techniques[13,15] allow spatial resolution.

PLASMONS IN QUASI-2D ELECTRON SYSTEMS

The plasmons of an electron gas are the self-sustaining collective carrier oscillations of the electrons. In a single two-dimensional electron gas (2DEG), the plasmon energy varies as a function of in-plane wave-vector $k_{\parallel}$ in the long-wavelength high-frequency approximation[8] as $k_{\parallel}^{1/2}$. When there are more than one parallel two-dimensional electron gas layers, the electron sheets are coupled together by the Coulomb interaction, which causes the plasmon branches for the electrons in each well to be coupled. So a modulation doped quantum well sample with N periods and a single occupied subband in each well will have a set of N $intra$subband plasmon modes[4,10,11] - collective excitations in the plane of the electron sheets which set up charge distributions along the layer. However, an accurate description of plasma properties must go beyond the 2DEG approach and take into account quantum size effects from the carriers being confined within wells of finite width.[3,6] For the modulation-doped quantum-well system studied here, the GaAs wells are sufficiently far apart for there to be no overlap between the electron states in neighbouring wells, although they are of course still coupled by the relatively weaker long-range Coulomb interaction. With M occupied subbands per well there is a further strong intra-well Coulomb interaction due to the overlap of electron wave functions which splits the plasmon modes into M groups. Thus there will be NxM $intra$subband plasmon eigenmodes.[7] In addition, $inter$subband plasmons, which are collective intersubband excitations with an induced charge density normal to the plane of the layer, will also exist. These are not dispersive at very long wavelength and are depolarisation-shifted to a frequency higher than the intersubband transition energy. Das Sarma has noted that there is a coupling between the intersubband and intrasubband plasmons.[3] The intersubband plasmon is only affected by mode-coupling effects as $O(k_{\parallel}^2)$ but the intrasubband plasmon frequency is affected in the leading order, and if the two modes cross they will split. We show clear evidence of this coupling of the longitudinal and the transverse plasmon excitations, both in the frequencies as well as in the intensities of the Raman spectra.

The dispersion of plasmons in 2DEG systems can be probed by electronic resonant Raman scattering measurements at liquid Helium temperatures.[9-11] There is no conservation of $k_{\perp}$ in the case of light scattering from a sample of only a few quantum wells, since there is no translational symmetry perpendicular to the sample-growth direction. So light coupling to all in-plane plasmons is possible. The plasma dispersion is measured as a function of $k_{\parallel}$ by varying the angles of the incident and scattered light with respect to the plane of the quantum wells.[11]

CALCULATIONS OF PLASMONS

The plasmon frequencies of a solid can be determined from calculations of the dielectric response within the random phase approximation (RPA), which is valid for the long wavelengths under consideration. This theory is well known and documented.[16] The plasmons are self-sustaining oscillations of the electrons in the solid and will therefore produce fluctuations of the total potential, in the absence of any applied field. If an external potential $\phi^{ext}(r,\omega)$ is applied, charges and currents will be induced, leading to a new total potential $\phi^{tot}(r,\omega)$ in the solid.

$$\phi^{tot}(r,\omega) = \int \varepsilon^{-1}(r,r',\omega) \ \phi^{ext}(r',\omega) \ dr' \tag{1}$$

Hence the poles of the inverse dielectric response function, $\varepsilon^{-1}(r,r',\omega)$, will give the plasmon energies. The inverse dielectric function is related to the density-density correlation function $D(r,r',\omega)$ by,

$$\varepsilon^{-1}(r,r',\omega) = \delta(r-r') + \int v(r-r'') \ D(r'',r',\omega) \ dr'' \tag{2}$$

where $v(r-r')$ is the Coulomb interaction. We can also express the dielectric response in terms of the polarisability $P(r,r',\omega)$, the response of a non-interacting electron gas, which is easily obtainable.

$$\varepsilon(r,r',\omega) = \delta(r-r') - \int P(r,r'',\omega) \ v(r''-r') \ dr'' \tag{3}$$

$$P(r,r',\omega) = \sum_{\kappa \ \kappa'} \frac{2 \ (n_\kappa - n_{\kappa'})}{E_\kappa - E_{\kappa'} - \hbar\omega - i\delta} \ \psi_\kappa(r) \ \psi_{\kappa'}^*(r) \ \psi_\kappa^*(r') \ \psi_{\kappa'}(r') \tag{4}$$

where $\psi_\kappa(r)$ are the single non-interacting electron wave functions with both wave vector and subband denoted by index κ, and n_κ are the Fermi occupation factors.

For calculations at absolute zero, assuming a parabolic conduction band and no scattering, an analytical expression for the polarisability can be used employing the envelope function approximation[17] for the single-electron wave functions $\psi_\kappa(r)$. Then, using the real-space inversion formalism of King-Smith and Inkson,[18] an expression for $\varepsilon^{-1}(z,z',k_{\parallel},\omega)$ can be readily obtained in terms of the envelope wave functions of the quasi-2DEGs, where $k_{\parallel}$ is the in-plane wave vector. The use of the *full* RPA expression for the dielectric response[7] employing self-consistently determined envelope wave functions is most important for a complete analysis of experimentally obtained plasmon dispersions, as quantum-size effects can be very important and long-wavelength, high frequency approximations for the polarisability, much favoured in the literature[4,10], are not sufficient.[7]

The interaction term $v(z,z',k_{\parallel})$ is of the form

$$v(z,z',k_{\parallel}) = \frac{2\pi e^2}{\varepsilon \ k_{\parallel}} (\ e^{-k_{\parallel}|z - z'|} + \alpha \ e^{-k_{\parallel}|z + z'|} \ ) + \frac{\partial v_{xc}[n(z)]}{\partial n} \delta(z-z') \tag{5}$$

$$\alpha = \frac{\varepsilon - \varepsilon_0}{\varepsilon + \varepsilon_0} \tag{5a}$$

The first two terms are contributions to the potential from the Coulomb interaction, the second term allowing for the image potentials experienced by each electron from the air-interface of the sample surface.[5] The third term allows for exchange and correlation effects[19,20], where $v_{xc}[n(z)]$ is the exchange-correlation potential within the local density approximation (LDA). Image potential effects arising from the small differences in dielectric constant between GaAs, AlGaAs and AlAs are ignored as differences between the dielectric constants of these materials are so small. The scalar dielectric constant ε takes into account the effect of the valence polarisability and of low-lying core energy levels. Interaction between phonons and electronic collective modes is included by using a frequency-dependent dielectric constant, with ω_L and ω_T the frequencies of the LO and TO phonons, respectively.

$$\varepsilon(\omega) = \varepsilon(\infty) \left(\frac{\omega^2 - \omega_L^2}{\omega^2 - \omega_T^2} \right) \tag{6}$$

Plasmon modes at energies and $k_{||}$ values outside the ranges of electron-hole excitations (i.e. those regions where single-particle excitations can occur) are infinitely long-lived and are determined as the poles of the determinant of $\varepsilon^{-1}(z,z',k_{||},\omega)$. Within the regions of electron-hole excitations, where plasmon modes are Landau-damped, the zeros of the determinant of ε^{-1} become complex. The plasmon modes are then given by the Lorentzian quasi-particle peaks in the imaginary part of ε^{-1} and hence in $-Im[D(z,z',k_{||},\omega)]$, the imaginary part of the density-density correlation function.[7] This agrees well with identifying the peaks in the Raman spectra as plasmon modes, since for incident light of wave vector k and decay length λ in the solid, the Raman scattering intensity, I, is given by[20,21]

$$I \propto \int -Im[D(z,z',k_{||},\omega)]\, e^{-2ik(z-z')}\, e^{-(z+z')/\lambda}\, dz\, dz' \tag{7}$$

To obtain envelope wave functions and confinement energies in modulation-doped quantum wells, we solve the effective-mass Schrödinger equation and Poisson equation self-consistently[22], using a variational method of Ekenberg,[23] which is a modification of the methods of Altarelli[24]. Exchange and correlation contributions to the potential are included within the LDA using the parametrization given by Gunnarson and Lundqvist,[25] also used in calculations of the dielectric response.

$$v_{xc}[n(z)] = -\frac{2}{\pi \, \alpha \, r_s} \left(\frac{9\pi}{4} \right)^{1/3} \left[1 + 0.0545 r_s \log_e \left(1 + \frac{11.4}{r_s} \right) \right] Ry^* \tag{8}$$

where

$$r_s \, a_B^* = \left(\frac{3}{4\pi n(z)} \right)^{1/3} \tag{8a}$$

and a_B^* and Ry^* are the effective Bohr radius (99.9Å) and Rydberg constant (5.74meV), respectively, for GaAs.

The variational approach adopted employs symmetric and antisymmetric pairs of Gaussians as a set of basis functions $f_i(z)$ for the subband wave functions $\phi^v(z)$.

$$f_i^s(z) = A_i^s \left[e^{-a_i (z - b_i)^2} + e^{-a_i (z + b_i)^2} \right] \qquad \text{symmetric (9)}$$

$$f_i^a(z) = A_i^a \left[e^{-a_i (z - b_i)^2} - e^{-a_i (z + b_i)^2} \right] \qquad \text{antisymmetric (10)}$$

The problem is split into three parts, left barrier, quantum well and right barrier, with interface functionals imposing the envelope function boundary conditions[17,26] at the two interfaces, these being:

1. Continuity of the wavefunction $\phi^v(z)$ at an interface.

2. Continuity of $\frac{1}{m^*} \frac{d\phi^v}{dz}$ where m* is the effective mass.

If N gaussian functions are used, the solution of Schrödinger's equation is then reduced to a 6Nx6N matrix, generalized eigenproblem. Typically, five pairs of basis functions are used, this being sufficient to give good eigenvalues. Note that we have previously found good agreement between results obtained by the variational method and by direct integration.[7]

CALCULATION OF THE DIELECTRIC RESPONSE OF A SINGLE QUANTUM WELL

In the present section we illustrate the calculation of the dielectric response to obtain the plasmon dispersion for the case of a *single* quantum well, modulation-doped on both sides.[7] We demonstrate the calculations here for a structure similar to those we have investigated experimentally. The well is 500Å wide, has spacer layers of 100Å $Al_{0.3}Ga_{0.7}As$ at the normal interface and 50Å AlAs at the inverted interface, and we assume has a total carrier concentration of $1.8x10^{12}$ cm^{-2}. Self-consistent electron subband calculations give three occupied subbands for this structure, their wavefunctions and the self-consistent potential being shown in Fig. 1.

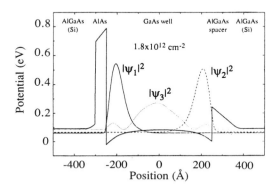

Fig. 1. Self-consistently determined potential and probability density distributions $|\psi|^2$ for a single 500Å-wide quantum well, modulation doped on both sides, with a carrier concentration of $1.8x10^{12}$cm^{-2}. We show the probability densities for the three occupied subbands (1,2,3).

In Fig. 2a we show the calculated plasmon dispersion of this single quantum well structure. As there are three occupied subbands in the quantum well, there are three intrasubband plasmon modes (labeled A,B and C in Fig. 2a) resulting from the strong coupling between the different subbands. For the highest frequency mode (C), the dispersion varies as $k_\parallel^{1/2}$, electrons in all the subbands moving in phase. The other two modes (A and B) are so-called acoustic plasmons[27] and disperse more like $k_\parallel$, electrons in different subbands moving in antiphase to one another. We show, in Fig. 3a, the carrier-density modulations parallel to the plane of the well for the two highest frequency modes (B and C) and in, Fig. 3b, the electrical potential profiles corresponding to the coupled plasmon modes. As all subbands are in phase for the highest frequency plasmon (C), the electric potential outside the well is greatest for this mode. Thus, if we bring several such quantum wells together, the coupling of plasmons between wells due to the long-range Coulomb potential is strongest for this mode, resulting in a strong splitting of plasmon modes. This effect is less for the lower frequency plasmons where the electric field outside the well is correspondingly smaller. Note in Fig. 3(b) that, for the lowest energy mode A, the carriers move in such a way that the potential outside the well is zero. Thus there can be no Coulomb coupling between wells for mode A. We show in, Fig. 2(b), the calculated plasmon dispersion for a five-quantum well structure. Mode A is now the degenerate set of modes P, B is the group of modes Q, and C has split into modes 1-5.

The shaded regions in Fig. 2 denote the single-particle excitation regimes where Landau damping can occur. As previously stated, in these regions we obtain the plasmon spectra by calculating the imaginary part of the density-density correlation function $-Im[D(z,z,k_\parallel,\omega)]$. As shown in Fig. 4, it has peaks at those frequencies where long-lived Landau-damped plasmons exist. Of particular interest is the presence of a Fano-type

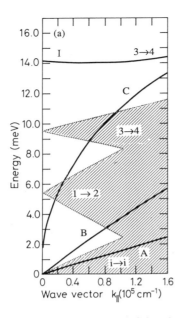
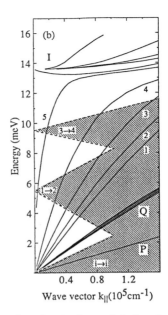

Fig. 2. (a) Full RPA calculation of the plasmon modes of a single modulation-doped quantum well. As there are three filled subbands in the quantum well, there are three coupled *intra*subband plasmon modes A,B,C. There is also one *inter*subband plasmon I from the 3→4 transition. The shaded areas show the intrasubband and the 1→2 and 3→4 single-particle excitation regimes where Landau damping occurs.
(b) Calculated plasmon dispersion (thin solid lines) for a multi-quantum-well structure with five periods. The inter-well interaction splits mode C of the single well into modes 1,2,3,4,5. Mode B is split into group Q and mode P is a degenerate set of plasmons corresponding to mode A.

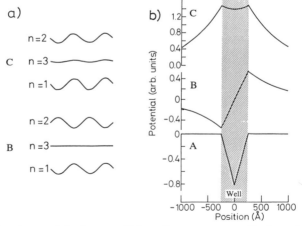

Fig. 3. (a) Charge-density modulation parallel to the quantum well plane for the plasmon modes B and C in Fig. 2(a). We do not show the result for the Landau damped mode A. (b) The electrical potential associated with plasmon modes A,B,C of the single quantum well. Here we have taken the approximation that the charge densities are infinitely thin layers, the two lowest subbands are located at the "inverted" and "normal" interfaces of the well with equal populations $n_1 = n_2$, and the third subband is located in the middle of the well with population $n_3 \ll n_1$. The shaded area shows the thickness of the quantum well.

548

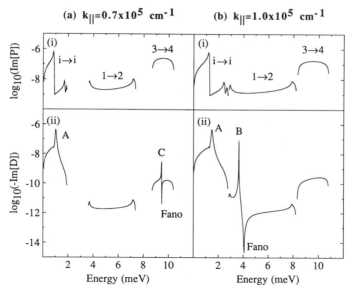

Fig. 4. Imaginary part of the density-density correlation function -Im[D(z,z,k$_{||}$,ω)] for a single quantum well with three filled subbands, calculated for in-plane wave vectors (a) $k_{||}=0.7\times10^5$cm^{-1} and (b) $k_{||}=1.0\times10^5$cm^{-1}. This demonstrates the nature of the plasmon resonances within the regions of the single-particle excitation continua. (i) The imaginary part of the RPA polarisability P - the density-density correlation function for vanishing interaction between electrons. This case corresponds to the single-particle excitations. (ii) -Im[D]. Fano-type interferences labeled Fano.

interference[28] for these Landau-damped plasmons, due to the interaction of the discrete plasmon mode with the continuum of single-particle excitations. As the Raman intensity is directly related to -Im[D(z,z',k$_{||}$,ω)], it may well be possible to observe this effect.

DETERMINATION OF THE ELECTRONIC STRUCTURE OF MODULATION-DOPED QUANTUM WELLS FROM RAMAN MEASUREMENTS OF THE PLASMON DISPERSION

In the present section we show that electronic Raman scattering measurements of the plasmon dispersion are a valuable tool for the complete determination of the electronic structure of modulation-doped quantum wells. A layered electron gas comprising a total of N 2D-electron subbands will yield N coupled intrasubband plasmons.[7] Thus, if all these plasmons are measured, a complete characterization of the multi-quantum well structure is possible.[29] Alternative characterization methods, such as Shubnikov-de Haas measurements, can be used to determine the carrier densities of multiple subbands in a *single* modulation-doped quantum well if the densities are sufficiently different. However, for *multiple* quantum wells, signals from subbands with similar densities in the same or in different wells are very difficult to separate.

Fig. 5 illustrates our technique schematically. Plasmon energies are measured from electronic Raman scattering measurements for varying in-plane wavevectors k$_{||}$. The plasmon dispersion is then calculated from the RPA dielectric response, including self-consistent electronic subband calculations. Experimental and calculated plasmon dispersions are then compared and parameters varied until agreement between theory and experiment is obtained, yielding carrier concentrations and subband wave functions and energies. When analysing an experimental dispersion, the calculated plasmon energies are not very sensitive to small changes in the form of the electron wave functions, although such quantum size effects are crucial in fitting the data. However, the dispersion of the intrasubband plasmon modes is *very* sensitive to subband carrier concentrations. Hence an initial 'guess' at the subband structure can be made from self-consistent calculations. Then the carrier densities and intersubband energies can be 'fine-tuned' to fit the experimental results. Note that if we know the carrier densities of two subbands in the same quantum well, then we know their intersubband energy (assuming a parabolic conduction band).

$$E_f - E^v = \frac{\hbar^2 \pi N^v}{m^*} \qquad \text{hence} \qquad E^v - E^{v'} = \frac{\hbar^2 \pi}{m^*} (N^{v'} - N^v) \qquad (11)$$

where E_f is the Fermi energy and E^v and N^v are the confinement energy and carrier density, respectively, for subband v.

When we start tending towards a reasonable fit, self-consistent calculations can be performed again, varying parameters of large uncertainty such as dopant concentrations, to obtain a good agreement between calculated and experimentally obtained plasmon dispersions.

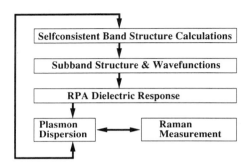

Fig. 5. Schematic flow diagram of the technique employed in the present work.

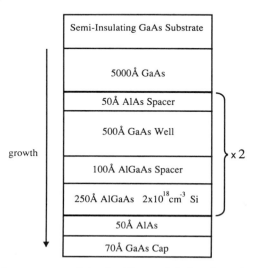

Fig. 6. Schematic of the structure of the double modulation-doped quantum well structure discussed in the text.

RAMAN DETERMINATION OF THE ELECTRONIC STRUCTURE OF A DOUBLE QUANTUM WELL

We demonstrate the above technique by considering a representative GaAs/$Al_{0.3}Ga_{0.7}As$ sample grown by molecular beam epitaxy (MBE) consisting of two 500Å-wide modulation-doped quantum wells.[29] The structure is shown in Fig. 6. The two wells have different environments, the well nearest the substrate (hereafter the 'substrate well') being only doped on one side and thus having an electronic structure similar to that of a heterojunction. The second well (hereafter the 'surface well') is doped on both sides, although surface states may completely deplete the doped layer on the surface side of the well, thus affecting the electronic structure of the quasi-2DEG in the well.

In Fig. 7 we show the Raman spectra of the plasmon modes of this sample, measured for different in-plane wave vectors $k_\parallel$ at a laser energy close to the E_0 resonance of the quantum well. There are four dispersive plasmons (labelled P,Q,1,2) and two less dispersive modes (R and S). The plasmon energies are shown as a function of $k_\parallel$ in Fig. 8, the point size indicating the Raman signal intensity.[30] The four dispersive *intra*subband plasmons indicate the presence of four 2D-electron subbands in the structure. As the asymmetric substrate well is likely to have only one occupied subband,[31] we can infer that there are three occupied subbands in the surface well. Modes 1,2 are the plasmons coupled by the inter-well Coulomb interaction, where all charges within each well are moving in phase. From the dispersion of these modes we can determine the total carrier concentrations for the surface and substrate wells to be $1.6 \times 10^{12} cm^{-2}$ and $6.5 \times 10^{11} cm^{-2}$, respectively. These values can be compared with a total carrier concentration for the structure, obtained from a Hall measurement with the sample illuminated, of $2.0 \times 10^{12} cm^{-2}$. The intrasubband modes P,Q arise from splitting of the intrasubband collective excitations in the surface well due to the strong intra-well interaction. Analysis of these modes thus enables one to determine the occupancies for the different subbands in the surface well. The non-dispersive modes R,S are *inter*subband plasmons due to collective excitations in the surface well.

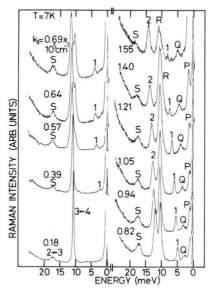

Fig. 7. Raman spectra of the plasmon modes of two modulation-doped quantum wells, for different in-plane wavevectors $k_\parallel$. Letters (P,Q,R,S) and numbers (1,2) refer to the different types of plasmon mode discussed in the text.

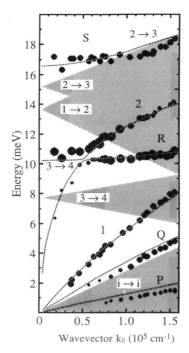

Fig. 8. Experimental points and full RPA calculation (dashed lines) of plasmon modes, as a function of the in-plane wavevector $k_{\parallel}$, of the two-quantum-well sample discussed in the text. Point size indicates relative Raman signal strength for each plasmon resonance (displaying signal strength as area of points is a graphical method due to J. C. Maan[30]). Modes P,Q,1,2 are *intra*subband plasmons. R,S are *inter*subband collective excitations. The shaded areas show the intrasubband and the $3{\to}4$, $1{\to}2$ and $2{\to}3$ single-particle excitation continua where Landau damping occurs.

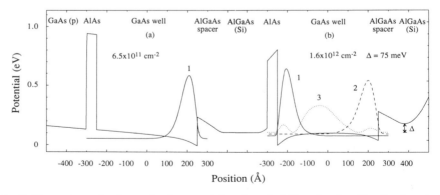

Fig. 9. Self-consistently determined potential and probability density distributions $|\psi|^2$ for the modulation-doped two-quantum-well sample. We show (a) in the 'substrate well' the probability density for the ground state (1). In the 'surface well' (b), probability densities are shown for the three occupied subbands (1,2,3).

Taking the total concentrations of 1.6×10^{12} cm^{-2} and 6.5×10^{11} cm^{-2}, the subband wave functions and energies for the two wells were calculated, the probability density distributions $|\psi|^2$ and the potential for the structure shown in Fig. 9. A band-bending offset Δ was used as a fitting parameter to describe the effect of surface depletion on the first well. The plasmon dispersion could then be determined from calculations of the full RPA dielectric response, utilising the self-consistently determined wave functions and energies with $\Delta = 75$ meV. The calculated plasmon dispersion (dotted lines) is shown in Fig. 8, together with that obtained by experiment (full experimental points), and we can assign the intersubband plasmons R and S as the collective excitations between the third and fourth and the second and third subbands, respectively. The intersubband plasmon between the first and second subbands does not occur because the spatial separation between these two states is too great since they are localised at opposite interfaces. The electron subband energies and carrier concentrations determined by this technique are given in Table I.

Table I. Subband energies and subband occupations for the present double-quantum-well sample under saturated illumination. (Energies are with respect to the first subband.)

		Subband (ν)	Electron density ($\times 10^{11}$ cm^{-2})	Energy (meV)
1.	Surface Well: 1.6×10^{12} cm^{-2}	1	9.27	0.0
		2	5.48	13.65
		3	1.25	28.9
		4	-	36.6 [a]
2.	Substrate Well: 6.5×10^{11} cm^{-2}	1	6.5	0.0
		2	-	32.1

[a] The energy of the first excited state ($\nu=4$) in the surface well was calculated to be 39.3 meV. The value given was chosen to give a $3 \rightarrow 4$ transition energy of 7.7 meV.

Of particular interest is the dispersion of the $2 \rightarrow 3$ intersubband mode S for large in-plane wave vectors. The intersubband plasmon frequency $\omega_{\nu\nu'}$ for a transition from subband ν to ν' can be approximately given for small in-plane wave vector by,

$$\hbar \omega_{\nu\nu'} = \left[(E^\nu - E^{\nu'})^2 + W_p^2 \right]^{1/2} \tag{12}$$

where the depolarization shift W_p is given by

$$W_p^2 = \frac{4\pi e^2}{\varepsilon} (E^\nu - E^{\nu'})(N^{\nu'} - N^\nu) \left[v(k_{||}) \right]_{k_{||} \rightarrow 0} \tag{13}$$

and $v(k_{||})$ is a matrix element term.

Thus there are two influences which will tend to reduce the depolarization shift of the $2 \rightarrow 3$ plasmon (S) compared to the $3 \rightarrow 4$ plasmon (R):
(a) The wave function overlap is smaller for $2 \rightarrow 3$ as the second subband is localised at the normal interface, whereas 3 and 4 are localised over the whole well.
(b) As $2 \rightarrow 3$ is from one occupied subband to another occupied subband, and the depolarization shift varies as the difference in the carrier densities of each subband, it is reduced with respect to a similar transition to an empty subband.
The net result is that the $2 \rightarrow 3$ intersubband collective excitation approaches its corresponding single-particle excitation regime at small in-plane wave vector. The plasmon then follows the

edge of the single-particle regime, giving the mode some dispersion, before finally entering and being strongly Landau-damped.

DETERMINATION OF CARRIER DENSITY IN A HETEROJUNCTION FROM RAMAN MEASUREMENTS OF THE INTRASUBBAND PLASMON

Electronic Raman scattering from plasmons in modulation-doped *multi*-quantum-well structures has been extensively studied in recent years.[7,9-11,29] In such structures there are a large number of adjacent 2DEG sheets, leading to complicated plasmon dispersions, as described in previous sections. For a structure containing just one sheet of electrons, there will be just one intrasubband plasmon mode with a dispersion that varies as $k_\parallel^{1/2}$. Raman spectra of intrasubband excitations have previously been obtained for an asymmetric quantum well doped on one side,[32] where the presence of an undoped AlGaAs barrier at the inverted interface ensures the confinement of photo-created holes in the same region as the confined electrons, thus enabling resonant Raman processes to occur. Here we present Raman scattering measurements on a new type of heterojunction,[33] the structure of which we illustrate schematically in Fig. 10. A δ-doped layer of Be acceptors at a well-defined distance z_0 from the interface makes possible the enhancement of resonant Raman scattering.

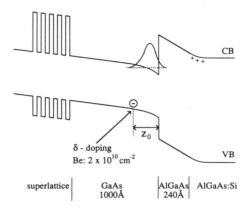

Fig. 10. The conduction and valence bands for the heterojunction described in in the text. There is a δ-layer of acceptors at a well-defined distance z_0 from the GaAs/AlGaAs interface.

In Fig. 11 we present Raman spectra obtained in parallel polarisation close to the E_0 resonance, measured as a function of the in-plane wave vector $k_\parallel$. The structure (sample #6639) is that of a modulation-doped heterojunction with a 1000Å 'active' GaAs layer between a GaAs/AlAs superlattice and the heterointerface. The acceptor δ-layer, doped with 2×10^{10} cm^{-2} of Be, is positioned at $z_0 = 250$Å from the GaAs/Al$_{0.33}$Ga$_{0.67}$As interface. We can infer that the dispersive mode present, which follows a square-root type dispersion, is the intrasubband plasmon mode of the 2DEG. Part of the width of the Raman signal is due to the presence of two weaker non-dispersive modes in the same energy range, which limits the observation of the plasmon to low $k_\parallel$ when the plasmon signal becomes smaller. The origin of these additional peaks, which may be seen more clearly further away from the E_0 resonance, has not yet been determined. In Fig. 12, we show the dispersion of the plasmon for the above sample and also one of a similar structure (#6505) with a Be δ-layer at $z_0 = 300$Å from the interface. To these experimentally determined dispersions we have calculated fits for the plasmon modes for carrier densities of 2.6×10^{11} cm^{-2} (#6639) and 3.4×10^{11} cm^{-2} (#6505). These can be compared with respective carrier densities, obtained from Hall measurements, of 4.5×10^{11} cm^{-2} (#6639 at 4.2K) and 5.2×10^{11} cm^{-2} (#6505 at 77K). The discrepancies between Hall and Plasmon measurements of carrier concentration are most probably due to differences in temperature and illumination.

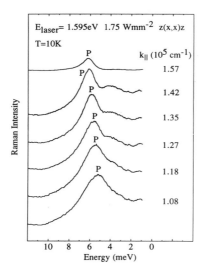

Fig. 11. Raman spectra of the single intrasubband plasmon mode of the heterojunction, as a function of in-plane wave vector $k_{\parallel}$.

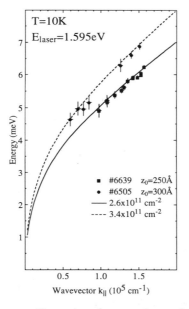

Fig. 12. Experimental plasmon dispersion for two heterojunctions #6639 and #6505. Plasmon modes calculated to fit the experimental points make possible a determination of sheet carrier density.

As the dispersions determined from Raman scattering match perfectly the expected square-root-like behaviour, we can infer that there is little or no parallel conduction present in the AlGaAs barrier. The plasmon dispersion obtained from an asymmetric quantum well which we studied separately (results not shown here) was at a much lower energy than expected and did not follow the expected behaviour for a single 2DEG. This was found to be due to the presence of a second quasi-2DEG of large carrier density in the AlGaAs barrier which, due to the Coulomb interaction, resulted in the plasmon mode being reduced in energy and having a more acoustic-like behaviour. Thus Raman determination of the plasmon dispersion is a sensitive probe for parallel conduction.

NON-PERSISTENT OPTICAL CONTROL OF ELECTRON DENSITY

It has been shown[34,35] that the carrier concentration of the 2DEG at a modulation-doped heterointerface can be reduced with increasing intensity of illumination above the AlGaAs barrier band-gap. This effect, as explained by Chaves *et al*,[34] is due to the photo-excitation of electron-hole pairs across the AlGaAs band gap. We illustrate the mechanism schematically in Fig. 13. Following the creation of the electrons and holes (1), they are separated by the built-in electric field, the electrons residing in the Si-doped AlGaAs, whereas the holes move to the more favourable potential in the GaAs (2) where they recombine with electrons from the quasi-2DEG, thus reducing the 2DEG concentration (3). The effective transfer of electrons from the 2D-channel to the Si donors shifts the quasi-Fermi levels in the different regions relative to one another. In the steady state the electrons then tunnel through the barrier to the 2D-channel (4). This is a dynamic effect entirely different in origin from the well known persistent photoconductivity.

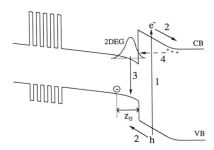

Fig. 13. Schematic diagram of conduction and valence bands of a heterojunction, illustrating the mechanism for the depletion of electrons from the 2D channel under illumination. (1) Electron-hole pairs are created across the AlGaAs band-gap. (2) Electrons and holes are separated by an in-built electric field. (3) Electrons from the 2D channel recombine with holes. (4) In the steady state, electrons tunnel through the barrier to the 2D channel.

We present here a direct observation of this phenomenon, by measuring the change of carrier concentration from the variation of the plasmon energy with varying illumination. Note that Shubnikov-de Haas measurements of this effect were reported to be unsuccessful in Ref. 34. In addition to the 'probe' beam, of energy close to the E_0 resonance, for the Raman detection of plasmons, a 'pump' spot of 647.1nm illumination was focussed onto the sample. In Fig. 14, we show Raman spectra for varying power densities of pump illumination (background luminescence due to the pump has been subtracted for clarity). A clear shift of the plasmon peak to lower energy can be observed for increasing pump power, and indeed this effect is in no way persistent, in accordance with the model proposed by Chaves *et al*. In Fig.15 we plot the change in carrier concentration versus power density, as

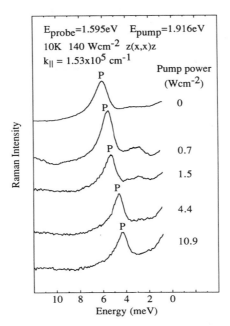

Fig. 14. Raman spectra of the plasmon mode P in the heterojuction for different pump-power densities. The in-plane wave vector $k_{||}=1.53\times10^5$ cm^{-1}. There is a strong shift to lower energy for increasing pump-power density, corresponding to a reduction in carrier concentration.

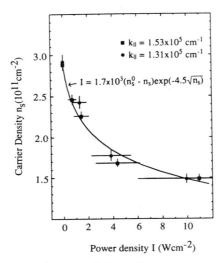

Fig. 15. Carrier concentration, determined from plasmon energy, plotted versus pump-power density I. This has been fitted using the simple model proposed in Ref. 34.

determined from the plasmon energy, using only the simple long-wavelength $k_{||}^{1/2}$ approximation of the plasmon dispersion - this will overestimate the determination of carrier density by up to 10%. Power densities have been corrected to account for reflection and change of sample angle with pump illumination when varying $k_{||}$.

From the model of Ref. 34, we can relate the pump-power density I to the carrier concentration n_s:

$$I = A \, (n_s^0 - n_s) \, \exp[-B n_s^{1/2}] \qquad (14)$$

The rate of return of electrons from the neutralized donors will be proportional to the pump-power density. It will also vary linearly with the number of electrons out of equilibrium and exponentially with the square root of the barrier height seen at the energy of the donor level. This barrier height is taken to be proportional to $n_s^{1/2}$. Using equation (14), we have obtained a fit to the experimental variation of carrier concentration with power density.

CONCLUSIONS

We have shown that electronic Raman scattering measurements of the plasmon dispersion, in conjunction with calculations of the dielectric response, can be used to determine the electronic structure of complicated quasi-2D systems that are difficult to characterize from Shubnikov-de Haas measurements. Moreover, this method is contactless and so should be of great use for topographic characterization and for the characterization of novel one-dimensional structures, where sample geometry may limit the use of methods employing contacts. For a single electron sheet, this light scattering characterization technique can be used to determine the carrier density of the quasi-2DEG within an error of 2×10^{10} cm^{-2}. Using a Raman determination of carrier concentration from plasmon energy, we have observed directly an optically controlled, charge-transfer process that is different from the persistent photoconductivity effect.

ACKNOWLEDGEMENTS

We would like to thank R. D. King-Smith and B. Jusserand for valuable discussions and support. This work was supported by the SERC (United Kingdom).

REFERENCES

* Present address: Hitachi Cambridge Laboratory, Hitachi Europe Ltd.,
Cavendish Laboratory, Madingley Road, Cambridge CB3 0HE, United Knigdom.

1. S. Das Sarma and J.J. Quinn, Phys. Rev. B **25**, 7603 (1982).
2. W. L. Bloss, J. Vac. Sci. Tech. A **2**, 519 (1983).
3. S. Das Sarma, Phys. Rev. B **29**, 2334 (1984).
4. J. K. Jain and P. B. Allen, Phys. Rev. Lett. **54**, 2437 (1985).
5. J. K. Jain and P. B. Allen, Phys. Rev. B **32**, 997 (1985).
6. J. K. Jain and S. Das Sarma, Surf. Sci. **196**, 466 (1988).
7. G. Fasol, R. D. King-Smith, D. Richards, U. Ekenberg, N. Mestres and K. Ploog, Phys. Rev. B **39**, 12695 (1989).
8. F. Stern, Phys. Rev. Lett. **18**, 546 (1967).
9. D. Olego, A. Pinczuk, A. C. Gossard and W. Wiegmann, Phys. Rev. B **25**, 7867 (1982).
10. A. Pinczuk, M. G. Lamont and A. C. Gossard, Phys. Rev. Lett. **56**, 2092 (1986).
11. G. Fasol, N. Mestres, H. P. Hughes, A. Fischer and K. Ploog, Phys. Rev. Lett. **56**, 2517 (1986).
12. A. F. J. Levi, J. R. Hayes and R. Bhat, Appl. Phys. Lett. **48**, 1609 (1986).

13. G. Abstreiter, this volume.
14. T. Zettler, C. Peters, J. P. Kotthaus and K. Ploog, Phys. Rev. B **39**, 3931 (1989).
15. T. Egeler, S. Beeck, G. Abstreiter, G. Weimann and W. Schlapp, Superlattices and Microstructures **5**, 123 (1988)
16. D. Pines, 'Elementary Excitations in Solids', Benjamin, New York (1963).
17. G. Bastard, Phys. Rev. B **24**, 5693 (1981).
18. R. D. King-Smith and J. C. Inkson, Phys. Rev. B **33**, 5489 (1986).
19. T. Ando, Z.Physik B **26**, 263 (1977).
20. S. Katayama and T. Ando, J. Phys. Soc. Jap. **54**, 1615 (1985).
21. P. Hawrylak, J-W. Wu and J. J. Quinn, Phys. Rev. B **32**, 5169 (1985).
22. T. Ando, A. B. Fowler and F. Stern, Rev. Mod. Phys. **54**, 473 (1987).
23. U. Ekenberg, Surf. Sci. **229**, 419 (1990).
24. M. Altarelli, Phys. Rev. B **28**, 842 (1983).
25. O. Gunnarsson and B. I. Lundqvist, Phys. Rev. B **13**, 4274 (1976).
26. D. J. BenDaniel and C. B. Duke, Phys. Rev. **152**, 683 (1966).
27. S. Das Sarma and A. Madhukar, Phys. Rev. B **23**, 805 (1981).
28. U. Fano, Phys. Rev. **124**, 1866 (1961).
29. D. Richards, G. Fasol and K. Ploog, Appl. Phys. Lett. **56**, 1649 (1990).
30. J. C. Maan, Private communication
31. F. Stern and S. Das Sarma, Phys. Rev. B **30**, 840 (1984).
32. B. Jusserand, D. R. Richards, G. Fasol, G. Weimann and W. Schlapp, Surf. Sci. **229**, 394 (1990).
33. I. V. Kukushkin, K. von Klitzing, K. Ploog and V. B. Timofeev, Phys. Rev. B **40**, 7788 (1989).
34. A. S. Chaves, A. F. S. Penna, J.M. Worlock, G. Weimann and W. Schlapp, Surf. Sci. **170**, 618 (1986).
35. B. Jusserand, J. A. Brum, D. Gardin, H. W. Liu, G. Weimann and W. Schlapp, Phys. Rev. B **40**, 4220 (1989).

MICRO-RAMAN SPECTROSCOPY FOR LARGE IN-PLANE WAVE

VECTOR EXCITATIONS IN QUANTUM-WELL STRUCTURES

G. Abstreiter, S. Beeck, T. Egeler, and A. Huber

Walter Schottky Institut
Techn. Univ. München
D-8046 Garching

INTRODUCTION

Inelastic light scattering has been used extensively in the past ten years to study phonon properties and electronic excitations in heterostructures, quantum wells and superlattices.[1] Most of the experiments were performed in backscattering geometry normal to the layers or interfaces. This limits the possibilities of exciting in-plane excitations. Due to the high refractive index of GaAs the accessible in-plane wave vector is $q_{\parallel} \lesssim 1.7 \times 10^5 \, \text{cm}^{-1}$ in the usual backscattering geometry with the exciting laser wavelength in the red spectral range. The dispersion of plasmons in layered 2D-electron systems has been studied, for example, in this limited wave-vector range.[2] A first extension of the in-plane wave-vector was achieved by a 90° scattering geometry.[3] The dispersion of interface phonon-polaritons in a GaAs/AlAs heterostructure has been studied by Nakayama et al.,[4] also in a quasi-backscattering geometry from the (100) surface. Otherwise excitations with large in-plane wave-vectors have been observed under resonance conditions due to wave vector non-conservation.[5,6,7] This, however, does not allow the extraction of detailed information on the dispersion of in-plane excitations.

Recently the accessible wave-vector range was extended by two new light-scattering techniques, namely micro-Raman spectroscopy [8,9] and grating-assisted Raman scattering.[9,10,11] Also, first results on wire-structured electron systems in GaAs quantum wells have been achieved.[9,12,13] In the present contribution we concentrate on the application of micro-Raman scattering to in-plane excitations in GaAs quantum-well structures. After the description of the experimental technique, which also allows low-temperature experiments with high spatial resolution, we present the first results on the dispersion of interface phonons in GaAs/AlAs super-lattices. Finally we briefly discuss single-particle and plasmon excitations in modulation-doped multi-quantum wells excited with in-plane scattering wave vectors up to $7 \cdot 10^5 \, \text{cm}^{-1}$.

Light Scattering in Semiconductor Structures and Superlattices
Edited by D.J. Lockwood and J.F. Young, Plenum Press, New York, 1991

561

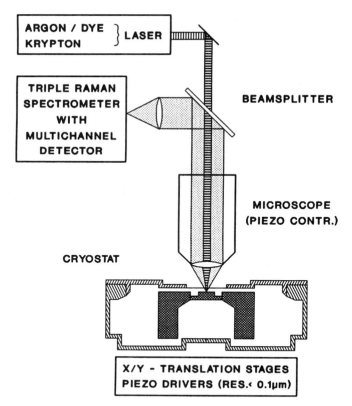

Fig. 1. Schematic experimental set-up of the micro-Raman spectrometer.

EXPERIMENTAL SET-UP

The micro-Raman spectrometer is shown schematically in Fig. 1. It consists of ion lasers (Argon, Krypton) and a dye laser which is used especially for resonant excitation, a triple grating spectrometer with multichannel detection, a small He flow-through cryostat where the sample can be mounted on a cooled copper block close to the window, and a microscope which is used to focus the exciting laser on the sample. The high-resolution microscope objective has a working distance of 8 mm. It is mounted on a high-precision, piezo-controlled translation stage in order to achieve a well defined focal position. The cryostat with the sample is connected to x/y-translation stages with piezo drivers (positioning accuracy $< 0.1\ \mu m$) which enables us to scan the sample in the submicron range. It is also possible to mount the sample on this stage at room temperature without the cryostat, which leads to a slightly better resolution because the extra cryostat window is then removed. The base sample temperature achievable in the cryostat is about 5 K, which allows resonant electronic light scattering. The Raman scattered light is collected by the same microscope objective and reflected by a beamsplitter into the spectrometer, which contains three gratings to obtain a strong reduction in stray laser light and higher spectral resolution. A Peltier cooled multichannel detector allows fast data acquisition. We have measured the spatial resolution by focusing the laser onto the cleavage plane of a p-n heterojunction GaAs/(AlGa)As structure. Monitoring the induced local photo-current while shifting the heterojunction across the focused laser gives a direct image of the laser spot profile. Fig. 2 shows spot

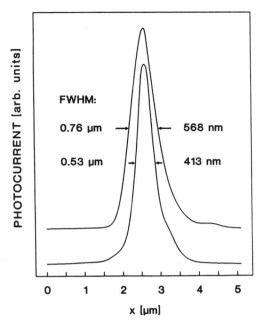

Fig. 2. Laser spot profiles in dependence on the
wavelength for a microscope objective
with high resolution (N.A. = 0.75).

profiles for two laser wavelengths. The small dimensions of this detector also
permit us to examine the influence of the cryostat window on the spatial resolution, which is found to be negligible for window thicknesses of 0.2 mm. This
high resolution of the micro-Raman set-up enables us to apply new scattering
geometries like backscattering from the (110) cleavage plane of GaAs/AlAs superlattices grown along (100). Selected examples are discussed in the next section.
It also allows us to analyze with high spatial resolution various properties, like
temperature under operation or built-in strain, of devices. Such experiments have
been reported, for example, in Refs. 15,16,17.

INTERFACE PHONONS IN GaAs/AlAs SUPERLATTICES

Semiconductor superlattices show a much larger number of phonon branches
as compared to the constituent materials due to the enlarged unit cell. For superlattices made of semiconductors with non-overlapping bulk optical phonon dispersions like GaAs and AlAs, the optical phonons are confined within one of the two
material layers. Due to the long-range macroscopic electric fields generated by
optical phonons in polar semiconductors, there appear in addition optical modes
which extend with decreasing amplitude from the interfaces into both neighbouring
layers. These interface (IF) phonons show, in contrast to the confined modes, a
large energy dispersion which is anisotropic with respect to the momentum direction, relative to the superlattice axis. Their energies are in the range between the
TO and LO phonon frequencies of the two bulk materials. The first observation of

IF phonons in short-period GaAs/AlAs superlattices by Raman spectroscopy revealed a broad peak filling almost the whole range between the TO and LO phonon bulk frequencies, especially in the AlAs frequency region.[5,6] This was explained by a resonantly enhanced impurity-induced breakdown of momentum conservation. The simultanous excitation of IF modes with many different momenta and their strong dispersion lead to the observed broad signals. Momentum conservation can be obtained by non-resonant excitation. In this case the possibilities of measuring the characteristic anisotropic IF phonon dispersion in backscattering from the sample surface are very limited due to the small values of in-plane momentum which can be achieved by changing the angle of incidence. An attempt to measure the dispersion in a GaAs/(AlGa)As superlattice using different laser excitation wavelengths provided only qualitative results.[18] The new scattering geometries which can be realized by micro-Raman spectroscopy are ideally suited for measuring the IF mode dispersion as they allow rotation of the scattering wave vector from parallel to perpendicular to the superlattice axis.

We have investigated a series of GaAs/AlAs superlattices grown on GaAs substrates with a common width of the GaAs layers of 100 Å and AlAs layer thicknesses ranging from 25 Å to 200 Å.[14] The number of periods varies from 50 to 90 to keep the total superlattice thickness at about 1 μm. Here we present, as an example, the results obtained from a sample with 75 Å wide AlAs layers. Fig. 3 shows Raman spectra for different values of momentum transfer parallel to the superlattice layers. The corresponding backscattering geometries are shown schematically in the insets. The spectra are excited at room temperature using the 568.2 nm Krypton laser line. The polarizations of incident and scattered light are parallel. For crossed polarization the superlattice phonons are much weaker or even absent. The dashed lines mark the bulk TO and LO phonon frequencies of GaAs and AlAs. At these energies phonons are observed in accordance with the selection rules for the different geometries. In addition, however, we observe forbidden GaAs LO phonons in backscattering from the (110) cleavage plane. This disappears if we shift the laser spot from the superlattice to the substrate region. We do not expect to observe a frequency shift due to the confinement as the layer thicknesses are rather large. The forbidden GaAs LO phonon, however, is a direct consequence of the confinement. The strength of the signal indicates that our laser spot is well focused in the range of the superlattice thickness of about 0.9 μm. This breaking of the selection rules in a superlattice was previously reported by J. Zucker et al.[19] in a 90° scattering experiment from a GaAs/(AlGa)As superlattice. It was explained by B. Jusserand and M. Cardona[20] in the following way: in our case, for parallel polarization, the selection rules permit the excitation only of phonons polarized in the (001) direction. The confined modes of the superlattice have quantized effective momenta along the superlattice axis which are integral multiples of π/d, where d is the layer thickness. In our experiment this momentum is larger than those transferred parallel to the layers in the scattering process. It therefore determines the LO character of the excited phonon.

For a finite in-plane momentum there appear two additional peaks (marked by arrows) between the TO and LO phonon energies of GaAs and AlAs, respectively. With increasing in-plane momentum they shift away from the bulk frequencies. Therefore these modes are attributed to IF phonons. For a quantitative comparison of the observed dispersion with theory we use the well known dielectric continuum model which describes the optical phonons using the macroscopic electric fields they generate in polar semiconductors. Confined and interface phonons are obtained

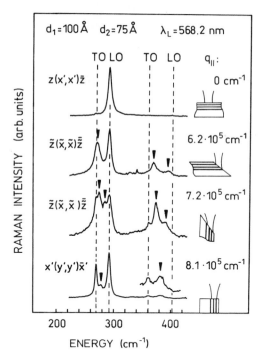

Fig. 3. Raman spectra of a GaAs/AlAs superlattice for different values of in-plane momentum transfer $q_{\parallel}$. The corresponding backscattering geometries are shown schematically in the inset. The IF modes are marked by arrows.

as solutions of the Maxwell equation $\mathrm{div}(\varepsilon E) = 0$. The superlattice structure is incorporated in the frequency-dependent bulk dielectric constant ε, which is spatially modulated with the period of the superlattice according to the different materials GaAs and AlAs. The dispersion of the bulk TO and LO phonon frequencies which enter into the dielectric constant is neglected for the calculation of interface phonons. The IF modes are then obtained from the implicit equation:[21]

$$\cos q_z(d_1 + d_2) = \cosh(q_{\parallel}d_1)\cosh(q_{\parallel}d_2) + 0.5(\varepsilon_1/\varepsilon_2 + \varepsilon_2/\varepsilon_1)\sinh(q_{\parallel}d_1)\sinh(q_{\parallel}d_2)$$

where the indices 1 and 2 distinguish the GaAs and AlAs layers. For each phonon momentum four IF modes are obtained, two of them situated in the energy range between the TO and LO phonons of GaAs and AlAs, respectively, as observed in the experiment. In Fig. 4 we present the calculated and the measured interface phonon

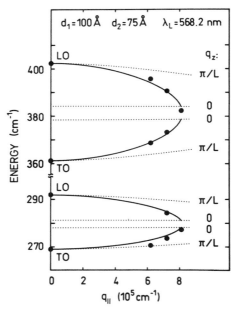

Fig. 4. Comparison of the measured IF mode dispersion (points) with the predictions of the dielectric continuum model. The allowed IF phonon bands extend between the dotted lines which give the dispersion for $q_z = 0$ and $q_z = \pi/L$, where $L = d_1 + d_2$ is the superlattice period. The solid line is the calculated dispersion for values $q_z = \sqrt{(4\pi n/\lambda)^2 - q_\parallel^2}$ as determined by the scattering geometry.

dispersions as functions of the in-plane momentum $q_\parallel$. The dotted lines give the calculated dispersion of the IF modes for a momentum q_z, perpendicular to the layers, equal to 0 and π/L, where L is the superlattice period. The IF phonon branches for intermediate values of q_z fill the energy range between these border lines. For the IF modes observed in the experiment, q_z is given by $q_z = \sqrt{(4\pi n/\lambda)^2 - q_\parallel^2}$. The solid line is the calculated dispersion for these momentum values, using the measured bulk phonon frequencies and the thicknesses of the layers, as determined from TEM images, as input. The energies of the observed IF modes, represented by points, are in good agreement with the calculated dispersion.

SINGLE-PARTICLE AND COLLECTIVE ELECTRONIC EXCITATIONS

The possibilities opened up by micro-Raman spectroscopy have also been exploited to investigate the characteristic in-plane momentum dependence of plasmons and single-particle excitations (SPE) in a layered two-dimensional electron gas (2DEG). In this case the low-temperature facility is indispensable for reducing electron scattering. The energy of a plasmon in a single 2DEG plane depends on the square root of the in-plane momentum. In a layered system the Coulomb coupling

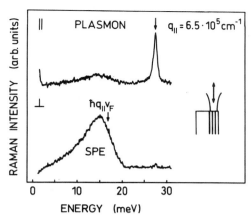

Fig. 5. Low temperature Raman spectra of a modulation-doped GaAs/(AlGa)As multi quantum well
structure, obtained in backscattering from the
(110) cleavage plane of the sample. The upper
spectrum shows the plasmon of the layered
electron gas, excited with parallel polarization
of incident and scattered light. For crossed
polarization in the lower spectrum the single
particle excitation is observed. It shows the
typical line shape for a 2DEG.

of modes in neighbouring planes results in a splitting of the 2D plasmon into a
plasmon band, depending on the number of layers. For an infinite system the 2D
plasmon energy is multiplied by a structure factor, which depends on the momentum of the excitation perpendicular to the layers.[22] Fig. 5 shows resonantly excited
Raman spectra of a modulation-doped GaAs/(AlGa)As multi-quantum-well structure
obtained in backscattering from the (110) cleavage plane. The collective plasmon
excitation is obtained with parallel polarization of incident and scattered light in
the upper spectrum, while the SPE in the lower spectrum requires crossed polarizations. As with IF phonons, only the plasmon branch is observed, which allows for
approximate conservation of the momentum perpendicular to the layers. Using backscattering from polished sample edges and different angles of incidence, the plasmon
dispersion was measured[8,11] from the range of quasi-2D behaviour for maximum
in-plane momentum down to the range of acoustic behaviour, investigated in previous experiments.[2] Good agreement with the calculated dispersion was obtained.
Due to the large in-plane momentum the quasi-2D plasmon in Fig. 5 is shifted to
energies large enough that the coupling to the LO phonons becomes appreciable.
A phonon-like coupled mode was observed, which was clearly shifted relative to
the LO phonon energy. SPE directly reflect the dielectric properties of a 2DEG.
There is no coupling between the electron gas planes because no charge density
fluctuations are generated. For the large in-plane momentum the 2D character of

the electron gas shows up directly in the line shape of the SPE spectrum. The peak is much sharper than in the 3D case due to the indentation of the low-energy flank which increases linearly for the 3D case. The SPE offers the possibility of determining the actual temperature of the electron gas from a fit of the spectrum to the Lindhard-Mermin dielectric function. An almost perfect fit is obtained for a temperature of 10 K. SPE spectra obtained in the different scattering geometries, along with calculated spectra, are discussed in Ref. 9.

CONCLUDING REMARKS

We have shown that micro-Raman spectroscopy allows the investigation of in-plane excitations in semiconductor heterostructures. The accessible wave-vector range is extended considerably. The high spatial resolution also makes it possible to perform optical studies of single micro-fabricated quantum wires and quantum dots. We expect stimulating new results in this field in the near future.

ACKNOWLEDGEMENTS

It is a pleasure to thank W. Ettmüller, H. Rothfritz, G. Tränkle, G.Weimann, and W. Schlapp, who provided the excellent samples necessary for this work. We acknowledge financial support from the Deutsche Forschungsgemeinschaft and from Siemens AG.

REFERENCES

1. For a review see "Light Scattering in Solids V, Superlattices and Other Microstructures," Topics in Applied Physics 66, M. Cardona and G. Güntherodt, eds., Springer-Verlag, Berlin Heidelberg (1989).
2. D. Olego, A. Pinczuk, A. C. Gossard, and W. Wiegmann, Phys. Rev. B 26, 7867 (1982).
3. R. Sooryakumar, A. Pinczuk, A. C. Gossard, and W. Wiegmann, Phys. Rev. B 31, 2578 (1985).
4. M. Nakayama, M. Ishida, and N. Sano, Phys. Rev. B 38, 6348 (1988).
5. R. Merlin, C. Colvard, M.V. Klein, H. Morkoc, A. Y. Cho, and A. C. Gossard, Appl. Phys. Lett. 36, 43 (1980).
6. A. K. Sood, J. Menedez, M. Cardona, and K. Ploog, Phys. Rev. Lett. 54, 2115 (1985).
7. A. Pinczuk, J. P. Valladares, D. Heitmann, A. C. Gossard, J. H. English, C. W. Tu, L. N. Pfeiffer, and K. West, Phys. Rev. Lett. 61, 2701 (1988).
8. T. Egeler, S. Beeck, G. Abstreiter, G. Weimann, and W. Schlapp, Superlattices and Microstructures 5, 123 (1989).
9. G. Abstreiter and T. Egeler, Proceedings of the 6 Int. Winterschool, Mauterndorf (1990), Springer Series in Solid State Sciences (in press).
10. T. Zettler, C. Peters, J. P. Kotthaus, and K. Ploog, Phys. Rev. B 39, 3931 (1989).
11. T. Egeler, G. Abstreiter, G. Weimann, T. Demel, D. Heitmann, and W. Schlapp, Surface Science 229, 391 (1990).
12. T. Egeler, G. Abstreiter, G. Weimann, T. Demel, D. Heitmann, P. Grambow, and W. Schlapp, Phys. Rev. Lett. 65, 1804 (1990).
13. J. S. Weiner, G. Danan, A. Pinczuk, J. Valladares, L. N. Pfeiffer, and K. West, Phys. Rev. Lett. 63, 1641 (1989).

14. A. Huber, T. Egeler, W. Ettmüller, H. Rothfritz, G. Tränkle, G. Weimann, and G. Abstreiter, Superlattices and Microstructures (in press), see also: A. Huber, Diplomthesis, Techn. Univ. Munich (1990).
15. S. Beeck, T. Egeler, G. Abstreiter, H. Brugger, P. W. Epperlein, D. J. Webb, C. Hanke, C. Hoyler, and L. Korte, in "ESSDERC '89," A. Heuberger, H. Ryssel, and P. Lange, eds., Springer, Berlin (1989) p. 508.
16. K. Brunner, G. Abstreiter, B. O. Kolbesen, and H. W. Meul, Appl. Surface Science 39, 116 (1989).
17. H. Brugger, this volume; see also S. Nakashima, this volume.
18. A. K. Arora, A. K. Ramdas, M. R. Melloch, N. Otsuka, Phys. Rev. B 36, 1021 (1987).
19. J. E. Zucker, A. Pinczuk, D. S. Chemla, A. Gossard, W. Wiegmann, Phys. Rev. Lett. 53, 1280 (1984).
20. B. Jusserand and M. Cardona, in ref. 1, p. 49.
21. R. E. Camley and D. L. Mills, Phys. Rev. B 29, 1695 (1984).
22. A. L. Fetter, Ann. Phys. (N.Y.), 88, 1 (1974).

INELASTIC LIGHT SCATTERING BY THE HIGH MOBILITY

TWO-DIMENSIONAL ELECTRON GAS

A. Pinczuk,[1] D. Heiman,[2] S. Schmitt-Rink,[1] C. Kallin,[3]
B. S. Dennis,[1] L. N. Pfeiffer,[1] and K. W. West[1]

(1) AT&T Bell Laboratories, Murray Hill, NJ 07974, USA
(2) MIT-National Magnet Laboratory, Cambridge, MA 02139, USA
(3) Dept. of Physics, McMaster Univ., Hamilton, ON L8S 4MI, Canada

ABSTRACT

We present a review of inelastic light scattering studies of two-dimensional electron systems in GaAs-AlGaAs heterostructures of ultrahigh electron mobility. In the case of intersubband excitations, between different confined states, the spectra show single-particle excitations in addition to collective spin-density and charge-density excitations. These results reveal exchange Coulomb interactions larger than previously anticipated. Light scattering spectra by large wave vector inter-Landau-level excitations of the 2D electron gas in a high perpendicular magnetic field display the excitonic binding and magnetoroton minima in the mode dispersion that are predicted by Hartree-Fock theories.

INTRODUCTION

Semiconductor heterojunctions, quantum wells and superlattices display new phenomena that appear because of the reduction in dimensionality of the electron states. A collection of free carriers in these artificial semiconductor structures has behaviors in common with those of an ideal two-dimensional (2D) electron gas.[1] The greatly enhanced carrier mobilities in modulation doped heterostructures stimulates research on the electron gas that is at the frontiers of condensed-matter physics.[2,3] In ultra-high mobility systems the combined effects of reduction in dimensionality and fundamental electron-electron interactions have direct manifestations in the energies and dispersions of the elementary excitations and give rise to remarkable phenomena like the fractional quantum Hall effect.

Resonant inelastic light scattering is a powerful method to study the elementary excitations of the 2D electron gas in semiconductors.[4,5] Interest in such light scattering studies was stimulated by a proposal of Burstein et al. presented at the 1978 International Conference on the Physics of Semiconductors.[6] This work considered results of resonant light scattering from bulk n-GaAs,[7,8] and pointed out that with resonant enhancement it should be possible to measure spectra of elementary excitations of free electrons confined in semiconductor space-charge layers. The first observations of resonant inelastic light

Light Scattering in Semiconductor Structures and Superlattices
Edited by D.J. Lockwood and J.F. Young, Plenum Press, New York, 1991

571

scattering from 2D electron systems in modulation doped GaAs-AlGaAs heterostructures were reported in 1979.[9,10]

Some of the most important applications of the light scattering method follow from polarization selection rules that allow separate measurements of the spectra of collective spin-density and charge-density excitations.[4-6,11] Recent work has also demonstrated that in ultrahigh mobility 2D electron systems single-particle excitations are also active in resonant light scattering spectra.[12] This unexpected result leads to unique quantitative determinations of the spacings between confined energy levels and fundamental electron-electron interactions. In the case of intersubband excitations the light scattering measurements have shown that exchange Coulomb interactions are larger than anticipated from previous work.

The case of the 2D electron gas in a strong perpendicular magnetic field is of great current interest. Here, the unique properties of the electron gas arise from quantization of the 2D kinetic energy states into discrete Landau levels in conjunction with long range electron-electron interactions. In principle, these interactions can be studied by measurements of elementary excitations associated with transitions between Landau levels.[13-17] However, the relevant range of wave vectors, $q > 10^6$ cm^{-1}, is not easily accessible in optical experiments. The resonant inelastic light scattering method may play an important role in these studies because the observation of large wave vector inter-Landau-level excitations was reported in spectra of the 2D electron gas in GaAs quantum wells.[18-20] The electron-electron interactions determined in these experiments are in excellent agreement with the predictions of current theories.

This paper presents a short review of the recent resonant inelastic light scattering studies of the ultra-high mobility 2D electron gas in modulation doped GaAs-AlGaAs quantum wells. In the next section we discuss the kinematics and selection rules of inelastic light scattering. The following three sections consider studies of intersubband excitations, the spectroscopy of inter-Landau-level excitations and present our concluding remarks.

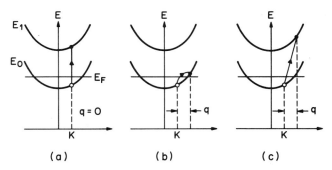

Fig. 1.   Single-particle transitions of the 2D electron gas in a quantum well. The two lowest subbands are shown. (a) Vertical intersubband transition. (b) Intrasubband transition. (c) Non-vertical intersubband transition. After Ref. 5.

LIGHT SCATTERING SELECTION RULES AND MECHANISMS

Figure 1 represents the transitions of free electrons between the single-particle states of a quasi-2D system in the absence of an external magnetic field. In this representation each of the confined states is associated with a subband in 2D wave vector space. Figure 1(a) shows a vertical intersubband transition between the lowest two subbands. For parallel subbands the transition energy is $E_{01} = E_1 - E_0$ independent of the in-plane wave vector $\vec{K}$. Figure 1(b) displays an intrasubband transition with an in-plane wave vector transfer q. Such transitions have a continuum of energies $\hbar^2(\vec{K} \cdot \vec{q} + q^2/2)/m^*$, where m^* is the effective mass of the carriers. In Fig. 1(c) we show a non-vertical intersubband transition. These excitations have a continuum of energies bounded by $E_{01} \pm qv_F$, where v_F is the Fermi velocity of the carriers. In a large perpendicular magnetic field all the transitions are discrete. They are intersubband transitions, inter-Landau-level transitions and combined transitions in which there is simultaneous change of Landau level and subband state.

The transitions shown in Fig. 1 can be classified according to the angular momenta of the electron-hole pair states. For electrons in conduction states of GaAs the orbital angular momentum is zero. In this case the particle-hole states in the transitions are singlets (J = 0) or triplets (J = 1). The singlets are charge-density excitations and the triplets are spin-density excitations. Within the effective-mass approximation the selection rules that apply in resonant light scattering by the 2D electron gas are similar to those of the bulk semiconductors.[6,20] Charge-density excitations are active in polarized spectra in which the incident and scattered light polarization are parallel. Spin-density excitations are active in depolarized spectra with orthogonal light polarizations.[11] These characteristic light scattering selection rules make possible unique determinations of electron-electron interactions.[4-6]

The observation of light scattering from low density systems requires large resonant enhancements of the intensities. These resonances occur at photon energies near the optical transitions between the states that contribute to the effective mass of the free carriers.[4-6] Much of the work in the field has been described by optical processes in which the photons couple directly to states of the Fermi sea.[4-6,21] This description explains experimental results from modulation doped GaAs quantum wells obtained with photon energies close to the $E_o + \Delta_o$ energy gap (~1.9 eV). A distinctive feature of this light scattering mechanism is the width of the profile of resonant enhancement, which is comparable to the Fermi energy.

In recent work carried out in high mobility single quantum wells the enhancement profiles show sharp and strong peaks which are characteristic signatures of excitonic resonances.[12,20,22] Figure 2(a) shows results for the profile of resonant enhancement of the lowest (32 meV) spin-density intersubband excitation. These results were obtained in a single modulation doped quantum well of width 200 Å.[22] They cover the energy range between 1.54 and 1.61 eV, which corresponds to optical transitions of above bandgap excitons of the GaAs quantum well. Surprisingly sharp peaks are seen when the incident (incoming resonance) or the scattered (outgoing resonance) photon energies are equal to the energies of excitons measured in photoluminescence-excitation (PLE) spectra. In this figure, the notation $E_n H_m (E_n L_m)$ refers to the exciton formed in the nth conduction subband and a heavy (light) hole in the mth valence subband. Figure 2(b) shows a schematic diagram of the one-side modulation doped single quantum well and the low-lying conduction and valence energy levels. In Fig. 2(a) the strong peak at 1.572 eV corresponds to a resonant enhancement when the laser photon energy is equal to that of the

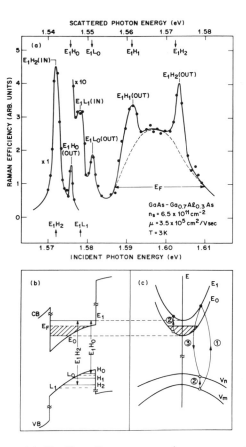

Fig. 2. (a) Profile of resonant enhancement of light scattering by the spin-density intersubband excitation. (b) Schematic structure of the quantum well. (c) Diagram of a resonant light scattering process. After Ref. 22.

E_1H_2 exciton. The associated outgoing resonance is observed at 1.604 eV. The E_1L_1 incoming resonance is also visible but much weaker. The other sharp structures in Fig. 2(a) are the outgoing resonances at E_1H_0, E_1L_0, and E_1H_1. The broad band centered at 1.595 eV is assigned to the conventional light scattering mechanisms with no excitonic intermediate states.

To explain the unexpected sharp incoming and outgoing resonances Danan et al.[22] proposed new inelastic light scattering processes by the electron gas. They are third-order processes in which the two optical transitions are associated with intermediate exciton states. Light scattering takes place because of coupling between the intermediate excitonic states and the Fermi sea. The exchange term of the Coulomb interactions are involved in light scattering by spin-density excitations. In the case of charge-density excitations the direct terms are involved. To consider the salient features of the new mechanisms it is convenient to describe the excitonic states as superpositions of electron-hole pair states. This simplifies the description of exciton scattering by the electron gas as single events in conduction or valence states. Figure 2(c) shows the sequence of three transitions for the case in which the hole in the valence subband makes the transition due to the Coulomb interaction. Danan et al.[22] have discussed the relevant features of this mechanism and have shown that it predicts sharp incoming and outgoing resonances, as found in the experiments. It was also pointed out that, as in the case of Raman scattering by optical phonons,[23] the differences in the intensities of incoming and outgoing resonances give information on the dominant contributions to the inelastic light scattering processes.

SPECTROSCOPY OF INTERSUBBAND EXCITATIONS

The capability to obtain separate spectra of spin-density and charge-density excitations has been a major factor in the impact of the light scattering method in research of the physics of the 2D electron gas in semiconductors. The energies of charge-density and spin-density excitations, the collective modes of the electron gas, are shifted by Coulomb interactions. Spin-density modes are shifted downwards by the exchange terms and the charge-density excitations are shifted upwards from the spin-density modes by the direct (or Hartree) terms. The light scattering determinations of the energies of the two collective modes, in conjunction with the measurements of the energies of single-particle excitations,[12] gives unique quantitative insights into the strengths of electron-electron interactions of 2D systems.

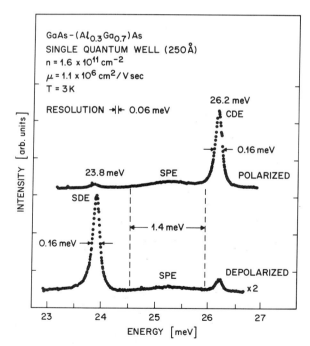

Fig. 3.   Light scattering spectra of vertical intersubband excitations. The peaks of collective spin-density excitations (SDE), collective charge-density excitations (CDE), and single-particle excitations (SPE) are shown. After Ref. 12.

The spectroscopy of intersubband excitations is an interesting example of the applications of the method in studies of electron-electron interactions. Because exchange interactions were expected to be small in GaAs,[1,6,11,24,25] the spin-density excitations were assumed to give single-particle transition energies.[4,5] However, recent work has shown that for intersubband excitations exchange interactions are more important than had been anticipated.[12] These experiments are carried out in single GaAs quantum wells of ultra-high electron mobility. The spectra are measured with a conventional backscattering geometry shown in the inset to Fig. 4. This geometry allows changes in the in-plane component of the scattering wave vector by changing the angle of incidence. For this experimental configuration we expect the excitations derived from the intersubband transitions shown in Figs. 1(a) and 1(c). The incident photon energies are resonant with sharp excitonic transitions of the GaAs quantum wells as discussed in the previous section.

Figure 3 shows high-resolution (0.06 meV) spectra in which the incident light was focussed to a spot about 50 μm in radius to minimize effects of inhomogeneous broadening. Typical incident power densities were about 1 W/cm². Optical multichannel detection was used to measure the relatively low-level signals. The sharp peaks in Fig. 3 are, to the best of our knowledge, the narrowest optical features reported for the electron gas in GaAs. The spectral lineshapes are not changed when the illuminated area is increased, indicating that inhomogeneous broadening is not significant in these measurements. The well-defined polarization selection rules identify the two peaks as due to collective spin-density (SDE) and charge-density (CDE) intersubband excitations.

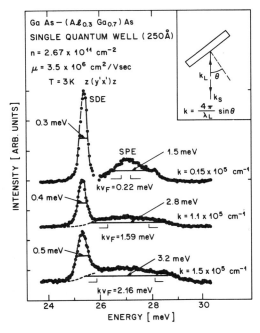

Fig. 4. Depolarized light scattering spectra of intersubband excitations for various scattering wavevectors k. Inset: scattering geometry and expression for k. After Ref. 12.

In addition to the SDE and CDE peaks, the spectra in Fig. 3 show unexpected bands labeled SPE. These features have been identified as single-particle excitations.[12] The assignment was made on the basis of studies of evolution of the SPE bands with changes of the in-plane scattering wave vector k. Results obtained from a sample with higher density are shown in Figs. 4 and 5. The inset to Fig. 4 gives the expression of k for the geometry of this experiment. The spectra in Fig. 4 reveal that the widths of the SPE bands have a pronounced dependence on k. The increase in the total widths of the SPE bands are comparable to $2kv_F$. Such behavior is expected for non-vertical single-particle intersubband transitions. As shown in the inset to Fig. 5, these transitions cover a continuum of energies. The assignment of the SPE features to single-particle intersubband excitations follows from the determination that at the larger values of k the widths of these bands are comparable to $2kv_F$. This assignment is also consistent with the strong Landau damping effects observed in the spectra of charge-density excitations seen in Fig. 5.

The simultaneous measurements of the energies of the two collective intersubband modes and of the spacing E_{01} between the lowest subbands (from the position of the SPE band) allow quantitative determinations of the strengths of Coulomb interactions from inelastic light scattering spectra. To carry out a quantitative analysis the light scattering intensities were written as[12]

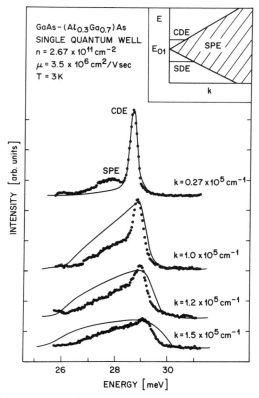

Fig. 5.   Polarized light scattering peaks of intersubband excitations at four values of the scattering wavevector. Inset: sketch of the k-dependence of long wavelength intersubband excitations. After Ref. 12.

$$I_j(q,\omega) \sim Im\; \chi_j(q,\omega) \qquad (1)$$

where $\chi_j(q,\omega)$ are the response-functions for the collective excitations. In a generalized random-phase approximation the response-functions are[24-26]

$$\chi_j(q,\omega) = \frac{\chi_0(q,\omega)}{1 - \gamma_j(q)\chi_0(q\omega)} \qquad (2)$$

where $\chi_0\;(q,\omega)$ is the intersubband susceptibility.[24-31] From Eqs. 1 and 2 we obtain the $q \sim 0$ collective mode energies

$$\omega_j^2 = E_{01}^2 + 2n\gamma_j(0)E_{01} \qquad (3)$$

In the case of spin-density excitations we have

$$\gamma_j \equiv \gamma_{SD} = -\beta_{01} \qquad (4)$$

where β_{01} is a positive parameter that enters in the exchange term of the Coulomb interaction. This term describes final-state, or excitonic, effects in intersubband excitations. For the charge-density modes we have

$$\gamma_j = \gamma_{CD} = [\alpha_{01}/\varepsilon(\omega)] - \beta_{01} \qquad (5)$$

where α_{01} accounts for the direct term and $\varepsilon(\omega)$ is the dielectric function of the polar lattice. The first term in the right side of Eq. (5) describes the depolarization field effect of intersubband excitations and their coupling to longitudinal optical phonons.[1,21,32,33]

Equations (3) - (5) provide a basis for the quantitative determination of strengths of Coulomb interactions from energies of intersubband excitations determined from inelastic light scattering spectra. This approach assumes wave vector conservation $(\vec{k}=\vec{q})$. To carry out the analysis α_{01} and β_{01} are written as

$$2n\;\alpha_{01} = \frac{\omega_{CD}^2 - \omega_{SD}^2}{E_{01}} \qquad (6)$$

$$2n\;\beta_{01} = \frac{E_{01}^2 - \omega_{SD}^2}{E_{01}} \qquad (7)$$

β_{01} and α_{01} are considered as parameters obtained from experiment.[12,34,35] The values of β_{01} and α_{01} and the ratio (β_{01}/α_{01}) in single GaAs quantum wells were recently determined as function of free electron density in the range $10^{11} < n < 5.5 \times 10^{11}$ cm^{-2}.[35] For densities $n < 3.5 \times 10^{11}$ cm^{-2} the ratio is $(\beta_{01}/\alpha_{01}) = 0.38$ independent of n, and at the largest density the ratio drops to a value of 0.23. These results suggest that exchange interactions are larger than the predictions of local-density-functional theory.[24,25] Better agreement with this theory has been reported by Gammon et al.[34]

The striking changes in the spectral lineshapes of the collective modes with changes of k were also interpreted with Eqs. (1) - (7). When k is increased the energies of the CDE overlap the continuum of single-particle intersubband excitations, as indicated in the inset to Fig. 5. At these values of k there is Landau damping due to decay of the collective modes into single-particle excitations. Landau damping explains the observed broadening of the spectra of charge-density excitations. The effect is not as marked for the spin-density excitations because these modes are further away from E_{01}. Recently the light scattering spectra have been calculated by solving numerically the Bethe-Salpeter equations including exchange interactions within the Hartree-Fock approximation.[20,36] These theoretical interpretations give excellent agreement with spectra of the collective modes. However, at this time there is no definitive interpretation for the mechanisms that lead to the appearance of the SPE bands in the light scattering spectra. Effects of residual disorder in the ultra-high mobility system could be among the mechanisms that explain the unexpected observations of single-particle excitations.

SPECTROSCOPY OF INTER-LANDAU-LEVEL EXCITATIONS

This section considers recent light scattering results that reveal the manifestations of electron-electron interactions in the dispersions of inter-Landau-level excitations. The energies of these excitations are written as

$$\omega(q,B) = \omega_c + \Delta(q,\omega) \tag{8}$$

where ω_c is the cyclotron frequency. Hartree-Fock calculations of the dispersions $\Delta(q,B)$ display characteristic "magnetoroton" minima at finite wave vectors $q > q_0 = 1/\ell_0$, where $\ell_0 = (\hbar c/eB)^{1/2}$ is the magnetic length.[14-17] The roton is due to the reduction at large wave vectors, $q \gg q_0$, of the excitonic binding between the electron in the excited Landau level and the hole in the lower Landau level.[14-17,37,38] These interactions play a leading role in the theories of elementary excitations of the incompressible fluid of the fractional quantum Hall effect. The magnetoroton minimum in the dispersion of intra-Landau level excitations, related to the gap of the fractional quantum Hall effect and the Wigner crystal is due to the excitonic attraction between fractionally charged

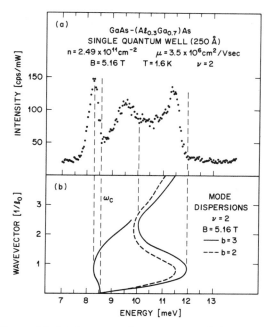

Fig. 6.   (a) Light scattering spectra of inter-Landau-level excitations.   (b) Calculated mode dispersions at $\nu = 2$ and two values of b. After Ref. 20.

quasiparticles. The strong correlations associated with the fractional quantum Hall effect are also expected to modify the dispersions of inter-Landau-level excitations.[16,17]

The observation of magnetoroton structure was reported in resonant inelastic light scattering spectra of inter-Landau-level excitations.[18,19,20] These experiments were carried out in modulation-doped GaAs-AlGaAs quantum wells and single heterojunctions. In the more interesting cases of extremely high electron mobility systems these measurements were made at integral values of the Landau level filling factor $\nu = 2\pi n \ell_0^2$. The spectra were interpreted in terms of critical points in the mode dispersions, where $(\partial\omega/\partial q) = 0$. The magnetorotons appear in the spectra as characteristic structure due to multiple critical points. These measurements are possible because of massive breakdown of wave vector conservation. The implied loss of translational invariance that allows the observation of modes with $q > q_0 \ 10^6$ cm^{-1}, much larger than k, was attributed to residual disorder.[18] This interpretation is consistent with the well-known reduction of screening of the disorder potential at the integral values of the Landau level filling factor.[39,40]

The initial results were obtained in multiple quantum wells and single heterojunctions. In the more interesting case of the ultrahigh mobility single heterojunctions comparison with theory is complicated by the overlap between inter-Landau-level and intersubband excitations. To avoid this difficulty the more recent results

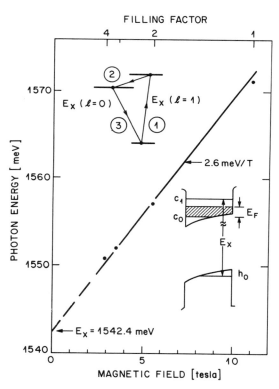

Fig. 7. $\hbar\omega_L$ of the maximum resonant enhancement. The lower inset shows B = 0 energy levels. The upper inset is a diagram of the light scattering process that explains the slope of the line and its B = 0 extrapolation. After Ref. 19.

were obtained in single quantum wells with electron mobilities comparable to those of low-disorder single heterojunctions. In these quantum wells the subband spacing can be adjusted to avoid the overlap with the cyclotron frequency.

Figure 6(a) displays results obtained at the filling factor $v = 2$. In these spectra the singlet and triplet excitations were not separated by polarization selection rules. The spectrum consist of a continuum where it is possible to identify several intensity maxima. The relative intensities of the structures in the spectra have a marked dependence on incident photon energy.[18,19] This is characteristic of spectra measured under strong and sharp resonant enhancements. Figure 6(b) shows calculated dispersions of inter-Landau-level excitations. A single mode approximation was used and two values of the finite-thickness parameter are considered.[20,41] The modes below ω_c are spin-density excitations (the $J_Z = 0$ component of the triplet) and those above are magnetoplasmons (singlet or charge-density excitations).

The results in Fig. 6 indicate that the multiple structures in the spectra are related to the critical points in the mode dispersions. The spectra in Fig. 6(a) have an onset followed by a maximum well below ω_c. This is evidence of the excitonic binding and magnetoroton minimum of spin-density inter-Landau-level excitations. The high energy maximum near 11.5 meV followed by a sharp cutoff is explained by the critical point in the dispersion of magnetoplasmons near $q = q_0 = 1/\ell_0$. The large scattering intensity between 9.5 and 11 meV appears to be caused by the superposition of the magnetoroton minimum of

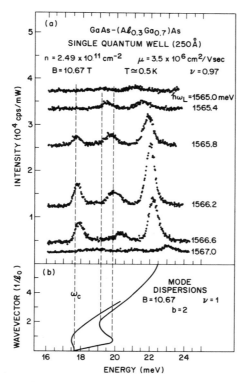

Fig. 8.   Light scattering spectra of inter-Landau-level excitations   at   four   different $\hbar\omega_L$. (b) Calculated mode dispersions at $v = 1$ and b = 2. After Ref. 20.

magnetoplasmons at $q \sim 2.5 q_0$ and a large density of states for larger values of q. We find relatively small (0.5 meV) differences between measured maxima in intensity and the calculated positions of critical points. The discrepancy could be explained by the strong field approximation ($\omega_c \gg e^2/\ell_0 \varepsilon_0$) used in the calculations. This approximation neglects the coupling to higher inter-Landau level transitions that at these relatively small fields reduces the energy of magnetoplasmons.

Figure 7 shows the energies of incident photons that correspond to the maxima in the profiles of resonant enhancement as a function of magnetic field. The four points shown correspond to the integer filling factors $v = 1,2,3,4$. Extrapolation to $B = 0$ gives the energy of the excitonic transition shown in the lower inset to Fig. 7. The slope of the line can be written as $(\ell + \frac{1}{2}) \omega'_c$, where ω'_c is the cyclotron energy for the optical reduced mass μ^*. Quantitative agreement with the measured slope, 2.6 meV/T, is obtained for $\ell = 1$ and $\mu^* = 0.064 \, m_0$. This result suggests the light scattering mechanism shown in the upper inset to Fig. 7. In these third-order processes the intermediate virtual transitions occur in the time sequence indicated by the numbers. Steps 1 and 3 are the optical transitions. The inter-Landau-level excitations are involved in step 2. Breakdown of wavevector conservation could occur in transitions between Landau levels broadened by disorder.

Figure 8(a) shows results obtained near $v = 1$ for several values of the incident photon frequency ω_L. In this case the resonance of the incident photon energy is with excitonic transitions that involve the second excited Landau level of the ground conduction subband.[18] The spectra were measured with the sample attached to the cold finger of a ^{3}He cryostat and with incident powers in the range of 10 microwatts. The intensities show a striking temperature dependence that is not observed for $v = 2$. At $T = 5K$ the intensity decreases by a factor of ~ 10, and the spectra cannot be measured above 10K. It is interesting that such temperature dependence is similar to that of the occupations of the two lowest spin-split Landau levels.[42] This is further evidence of the correlation between light scattering intensities and Landau level broadening. Figure 8(b) shows calculated mode dispersions that ignore the Zeeman energy in spin-density excitations. The figure displays the excellent agreement between the energies of the structures in the spectra and the calculated position of the critical points in the mode dispersions.

The spectra in Fig. 8(a) show an additional band, the strongest in the spectra, centered at 22 meV. At the present time the assignment of this band is not definitive. It is possible that it originates from the density of states at wavevectors $q \gg 1/\ell_0$. If this is the case, the shift from ω_c is a measure of exchange interactions.[14,37] It is also possible that the band at 22 meV is due to excitations at defects.

At non-integral values of the Landau level filling factor there is a marked decrease in the light scattering intensities by inter-Landau-level excitations measured in samples of ultrahigh electron mobility. This indicates that for partial occupation of the Landau levels the disorder potential required to relax the condition of vector conservation is effectively screened by the electron gas. In the case of modulation-doped single heterojunctions of very low electron density and high mobility we have been able to measure spectra of intersubband and inter-Landau-level excitations at filling factors $v < 1$. Figure 9 shows results obtained near $v = 0.24$ for several values of ω_L. The intensities of inter-Landau-level excitations are weak and, like the spectra in Fig. 6(a), they show additional structure at higher energies. The intersubband excitations are much more intense. At these filling factors the intersubband excitations display a remarkable temperature dependence for $v < 1/3$.[43] These unexpected results could be related to localization phenomena and phase transitions to insulating phases discovered near $v = 0.25$.[44]

CONCLUSION

The recent light scattering research described above has revealed the single particle and collective excitations of the 2D electron gas in GaAs-AlGaAs quantum wells and single heterojunctions. This work gives unique insights into many-body interactions in the 2D electron gas. The light scattering method is being applied in studies of fundamental interactions at high magnetic fields.[43] The large scattering intensities achieved in these experiments allow measurements with the incident light power levels compatible with low temperatures in the vicinity of 0.5K. These experiments explore the regimes of the fractional quantum Hall effect and the Wigner crystal. The light scattering method should be very effective in all reduced dimensionality situations and is currently used to investigate the elementary excitations of one-dimensional electron gases.[45,46]

ACKNOWLEDGEMENTS

We wish to thank L. Rubin and B. Brandt for valuable assistance at the Francis National Laboratory. This work was supported in part by NSF grants DMR-8807682 and DMR-8813164.

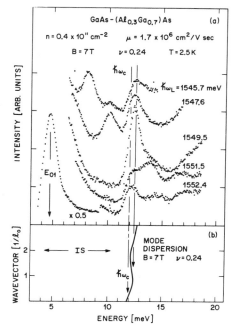

Fig. 9.　(a) Light scattering spectra from a low-density single heterojunction. (b) Calculated inter-Landau-level excitations in the Hartree-Fock approximation for $v = 0.24$ and $b = \infty$. IS is the range intersubband excitations. After Ref. 20.

REFERENCES

1. T. Ando, A. B. Fowler, and F. Stern, Rev. Mod. Phys. $\underline{54}$, 437 (1982).
2. H. L. Stormer, "Physics in a Technological World," A. P. French ed., American Institute of Physics, New York (1988).
3. D. C. Tsui and H. L. Stormer, IEEE J. Quantum Electron. $\underline{QE-22}$, (1988).
4. G. Abstreiter, R. Merlin, and A. Pinczuk, ibid page 1771.
5. A. Pinczuk and G. Abstreiter, "Light Scattering in Solids V," M. Cardona and G. Guentherodt eds., Springer-Verlag, Berlin-Heidelberg, (1989).
6. E. Burstein, A. Pinczuk, and S. Buchner, "Physics of Semiconductors 1978," B. L. H. Wilson ed., The Institute of Physics, London, (1979).
7. A. Pinczuk, L. Brillson, E, Anastassakis, and E. Burstein, Phys. Rev. Lett. $\underline{27}$, 327 (1971).
8. A. Pinczuk, G. Abstreiter, R. Trommer, and M. Cardona, Solid State Commun. $\underline{30}$, 703 (1979).
9. G. Abstreiter and K. Ploog, Phys. Rev. Lett., $\underline{42}$, 1308 (1979).
10. A. Pinczuk, H. L. Stormer, R. Dingle, J. M. Worlock, W. Wiegmann, and A. C. Gossard, Solid State Commun. $\underline{32}$, 1001 (1979).
11. D. C. Hamilton, and A. L. McWhorter, "Light Scattering Spectra of Solids," G. B. Wright ed., Springer-Verlag, New York, (1969).
12. A. Pinczuk, S. Schmitt-Rink, G. Danan, J. P. Valladares, L. N. Pfeiffer, and K. W. West, Phys. Rev. Lett. $\underline{63}$, 1633 (1989).
13. K. W. Chiu and J. J. Quinn, Phys. Rev. $\underline{B\,9}$, 4724 (1974).
14. C. Kallin and B. I. Halperin, Phys. Rev. $\underline{B\,30}$, 5655 (1984).
15. A. H. MacDonald, J. Phys. $\underline{C\,18}$, 1003 (1985).
16. A. H. MacDonald, H. C. A. Oji, and S. M. Girvin, Phys. Rev. Lett. $\underline{55}$, 2208 (1985).
17. P. Pietilainen and T. Chakraborty, Europhys. Lett. $\underline{5}$, 157 (1988).
18. A. Pinczuk, J. P. Valladares, D. Heiman, A. C. Gossard, J. H. English, C. W. Tu, L. Pfeiffer, and K. West, Phys. Rev. Lett. $\underline{61}$, 2701 (1988).
19. A. Pinczuk, J. P. Valladares, D. Heiman, L. N. Pfeiffer, and K. W. West, Surf. Sci. $\underline{229}$, 384 (1990).
20. A. Pinczuk, D. Heiman, S. Schmitt-Rink, S. L. Chuang, C. Kallin, J. P. Valladares, B. S. Dennis, L. N. Pfeiffer, and K. W. West, to be published in the Proc. of the 20th Int. Conf. on the Physics of Semiconductors.
21. E. Burstein, A. Pinczuk, and D. L. Mills, Surf. Sci. $\underline{98}$, 451 (1980).
22. G. Danan, A. Pinczuk, J. P. Valladares, L. N. Pfeiffer, and K. W. West, Phys. Rev. $\underline{B\,39}$, 5512 (1989).
23. J. Zucker, A. Pinczuk, D. S. Chemla, A. C. Gossard, and W. Wiegmann, Phys. Rev. Lett. $\underline{51}$, 1293 (1983).
24. T. Ando, J. Phys. Soc. Japan $\underline{51}$, 3893 (1982).
25. S. Katayama and T. Ando, J. Phys. Soc. Japan $\underline{54}$, 1615 (1985).
26. A. C. Tsellis and J. J. Quinn, Phys. Rev. $\underline{B\,29}$, 3318 (1984).
27. G. Eliasson, P. Hawrilak, and J. J. Quinn, Phys. Rev. $\underline{B\,35}$, 5569 (1987).
28. D. Dahl and L. J. Sham, Phys. Rev. $\underline{B\,16}$, 651 (1977).
29. S. Das Sarma, Appl. Surf. Sci. 11/12: 535 (1982).
30. L. Wendler and R. Pechstedt, Phys. Rev. $\underline{B\,35}$, 5887 (1987).
31. D. H. Ehlers, Phys. Rev. $\underline{B\,38}$, 9706 (1988).
32. A. Pinczuk, J. M. Worlock, H. L. Stormer, R. Dingle, W. Wiegmann, and A. C. Gossard, Solid State Commun. $\underline{36}$, 43 (1980).
33. A. Pinczuk and J. M. Worlock, Surf. Sci. $\underline{113}$, 69 (1982).
34. D. Gammon, B. V. Shannabrook, J. C. Ryan, and D. S. Katzer, Phys. Rev. $\underline{B\,41}$, 12311 (1990).

35. A. Pinczuk, B. S. Dennis, L. N. Pfeiffer, and K. W. West, unpublished.
36. S. L. Chuang and S. Schmitt-Rink, unpublished.
37. I. V. Lerner and Yu. E. Lozovik, Sov. Phys.-JETP 51, 5881 (1980).
38. Yu. A. Bychkov, S. V. Iordanskii, and G. M. Eliashberg, JETP Lett. 33, 153 (1981).
39. T. Ando and Y. Murayama, J. Phys. Soc. Japan 54, 1519 (1985).
40. S. Das Sarma and X. C. Xie, Phys. Rev. Lett. 61, 7381 (1988).
41. C. Kallin, in: "Interfaces, Quantum Wells and Superlattice," C. R. Leavens and R. Taylor eds., Plenum Press, New York, (1988).
42. B. B. Goldberg, D. Heiman, and A. Pinczuk, Phys. Rev. Lett. 63, 1102 (1989).
43. D. Heiman, A. Pinczuk, B. S. Dennis, L. N. Pfeiffer, and K. W. West, unpublished.
44. H. W. Jiang, H. L. Stormer, R. W. Willet, D. C. Tsui, L. N. Pfeiffer, and K. W. West Phys. Rev. Lett. 65, 633 (1990).
45. J. S. Weiner, G. Danan, A. Pinczuk, J. P. Valladares, L. N. Pfeiffer, and K. W. West, Phys. Rev. Lett. 63, 1641 (1989).
46. T. Egeler, G. Abstreiter, G. Weimann, T. Demel, D. Heitmann, and W. Schlapp, to be published in the Proc. of the 20th Int. Conf. on Physics of Semiconductors.

CONCLUDING REMARKS

M. Cardona

Max-Planck-Institut für Festkörperforschung
7000 Stuttgart 80
Federal Republic of Germany

I told you Tuesday an anecdote about Raman's visit to Italy. Anant Ramdas, one of his last students, could tell you many more. But inspite of being an egocentric, he did great many things. Not all of them right, as you know, but on the whole the balance for a man working in a country under colonial rule is pretty good. Maybe this says something about British colonial rule also. Which other country under colonial, or semicolonial rule has received a Nobel Prize in Science? And "semi" here brings to my mind Central America as much as Eastern Europe, the so-called banana and banana-less republics.

Well, Raman did many things. Yesterday when Djafari-Rouhani was talking about contributions from interface motion I remembered that the mechanism was discussed by Raman and is called the Raman-Nath effect. In fact, we know it is the dominant mechanism for Brillouin scattering at the surface of a metal such as aluminum, but it is rather small for a GaAs-air interface.

Back to Raman. He, of course, became a symbol. A symbol of hope to scientists in third world countries. Which other third world country has received a Nobel Prize in Science? Well, you can think about it. I'll come back to that later. Anyhow, Raman to get his Nobel Prize used natural resources: Sunlight. The Sun was soon replaced by the Toronto Arc, no doubt a Canadian conspiracy. The Sun did set within the British Empire while allegedly it never set in the Spanish Empire. The Toronto Arc was replaced in due course by the laser. Let me come back to Nobel Prizes and third world countries. Well, I have not checked the list since I have prepared these remarks here, but only one comes to my mind: Argentina has received two Nobel Prizes for work done there (I bet most of you did not know that). Of course, when the prizes were received, the Country was not third world but only 2-1/2, although when it comes to the third worldliness of Argentina it would be better to use irrational instead of rational numbers. The country is probably the only one in the world with a serious brain drain problem, the largest number of prominent emigré scientists per capita, present company included, and when you normalize it to the GNP at current rates of exchange it is even worse. Only when you normalize it to the inflation rate they do better. Why do I say that? because we have had the opportunity to see here that Argentine expatriates have made a formidable contribution to our field and also because Argentina is the conjugate of Canada with respect to equatorial reflection. In spite of its down-under position it has received as many Nobel Prizes in Science as Canada, namely two.

You already see that I am not going to review in detail the contributions to the conference but nevertheless I would like to mention a few important points. They will be biased by my own interest in phonons and electron-phonon interaction and will cover less

Light Scattering in Semiconductor Structures and Superlattices
Edited by D.J. Lockwood and J.F. Young, Plenum Press, New York, 1991

587

electronic excitations which, anyhow, were mainly presented this morning and I did not have time to include them in my talk.

We have beaten the subject of interface modes to death, I hope. I wish you pass the message to your colleagues who do transport and hot electrons. One point I would like to make concerns the confined modes. The dichotomy electrostatic vs. mechanical boundary conditions has been mentioned. Before it became clear that the <u>mechanical ones</u> have first rank this was settled experimentally by Sood and Menéndez in my lab in 1985 (see Fig. 1). The reason why this had not been done before is that near resonance, and everybody was trying to work near resonance to obtain strong signals, you see only $m =$ even modes, regardless of polarization direction. We now know that the anomalous cross-polarized scattering is due to a process involving inelastic scattering by a phonon and elastic scattering by a defect, probably interface roughness. The beautiful infrared work presented by Tilley is, to my knowledge, the first infrared work in which modes with $m > 1$ are seen. Since these modes correspond to $m = 3$, it also confirms the primacy of mechanical boundary conditions. Tilley also told us about virtual modes which, like virtue itself, disappear in an unexpected way, some place, somewhere. I nevertheless like to say that Raman is intrinsically more powerful for the study of microstructures than infrared. Besides the flexibility of polarization selection rules, it is hardly possible to see in infrared folded acoustic modes, since the difference in dynamical charge between GaAs and AlAs is very small. The ratio of photoelastic constants is, however, nearly as large as you want to make it because of resonances and antiresonances (I presume everybody knows that photoelastic constants exhibit both resonances and antiresonances near gaps!)

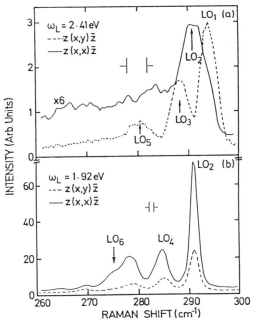

Fig. 1. Raman spectra of a GaAs-AlAs superlattice for polarized and depolarized scattering configurations in (a) off-resonance and (b) near resonance. The peak on the high frequency side of LO_6 is believed to be the GaAs-like interface mode. [From A.K. Sood, J. Menéndez, M. Cardona, and K. Ploog, Phys. Rev. Lett. <u>54</u>, 2111 (1985.]

And this brings me to the need for data base parameters for the bulk materials. Sometimes students ask me, but why work on bulk silicon and GaAs? Isn't everything known about these guys? Well, no. For instance in the visible the real parts of the photoelastic constants are poorly known. The imaginary parts even worse, in fact it became clear to me that few of the people here present were aware of the fact that they are nearly as important as the real parts.

I have made a career out of measuring such things when they were only of academic interest, and it has been a rewarding one. Evangelos Anastassakis and Eli Burstein have discussed the pioneering experiments of piezo-Raman coefficients which now, 22 years later, everybody is referring to and have become the basis of sophisticated strain characterization techniques. I could tell you anecdotes related to those measurements but I shall put them on file for the day when I get invited to give after dinner speeches. Many of these constants are now known thanks to the work of Evangelos and my group, but still many are still needed, especially those involving the change of LO-TO splittings with stress and the corresponding constants for II-VI compounds. In fact, I believe many of the numbers are used incorrectly since they correspond to LO phonons when they are used for TO and vice versa.

We have had lots of semantic discussions and it is good so language becomes sharper and we learn what we are talking about. Let me remind you of the motto of the Royal Academy of the Spanish Language: It cleanses, it polishes, and it brightens. Evangelos pointed out the inappropriateness of calling strain in [100] superlattices biaxial as done commonly. It is indeed biaxial in the [110] and lower symmetry cases. For [100] and [111], however, it is actually uniaxial or, as Evangelos' Hellenic erudition suggested, "bisotropic".

Well, we worry an awful lot about strain and the elastic constants of AlAs are not reliably known experimentally. The best values to use may actually be those recently calculated by Richard Martin! Actually total energy calculations have become so good, as Elisa Molinari has shown us, that it is often better to use theoretical instead of experimental values. This also applies to the lattice dynamics of AlAs.

The expansion coefficient of AlAs is not known to the accuracy required to calculate lattice mismatches, in fact people have even conjectured that the mismatch may even reverse sign at high temperatures!

There has been a lot of talk about mismatch strains but I was surprised that no one mentioned that these strains do not suffice to characterize the structure of a superlattice. In fact a number of additional internal super-unit-cell parameters are needed about which we have little and only unreliable information from theoretical models, none experimental. What are they? Evangelos and I discussed some time ago that every phonon of A_{1g} (fully symmetric) symmetry can be applied as a static distortion (frozen phonon) without lowering the symmetry of the structure. The amplitude of this "frozen phonon" is thus a free structural parameter. Imagine how many you have (see Fig. 2) even in a [001] $A_m B_n$ superlattice for large values of m and n! This situation becomes even worse if the superlattice has lower symmetry.

And back to piezo-Raman coefficients, how many of you know that they are terribly anisotropic even in cubic crystals like Ge, Si, GaAs? In fact, for these materials the singlet-doublet splitting reverses sign from a [001] to a [111] strain (see Evangelos talk). Curiously enough, in the case of diamond the sign remains the same for both directions, facts which are theoretically well understood. A point which has just been barely discussed is the dependence of band structure on the mismatch strain. Some parameters (e.g., at the top of the valence band) are well known. Others, relating to conduction bands and E_1 gaps, less so. Their measurement is a highly pedagogical exercise for training students and they may find that 30 years from now their names are profusely quoted in the literature.

Obviously time resolved measurements of various sorts, on excited electrons and non-equilibrium phonons, have reached a high degree of sophistication and credibility. I

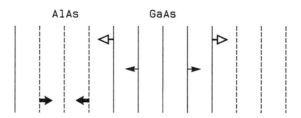

AlAs   GaAs

Fig. 2. Schematic diagram of a $(GaAs)_5(AlAs)_3$ superlattice. The three pairs of arrows represent free internal structural parameters.

mention work by Kash, Yu, Shen, Young and Ryan. Of particular interest, also technologically, are the values of the $\Gamma \to X$ and $\Gamma \to L$ electron-phonon coupling constants obtained from them. They seem to be converging to values within the limits of those predicted by theory although this theory could stand some improvement in sophistication. John Ryan reiterated his being a spectator in the "interface phonon" controversy. Given the high quality of his data, I hate to think what would happen if he were to become an actor!

Problems of localization and confinement kept coming back. It was nice to see that even when optical bands strongly overlap (e.g., GaSb/InAs) you can get strong confinement. Some people were asking for a definition of confined modes. I guess we can make up various such definitions but they will not enhance our understanding of the matter.

Localization in alloys and the whole problem of lattice dynamics, especially the **k** dependence of the modes, was discussed. Jeff Kash told us that in GaAlAs the optical phonons are not localized. I mentioned that in GeSi they are. Is the reason for the lack of localization in the former case percolation or the coherence imposed by the long-range electrostatic fields? Jusserand also discussed within the CPA the confined phonons of GaAlAs/GaAs superlattices. Another interesting problem involves confinement of electrons in minima above the lowest, in particular those which contribute to the E_1 transitions. Like in the case of overlapping phonon bands, confinement seems to exist. The theoretical underpinning is missing

I enjoyed listening to the beautiful structural work done at NRC Canada for Ge-Si superlattices in which they were even able to determine activation energies for dislocations. The problem of the metastable or interface stabilized Ge-Sn system has come to the fore only recently and is of interest because of potential applications in the range of direct gaps. I have been a fan of gray tin since 1960 when I probably was the second person in the world to grow single crystals of it (the first were grown by A. Ewald at Northwestern U.) and I could tell you many anecdotes of the so-called tin pest. In fact the first Raman measurements on gray tin were performed on polycrystalline powders made in my home freezer after rubbing a piece of white tin with InSb. This is the reason why I got into high T_c superconductivity. When I saw the first ceramic samples they looked like the old gray tin and I thought: if you can do Raman in one you can do it in the other.

We have seen various people, for instance Ramdas and Djafari-Rouhani, use phonon dispersion relations for imaginary k in order to explain behavior at interfaces. This brought back to my mind the time in 1965 when I was calculating electronic band structures for imaginary k in Buenos Aires with N. Majlis and M. Chaves. Shortly after there was a military takeover. The police entered the University and with truncheons destroyed the old Ferranti-Mercury tube computer. We had calculated the [111] and the [110] direction but we never got to the [100]. The referee of course wanted to see the [100] direction.

This is a NATO conference and we should be grateful to NATO. We have needed NATO's support and now NATO may need ours in the years ahead; they are up to some changes. I think now, in retrospect, most of us are thankful for the security and peace NATO has provided and for the bonds of friendship and cultural ties across the Atlantic. Military emphasis will decline but those bonds should survive and, in my opinion, even be strengthened.

For many years, when I have any time to kill in New York City, I go to the gardens of the United Nations where one can contemplate an inspiring Soviet bronze group based on Isaiah's prophecy: they shall bend their swords into plowshares and their spears into pruning hooks and of war there shall be no more. This prophecy, and the Soviet Statue, became the inspiration of protest groups in East Germany a few years ago. Tee-Shirts were even printed with the Soviet statue. They were immediately banned and people wearing them beaten up. May I prophesy that "we shall bend SDI lasers into Raman Spectrometers and night vision snooperscopes into detectors for infrared spectroscopy and of war there shall be only that with the editors of Physical Review Letters".

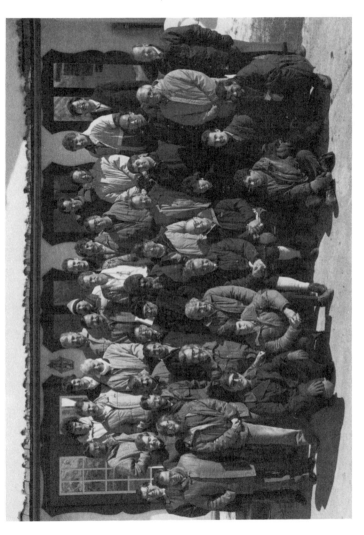

Standing (elevated): S. Das Sarma, J. Wagner, H. Brugger, G. Fasol, G. Kanellis, D. Tilley, R. Ellialtioglu, G.P. Schwartz, J.M. Worlock, A.P. Smart, R. Merlin, J.M. Calleja, A.V. Nurmikko, P.Y. Yu, J.F. Ryan, K.T. Tsen

Standing: B. Wakefield, A. Ercelebi, J. Menéndez, B. Jusserand, J.A. Kash, A. Aydinli, J. Sapriel, S. Nakashima, A.K. Ramdas, M.W.C. Dharma-wardana, J.C. Tsang, M.G. Cottam, L. Brey, Z.P. Wang, B. Djafari Rouhani, L. Dobrzynski

Sitting: D.J. Lockwood, A. Pinczuk, M. Cardona, E. Burstein, E. Molinari,

Squatting: E. Anastassakis, J.F. Young, L. Viña, P.M. Fauchet, G. Abstreiter

PARTICIPANTS

Professor G. Abstreiter
Walter Schottky Institut
Technische Universität Müchen
D-8046 Garching
Federal Republic of Germany

Professor E.M. Anastassakis
Physics Department
National Technical University
Zografou Campus
Athens 15773
Greece

Professor A. Aydinli
Department of Physics Engineering
Hacettepe University
06532 Beytepe-Ankara
Turkey

Professor L. Brey
Departamento de Fisica
Universidad Autonoma
Cantoblanco
E-28049 Madrid
Spain

Dr. H. Brugger
Daimler Benz AG
Research Center FAU
Wilhelm-Runge-Str. 11
D-7900 Ulm
Federal Republic of Germany

Professor E. Burstein
Physics Department
University of Pennsylvania
Philadephia, PA 19104
U.S.A.

Professor J.M. Calleja
Departamento de Fisica Aplicada C-IV
Universidad Autonoma de Madrid
Cantoblanco
E-28049 Madrid
Spain

Professor M. Cardona
Max-Planck-Institut für Festkörperforschung
Heisenbergstrasse 1
D-7000 Stuttgart 80
Federal Republic of Germany

Professor M.G. Cottam
Physics Department
University of Western Ontario
London, Ontario N6A 3K7
Canada

Professor S. Das Sarma
Dept. of Physics and Astronomy
University of Maryland
College Park, Maryland 20742-4111
U.S.A.

Dr. M.W.C. Dharma-wardana
Institute for Microstructural Sciences
National Research Council of Canada
Ottawa, Ontario K1A 0R6
Canada

Professor B. Djafari Rouhani
Faculté des Sciences et Techniques
Université de Haute Alsace
4, rue des Frères Lumière
F-68093 Mulhouse Cédex
France

Professor L. Dobrzynski
U.F.R. de Physique
Université des Sciences et Techniques
   de Lille Flandres Artois
Bâtiment P. 5
F-59655 Villeneuve d'Ascq Cédex
France

Professor R. Ellialtioglu
Faculty of Engineering and Science
Bilkent University
P.O.B. 8
06572 Maltepe-Ankara
Turkey

Dr. A. Ercelebi
Department of Physics
M.E.T.U.
06531 Ankara
Turkey

Dr. G. Fasol
Cavendish Laboratory
University of Cambridge
Madingley Road
Cambridge CB3 0HE
U.K.

Professor P.M. Fauchet
Dept. of Electrical Engineering
Princeton University
Princeton, NJ 08544
U.S.A.

Dr. P. Hawrylak
Institute for Microstructural Sciences
National Research Council of Canada
Ottawa, Ontario K1A 0R6
Canada

Dr. B. Jusserand
Centre National d'Études des
 Télécommunications
196 rue de Paris
F-92220 Bagneux
France

Professor G. Kanellis
Physics Department
University of Thessaloniki
54006 Thessaloniki
Greece

Dr. J.A. Kash
IBM Thomas J. Watson Research Center
P.O. Box 218
Yorktown Heights, NY 10589
U.S.A.

Dr. D.J. Lockwood
Institute for Microstructural Sciences
National Research Council of Canada
Ottawa, Ontario K1A 0R6
Canada

Professor J. Menéndez
Physics Department
Arizona State University
Tempe, AZ 85287
U.S.A.

Professor R. Merlin
Department of Physics
The University of Michigan
Ann Arbor, Michigan 48109-1120
U.S.A.

Dr. E. Molinari
Istituto di Acustica "O.M. Corbino"
Consiglio Nazionale delle Ricerche
Via Cassia, 1216
I-00189 Roma
Italy

Professor S. Nakashima
Department of Applied Physics
Osaka University
Suita
Osaka 565
Japan

Professor A.V. Nurmikko
Division of Engineering
Brown University
Providence, RI 02912
U.S.A.

Dr. A. Pinczuk
AT&T Bell Laboratories
600 Mountain Avenue
Murray Hill, NJ 07974-2070
U.S.A.

Professor A.K. Ramdas
Department of Physics
Purdue University
West Lafayette, IN 47907
U.S.A.

Dr. J.F. Ryan
Clarendon Laboratory
University of Oxford
Parks Road
Oxford, OX1 3PU
U.K.

Dr. J. Sapriel
Centre National d'Études des
 Télécommunications
196 Avenue Henri Ravera
F-92220 Bagneux
France

Dr. G.P. Schwartz
AT&T Bell Laboratories
Materials Research Laboratory
600 Mountain Avenue
Murray Hill, NJ 07974-2070
U.S.A.

Dr. A.P. Smart
Department of Electronics and
 Electrical Engineering
The University
Glasgow G12 8QQ
U.K.

Dr. D. Tilley
Department of Physics
University of Essex
Colchester C04 3SQ
U.K.

Dr. J.C. Tsang
IBM Thomas J. Watson Research Center
P.O. Box 218
Yorktown Heights, NY 10589
U.S.A.

Professor K.T. Tsen
Department of Physics
Arizona State University
Tempe, AZ 85287
U.S.A.

Professor L. Viña
Dept. de Fisica Aplicada, C-IV
Universidad Autonoma de Madrid
Cantoblanco
E-28049 Madrid
Spain

Dr. J. Wagner
Fraunhofer-Institut für
 Angewandte Festkörperphysik
Eckerstrasse 4
D-7800 Freiburg
Federal Republic of Germany

Dr. B. Wakefield
British Telecom Research Laboratories
Martlesham Heath
Ipswich IP5 7RE
U.K.

Dr. Z.P. Wang
Institute of Semiconductors
Academia Sinica
P.O. Box 912
Beijing
People's Republic of China

Dr. J.M. Worlock
Bell Communications Research
331 Newman Springs Road
Red Bank, NJ 07701-7020
U.S.A.

Dr. J.F. Young
Institute for Microstructural Sciences
National Research Council of Canada
Ottawa, Ontario K1A 0R6
Canada

Professor P.Y. Yu
Physics Department
University of California
Berkeley, CA 94720
U.S.A.

AUTHOR INDEX

Abstreiter, G., 561
Aers, G.C., 81
Anastassakis, E., 173
Arnot, H.E.G., 247

Baribeau, J.-M., 81, 197, 401
Baroni, S., 39
Beall, R.B., 461
Beeck, S., 561
Berdekas, D., 63
Brey, L., 525
Briones, F., 53
Brugger, H., 259
Burstein, E., 1, 441

Calle, F., 53, 353
Calleja, J.M., 53, 353
Cardona, M., 1, 19, 587
Chang, L.L., 353
Cottam, M.G., 311

Das Sarma, S., 499
Dempsey, J., 525
Dennis, B.S., 571
Dharma-wardana, M.W.C., 81
Djafari Rouhani, B., 139, 451
Dobrzynski, L., 451
Driel, H.M. van, 401
Dumelow, T., 461

Egeler, T., 561
Ekenberg, U., 543
Engelhardt, M.A., 33

Fasol, G., 543
Fauchet, P.M., 229
Florez, L.T., 477
Foad, M.A., 247
Foxon, C.T.B., 461
Freeouf, J.L., 159

Giannozzi, P., 39
Gironcoli, S. de, 39
Gopalan, S., 311

Halperin, B.I., 525

Hamilton, A.A., 461
Han, H.X., 79
Harbison, J.P., 477
Harris, J.J., 461
Hawrylak, P., 479
He, J., 123
Heiman, D., 571
Hilton, D., 461
Höchst, H., 33
Hong, M., 353
Huber, A., 561

Iyer, S.S., 159

Jiang, M.Y., 441
Johnson, N.F., 525
Jusserand, B., 103

Kallin, C., 571
Kanellis, G., 63
Kash, J.A., 367
Kelly, P.J., 401
Khourdifi, E.M., 139
Kim, D., 383
Kwok, S.H., 491

Li, G.H., 79
Liarokapis, E., 491
Lockwood, D.J., 1, 81, 197, 401, 451
López, C., 53

Maciel, A.C., 477
Maslin, K.A., 461
Mendialdua, J., 451
Menéndez, J., 33
Merlin, R., 491
Meseguer, F., 53, 353
Molinari, E., 39
Moore, K.J., 461

Nahory, R.E., 477
Nakashima, S., 291
Nurmikko, A.V., 341

Othonos, A., 401

Parker, T.J., 461
Pfeiffer, L.N., 571
Pinczuk, A., 1, 571
Ploog, K., 53, 491, 543

Ramdas, A.K., 323
Richards, D., 543
Rodriguez, A., 451
Rodriguez, S., 323
Rothwell, W.J., 257
Ryan, J.F., 421, 477

Samson, B., 461
Sankey, O.F., 393
Sapriel, J., 123
Schmitt-Rink, S., 571
Schwartz, G.P., 219
Sinha, K., 33
Smart, A.P., 247
Smith, S.R.P., 461
Sotomayor Torres, C.M., 247

Tatham, M.C., 421, 477
Tejedor, C., 53
Tilley, D.R., 461
Tsang, J.C., 159
Tsen, K.T., 393

Viña, L., 53, 353

Wagner, J., 275
Wakefield, B., 257
Wang, Z.P., 79
Watt, M., 247
West, K.W., 571
Wilkinson, C.D.W., 247
Worlock, J.M., 477

Yoshino, J., 353
Young, J.F., 1, 401
Yu, P.Y., 383

SUBJECT INDEX

α-Sn, 34-35
α-SnGe, *see* GeSn

Acousto-optic deflection, 123, 135-138
Activation energy, 197, 211-212, 216
AlAs, 43-45
AlSb , 188, 223-224
Anti-Stokes/Stokes intensities, 314, 333-334
  temperature determination from, 10, 261,
    387-391, 467
  time-resolved, 10, 368-374, 402-419,
    429-438
Asymmetric quantum wells, 441, 444-447
Atomic layer superlattices, 212-216

Boltzmann equation, 402-411
Bond polarizability model, 64, 76-78, 82
Brillouin scattering, 1-4, 150, 312

Cation mixing, 39, 50
CdTe, 341-350, 353-364
CdMnTe, 316, 324-335
CdTe/CdMnTe, 353-364
CdTe/MnTe, 341-348
CdZnSe/CdMnSe, 334-338
Characterization
  of composition, 197, 200-201
  of crystal orientation, 293-298
  of epitaxial quality, 197, 212-216
  of free carrier concentration, 298-304,
    556-558
  of layer thickness, 197, 200
  of mobility, 299-301
  of strain relief, 197-211, 223-224, 226
Charge density excitation, 496, 500-521,
  571-583
Coherent potential approximation, 110
Coulomb interactions, 480-481
Coupled oscillator model, 75
CrBr$_3$, 312-314
Critical thickness, 266, 271
Cu, 457-459

Damage, 307
Damping, 300, 303
  acoustic, 131-135

Damping (continued)
  Landau, 543-549, 552, 577, 578
  optical, 131-135
Defects, 275-288
Delta doping, 276, 284-288, 470-475, 477,
  554
Diamond, 188
Dielectric function, 463-474, 545, 568, 578
Diffraction of light by ultrasound, 123,
  135-137
Dispersion, 590
  in-plane, 30
  interface, 27-30, 423, 561, 566
  intersubband, 577
  phonon, 21-30, 43-50, 63-78, 79-80, 112,
    119, 124-133, 142-149, 231, 233,
    249, 267, 325, 423
  plasmon, 505-515, 545-558
  three dimensional calculations in InAs/GaSb,
    64-68
Domains, electric field, 491-496
Dynamic charge transfer, 543

Effective medium theory, 465
Elastic constants, 452-454
Electric field induced Raman scattering, 270
Electron-phonon excitations, 4
Electron-phonon interaction, 7, 10
  deformation potential, 353, 363, 401-419,
    425
  cooling, 368-374, 389-391
  in GaP, 292
  intersubband, 421-438
  intervalley, 375-380, 383, 406-411
  intrasubband, 421-438
  intravalley, 331, 342, 363, 368-374, 383,
    406-411
Electron-electron interactions, 379-380,
  479-489, 525-539, 571-583
Electron paramagnetic resonance, Raman, 334
Ellipsometry, spectroscopic, 160-170
Energy gaps in Si/Ge, 159, 163-164, 170
EuS, 316
Excitations
  collective, 500-521, 567, 571-583
  electronic, 3-4, 8, 543-558, 571-583

Excitations (continued)
  intersubband, 11, 477, 491-496, 500-521, 571-583
  non-equilibrium, *see* Non-equilibrium excitations
  single particle, 500-521, 567, 571-583
Excitons, 330, 354-364, 428, 435, 441-449, 479-489, 573
  coupling to phonons, 341-350, 393-399

Fano resonance, 543-549
Far infrared, 461-475
Fe, 457, 459
Forbidden or non-allowed scattering, 269, 271, 276, 407, 415-418, 435-436
Fourier transform infrared spectroscopy, 461-475

GaAlAs, 110-113, 368-374
GaAs, 43-45, 188, 249, 254, 276-288, 367-380, 383-391, 470-474, 477
GaAs/AlAs, 19, 23-26, 29-30, 39-51, 53-60, 79-80, 103-120, 123-158, 465-470, 561-566, 588
GaAs/AlGaAs, 80, 111-113, 260-263, 305, 393-399, 421-438, 441-449, 454-457, 461-465, 491-496, 499-521, 543-558, 567, 571-583
GaAs/Ge, 269-271
GaP, 188, 298-304
GaSb, 63-65, 188
GaSb/AlSb, 219-224
Ge, 188, 235, 401-419
GeSi, *see* SiGe
GeSn, 33-37, 219, 224-227
Green function method, 139-140, 311-320, 482, 501

Historical review of light scattering from semiconductors, 1-12
Hot phonons, 368-374, 383-391

Impurities, 275-288
InAs, 63-65, 188
InAs/GaSb, 63-78
InP, 188
InSb, 188
Infrared spectroscopy, 425, 461-475, 526-539, 588
Interface
  defects, 139, 148-149
  disorder, 91-101, 213-214
  phonons, *see* Phonons
  roughness, 103-120, 588

Landau level excitations, 3, 55, 579-583
Lattice dynamics, *see* Phonons
Lattice mismatch, 176, 325-326, 341, 353
Lifetimes, phonon, 368-372, 401-419
Lineshapes, phonon, 229-243
Luminescence, *see* Photoluminescence

Magnetic excitations,
  bulk, 316-320, 331-336
  surface, 316-320
Magnetic fields, 53-60, 329-336, 342-345, 353-364, 529-531
Magnetic semiconductors, 311-320
  dilute, 311, 323-338, 344, 353
Magnetostatic modes, 314
Mapping, 260-263, 305-308 (*see also* Micro-Raman)
Many body effects, 479-489, 500, 583
  fermion-boson exchange, 483-489
  phase space filling, 485, 489
Metastable structures, 198, 219, 224-227
Micro-Raman, 257, 259-263, 291-308, 561-568
Microcrystals, 229-243
Misfit dislocations, 204, 207-208, 211, 223
MnTe, 341-350
Modulation doped, 543-558, 567, 571-583

Nanocrystals, 229-243
Non-equilibrium excitations
  electrons, 9-12, 367-380, 383-391, 421-438
  phonons, 9-12, 367-380, 383-391, 401-419, 432-438, 589
Nb, 457-459
Non-linear optics, 441-449, 479-489

One-dimensional electron gas, 9
Optical Bloch equations, 479-483

Parabolic quantum wells, 499, 521, 525-539
Phonon deformation potentials, 173, 175, 188, 192-193, 210, 220-221, 223, 226
Phonon dynamics, 368-372, 411-419
Phonon localization, 374
Phonons, 305-308, 587-590
  acoustic, 21-30, 64-75, 79-81, 123-158, 198-200, 324-327, 451-459
  confined, 21-27, 87, 103-120, 231-232, 236, 238, 248-249, 277, 328, 330, 423, 433-435, 467, 564
  eigen vectors, 23, 26, 29, 50, 68-75, 118-120, 176-177, 462, 482
  folded, 65-75, 79-80, 83, 127, 198-200, 267, 269, 326, 329, 451-459
  interface, 27-30, 39, 47, 65-75, 146-148, 160, 164, 303, 325, 330, 333, 355, 360, 423, 425, 433-435, 438, 563-566
  local modes, 275-288
  longitudinal optic, 21-30, 39-51, 53-60, 197-216, 263-272, 276-280, 298-304, 341-350, 353-364, 368-374, 383-391, 395, 401-419, 421-438, 493-496, 515-519, 578
  non-equilibrium, 9-12, 367-380, 383-391, 401-419, 432-438, 589
  optic, 2-12, 33-37, 65-101, 103-120, 159-172, 229-243, 248-254, 324-331, 461-475

Phonons (continued)
  strain shift of, 89-91, 173-193,
    219-227
  transverse optic, 21-30, 39-51, 261, 270
  two-mode behaviour, 80, 264, 354
Photoelastic coefficients, 130-131, 149,
    156, 158, 192
Photoluminescence, 54, 57, 247, 250-253,
    329-332, 353-364, 477, 494-496
  excitation spectra, 54-58, 353-364, 573
  hot, 56, 342-347, 375-380, 384-386,
    434
  at neutral acceptors, 375-380
Photoreflectance, 342
Picosecond spectroscopy, 10-12
Plasmons, 3, 8-9, 11, 30, 280, 298-304,
    386, 426-429, 437, 499-521, 528,
    543-558, 567-568
Point groups of superlattices
  diamond, 19-20
  zinc blende, 19
Polaritons, 3-5, 424, 448, 463, 465,
    469-470
Polytypes of SiC, 292, 306

Random-phase approximation, 499-521,
    549
Reactive ion etching, 247-248, 254, 276,
    279-280
Resonance Raman, 53-60, 162-171,
    282-288, 330, 333, 342, 344,
    347-349, 353-364, 421-438, 449,
    477, 544
Resonant tunneling, 491-496

Sapphire, 188
Schottky diode, 247, 280
Second harmonic generation, 441
Second-order Raman, 2, 237, 417
Selection rules, 34, 230, 235, 237-238,
    248, 264, 394, 407, 419, 424,
    442-445, 564, 572-573, 576, 581,
    588
Si, 188, 233-243, 295-298, 307
SiC, 292, 306
SiGe, 19-22, 27-28, 81-101, 131-134,
    159-172, 197-216, 264-272,
    411-419
SiGe/GaAs, 271-272
Single heterostructure, 502-506, 543,
    580-583

Single heterostructure (continued)
  self-consistent subband calculations,
    547-550
Single particle excitation, see Excitations
Space groups of superlattices
  diamond, 19-20
  zinc blende, 19, 65
Spin density excitation, 496, 500-521,
    571-583
Spin-flip Raman, 3, 335-336
Strain, 81, 89-101, 159-193, 197-211, 216,
    219-227, 229, 238-240, 263-267,
    325, 341, 347-348, 353, 360, 412,
    467, 589
Stress, see Strain
Submicron resolution, 562
Sub-picosecond spectroscopy, 383-391
Sum and difference frequency generation,
    441, 446-449
Surface modes, 139, 142-145, 229, 240,
    311-320, 457-459, 463, 465, 469,
    519-520

Tensor, strain, 175-178, 183
Thin films and multilayers, 311-321
Three wave mixing, 441-449
Time-resolved spectroscopy, 9-12, 347,
    368-374, 383-391, 393-399,
    401-419, 421-438, 589
Transfer matrix method, 139-140
Transport properties of excitons, 393-399
Two-dimensional electron gas, 8-9, 499,
    571-583
II-VI materials, 247-254, 323-338, 341-350,
    589

Ultrasound, 123, 135-138
Unit cell, n-layer, 451-459

Valence overlap shell model, 63-64
Virtual excitations, 479-489

Wire, quantum, 254, 499-521, 543

X-ray diffraction, 192, 197-216, 223, 225,
    293, 354

ZnSe, 188, 250-254
ZnSe/CdMnSe, 334-337
ZnTe, 341-350
ZnTe/MnTe, 341-350